Many-Electron and Multiphoton Atomic Processes: A Tribute to Miron Amusia

Many-Electron and Multiphoton Atomic Processes: A Tribute to Miron Amusia

Editors

Anatoli Kheifets
Gleb Gribakin
Vadim Ivanov

MDPI • Basel • Beijing • Wuhan • Barcelona • Belgrade • Manchester • Tokyo • Cluj • Tianjin

Editors
Anatoli Kheifets
Australian National University
Australia

Gleb Gribakin
Queen's University Belfast
UK

Vadim Ivanov
Peter the Great St. Petersburg Polytechnic University
Russia

Editorial Office
MDPI
St. Alban-Anlage 66
4052 Basel, Switzerland

This is a reprint of articles from the Special Issue published online in the open access journal *Atoms* (ISSN 2218-2004) (available at: https://www.mdpi.com/journal/atoms/special_issues/ManyElectronMultiphotonAtomicProcesses_ATributeMironAmusia).

For citation purposes, cite each article independently as indicated on the article page online and as indicated below:

LastName, A.A.; LastName, B.B.; LastName, C.C. Article Title. *Journal Name* **Year**, *Volume Number*, Page Range.

ISBN 978-3-0365-6598-9 (Hbk)
ISBN 978-3-0365-6599-6 (PDF)

Cover image courtesy of Gleb Gribakin

Contents

atoms MDPI

Editorial

"Atoms" Special Issue (Many-Electron and Multiphoton Atomic Processes: A Tribute to Miron Amusia)

Anatoli S. Kheifets [1,*], Gleb Gribakin [2] and Vadim Ivanov [3]

1 Research School of Physics, The Australian National University, Canberra, ACT 2601, Australia
2 School of Mathematics and Physics, Queen's University Belfast, Belfast BT7 1NN, UK; g.gribakin@qub.ac.uk
3 Department of Physics, Peter the Great St. Petersburg Polytechnic University, 195251 St. Petersburg, Russia; ivanov@physics.spbstu.ru
* Correspondence: a.kheifets@anu.edu.au

The late Professor Miron Amusia was a key figure in theoretical atomic physics on the international stage for more than five decades. His main achievement was the discovery of a collective nature of atomic processes and the role played by many-electron correlations, which have a profound effect on atomic interactions with radiation [1] and matter [2]. Amusia was one of the pioneers in the application of many-body theoretical methods in atomic physics, which offered a universal approach for studying a wide range of processes. It also led to the creation of an unparalleled suite of codes that enabled early calculations of many such processes: photoionization, electron and positron scattering, Auger decays, post-collision interaction, multi-hole decays, and many others.

Amusia's work in atomic physics began in 1966–1967 at the Ioffe Institute in St Petersburg, Russia. There, after discussions with experimentalists [3], severe limitations of the single-electron approximation in atoms became clear. A fundamental result by Amusia from those and the next few years was a theoretical justification and computational implementation of the Random Phase Approximation with Exchange (RPAE)—the first self-consistent (and gauge invariant) method capable of accounting for collective, many-electron effects in atomic photoionization [4–6]. That period saw the creation of the first computer codes of what later became the ATOM suite of programs for atomic calculations. It was followed by a rapid expansion of the community of theorists involved in atomic calculations, extension of the work to new processes and targets, and emergence of international contacts and collaborations (something that was by no means encouraged in the USSR at the time). These collaborations grew to establishing productive ties with leading atomic physics groups in Serbia, Germany, the United States, and Israel. Miron Amusia held invited positions with these groups and was a frequent speaker at conferences and seminars.

For many years, Amusia led a very active weekly atomic physics seminar at the Ioffe Institute. In addition to local speakers, these two-hour seminars saw many invited guests, theorists and experimentalists alike, from around the globe. Suitably qualified graduate students were also involved, first as part of the audience, and later, reporting on their progress, which provided them with an invaluable exposure to sometimes heated but always genuinely interesting and insightful discussions. Later, many of Amusia's pupils created their own groups, often continuing and expanding the work they started in that creative and stimulating environment. Some of their work is represented in this Special Issue, which is dedicated to the memory of Professor Miron Amusia, eminent physicist and great human being.

The papers in this Special Issue reflect the lasting legacy of Professor Amusia. His seminal ideas have been developed by his many long-term collaborators, colleagues, and former students. A special emphasis is on many-electron atomic processes driven by radiation and a charged particle impact. Radiation sources include conventional and free-electron lasers taken to extreme intensities where strongly non-linear multiphoton processes take place.

Citation: Kheifets, A.S.; Gribakin, G.; Ivanov, V. "Atoms" Special Issue (Many-Electron and Multiphoton Atomic Processes: A Tribute to Miron Amusia). *Atoms* **2023**, *11*, 18. https://doi.org/10.3390/atoms11020018

Received: 16 January 2023
Accepted: 17 January 2023
Published: 20 January 2023

It is fitting to first mention the topical review by Chernysheva and Ivanov [7]. Their contribution describes the ATOM program suite and its extensive use for various computational experiments. This suite of codes summarizes many years of computational developments that were led by Miron Amusia and carried out by his group at the Ioffe Institute in St. Petersburg, Russia [8]. The programs included in the ATOM suite are designed to study the electronic structure, transition probabilities, and cross sections of various processes in many-electron atoms. The main numerical methods are presented for taking into account many-electron correlations and determining their role in photoionization, elastic and inelastic particle scattering, decay of vacancies, and several other processes. The most significant results obtained with the ATOM software are highlighted.

One of the key computational developments of the ATOM system was an efficient numerical solution of the Hartree–Fock (HF) equations for many-electron atoms, making it a standard starting point for higher-level calculations. This development has aided and stimulated many other fields of computational atomic physics. Bray et al. [9] have utilized the HF method to take the convergent close-coupling (CCC) method beyond the simplest hydrogen and helium atoms [10,11]. The CCC method was initially developed to describe electron scattering on atomic hydrogen and hydrogenic ions, such as He^+. The latter allows implementation of double photoionization (DPI) of the helium atom. For more complex single-valence-electron atomic and ionic targets, the direct and exchange interaction with the inner electron core needs to be taken into account. For this purpose, the self-consistent field and frozen-core HF computer codes from the ATOM system have been adopted. The utility of the HF technique is demonstrated by examples of electron scattering on Li and the DPI of the H^- and Li^- ions. The authors also discuss the possibility of running modern computer infrastructure associated with the CCC code directly via the Atomic, Molecular and Optical Science Gateway.

The topical review by Shaginyan et al. [12] describes the peculiar physics of heavy-fermion metals, a topic that Amusia actively contributed to over the past decade [13–15]. The present review considers the topological fermion condensation quantum phase transition that leads to flat bands and elucidates the special behavior of heavy-fermion metals, not exhibited by common metals described by the Landau Fermi-liquid theory. The authors bring together theoretical considerations and experimental data on heavy-fermion metals, which demonstrate peculiar thermodynamic, transport, and relaxation properties.

Recent progress in low-energy electron elastic collisions with multi-electron atoms and fullerene molecules is reviewed by Msezane et al. [16]. The authors describe the application of the Regge pole analysis to this process, treating the fullerene molecules as "big atoms" [17]. The authors demonstrate the sensitivity of the Regge-pole-calculated Ramsauer–Townsend minima and shape resonances to the electronic structure and dynamics of the actinide atoms, and their use as a novel and rigorous approach to validation of recent experimental observations.

Nora Berrah [18] provides a perspective on probing fullerene molecules using free-electron lasers (FELs). Non-linear processes in the interaction of atoms and molecules with intense EUV and X-ray fields have been the subject of Amusia's keen interest [19]. Ultra-short and ultra-intense FELs have allowed molecular research in a new photon-energy regime. Illuminated from within by the flow of photoelectrons, the fullerenes reveal fine details of the structural and electronic properties. FELs have allowed the study of the response of fullerenes to X-rays, which includes femtosecond multiphoton processes, as well as time-resolved ionization and fragmentation dynamics.

Grundmann et al. [20] put the process of quasi-free photoionization under the reaction microscope. The novel quasi-free mechanism (QFM) of single-photon double ionization of the helium atom was predicted theoretically by Amusia and co-authors [21], to be confirmed experimentally nearly 50 years later in the group of Experimental Atomic Physics at the Goethe University Frankfurt [22,23]. The work by Grundmann et al. [20] provides new insights into the elusive QFM photoionization. They found a distinct four-fold symmetry in the angular emission pattern of QFM electrons from the He atom and H_2 molecule. Furthermore, they provided experimental evidence that during the quasifree photoionization, the photon momentum is not imparted onto the centre of mass, in contrast to the single ionization and double ionization mediated by the shake-off and knock-out mechanisms.

Horst Schmidt-Böking [24] describes his personal encounters with Professor Amusia, which led to a patented proposal of a new and efficient method of energy storage. The method is based on producing long-life multiply-excited spin-polarized atoms or ions, whose decay is strongly delayed or even blocked by the intra-ionic magnetic stabilization. Specific configurations with huge internal magnetic fields capture only spin-polarized electrons in collisions with spin-aligned atomic hydrogen gas targets.

The work by Lagutin et al. [25] exploits the use of super-intense free-electron laser sources driving atomic photoionization in a strongly non-linear multiphoton regime. The authors studied sequential two-photon double ionization of the Ar atom, with a focus on the role of electron correlations in this process. They demonstrated a strong dependence of the low-energy part of the photoelectron spectrum on both the photon energy and the flux of the exciting beam.

Dolmatov and Manson [26] explore photodetachment of giant and nested fullerene anions. Miron Amusia contributed vastly to the study of the interaction of particles and light with fullerenes and endo-fullerenes [17,27,28]. The negative molecular ions $(C_N@C_M@\ldots)^-$ are formed by adding an electron to several nested fullerene cages, where the attached electron is captured into the s-wave ground state. The authors gain insight into the changes in photodetachment of this valence electron as a function of the different geometries and potentials of the various underlying fullerenes, depending on their increasing size and packing.

Gregg and Gribakin [29] calculate low-energy positron-atom scattering, taking into account meaning is retained strong electron–positron correlations, including the effect of virtual positronium formation. Amusia and co-workers were probably the first to recognize the importance of this effect [30]. The aim of the present study is to find the most computationally economical way of accounting for correlations using a square-integrable wave-function basis. As a demonstration of the utility of their method, the authors calculate the phaseshift and the annihilation rate parameter Z_{eff}, which are found to be in good agreement with other benchmark calculations.

Kheifets [31] considers, theoretically, the shake-off process in non-sequential single-photon double-ionization of closed-shell atomic targets. This process is facilitated by a sudden re-arrangement of the residual ionic core which shakes off an extra electron into continuum. Amusia and Kheifets [32,33] introduced the Green's function formalism to describe the effect of many-electron correlations on the ionization spectra of atoms. This formalism is used in the present work. It is validated by making a comparison with more elaborate techniques, such as convergent and time-dependent close coupling.

Deshmukh and Manson [34] consider photoionization of atomic systems using the relativistic random-phase approximation. In this article, the historical reasons behind the term "random-phase approximation" (RPA) are revisited. A brief introduction to the relativistic RPA (RRPA) is provided, illustrated by a number of applications.

Vinbladh et al. [35] present a theory of two-photon above-threshold ionization and its application to heavy atoms in attosecond science. They employ the Dirac–Fock formalism and account for many-body effects using the relativistic random-phase approximation. Strong relativistic effects are revealed close to ionization thresholds and Cooper minima, predicting differences in the fine-structure level delays as large as tens of attoseconds.

Kabachnik and Sazhina [36] describe theoretically the spin polarization effects in XUV photoionization of atoms dressed with an optical laser field. They show how different photoelectron spectral lines originating from the ionizing XUV radiation and supplemented by the optical dressing display various degrees of spin polarization.

Popova et al. [37] explore theoretically various spectroscopic peculiarities of the Ne photoionization, focusing on the Cooper minima and auto-ionizing resonances. They use the R-matrix approach to calculate the photoionization cross sections for metastable and dipole-allowed excited states.

Kornev et al. [38] consider theoretically the process of X-ray bremsstrahlung in electron scattering from noble-gas atoms. The calculated isochromatic (i.e., fixed-photon energy) spectra as functions of the electron energy, are consistent with the absolute values of the experimental differential cross sections.

Yarzhemsky and Teterin [39] describe the effects of many-electron correlation on the formation of complex satellite structures in photoelectron spectra. In the absence of correlations, the spectra contain only single lines corresponding to one-hole states. The theoretical results obtained for satellites and low-energy Auger lines in Ne, Co, and Th atoms are found to be in agreement with the experiment.

Fritzsche and Böning [40] present the Jena Atomic Calculator, a novel computational tool for studying relativistic atomic structure and dynamics. They illustrate its use by evaluating the above-threshold ionization process for many-electron targets in the strong-field regime. The authors also discuss how this approach can be extended to incorporate re-scattering and high-harmonic generation.

Liverts and co-authors [41,42] focus their attention on spectroscopically accurate calculations for two-electron atomic systems. Such systems are a testbed for various computational techniques aiming to describe many-electron correlations [43,44]. The authors propose a compact, yet very accurate method for computing S-wave functions as linear combinations of a few single exponentials. They further employ the Fock expansion to calculate high-order angular coefficients.

Amusia, Baltenkov, and Woiciechowski [45] study the angular-dependent time delay in the low-energy electron elastic scattering by spherical targets. Specific features of both angular and energy dependencies of the time delay are discussed in detail. Examples of the hard-sphere and delta-shell potential well are considered for illustration.

In conclusion, this brief Editorial summarizes only a fraction of efforts of the followers, colleagues and former students of Miron Amusia, who developed his seminal ideas in various areas of computational and theoretical atomic physics. The present collection of works demonstrates vividly that the ideas and methods pioneered by Amusia are still relevant and fruitful, and continue to impact on this field of research.

Author Contributions: All authors contributed equally to writing this Editorial. All authors have read and agreed to the published version of the manuscript.

Conflicts of Interest: The authors declare no known conflicts of interest.

References

1. Amusia, M.Y. *Atomic Photoeffect*; Plenum Press: New York, NY, USA, 1990.
2. Amusia, M.Y.; Cherepkov, N.A. Many-electron correlations in scattering processes. *Case Stud. At. Phys.* **1975**, *5*, 47–179.
3. Aleksandrov, A.P.; Artsimovich, L.A.; Afrosimov, V.V.; Gaev, B.A.; Tuchkevich, V.M.; Flerov, G.N. Nikolai Vasil'evich Fedorenko (Obituary). *Sov. Phys. Uspekhi* **1973**, *15*, 530–531. https://doi.org/10.1070/PU1973v015n04ABEH005027.
4. Amusia, M.; Cherepkov, N.; Sheftel, S. Many-body correlations and the photoeffect. *Phys. Lett. A* **1967**, *24*, 541–542. https://doi.org/10.1016/0375-9601(67)90827-4.
5. Amusia, M.Y.; Cherepkov, N.A.; Chernysheva, L.V. Cross section for the photoionization of noble-gas atoms with allowance for multielectron correlations. *JETP (Sov. Phys.)* **1971**, *33*, 90.
6. Amusia, M.; Ivanov, V.; Cherepkov, N.; Chernysheva, L. Interference effects in photoionization of noble gas atoms outer s-subshells. *Phys. Lett. A* **1972**, *40*, 361–362. https://doi.org/10.1016/0375-9601(72)90531-2.
7. Chernysheva, L.V.; Ivanov, V.K. ATOM Program System and Computational Experiment. *Atoms* **2022**, *10*, 52. https://doi.org/10.3390/atoms10020052.
8. Amusia, M.I.; Chernysheva, L.V. *Computation of Atomic Processes: A Handbook for the ATOM Programs*; Institute of Physics Publishing Ltd.: Bristol, UK; Philadelphia, PA, USA, 1997.
9. Bray, I.; Weber, X.; Fursa, D.V.; Kadyrov, A.S.; Schneider, B.I.; Pamidighantam, S.; Cytowski, M.; Kheifets, A.S. Taking the Convergent Close-Coupling Method beyond Helium: The Utility of the Hartree-Fock Theory. *Atoms* **2022**, *10*, 22. https://doi.org/10.3390/atoms10010022.
10. Bray, I.; Fursa, D.; Kadyrov, A.; Stelbovics, A.; Kheifets, A.; Mukhamedzhanov, A. Electron- and photon-impact atomic ionisation. *Phys. Rep.* **2012**, *520*, 135. https://doi.org/10.1016/j.physrep.2012.07.002.

11. Bray, I.; Fursa, D.V.; Kheifets, A.S.; Stelbovics, A.T. Electrons and photons colliding with atoms. *J. Phys. B* **2002**, *35*, R117–R146. https://doi.org/10.1088/0953-4075/35/15/201.
12. Shaginyan, V.R.; Msezane, A.Z.; Japaridze, G.S. Peculiar Physics of Heavy-Fermion Metals: Theory versus Experiment. *Atoms* **2022**, *10*, 67. https://doi.org/10.3390/atoms10030067.
13. Amusia, M.; Shaginyan, V. Quasiparticle dispersion and lineshape in a strongly correlated liquid with the fermion condensate. *Phys. Lett. A* **2000**, *275*, 124–130. https://doi.org/10.1016/S0375-9601(00)00569-7.
14. Shaginyan, V.; Amusia, M.; Msezane, A.; Popov, K. Scaling behavior of heavy fermion metals. *Phys. Rep.* **2010**, *492*, 31–109. https://doi.org/10.1016/j.physrep.2010.03.001.
15. Amusia, M.Y.; Popov, K.G.; Shaginyan, V.R.; Stephanovich, V.A. Theory of heavy-fermion compounds. *Springer Ser. Solid-State Sci.* **2014**, *182*, 33.
16. Msezane, A.Z.; Felfli, Z. Recent Progress in Low-Energy Electron Elastic-Collisions with Multi-Electron Atoms and Fullerene Molecules. *Atoms* **2022**, *10*, 79. https://doi.org/10.3390/atoms10030079.
17. Amusia, M. Fullerenes and endohedrals as "big atoms". *Chem. Phys.* **2013**, *414*, 168. https://doi.org/10.1016/j.chemphys.2012.08.007.
18. Berrah, N. Probing C_{60} Fullerenes from within Using Free Electron Lasers. *Atoms* **2022**, *10*, 75. https://doi.org/10.3390/atoms10030075.
19. Berrah, N.; Bozek, J.; Costello, J.; Düsterer, S.; Fang, L.; Feldhaus, J.; Fukuzawa, H.; Hoener, M.; Jiang, Y.; Johnsson, P.; et al. Non-linear processes in the interaction of atoms and molecules with intense XUV and X-ray fields from free electron lasers. *J. Mod. Opt.* **2010**, *57*, 1015–1040. https://doi.org/10.1080/09500340.2010.487946.
20. Grundmann, S.; Trinter, F.; Fang, Y.K.; Fehre, K.; Strenger, N.; Pier, A.; Kaiser, L.; Kircher, M.; Peng, L.Y.; Jahnke, T.; et al. Quasifree Photoionization under the Reaction Microscope. *Atoms* **2022**, *10*, 68. https://doi.org/10.3390/atoms10030068.
21. Amusia, M.Y.; Drukarev, E.G.; Gorshkov, V.G.; Kazachkov, M.O. Two-electron photoionization of helium. *J. Phys. B* **1975**, *8*, 1248.
22. Schöffler, M.S.; Stuck, C.; Waitz, M.; Trinter, F.; Jahnke, T.; Lenz, U.; Jones, M.; Belkacem, A.; Landers, A.L.; Pindzola, M.S.; et al. Ejection of Quasi-Free-Electron Pairs from the Helium-Atom Ground State by Single-Photon Absorption. *Phys. Rev. Lett.* **2013**, *111*, 013003.
23. Grundmann, S.; Trinter, F.; Bray, A.W.; Eckart, S.; Rist, J.; Kastirke, G.; Metz, D.; Klumpp, S.; Viefhaus, J.; Schmidt, L.P.H.; et al. Separating Dipole and Quadrupole Contributions to Single-Photon Double Ionization. *Phys. Rev. Lett.* **2018**, *121*, 173003.
24. Schmidt-Böcking, H.; Gruber, G. On Producing Long-Lived Spin Polarized Metastable Atoms: Feasibility of Storing Electric Energy. *Atoms* **2022**, *10*, 76. https://doi.org/10.3390/atoms10030076.
25. Lagutin, B.M.; Petrov, I.D.; Sukhorukov, V.L.; Kilin, V.A.; Novikovskiy, N.M.; Demekhin, P.V.; Ehresmann, A. Electron Correlations in Sequential Two-Photon Double Ionization of an Ar Atom. *Atoms* **2022**, *10*, 139. https://doi.org/10.3390/atoms10040139.
26. Dolmatov, V.K.; Manson, S.T. A Glimpse into Photodetachment Spectra of Giant and Nested Fullerene Anions. *Atoms* **2022**, *10*, 99. https://doi.org/10.3390/atoms10040099.
27. Amusia, M.; Baltenkov, A.; Krakov, B. Photodetachment of negative $C_{60}-$ ions. *Phys. Lett. A* **1998**, *243*, 99–105. https://doi.org/10.1016/S0375-9601(98)00158-3.
28. Amusia, M.Y.; Baltenkov, A.S.; Chernysheva, L.V.; Felfli, Z.; Msezane, A.Z. Dramatic distortion of the 4*d* giant resonance by the C_{60} fullerene shell. *J. Phys. B* **2005**, *38*, L169.
29. Gregg, S.; Gribakin, G. Calculation of Low-Energy Positron-Atom Scattering with Square-Integrable Wavefunctions. *Atoms* **2022**, *10*, 97. https://doi.org/10.3390/atoms10040097.
30. Amusia, M.Y.; Cherepkov, N.A.; Chernysheva, L.V.; Shapiro, S.G. Elastic scattering of slow positrons by helium. *J. Phys. B* **1976**, *9*, L531.
31. Kheifets, A.S. Shake-Off Process in Non-Sequential Single-Photon Double Ionization of Closed-Shell Atomic Targets. *Atoms* **2022**, *10*, 89. https://doi.org/10.3390/atoms10030089.
32. Amusia, M.Y.; Kheifets, A.S. Effect of correlations on the photoelectron spectrum of atom. *Sov. Phys.-JETP* **1984**, *59*, 710–715.
33. Amusia, M.Y.; Kheifets, A.S. On our ability to measure the singly ionized rare-gas spectroscopic factors using the (γ,e) and (e,2e) reactions. *J. Phys. B* **1985**, *18*, L679–684.
34. Deshmukh, P.C.; Manson, S.T. Photoionization of Atomic Systems Using the Random-Phase Approximation Including Relativistic Interactions. *Atoms* **2022**, *10*, 71. https://doi.org/10.3390/atoms10030071.
35. Vinbladh, J.; Dahlström, J.M.; Lindroth, E. Relativistic Two-Photon Matrix Elements for Attosecond Delays. *Atoms* **2022**, *10*, 80. https://doi.org/10.3390/atoms10030080.
36. Kabachnik, N.M.; Sazhina, I.P. Spin Polarization of Electrons in Two-Color XUV + Optical Photoionization of Atoms. *Atoms* **2022**, *10*, 66. https://doi.org/10.3390/atoms10020066.
37. Popova, M.M.; Kiselev, M.D.; Burkov, S.M.; Gryzlova, E.V.; Grum-Grzhimailo, A.N. Spectroscopic Peculiarities at Ionization of Excited States of Ne: Cooper Minima and Autoionizing Resonances. *Atoms* **2022**, *10*, 102. https://doi.org/10.3390/atoms10040102.
38. Kornev, A.S.; Zon, B.A.; Chernov, V.E.; Amusia, M.Y.; Kubelík, P.; Ferus, M. Absolute Double-Differential Cross Sections of Ultrasoft Isochromatic X-ray Radiation in Electron Scattering on Atoms. *Atoms* **2022**, *10*, 86. https://doi.org/10.3390/atoms10030086.
39. Yarzhemsky, V.G.; Teterin, Y.A. Satellite Excitations and Final State Interactions in Atomic Photoionization. *Atoms* **2022**, *10*, 73. https://doi.org/10.3390/atoms10030073.
40. Fritzsche, S.; Böning, B. Strong-Field Ionization Amplitudes for Atomic Many-Electron Targets. *Atoms* **2022**, *10*, 70. https://doi.org/10.3390/atoms10030070.

41. Liverts, E.Z.; Barnea, N. Accurate Exponential Representations for the Ground State Wave Functions of the Collinear Two-Electron Atomic Systems. *Atoms* **2022**, *10*, 4. doi:10.3390/atoms10010004.
42. Liverts, E.Z.; Krivec, R. Fock Expansion for Two-Electron Atoms: High-Order Angular Coefficients. *Atoms* **2022**, *10*, 135. doi:10.3390/atoms10040135.
43. Amusia, M.Y.; Drukarev, E.G.; Liverts, E.Z.; Mikhailov, A.I. Effects of small recoil momenta in one-photon two-electron ionization. *Phys. Rev. A* **2013**, *87*, 043423.
44. Amusia, M.Y.; Drukarev, E.G.; Liverts, E.Z. Small recoil momenta double ionization of He and two-electron ions by high energy photons. *Eur. Phys. J. D* **2020**, *74*, 173.
45. Amusia, M.Y.; Baltenkov, A.S.; Woiciechowski, I. Time Delay in Electron Collision with a Spherical Target as a Function of the Scattering Angle. *Atoms* **2021**, *9*, 105. https://doi.org/10.3390/atoms9040105.

atoms MDPI

Review

ATOM Program System and Computational Experiment

Larissa V. Chernysheva [1,*] and Vadim K. Ivanov [2,*]

[1] Sector of Solid State Theory, A.F. Ioffe Physical-Technical Institute RAS, 194021 St. Petersburg, Russia
[2] Department of Physics, St. Petersburg Polytechnic University of Peter the Great, 195251 St. Petersburg, Russia
* Correspondence: larissa.chernysheva@mail.ioffe.ru (L.V.C.); ivanov@physics.spbstu.ru (V.K.I.)

Abstract: The article is devoted to a brief description of the ATOM computer program system, designed to study the structure, transition probabilities and cross sections of various processes in multielectron atoms. The theoretical study was based on the concept of a computational experiment, the main provisions of which are discussed in the article. The main approximate methods used in the system of programs for taking many-electron correlations into account and determining their role in photoionization processes, elastic and inelastic electron scattering, the decay of vacancies, and many others are presented. The most significant results obtained with this software are listed.

Keywords: atom; RPAE; photoionization; electron-impact ionization; ionization of the inner shells; decay of vacancies; computer

Citation: Chernysheva, L.V.; Ivanov, V.K. ATOM Program System and Computational Experiment. *Atoms* **2022**, *10*, 52. https://doi.org/10.3390/atoms10020052

Academic Editor: Yew Kam Ho

Received: 16 April 2022
Accepted: 18 May 2022
Published: 24 May 2022

1. Introduction

In the theoretical study of multielectron atomic systems, software plays an important role because it is necessary for calculating the energy and spatial structure and characteristics of various processes occurring in atoms. Calculations in the framework of relatively simple one-electron models did not lead to agreement between calculations and experimental data, despite the fact that the pair interaction between electrons is the well-known *Coulomb interaction* [1–7]. In the 1960s–1970s, it became completely clear that in order to describe the structure and reaction of atoms to external influences, it is necessary to go beyond single-electron concepts and take collective multielectron effects into account [3–7]. At the same time, in the theoretical group of the St. Petersburg Ioffe Institute, under the guidance of Prof. Miron Ya. Amusia, began to create a package of applied programs known as ATOM [8–10], which was improved in subsequent years and aided a huge number of direct calculations that were carried out in atomic systems to describe experimental data and predict new many-particle effects. These calculations laid the foundation for the development of the concept of a computational experiment.

Theoretical studies of many-body systems, as a rule, are accompanied by the need to take many-body interactions into account. At the same time, the determination of the spatial and energy structure, as well as the response of such systems to external influences, faces significant computational difficulties, especially in cases where perturbation theory cannot be consistently applied [7–11]. The theory of atoms and atomic particles is also facing such difficulties despite the fact that the pair interaction between electrons has been well-known to physicists for more than a century. Nevertheless, a significant development of methods for describing multielectron atoms and their interactions with external fields began in the second half of the last century when the apparatus of many-body theory was developed, great computational capabilities appeared, and experimental studies of atomic systems began to be widely carried out [8–10]. These experiments showed that the simplest description of atoms based on single-particle models—the Hartree–Fock (HF) approximation, in particular [12]—is unable to describe the characteristics and properties of the behavior of atoms in cases of their interactions with electromagnetic fields and other particles.

One of the most successful theoretical approaches to describing the structure and processes in atoms are approximations based on many-body theory, particularly including perturbation theory [13] and random phase exchange approximation (RPAE) [3,5,6]. It was the RPAE method, developed as the 1960s turned into the 1970s, that first made it possible to describe the photoionization cross sections of multielectron atoms and to show the decisive role of many-particle (collective) interactions in atomic systems. The big advantage of the RPAE is the self-consistency of this approximation, within which the main general laws of the processes of the ionization and excitation of electrons, such as the oscillator sum rule and the condition of gradient invariance, are satisfied [7,8]. Subsequently, the application of this approximation made it possible to describe and explain numerous experimental data and predict many-electron effects that manifest themselves in the processes of the ionization and excitation of atomic systems. It is these calculations that have stimulated numerous theoretical and experimental studies of many-particle processes in various research centers and laboratories.

The theoretical group created by Prof. Miron Ya. Amusia has developed and described the concepts of a computational experiment that is currently considered to be a new methodology and technology of scientific research [14,15]. The implementation of this concept makes it possible to use the capabilities of computers in combination with existing traditional research methods, creating a new style of research work that combines the work of theoreticians, experimenters and computer programmers.

The purpose of this work was to present a brief description of the ATOM program system and its capabilities, as well as a generalization of the results obtained in atomic physics using the developed software. In particular, this paper describes a computational experiment on the development and study using the ATOM system of theoretical models for studying the structure of atoms, simple molecules, endohedral atoms and their interactions with external fields.

2. Software for the Computational Experiment

The main task of the computational experiment is to develop numerical models for the specified range of problems within the framework of a pre-selected theoretical approach. The Hartree–Fock (HF) method is taken as the initial approximation, and many-electron correlations in atoms are taken into account in the framework of the RPAE or its simplified modifications [8–10].

The theoretical model of any process under study in the chosen approximation should provide a satisfactory mathematical description of the experiment. First of all, this model includes the choice of the wave functions of atomic electrons in both the ground and excited states, as well as additional external particles participating in the process. The wave functions of the ground state of an atom are determined in the HF approximation. The functions of excited states in the HF method can be found in a self-consistent field or in the field of the "frozen" atomic core for electrons. The choice of the one-particle approximation is determined by both the problem under consideration and the role of many-electron interactions, some of which can be taken into account by the choice of one-particle wave functions. In addition, not only electrons but also other particles can act as additional external particles, e.g., mesons and positrons. The single-particle transition amplitudes are determined in terms of wave functions, which makes it possible to calculate the characteristics of atoms or the probabilities of processes in the HF approximations. Then, the matrix elements of the interaction between electrons in various processes are calculated, and this makes it possible to find the characteristics of atoms or the probabilities of processes while considering multielectron correlations, RPAE in particular [8,9].

The ATOM system was created as a result of many years of research into the structure and processes in multielectron atoms; it was designed to calculate the characteristics of atoms, endohedral fullerenes, and diatomic molecules, as well as the probabilities of the processes of interaction of electrons, photons and other particles, on a computer [3–10,16,17].

Let us briefly list some main characteristics of atoms and processes that are determined within the framework of this system:

- Wave functions of atoms in the ground and excited states in the HF and Hartree–Fock–Dirac approximations.
- Amplitudes and cross sections of photoionization (including the strength of oscillators of discrete transitions) of atoms with filled and half-filled shells while taking the interaction between electrons of one, two or more shells into account.
- Atom polarizability.
- Characteristics of the angular distribution of photoelectrons and secondary electrons both in the dipole and outside the dipole approximation.
- Parameters of the spin polarization of photoelectrons.
- Scattering cross sections of fast electrons through the generalized oscillator strengths, taking the influence of electrons in one, two or more shells into account.
- Angular distribution of secondary electrons arising from the inelastic scattering of fast particles on atoms.
- Phases and cross sections of the elastic and inelastic scattering of particles (electrons, positrons, and mesons) of low and medium energies on atoms.
- Photoabsorption cross sections with allowance for the decay of vacancies and the inelastic scattering of a photoelectron.
- Cross sections of the ionization and excitation of an atom by electron impact.
- Probabilities of the single-electron and double Auger decay of vacancies in atoms.
- Probabilities of the one-photon decay of one- and two-hole states.
- Characteristics of bremsstrahlung of high and intermediate energy incident particles.
- Characteristics of capture of mu-mesons by atoms.
- Amplitudes, photoionization cross sections, and angular anisotropy parameters of endohedral atoms and the decay of vacancies in such atoms.
- Characteristics of the inelastic scattering of fast electrons on endohedral atoms.
- Characteristics of photoabsorption processes in negative and positive ions.

The creation of theoretical models is only the first step in a computational experiment. On their basis, numerical models are constructed and make it possible to obtain an approximate solution of the initial problems on a computer with the required accuracy. The wave functions of an atom in the HF approximation are represented as a product of the radial, angular and spin parts. Radial functions are the solution of the HF equation (or system of equations) through the method of successive approximations. Integration over the angular variables of the wave functions and summation over the spins are carried out analytically and enter the expression for the matrix elements, which are the results of solving integral equations in RPAE [9,10]. The multidimensional integrals in these equations are reduced to one-dimensional ones after separating the corner parts and integrating over the angle variables. The RPAE equations are transformed into a system of algebraic equations, the solutions of which are reduced to the inversion of matrices. The expression for the amplitudes of physical processes, defined in terms of matrix elements in RPAE, often contains summation and integration over intermediate states and may have singularities due to energy denominators [9,10]. The integrals of the emerging singularities are analytically calculated. The resulting expression is "matched" with the contribution of non-singular regions, which is found as a result of numerical integration. In this case, the presence of an imaginary additive in the energy denominator of the calculated amplitudes leads to the fact that the process amplitude has both real and imaginary parts.

The numerical solution of the problem, which makes it possible to study the chosen physical process, is sought without separation from physical considerations, which often lead to model simplification. For example, when discretizing the original model, an important element of the calculation is the choice of a sufficiently high upper limit; up to this limit, instead of integration over the continuous spectrum, summation is carried out on a computer [8,9]. The practice of calculations shows that in the physics of the atom, one has to deal with matrix elements that rather quickly decrease as the energies of the states entering into them increase. In addition, to improve the accuracy of calculations, it is advisable to choose sampling points so that the largest number is in the region of relatively low energies E. This condition is ensured by using the electron momentum $p = E^{1/2}$ as an integration variable.

3. Organization of the Computational Experiment

The optimal organization of the computational process depends on the choice of theoretical and technical characteristics of computational algorithms. The theoretical parameters of the algorithms are related to the formulation of the problem. They include a formal description of the problem, a solution method, the algorithm itself, and its implementation in the chosen language. Technical characteristics depend on the computer used and include system tools, the capabilities and features of the algorithmic languages used, the form of presentation, and the storage of initial information.

The choice of strategy in the development of the algorithm itself includes a number of issues. First, in all possible cases, the complex initial problem is divided into a number of simple ones that are easier to implement on a computer. Due to the allocation of subtasks into independent modules, such as the calculation of dipole and Coulomb matrix elements, integration with a pole, matrix inversion, the problem of using them in the study of new physical processes is simplified. Furthermore, in each of the subtasks, the necessary algebraic simplifications are carried out. In particular, the change of variable required in the calculation of wave functions is carried out in the calculation of all characteristics of atoms, since the matrix elements used in them are expressed in terms of wave functions. Access to auxiliary (intermediate) quantities is as important for a theoretical physicist as measurements are for an experimenter in an experiment. In the processes under study, auxiliary quantities usually refer to the matrix elements that are used to obtain qualitative estimates.

One of the components of the algorithm is the analysis of the accuracy of calculations. The accuracy of theoretical models is generally unknown, and it is established with the help of estimates obtained as a result of roughly approximate calculations. The accuracy of computer calculations can be very high. It is advisable to use an accuracy that it is somewhat higher than the expected error of the physical result itself. Often the error of intermediate results is more important, since the value obtained in a physical experiment is found as the difference of large numbers. In addition, in the process of computing, the error can accumulate, and, as a rule, it is difficult to evaluate it. Therefore, the final result on a computer often does not require a level of accuracy as high as that needed for intermediate data and that realized at intermediate stages. All this is considered when choosing numerical methods for solving problems.

The development of computational algorithms is the second stage of the computational experiment. In the next steps, computer programs are compiled that implement the selected algorithms. The ATOM system uses numerical methods for the theoretical study of the structure of complex atoms and the processes occurring with their participation [11]. The high-level language Fortran, which is the main language in many physical calculations, is used to write the algorithms. To facilitate the implementation of the ATOM system on other computers, Fortran does not use the features and extensions of languages implemented in some translators. Algorithms are recorded in accordance with the respective technology. Algorithms are designed in the form of modules, which are divided into three groups. Modules of the first type are procedures without formal parameters and include the description of variables, the input and printing of initial data, the description of the algorithm containing printing of intermediate values, and the output of results. These are executive programs, each of which solves an independent physical problem. Modules of the second type (specialized) are programs or function procedures with formal parameters. They contain descriptions of the variables and the algorithm. Modules of the third type (service) contain descriptions of variables, the input and printing of initial data, or the printing of results. These modules are used in the development and assembly of modules of the first type. Modules of the fourth type (generic) implement standard mathematical methods. For clarity and ease of finding errors, variables in modules are grouped according to their purpose; variables common to all modules are assigned permanent identifiers, and the names of the variables are usually abbreviated names of the physical processes under study. The input of initial data is accompanied by the printing of all physical quantities. The same is done for intermediate values and program results. Detail printing turns a software module into a theoretical physicist's tool; it plays the same role as diagnostics in a natural experiment for an experimenter.

The ATOM system includes an application program (AP) and a database (DB). The AP contains more than 50 executive modules (according to the number of physical tasks to be solved), more than 10 service modules, more than 70 specialized modules and more than 16 generic modules, and it has a hierarchical structure. During its development, the following requirements were taken into account: the modular principle of organization, the constant expansion of the system's capabilities, the convenience of users, ease of implementation, and the use of the system as part of the software for the computing network for collective use. The ATOM system was built with basic tools, so it is easily implemented and widely used in many places. For each atom and each physical process under study, the DB contains the required wave functions, input and output physical characteristics.

The input language of the ATOM system belongs to the class of task languages that allow for a wide class of users who do not have special training in programming to work.

An important stage of the computational experiment is the implementation of computer calculations, during which the capabilities of the created programs are constantly expanding. At present, the ATOM system enables the solving of the following problems [8,9]. In the Hartree and HF approximations, one can obtain various wave functions, namely wave functions of the ground state of an atom, wave functions of excited states consistent with the functions of the ground state, wave functions of excited states in a continuous spectrum for given energies in a fixed field of an atom with or without orthogonalization to the wave functions of the ground state, wave functions of excited states in the discrete spectrum for given values of the principal quantum number in a fixed core field with or without orthogonalization to the wave functions of the ground state, and wave functions of the mu-meson and positron.

Because it has a set of necessary wave functions in the DB, the ATOM system allows one to determine, in the HF and RPAE approximations, matrix elements within one or two transitions, each of which is characterized by one wave function of the ground (hole) state and a set of wave functions of excited (partial) state discrete and continuous spectra for a finite series of energies, namely dipole matrix elements of the length and velocity form, the Coulomb matrix of effective interaction, matrix elements of the terms of the expansion of a plane wave in a series in terms of Legendre polynomials, and Coulomb matrix elements such as "three particles–one hole", "two particles–two holes", and "three holes–one particle". The resulting matrix elements are the basis for studying a number of processes.

Using the example of calculations of the photoionization processes in the framework of the RPAE, we present the main stages of calculations. First, the wave functions of the ground and excited states are determined in the HF approximation. Using the obtained wave functions, the dipole matrix elements of phototransitions to discrete states and states of the continuous spectrum are calculated in the zeroth approximation. In the next stage, the matrix elements of the Coulomb interaction between all the HF states involved in the process, both real and virtual, are calculated. Furthermore, integral equations are solved in the program to determine the amplitudes and cross sections of phototransitions while taking multielectron correlations within the framework of the RPAE into account.

Thus, the analytical expression for the photoionization amplitude upon the absorption of a quantum with frequency ω, considering intra- and intershell interactions, can be represented in the following symbolic form [9,11]:

$$\langle \nu | \hat{D}(\omega) | i \rangle = \left\langle \nu \middle| \hat{d} \middle| i \right\rangle + \left(\sum_{\substack{k_2 > F \\ k_1 < F}} - \sum_{\substack{k_1 > F \\ k_2 < F}} \right) \frac{\langle k_2 | \hat{D}(\omega) | k_1 \rangle \langle \nu, k_1 | \hat{U} | i, k_2 \rangle}{\omega - E_{k_2} + E_{k_1} + i\delta(1 - 2n_{k_2})} \quad (1)$$

Here, $\left\langle \nu \middle| \hat{d} \middle| i \right\rangle$ and $\langle \nu | \hat{D}(\omega) | i \rangle$ are the dipole matrix elements of the transition between the $|i\rangle$-initial and $\langle \nu|$-final states in the HF and RPAE approximations, respectively. The matrix element of the interaction between electrons involved in the transition between intermediate $|k_1\rangle$ and final $|k_2\rangle$ states is determined by the sum of the direct and exchange matrix elements.

$$\langle \nu, k_1 | \hat{U} | i, k_2 \rangle = \langle \nu, k_1 | \hat{V} | i, k_2 \rangle - \langle \nu, k_1 | \hat{V} | k_2, i \rangle \quad (2)$$

where $\langle \nu, k_1 | \hat{V} | i, k_2 \rangle$ is the matrix element of the Coulomb interaction. Summation (integration) over the intermediate states $|k_1\rangle$ and $|k_2\rangle$ with energies E_{k_1} and E_{k_2} involved in the process above and below Fermi level F is carried out in both a time-direct process and a time-reversible process; n_k is the Fermi step $n_k = \begin{cases} 0\, k > F \\ 1\, k < F \end{cases}$.

When only intrashell correlations are taken into account, summation in (1) is carried out only over intermediate excited states of electrons in one shell under consideration. When intershell interactions are taken into account in the summation, the correlation term takes the transitions of electrons in different shells involved in the process into account.

As a result of Solution (1), we obtain the transition amplitude in the RPAE approximation, which is substituted into the formulas for determining the partial or total photoionization cross sections ($\hbar = m_e = e = 1$) [11]:

$$\sigma_{i \to \nu}(\omega) = \frac{4\pi^2}{\omega c} \int |D_{i\nu}(\omega)|_2 \delta(E_\nu - E_i - \omega) d\nu \quad (3)$$

where $\omega_{\nu i} = E_v - E_i$ is the transition energy and ν is the total set of quantum numbers that characterize the final state.

4. Interference Effects in the Processes of Photoionization of Atoms

The photoionization cross section (oscillator strengths) and the anisotropy coefficients of the angular distribution of photoelectrons are expressed in terms of the dipole matrix elements of the coordinate or momentum and are determined by considering the interactions of all electrons of the shell under study with each other, as well as intershell and intersubshell interactions.

Within the framework of the one-particle HF approximation, it was not possible to obtain agreement between the theory and the available experimental data, on photoabsorption in atoms in particular. In addition, the results of calculations in the HF approximation did not obey the conditions of gauge invariance, since the photoionization cross sections obtained with different transition dipole operators gave different results. It became obvious that it is necessary to go beyond the HF approximation when describing photoionization processes. This means, by definition, that it is necessary to consider the many-electron correlations caused by the part of the electron–electron interaction that is neglected in determining the self-consistent mean field.

The correlation interaction between the electrons of an atom can be taken into account using a number of theoretical methods (review [18]). Those that use the apparatus of the theory of many bodies and apply the diagram technique, namely RPAE [5,11,19] and many-particle perturbation theory (MPT) [20], are widespread. Based on the HF approximation as the zeroth approximation, the apparatus of many-body theory makes it possible to represent the mechanism of any process under consideration in the lowest nonvanishing order of perturbation theory in terms of interelectronic interaction and to present corrections to it in higher orders.

The development of the RPAE to account for many-electron correlations in atoms was the next step after the HF approximation in the creation of self-consistent approximations in many-body theory. From a many-body perspective, RPAE takes an infinite sequence of perturbation series terms of a certain class (class of diagrams) that contribute the most in each order of perturbation theory into account. The self-consistency of the approximation and its advantage also lie in the fact that, within its framework, one can obtain the gauge invariance of the obtained photoionization cross sections, namely the equality of the results in calculations with different types of dipole operators (in contrast to the HF approximation).

The first natural step in considering the correlation interaction, sometimes called residual, was to consider it between the electrons of only one shell, since the latter are well-separated spatially and energetically from the electrons of other shells. In application to the study of the processes of the ionization and excitation of atoms, correlations were first successfully taken into account in the framework of the RPAE [5,11]. In these calculations, it turned out to be sufficient to take the residual interaction between the electrons of the ionizable shell (intrashell interaction) into account in order to obtain satisfactory agreement with the experimental data on the total photoionization cross sections in the photon energy range from ionization thresholds to several hundreds of electron volts. It turned out that the residual interaction between electrons is essential for all outer and intermediate multielectron shells (p^6, d^{10}, and f^{14}), which make the largest contribution to the total photoionization cross section.

Intrashell correlations can significantly change the value of the photoionization cross section but usually do not lead to its qualitative changes depending on the energy. Intershell interaction, on the contrary, often leads to qualitative changes.

In partial ionization cross sections of few-electron shells whose electrons participate in relatively "weak" transitions, the role of intrashell correlations is usually small. However, when describing the partial cross sections for the ionization and excitation of such electrons, it is necessary to take intershell correlations (intershell effects) into account, since few-electron shells, e.g., ns^2, are subject to a strong screening effect of the surrounding many-electron shells. In other words, their behavior is completely collectivized and determined by the surrounding many-electron shells. Although the atomic shells are well-separated from each other spatially and energetically, considering the connection of electrons in different shells turns out to be very important in describing a number of dynamic processes in the atom. Intershell interactions manifest themselves most strongly in the ionization cross sections of few-electron shells, total ionization cross sections at the thresholds of inner shells, the angular distribution and polarization of photoelectrons, the decay of vacancies in inner shells, and photoelectron spectra. The prediction of a significant effect of many-electron shells on few-electron shells [7,11,21], the complete loss of their individuality by the latter, and their consequent collectivization served as the impetus for a wide experimental and theoretical study of the manifestations of intershell interactions.

Intershell interactions most clearly manifest themselves in the study of the photoionization cross sections of $5s^2$ shells in atoms whose electrons have completely lost the features of individual behavior. In this case, it is appropriate to speak of the collectivization of $5s^2$ electrons under the influence of the surrounding $4d^{10}$ and $5p^6$ multielectron shells. Under the influence of external electrons, a minimum appears in the partial cross section for the photoionization of $5s$ electrons, followed by a maximum under the influence of electrons from the inner shell. Such collectivization is typical for the $5s^2$ shells of many elements, beginning with Cd (Z = 48). However, the dependence of the photoionization cross section on energy undergoes successive changes with increasing nuclear charge Z. Thus, when going from Xe to La, the interference minimum shifts to the region of the discrete excitation spectrum. It should be noted that the collectivization of ns^2 electrons was initially theoretically predicted and then confirmed by experimental measurements.

The study of the angular distribution and polarization of photoelectrons in principle provides more detailed information about dipole transitions in atoms and the effect of many-electron correlations, since the parameters describing these processes are determined by the transition amplitudes together with the scattering phases of a photoelectron in the ion field. Thus, the angular distribution of photoelectrons knocked out of a shell with quantum numbers n, l when an unpolarized atom is irradiated with unpolarized light is determined by the expression [7–9]:

$$\frac{d\sigma_{nl}}{d\Omega} = \frac{\sigma_{nl}(\omega)}{4\pi}\left[1 - \frac{1}{2}\beta_{nl}(\omega)P_2(\cos\theta)\right] \tag{4}$$

where $P_2(\cos\theta)$ is the Legendre polynomial, $d\Omega$ is the element of the solid angle of emission of a photoelectron, and $\sigma_{nl}(\omega)$ is the total photoionization cross section of the nl-shell (obtained from (1)–(3)). The angular anisotropy parameter $\beta_{nl}(\omega)$ is expressed in terms of dipole matrix elements and photoelectron scattering phases with angular momenta $(l \pm 1)$.

The intrashell interaction, as a rule, does not lead to qualitative changes in the dependence of the anisotropy parameter in comparison to a single-particle calculation. The situation is different when the intershell interaction is taken into account: the transition amplitude can acquire additional zeros, maxima, and minima, which is reflected in quantitative and qualitative changes in the partial photoionization cross sections and angular distributions of photoelectrons [7,11]. A striking example is the behavior of the angular anisotropy parameter of electrons in the $5p^6$ shell [22,23].

The study of the polarization of photoelectrons also makes it possible to obtain even more detailed information about the behavior of dipole amplitude. The experimental determination of partial cross sections, angular distributions of photoelectrons, and their polarization forms a so-called full quantum-mechanical experiment, which makes it possible to measure all the amplitudes characterizing photoionization, including their real and imaginary parts. Calculations have shown that the degree of polarization of photoelectrons as a function of energy is very sensitive to variations in the amplitudes of dipole transitions and, consequently, to manifestations of intershell interactions [11,24].

Many-electron effects are often more significant in negative atomic ions [11,25] than in neutral atoms, since the interaction between outer electrons is relatively stronger due to more complete screening of the Coulomb field of the nucleus. The formation of negative ions mainly occurs due to the polarization attractive interaction between an additional electron and a neutral atom. Therefore, in determining the cross sections for the photodetachment of electrons from negative ions, in addition to RPAE correlations, it is necessary to take the polarization potential into account. The latter can be conveniently achieved by redefining the wave functions of the additional electron: instead of the HF functions, use Dyson orbitals, which take the self-energy parts of the external electron into account [25].

Let us list the most important results obtained in the study of many-electron correlations in the processes of photoionization and photoexcitation. Most of the results were obtained using the ATOM system [7–11,16,17,24].

1. It has been demonstrated that giant resonances in the photoionization cross sections of a number of atoms, xenon (Xe) and its neighbors such as iodine (I), cesium (Cs) and lanthanides in particular, are completely multielectron in nature. They are analogues of plasmons in solids, and at least all ten electrons of the $4d^{10}$ subshell participate in their formation in atoms.
2. It was predicted that the action of multielectron shells qualitatively changes the photoionization cross section of few-electron subshells, leading to a new continuous spectrum, the so-called *interference resonances* [6,11]. These resonances are a direct

consequence of the interaction between electrons that belong to different subshells or even shells, and they have indeed been observed in many atoms.

3. It has been shown that the interelectronic interaction and the correlation effects caused by it are sharply manifested not only in the photoionization cross sections but also in the angular distributions of photoelectrons [22] and spin orientation [24].
4. It has been demonstrated that multielectron correlations significantly affect the photoionization cross sections of outer and intermediate atomic subshells in a very wide frequency range. These effects are especially strong near the subshell ionization thresholds, but they are quite noticeable at relatively high photon frequencies of up to 1.5 or 2 keV and even above the threshold values [7,9,11].
5. Researchers have identified specific examples of the strong interaction of electrons belonging to two or even three different subshells, which leads to strong changes in their cross sections far from any ionization threshold, including the formation of completely new maxima [21,26].
6. It has been shown that along with the main line, which corresponds to the removal of an electron from a given subshell, there are satellite and shadow lines of a pure many-electron nature in the photoelectron spectrum. They appear at any incoming frequency of photons with the same strength relative to the main line. Quite often, the strength of these lines is high, which is a direct manifestation of the very important role played by the interelectronic interaction [27].
7. It has been demonstrated that the interelectronic interaction can in some cases be so strong that single-electron lines completely disappear [27] This effect, called the "melting" of electron shells, has been observed in Xe and some of its neighbors. In atomic physics, this effect is less common than other manifestations of correlations, but it is of great importance as an example of the possible power of the interelectronic interaction.
8. It has been shown that one photon, which can interact with only one electron, is able to simultaneously remove two or even more electrons from an atom. This process only occurs due to the presence of interelectronic interactions when the electron of the inner shell is removed from the atom and is accompanied by the Auger effect [7,9,16]. One photon can simultaneously remove two or even more electrons from the outer subshell, even if the photon frequency is below the intermediate photoionization threshold.
9. The autoionization of the continuous spectrum was predicted [25,28] due to the existence of relatively narrow resonances in the photoabsorption cross section of negative ions that arise as a result of the strong interaction of electrons belonging to two outer subshells.
10. Researchers have identified strong autoionization resonances that arise due to the effective interaction between the discrete excitation "two electrons–two vacancies" from one subshell with the continuum "one electron–one vacancy" of the other [29]. Therefore, almost everywhere, with the exception of the immediate vicinity of the first ionization potential, the photoionization cross section has a fine structure consisting of narrow autoionization resonances.
11. Researchers have predicted unexpectedly large low-energy non-dipole corrections in the angular anisotropy of photoelectrons, which lead to the creation of resistance currents that are quite observable macroscopic joint effects during the photoionization of atomic gases [30].

The intershell interactions in atoms with open *np*- and *nd*-shells have been theoretically and experimentally studied to a lesser extent than in atoms with filled shells. The reasons for this are not only computational difficulties but also experimental ones associated with obtaining these atoms in a vapor state. There are a number of calculations of the photoionization cross sections for atoms of groups VI and VII of the periodic system in the single-particle approximation while taking correlations into account. These calculations indicate that the correlations between electrons—both intrashell and intershell—should be no less in these atoms, and sometimes even more, than in atoms neighboring in the periodic system with filled shells.

5. Many-Electron Effects in Electron-Impact Ionization Processes

The study of the ionization of atoms by fast electrons (or other particles) makes it possible to trace the dependence of intershell interactions on the momentum q and angular momentum Δl transferred during scattering. The transfer of various moments to an atom leads to (along with dipole moments) monopole, quadrupole, and other transitions in the atom, thus making it possible to find out the role of the intershell interaction components of different multipolarities. The cross section for inelastic scattering of fast electrons on atoms is determined in terms of the generalized oscillator strengths (GOS), which are calculated for transitions of different multipolarities and describe the reaction of an atom to the transfer of momentum and energy to it. The differential cross section for inelastic scattering of fast electrons is proportional to the density of the GOS.

In the limiting case of the transferred momentum, only the dipole component of the interaction "survives"; therefore, in ionization by fast electrons, the influence of intershell correlations manifests itself similarly to that in the process of photoionization. Following the change in the influence of the intershell interaction with increasing momentum transfer q using the dipole-density component GOS as an example, we note that with increasing q, the influence of the outer shells on the ionization of the deeper ones decreases. The reason for this is that as q increases, the incident electron penetrates deeper and deeper into the atom and the effective radius r of the interaction with it finally becomes smaller than the radius of the outer shell. The outer electrons shield the inner shell from the impact of the incident electron to a lesser extent [11,16].

The differential cross section for inelastic scattering of fast electrons is proportional to the density of the GOS $\partial f(\omega, q)/\partial\omega$ and can be written as [31]

$$\frac{d^2\sigma}{d\omega\, d\Omega} = \frac{4\pi}{\omega E}\frac{\partial f(\omega, q)}{\partial\omega}\frac{d\ln q^2}{d\Omega} \tag{5}$$

where E is the energy of the incident electron; ω and q are the energy and momentum, respectively, transferred to the atom during scattering; and $d\Omega = 2\pi \sin\theta d\theta$ is the element of the solid angle into which the incident electron was scattered. When many-electron correlations are taken into account, instead of the single-electron matrix element included in (5), the corresponding matrix element—which is determined by expressions similar to those written for the dipole component of the photoionization amplitude—is substituted.

The differential cross section is proportional to the total density GOS. For small transferred momenta q, the GOS is determined by the contribution of the dipole component. As q increases, the contribution of monopole and especially quadrupole transitions increases, for which the influence of surrounding shells is smaller than for dipole ones.

As q increases, the influence of the outer shells on the ionization of the deeper ones decreases. The aforementioned reason for this is that as q increases, the incident electron penetrates deeper and deeper into the atom and the effective radius $r_э$ of the interaction with it becomes smaller than the radius of the outer shell. The outer electrons shield the inner shell from the action of the incident electron to a lesser extent [32].

On the contrary, the influence of inner shells on the ionization of electrons from outer shells can remain significant even at sufficiently large transferred momenta q. Moreover, since the rate of decrease in the ionization amplitude is determined by the product $qr_э$ and the radius of the inner shell is less than the radius of the outer one, the contribution of the direct amplitude usually decreases with increasing q faster than the correlation one. Thus, the relative role of internal electrons can increase.

As in the process of photoionization, the scattering cross section for fast electrons with ionization of outer s-electrons has a collective character. The influence of the surrounding multielectron shells on their ionization leads to qualitative changes in the dependence of the cross section on the transferred energy and momentum [11,32].

When studying the scattering of slow electrons by atoms, information about the role of many-electron correlations and the probability of the process can be obtained by determining the self-energy part of a hole or a particle of a single-particle Green's function in a simplified RPAE.

The cross section for elastic scattering of electrons with energy E is expressed in terms of the scattering phases $\delta_l(E)$ of partial waves with moment l. Using the ATOM system, RPAE corrections $\Delta\delta_l(E)$ to the HF phases of elastic scattering $\delta_l^{HF}(E)$ are calculated [9,11].

$$e^{i\Delta\delta_l(E)} \sin \Delta\delta_l(E) = -\pi \langle El | \hat{\Sigma}(E) | El \rangle \quad (6)$$

where $\langle El | \hat{\Sigma}(E) | El \rangle$ is the matrix element of the polarization interaction of the incident electron with the atom. The self-energy part of the Green's function depends on the energy of the incident electron and describes the nonlocal interaction between the incident electron and the electrons of the atom. This approach first made it possible to describe experimental data on the elastic scattering of slow electrons by a significant number of atoms with high accuracy and without using the phenomenological polarization potential. The total cross section of inelastic scattering is expressed in terms of the imaginary part of the phase shifts (6) $\mathrm{Im}\delta_l(E)$.

The same methods were applied to the description of the elastic scattering of slow positrons by atoms [11,33]. In contrast to the scattering of electrons, in this problem, on the one hand, it is not necessary to consider the exchange interaction, but it was necessary to consider the formation of a bound state, such as positronium, which complicates its solution.

Below, we present the most important results of the study of electron or positron scattering, both elastic and inelastic [8–11,16]. Some of the obtained results are also important for understanding the collisions of atoms involving heavy charged particles, such as protons and μ-mesons.

1. It was demonstrated that in the cross section of the elastic scattering of electrons on atoms, there is a Ramsauer minimum that arises due to the action on the incoming electron, along with the self-consistent HF potential and the polarization potential. This potential has a purely many-electron nature. This leads to Ramsauer minima in the electron scattering cross sections, not only on atoms of the noble gases Ar, Kr, and Xe but also on alkaline earth elements such as Ca. In the latter atoms, the Ramsauer minima have very high energies [11,25,34].
2. The polarization potential turned out to be strong enough to form stable negative ions of a number of atoms with filled subshells, alkaline earth elements in particular [25,34,35], although with a very low binding energy. These negative ions have been observed in experiments.
3. The important role of the polarization potential in the elastic and inelastic scattering of slow positrons by atoms was investigated and discovered. It has been demonstrated that the possibility of an incoming positron to form a bound state (called positronium) with an external atomic electron during scattering, greatly affects polarization potential [11,36]. Accounting for this temporary formation of positronium can lead to bound states of a positron with an atom, which is a completely new kind of positively charged ion [37].
4. The temporary formation of positronium during the elastic scattering of positron atoms can, in principle, lead to a qualitative modification of the polarization potential, i.e., it can become repulsive instead of always being attractive [38]. This reversal of the sign of the potential explains the qualitative difference between the low-energy elastic scattering cross sections of a positron on He and a positron on Li and also why the former is orders of magnitude smaller than the latter.
5. It has been demonstrated that many-electron correlations play a very important role in inelastic collisions for practically any projectile energy. The cross section of inelastic scattering is strongly affected by not only such collective excitations as dipole giant resonances but also multielectron nondipole excitations [10,11].

6. It was shown that for incident particles with energies close to the excitation thresholds of the intermediate or inner shell, the spectra of the inelastically scattered projectile are strongly modified due to the Auger decay of the created vacancy. This effect is called the post-collision interaction [39]. The many-particle theory of this effect can be found in [11,40].
7. A new mechanism for the generation of continuous spectrum electromagnetic radiation in inelastic collisions in atoms was proposed. This radiation, called atomic or polarization bremsstrahlung, is mainly due to the dipole deformation of the target atom during the collision. This radiation is strongly influenced by the interaction between atomic electrons and the collective effects caused by it [41].

6. Ionization of the Inner Shells of the Atom and Decay of Vacancies

Collective effects during the photoionization of inner shells near their threshold are more complex than those of outer shells. Along with the forced joint ordered motion of electrons of one or several neighboring shells occurring under the action of an external electromagnetic field, various relaxation processes also take place. Relaxation, or rearrangement, is a complex dynamic process that reflects the reaction of atomic electrons to the appearance of a vacancy in one of the shells and its subsequent decay. Accounting for the rearrangement leads to changes in the photoionization amplitudes and the interaction between electrons.

The simplest method that considers the reaction of atomic electrons to a vacancy appearing in an atom after photoabsorption is the "static rearrangement" approximation. In this case, during the decay time of the hole, the photoelectron with low energy does not have time to move far enough from the remaining ion and almost immediately moves into the field of the ion field that has already been rearranged due to decay [7,9].

The "static" rearrangement approximation becomes inapplicable when the lifetime of a vacancy T_{nl} in a shell with quantum numbers nl is comparable to the photoelectron escape time t from an atom. In this case, it is necessary to take the dynamics of the process into account. During Auger decay, the new field acting on a photoelectron corresponds to a charge one greater than the initial field formed during the absorption of a quantum. As a result of increased attraction, the slow electron in the new field has an energy lower than that which it would have if the decay of the vacancy was neglected. The released energy is carried away by a fast Auger electron. This phenomenon, associated with the redistribution of energy between a photoelectron and an Auger electron, is a strong correlation effect called the post-collision interaction (PCI) [39,40]. This effect significantly changes the amplitude and, accordingly, the photoionization cross section, as well as the energy distribution of Auger and photoelectrons. Recently, the effects of PCI have been intensively studied both experimentally and theoretically. An analytical study of the amplitude of the photoprocess, taking the PCI into account, showed a redistribution of energy between electrons: a fast electron is accelerated while a slow one is decelerated [40]. The analysis of this correlation effect is simplified when the width of the deep vacancy is not too large and (as a result of its decay) a sufficiently fast electron is formed, so that the interaction with a slow photoelectron can be neglected.

Many-electron correlations are clearly manifested in the decay of vacancies formed during the interaction of photons, electrons, or positrons with atoms. Decay with the emission of even one electron (Auger decay) is in itself a manifestation of the interelectronic interaction. However, in some decay processes, the role of interaction with other electrons of the atom (many-electron correlations), which are not directly involved in the decay, is especially important.

Through the matrix element of the energy proper part of the single-particle Green's function, taken between the wave functions of the occupied states, the shift of the ionization potential is expressed in comparison with its HF value due to the correlation interaction of atomic electrons. The imaginary part of this matrix element gives the total width of the hole level with respect to the Auger decay. In addition to direct Auger decay, in which another atomic electron is removed as a result of the transition of an electron from an occupied level to a free one, a more complex process (which proceeds due to the effective interaction of electrons and is taken into account in RPAE) is also investigated.

The main results obtained in the study of the decay of vacancies are presented below.

1. Taking the interaction between the electrons of an atom into account, it was shown that the probability of both radiative and non-radiative—i.e., flowing with the emission of electrons (Auger electrons)—decays can be significantly reduced or increased. Owing to correlations, radiative decay can be completely blocked, which is called the *radiative self-blocking of vacancies* [11].
2. It has been demonstrated that a state with one vacancy can decay upon the simultaneous emission of one electron and one photon, several electrons, or several photons [42]. At the same time, states with two vacancies can decay upon the emission of one photon or electron [16,17].
3. It was shown that the energy of an Auger electron leaving an atom after the decay of an internal vacancy shifts towards higher energies when a vacancy is created near the threshold of its formation. This shift increases with a decrease in the energy of the photoelectron that leaves the atom when an internal vacancy is created. This shift is a manifestation of the interaction after the collision [11,16].
4. The increase in the energy of the Auger electron mentioned in the previous paragraph can be so large that a slow photo- or inelastically scattered electron does not actually have enough energy to leave and remains in the atom because it is intercepted by one of the higher excited levels of the residual ion.
5. The creation of the innermost shell vacancies leads to avalanches of secondary electrons. Most of them are formed as a result of multistage Auger decay. However, multielectron Auger processes also play an unexpectedly large role in the creation of these avalanches.

7. Conclusions

With the help of the developed ATOM software, a huge number of calculations of various atomic processes have been carried out, in most of which multielectron processes play a decisive role. These calculations can be considered the result of a computational experiment in which many-particle effects are studied for a well-defined and well-known pair interaction between electrons. Naturally, in this short review of the obtained results and the possibilities of applying the presented experiment, not all processes associated with the reaction of the electronic system of atoms to external influences are discussed.

In the processes mentioned above, relativistic effects that have little impact on the partial photoionization cross sections of individual shells were neglected. However, when describing photoprocesses in atoms, such characteristics that are directly related to relativistic corrections to the interaction are also determined. One of them is the parameter called the *"branching ratio"* [18], which characterizes the relative probability of the ionization of shell sublevels with different total angular momenta j. The presence of a spin in an electron leads to the fact that during the photoionization of shells, the remaining ion can be in states that differ in the total momentum $j = l \pm 1/2$.

In the process of the photoionization of internal atomic shells with a total moment $j > 1/2$ ($l > 0$), the resulting ions have a certain alignment (the predominant orientation of the total moment j) along the direction of the incident photon beam. This alignment, which arises as a result of the uneven population of states with different projections of the total momentum of the ion M, manifests itself in the anisotropy of the angular distribution of emission photons or Auger electrons emitted during the decay of a vacancy. The alignment of ions depends on the squares of the dipole amplitudes in a different way than the photoionization cross section, and the measurement of the angular anisotropy of electrons or photons therefore also provides additional independent information about the photoionization process.

The final stage of the computational experiment is the analysis of the results and their comparison with the experimental data. The ATOM system has turned out to be very effective in solving a wide class of problems in the study of atomic processes, including processes involving atomic ions, endohedral fullerenes, diatomic molecules, and atomic clusters, as well as processes in a number of problems in solid state physics.

The ATOM system was created on the initiative and with the active participation of Prof. Miron Ya. Amusia. It was his ideas that led to the creation of the self-consistent RPAE approximation, which was the next step in many-body theory after the HF approximation. The development of the proposed approach and the creation of the ATOM complex of computational programs continued for more than a year of research in the field of physics of multielectron atoms. The development of such systems will make it possible to carry out mass molecular calculations, as well as calculations of atoms placed in strong electric or magnetic fields. The obtained results will become the theoretical base, the starting point for comparing theory and experiment. Thus, the area of applicability of the approach that a group of theoretical physicists under the leadership of Miron Ya. Amusia developed in the theory of the atom will be significantly expanded, providing experimenters with theoretical results of initial approximations that have a relatively high accuracy "on average".

The results of a computational experiment using the ATOM system can be found in supplementary materials and in a number of our monographs [7–11,16,17,19,43].

Supplementary Materials: The following supporting information can be downloaded at https://www.mdpi.com/article/10.3390/atoms10020052/s1. List of main publications in which the results were obtained using the ATOM program system.

Author Contributions: Conceptualization, L.V.C. and V.K.I.; methodology, V.K.I.; software, L.V.C.; writing—original draft preparation, L.V.C. and V.K.I.; writing—review and editing, L.V.C. and V.K.I. All authors have read and agreed to the published version of the manuscript.

Funding: This research received no external funding.

Data Availability Statement: Not applicable.

Acknowledgments: We consider it our duty to thank Kheifetz A. for his advice and help in preparing our manuscript for publication.

Conflicts of Interest: The authors declare no conflict of interest.

References

1. Lukirskii, A.P.; Britov, I.A.; Zimkina, T.M. Photoionization of the 4d-electrons in xenon. *Opt. Spectrosc.* **1964**, *17*, 438. (In Russian)
2. Ederer, D.L. Photoionization of the 4d-electrons in xenon. *Phys. Rev. Lett.* **1964**, *13*, 760.
3. Amusia, M.Y.; Cherepkov, N.A.; Chernysheva, L.V. Many-electron correlations in photoabsorption in the M-shell of Ar. *Phys. Lett. A* **1970**, *31*, 553. [CrossRef]
4. Wendin, G.J. Collective effects in atomic photoabsorption spectra. III. Collective resonance in the $4d^{10}$ shell in Xe. *Phys. B* **1971**, *4*, 1080.
5. Amusia, M.Y.; Cherepkov, N.A.; Chernysheva, L.V. Cross section for the photoionization of noble gas atoms with allowance for multielectron correlations. *Zh. Exp. Theor. Fiz.* **1971**, *60*, 160.
6. Amusia, M.Y.; Ivanov, V.K.; Cherepkov, N.A.; Chernysheva, L.V. Interference effects of noble-gas atoms outer s-subshell. *Phys. Lett. A* **1972**, *40*, 361.
7. Amusia, M.Y. *Atomic Photoeffect*; Plenum Press: New York, NY, USA; London, UK, 1990; p. 303.
8. Amusia, M.Y.; Chernysheva, L.V. *Automatic System of Atomic Structure Investigations*; Nauka: Leningrag, Russia, 1983; p. 180. (In Russian)
9. Amusia, M.Y.; Chernysheva, L.V. *Computation of Atomic Processes*; "Adam Hilger" Institute of Physics Publishing: Bristol, PA, USA, 1997; p. 253.
10. Amusia, M.Y.; Chernysheva, L.V.; Semenov, S.K. *ATOM-M. Algorithms and Programs for the Study of Atomic and Molecular Processes*; Nauka: St. Petersburg, Russia, 2016; p. 551. (In Russian)
11. Amusia, M.Y.; Ivanov, V.K.; Cherepkov, N.A.; Chernysheva, L.V. *Processes in Many-Electron Atoms*; Nauka: St. Petersburg, Russia, 2006; p. 325. (In Russian)
12. Froese-Fischer, C. *The Hartree—Fock Method for Atoms*; Wiley (Interscience): New York, NY, USA, 1977; p. 308.
13. Kelly, H.P. *Atomic Inner-Shell Processes*; Crasemann, B., Ed.; Academic Press: New York, NY, USA, 1963; p. 331.
14. Jablon, C.; Simon, J.C. *Applications de Modeles Numeriques en Physique*; Birkhauser: Basel, Switzerland, 1978; p. 283.

15. Samarskiy, A.A. *Mathematical Modeling and Computational Experiment*; No. 5; Bulletin of the the Academy of Science USSR: Moscow, Russia, 1979; pp. 38–49.
16. Amusia, M.Y.; Chernysheva, L.V.; Yarzhemsky, V.G. *Photon Absorption, Electron Scattering, Vacancy Decay: Atomic Data*; Nauka: St. Petersburg, Russia, 2010; p. 314. (In Russian)
17. Amusia, M.Y.; Chernysheva, L.V.; Yarzhemsky, V.G. *Handbook of Theoretical Atomic Physics*; Springer: Berlin, Germany, 2012; p. 820.
18. Starace, A.F. *Handbuch der Physik*; Berlin: Springer: Berlin, Germany, 1982.
19. Amusia, M.Y.; Cherepkov, N.A. *Case Studies in Atomic Physics*; North-Holland Publishing Company: Amsterdam, The Netherlands, 1975; Volume 5, pp. 47–179.
20. Kelly, H.P.; Carter, S.L. Many Body perturbation calculations of the interaction of atoms with electromagnetic radiation. *Phys. Scr.* **1980**, *21*, 448. [CrossRef]
21. Amusia, M.Y.; Ivanov, V.K.; Cherepkov, N.A.; Chernysheva, L.V. Intershell and Intersubshell Effects in Atomic Photoionization. *Zh. Exp. Theor. Fiz.* **1974**, *66*, 1537.
22. Amusia, M.Y.; Ivanov, V.K. The Peculiarities of Photoelectron Angular Distribution and Ionization Cross Section of 5p subshell in Xe. *Phys. Lett. A* **1976**, *59*, 194–196.
23. Torop, L.; Morton, J.; West, J.B. The angular distribution of photoelectrons from xenon. *J. Phys. B* **1976**, *9*, 2035–2041. [CrossRef]
24. Cherepkov, N.A. Spin polarization of photoelectrons ejected from unpolarised atoms. *J. Phys. B* **1979**, *12*, 1279. [CrossRef]
25. Ivanov, V.K. Many-body effects in negative ion photodetachment. *J. Phys. B At. Mol. Opt. Phys.* **1999**, *32*, R67–R101. [CrossRef]
26. Amusia, M.Y.; Ivanov, V.K.; Chernysheva, L.V. Photoproduction of Kr+ and Xe+ Ions in the Vicinity of the Outer d-subshell Threshold. *Phys. Lett. A* **1973**, *43*, 243. [CrossRef]
27. Amusia, M.Y. *Advances in Atomic and Molecular Physics*; Bates, D., Ed.; Academic Press: Cambridge, MA, USA, 1981; Volume 17, pp. 1–54.
28. Amusia, M.Y.; Gribakin, G.F.; Ivanov, V.K.; Chernysheva, L.V. Many-electron Correlations in the Negative Ion Photodetachment. *J. Phys. B At. Mol. Opt. Phys.* **1990**, *23*, 385. [CrossRef]
29. Ehresmann, A.; Vollweiler, F.; Schmoranzer, H.; Sukhorukov, V.L.; Lagutin, B.M.; Petrov, I.D.; Menzel, G.; Schartner, K.-H. Photoionization of Kr 4s. III. Detailed and extended measurements of the Kr 4s-electron tonization cross section. *J. Phys. B At. Mol. Opt. Phys.* **1994**, *27*, 1489. [CrossRef]
30. Amusia, M.Y.; Baltenkov, A.S.; Felfli, Z.; Msezane, A.Z. Large Non-Dipole Correlations Effects near Atomic Photoionization Threshold. *Phys. Rev. A* **1999**, *59*, R2544–R2547. [CrossRef]
31. Inokuti, M. Inelastic Collisions of fast charged Particles with Atoms and Molecules—The Bethe Theory Revisited. *Rev. Mod. Phys.* **1971**, *43*, 297. [CrossRef]
32. Amusia, M.Y.; Ivanov, V.K.; Sheinerman, S.A. Intershell correlations in the Inelastic Scattering of fast Electrons on the Outer Subshells in Ar and Xe. *J. Phys. B* **1976**, *9*, 1537. [CrossRef]
33. Amusia, M.Y.; Cherepkov, N.A.; Chernysheva, L.V. Elastic scattering of slow positrons on Atoms. *JETP* **2003**, *124*, 1–9. [CrossRef]
34. Gribakin, G.F.; Gul'tsev, B.V.; Ivanov, V.K.; Kuchiev, M.Y. Interaction of an alkaline-earth atom with an electron: Scattering, negative ion and photodetachment. *J. Phys. B At. Mol. Opt. Phys.* **1990**, *23*, 4505. [CrossRef]
35. Ivanov, V.K. Theoretical studies of photodetachment. *Radiat. Phys. Chem.* **2004**, *70*, 345–370. [CrossRef]
36. Amusia, M.Y.; Cherepkov, N.A.; Chernysheva, L.V.; Shapiro, S.G. Elastic Scattering of Slow Positrons be helium. *J. Phys. B At. Mol. Opt. Phys.* **1976**, *9*, L531–L534. [CrossRef]
37. Dzuba, V.A.; Flambaum, V.V.; Gribakin, G.F.; King, W.A. Many-body calculations of positron scattering and annihilation from noble-gas atoms. *J. Phys. B At. Mol. Opt. Phys.* **1996**, *29*, 3151. [CrossRef]
38. Amusia, M.Y.; Taulbjierg, K. Abstracts of Papers. In Proceedings of the XIX International Conference on the Physics of Electronic and Atomic Collisions, Whistler, BC, Canada, 26 July–1 August 1995; Volume 1, p. 364.
39. Read, F.H. Displaced electron energies and the "shake-down" effect. *Radiat. Res.* **1975**, *64*, 23. [CrossRef]
40. Amusia, M.Y.; Kuchiev, M.Y.; Sheinerman, S.A. Near-Threshold Effects in the Atomic Ionization Processes (Post-Collision Interaction). *J. Exp. Theor. Phys. (USSR Acad. Sci.).* **1979**, *76*, 470–481.
41. Amusia, M.Y.; Buimistrov, V.M.; Zon, B.A.; Tsitovich, V.N. *Polarization Bremsstrahlung*; Plenum Press: New York, NY, USA; London, UK, 1992.
42. Amusia, M.Y.; Lee, I.S. Radiative Semi-auger Decay in Atoms. *Phys. Scr.* **1992**, *41*, 23–27. [CrossRef]
43. Amusia, M.Y.; Chernysheva, L.V. *Computation of Atomic and Molecular Processes. Introducing the ATOM-M Software Suite*; Springer Series on Atomic, Optical, and Plasma Physics; Springer: Berlin/Heidelberg, Germany, 2021; Volume 117, pp. 1–456.

atoms

MDPI

Article

Taking the Convergent Close-Coupling Method beyond Helium: The Utility of the Hartree-Fock Theory

Igor Bray [1,*], Xavier Weber [1], Dmitry V. Fursa [1], Alisher S. Kadyrov [1], Barry I. Schneider [2], Sudhakar Pamidighantam [3], Maciej Cytowski [4] and Anatoli S. Kheifets [5]

1 Curtin Institute for Computation and Department of Physics and Astronomy, Curtin University, Perth, WA 6845, Australia; xavier.weber@student.curtin.edu.au (X.W.); D.Fursa@curtin.edu.au (D.V.F.); A.Kadyrov@curtin.edu.au (A.S.K.);
2 National Institute of Standards and Technology, Gaithersburg, MD 20899, USA; barry.schneider@nist.gov
3 Cyberinfrastructure Integration Research Center, Department of Chemistry, Indiana University, Bloomington, IN 47408, USA; pamidigs@iu.edu
4 Pawsey Supercomputing Centre, 1 Bryce Ave., Kensington, WA 6151, Australia; maciej.cytowski@pawsey.org.au
5 Research School of Physics, The Australian National University, Canberra, ACT 2601, Australia; a.kheifets@anu.edu.au
* Correspondence: I.Bray@curtin.edu.au

Abstract: The convergent close-coupling (CCC) method was initially developed to describe electron scattering on atomic hydrogen and the hydrogenic ions such as He^+. The latter allows implementation of double photoionization (DPI) of the helium atom. For more complex single valence-electron atomic and ionic targets, the direct and exchange interaction with the inner electron core needs to be taken into account. For this purpose, the Hartree-Fock (HF) computer codes developed in the group of Miron Amusia have been adapted. In this brief review article, we demonstrate the utility of the HF technique by examples of electron scattering on Li and the DPI of the H^- and Li^- ions. We also discuss that modern-day computer infrastructure allows the associated CCC code, and others, to be readily run directly via the Atomic, Molecular and Optical Science Gateway.

Keywords: electron–atom scattering; atomic photoionization; many-electron correlation; electronic structure

PACS: 32.80.Rm; 32.80.Fb; 42.50.Hz

arXiv: 2108.02392

Citation: Bray, I.; Weber, X.; Fursa, D.V.; Kadyrov, A.S.; Schneider, B.I.; Pamidighantam, S.; Cytowski, M.; Kheifets, A.S. Taking the Convergent Close-Coupling Method beyond Helium: The Utility of the Hartree-Fock Theory. *Atoms* **2022**, *10*, 22. https://doi.org/10.3390/atoms10010022

Academic Editor: Grzegorz Piotr Karwasz

Received: 19 January 2022
Accepted: 5 February 2022
Published: 11 February 2022

1. Introduction

Collisions between particles on the atomic scale are ubiquitous throughout the universe. Our interest is in the collisions of fundamental particles such as electrons, positrons, photons, and protons with atoms and molecules. The field relies on strong interactions between experimental and theoretical approaches to the collision problems. The expectation is that experiments provide benchmark measurements for theorists to test their models, and when deemed sufficiently accurate the models provide extensive data for use in applications. The latter include astrophysics, fusion, lighting, nanolithography, and medical imaging and therapy. Accordingly, it is of great concern whenever there are substantial discrepancies between theory and experiment that are not understood. One such case was the discrepancy for the angular correlation parameters in the fundamental Coulomb three-body collision problem of e-H excitation of the 2p state [1,2].

From the theoretical side, the difficulties associated with calculating e-H scattering relate to the fact that the target has a countably infinite number of discrete states, an uncountably infinite continuum, and that the Coulomb interaction extends to infinite distances for charged particles. The convergent close-coupling (CCC) method [3,4] was developed in response to the abovementioned discrepancy by providing a systematic and mathematically rigorous approach to the underlying computational difficulties. By expanding the target wave-functions in a truncated complete Laguerre basis the target is represented with a finite number of target states N. Furthermore, due to the exponential falloff of the basis, the interactions do not extend to infinite distances. The problem is then reduced to obtaining convergence in the scattering amplitudes of interest with increasing N. Despite demonstrating such convergence, the CCC method was unable to resolve the discrepancy with the experiment, and yielded results similar to other sophisticated theoretical approaches to the problem [4]. However, it was able to yield excellent agreement with the measurements of the total ionization cross section and its spin asymmetry [5], which is a particularly stringent test of any theory.

While atomic hydrogen is the ideal starting point for testing any electron–atom scattering theory, it is not ideal from the experimental point of view. Alkali atoms such as Li and Na are easier to prepare in the laboratory, and yet their chemistry is very similar to that of H, having just one valence electron. This allowed for some very accurate measurements of e-Li [6–8], e-Na [9–12], and e-K [13–15] scattering systems. So the next stage in trying to understand the discrepancy for the e-H system was to extend the CCC method to the alkali atoms. The group of Miron Amusia generously provided the Hartree-Fock computational code [16,17] for a self-consistent treatment of the core electrons. This allowed the reduction of the electron–alkali atom collision problem to also be a Coulomb three-body problem, albeit with some more complicated nonlocal potentials [18]. The Hartree-Fock code also enabled the extension of the two-electron CCC code [19] to quasi two-electron targets such as alkaline earth metals [20,21]. Reviews of the development of the CCC method, including application to differential ionization, are available [22–25].

Unlike the case of the e-H system, the agreement between the CCC calculations and experiment for the electron–alkali collisions was spectacular, so much so, that the e-H collision system was remeasured by other groups, who found excellent agreement with the CCC calculations [26,27]. Subsequently, errors in the original measurements were identified [28].

A similar development track was taken by DPI on He, where the final state is determined from the e-He^+ collision system. Here too. theoretical predictions did not always agree with the initial experiment. Mergel et al. [29] measured the DPI of He using circularly polarized light. However, the corresponding CCC calculations did not yield agreement with the measurements [30]. Subsequently, Achler et al. [31] revisited the experiment and found excellent agreement with the CCC calculations. The CCC method was then extended to describe double ionization of He by electron impact (the so-called (e,2e) reaction [32,33]). Then an extension was made to the process of two-photon double ionization of He [34]. The utility of the HF theory was instrumental to describe the valence-shell DPI of alkaline–earth metal atoms [35]. These calculations were later found in good agreement with experiments [36]. Most recently, the CCC technique aided by the HF theory was applied to time-resolved atomic photoemission. The time delay in photoemission, expressed via the phase of the complex ionization amplitude, became experimentally accessible [37,38]. While the measurements are restricted at present to single active electron targets, theoretical predictions for two-electron targets have been made [39,40]. Most recently, photodetachment time delay was analyzed to reveal the implications of the fundamental threshold laws [41].

Here we shall demonstrate the agreement between theory and experiment by focusing on just the simplest electron–alkali collision system, that of e-Li scattering. This collision system is also a key component in calculating the DPI of Li^-, as upon single or double photoionization the e-Li wave-function corresponds to the final state of Li^-. The CCC computer codes utilize modern computational infrastructure including massive parallelism and GPU acceleration and are readily accessible for execution via the Atomic, Molecular and Optical Science Gateway, https://amosgateway.org (accessed on 4 February 2022) [42].

2. Theory

The details of the implementation of the CCC theory to electron scattering on quasi one-electron targets, such as the alkalis, have been given by Bray [18]. Briefly, the core-electron wave-functions ψ_j of target T are solved for by utilizing the Self-consistent-Field Hartree-Fock (SCHF) equations [16]

$$\left(K + V^{\mathrm{HF}} - \varepsilon_j\right)\psi_j(\boldsymbol{r}) = 0, \qquad \psi_j \in T, \tag{1}$$

where

$$\begin{aligned} V^{\mathrm{HF}}\psi_j(\boldsymbol{r}) &= \left(-\frac{Z}{r} + 2\sum_{\psi_{j'} \in T}\int d^3r' \frac{|\,\psi_{j'}(\boldsymbol{r}')\,|^2}{|\,\boldsymbol{r}-\boldsymbol{r}'\,|}\right)\psi_j(\boldsymbol{r}) \\ &- \sum_{\psi_{j'} \in T}\int d^3r' \frac{\psi^*_{j'}(\boldsymbol{r}')\psi_j(\boldsymbol{r}')}{|\,\boldsymbol{r}-\boldsymbol{r}'\,|}\psi_{j'}(\boldsymbol{r}). \end{aligned} \tag{2}$$

The core-electron wave-functions are then used to define the Frozen-Core Hartree-Fock (FCHF) potential V^{FC} as

$$\begin{aligned} V^{\mathrm{FC}}\phi_j(\boldsymbol{r}) &= \left(-\frac{Z}{r} + 2\sum_{\psi_{j'} \in C}\int d^3r' \frac{|\,\psi_{j'}(\boldsymbol{r}')\,|^2}{|\,\boldsymbol{r}-\boldsymbol{r}'\,|}\right)\phi_j(\boldsymbol{r}) \\ &- \sum_{\psi_{j'} \in C}\int d^3r' \frac{\psi^*_{j'}(\boldsymbol{r}')\phi_j(\boldsymbol{r}')}{|\,\boldsymbol{r}-\boldsymbol{r}'\,|}\psi_{j'}(\boldsymbol{r}), \end{aligned} \tag{3}$$

where the notation C indicates the set of frozen-core wave-functions. The target wave-functions are then obtained from the effective one-electron Hamiltonian $K + V^{\mathrm{FC}}$ via

$$\left(K + V^{\mathrm{FC}} - \epsilon_j\right)\phi_j(\boldsymbol{r}) = 0. \tag{4}$$

The eigenstates can be obtained directly [17] or via diagonalization in some suitable basis. In the CCC method we do both, with the utilization of the Laguerre basis, which yields negative-energy eigenstates and a discretization of the target continuum.

With all of the potentials V defined, the close-coupling equations are formed in momentum space as coupled Lippmann-Schwinger equations directly for the transition amplitudes $\langle \boldsymbol{k}_f\phi_f|T_S|\phi_i\boldsymbol{k}_i\rangle$, as if the problem was a three-body one [4,18]

$$\langle \boldsymbol{k}_f\phi_f|T_S|\phi_i\boldsymbol{k}_i\rangle = \langle \boldsymbol{k}_f\phi_f|V_S|\phi_i\boldsymbol{k}_i\rangle + \sum_{n=1}^{N}\int d^3k \frac{\langle \boldsymbol{k}_f\phi_f|V_S|\phi_n\boldsymbol{k}\rangle\langle \boldsymbol{k}\phi_n|T_S|\phi_i\boldsymbol{k}_i\rangle}{E^{(+)} - \epsilon_n - k^2/2}, \tag{5}$$

where S is the total electron spin, and N is the number of Laguerre-based target states. We check for convergence in the required $\langle \boldsymbol{k}_f\phi_f|T_S|\phi_i\boldsymbol{k}_i\rangle$ by simply increasing N.

3. Results

3.1. Electron-Lithium Excitation

In the seminal publications of Bederson [43,44], the idea of perfect scattering experiments was introduced. In such experiments, the maximum amount of the underlying scattering information is measured. In the case of P-state excitation, the angular correlation parameters discussed above, or the equivalent Stokes parameters [45], which may depend on total electron spin [46], complement the differential cross sections in providing the extra experimental information to compare with the theory. In the following three figures, we examine e-Li excitation of the 2P state by presenting spin-dependent and -weighted results for the differential cross sections (DCS) and the three Stokes parameters. For each total electron spin S, at every scattering angle, there are two independent scattering (complex) amplitudes resulting in four independent theoretical predictions. While experiment [8] is available only for the spin-weighted parameters, it can be thought of as perfect for the spin-weighted case.

In addition to presenting the CCC calculations, which are convergent in the treatment of both the target discrete and the continuous spectrum, the CC calculations are also presented. These are convergent in the treatment of just the discrete spectrum; hence, the differences between CCC and CC highlight the importance of the inclusion of the target continuum on the transition of interest.

In Figure 1, we present the results for 7 eV e-Li excitation of the 2P state. We see that there is a substantial dependence of the results on the total electron spin S. The importance of the continuum is quite small at this energy. The agreement of the calculations with the experiment of Karaganov et al. [8] is quite extraordinary. Note that this experiment utilises the superelastic technique, which yields much better statistics than the traditional electron–photon coincidence experiments [1,2].

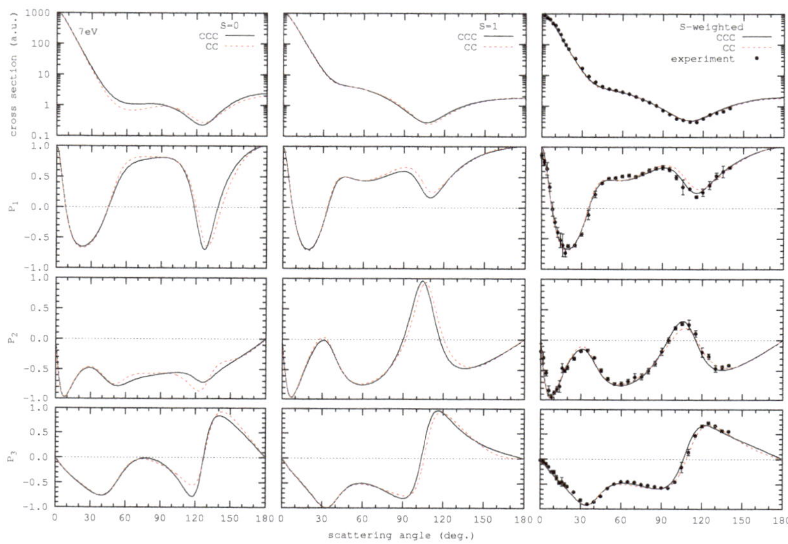

Figure 1. Differential cross section and (reduced) Stokes parameters for 7 eV electron-lithium 2P excitation. CCC results are convergent in the treatment of the Li discrete and continuous spectrum, whereas CC results are convergent in the treatment of the discrete spectrum only. Experiment is due to Karaganov et al. [8].

The results for 14 eV e-Li excitation of the 2P state are presented in Figure 2. This time we see substantial differences between the CCC and CC calculations, though mostly for the $S = 0$ DCS and Stokes parameters. These differences are much less visible when compared with the experiment for the spin-weighted Stokes parameters, where the agreement is again outstanding.

Lastly, the 22 eV e-Li excitation of the 2P state is presented in Figure 3. At this energy the inclusion of the target continuum is clearly important for $S = 0$, though this is less visible for the spin-weighted parameters. The agreement with the experiment is outstanding once more. It was such good agreement, first reported by [6], that others seriously questioned the accuracy of the corresponding e-H 2P excitation measurements.

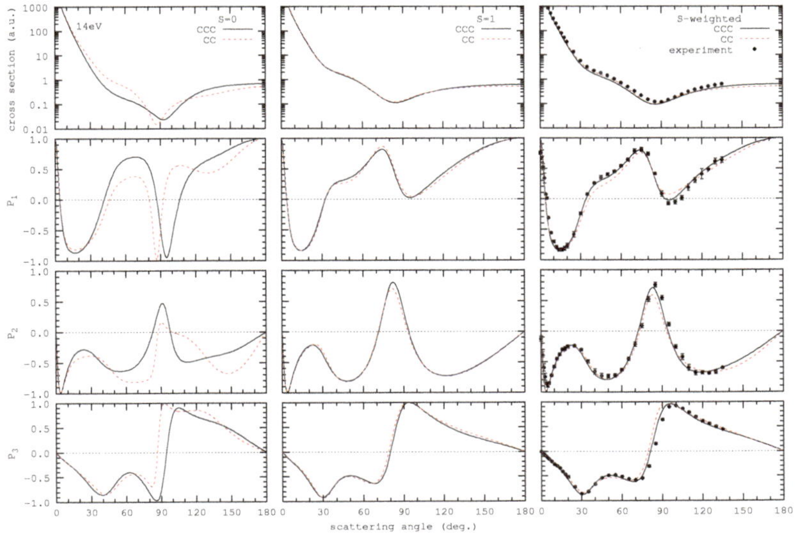

Figure 2. Same as for Figure 1, except for 14 eV.

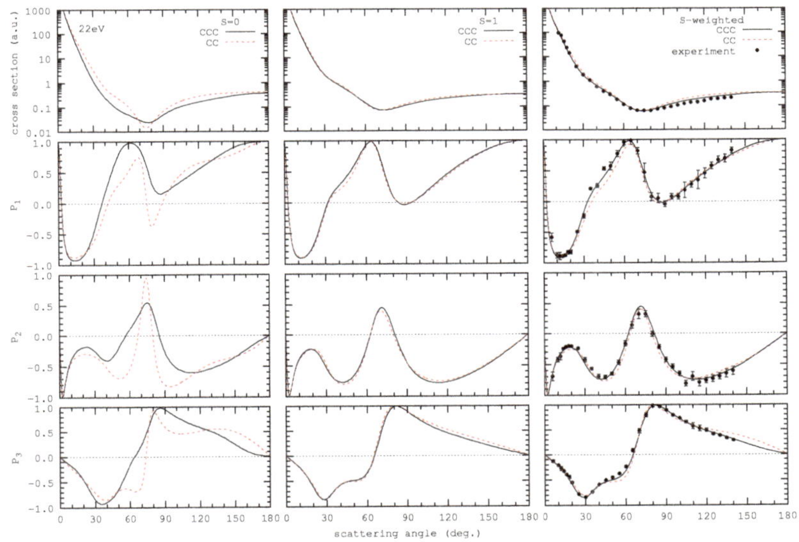

Figure 3. Same as for Figure 1, except for 22 eV.

3.2. Double Photoionisation

The knowledge of the scattering T-matrix allows expression of the two-electron dipole matrix element in the form of an integral [47]:

$$\begin{aligned} \langle \Psi_j^{(-)}(\boldsymbol{k})|\mathcal{D}|\Psi_0\rangle &= \langle \boldsymbol{k}^{(-)}\phi_j|\mathcal{D}|\Psi_0\rangle \\ &+ \sum_i \sum\!\!\!\!\!\!\int d^3\kappa \frac{\langle \boldsymbol{k}^{(-)}\phi_j|T|\phi_i\boldsymbol{\kappa}^{(+)}\rangle\langle \boldsymbol{\kappa}^{(+)}\phi_i|\mathcal{D}|\Psi_0\rangle}{E-\varepsilon_\kappa-\epsilon_i+i0}. \end{aligned} \quad (6)$$

Here, the bare dipole matrix element $\langle \boldsymbol{k}^{(-)}\phi_j|\mathcal{D}|\Psi_0\rangle$ is calculated between the two-electron initial state Ψ_0 and the final channels $\langle \boldsymbol{k}_b^{(-)}\phi_j|$. The initial state is expressed in the form of the multiconfiguration Hartree-Fock (MCHF) expansion

$$\Psi_0(\boldsymbol{r}_1,\boldsymbol{r}_2) = \sum_{l=0}^{l_{\max}} \sum_{m=n}^{n_{\max}} C_{ml}\,|\psi_{ml}(\boldsymbol{r}_1)\,\psi_{ml}(\boldsymbol{r}_2)\,:\,{}^1\mathrm{S}\rangle. \tag{7}$$

The MCHF orbitals $\psi_{ml}(\boldsymbol{r})$ are found in the frozen $1s^2$ core in the case of Li^- $(n=2)$, while this core is absent in the case of $\mathrm{H}^-(n=1)$. Only diagonal ml^2 terms are included in expansion (7), as is always the case for the closed-shell MCHF ground state. The coefficients C_{ml} in the MCHF expansion (7) are found by using the multiconfiguration Dirac-Fock computer code [48].

The cross section of a two-electron transition, measured as a function of the photon energy ω and corresponding to a particular state j of the remaining bound electron, is given by [49]:

$$\sigma_j(\omega) \quad = \quad \frac{4\pi^2}{\omega c} \sum\!\!\!\!\!\!\int_{m_j} d^3k\,|\langle \Psi_j^{(-)}(\boldsymbol{k})\,|\mathcal{D}|\,\Psi_0\rangle|^2 \delta(\omega - E + E_0)\,. \tag{8}$$

Here, $c \simeq 137$ is the speed of light in atomic units, while other fundamental constants are set to $e = m = \hbar = 1$. The final channels are separated into single and double ionization according to the energy ϵ_j, which is positive for the double ionized channels and negative for the singly ionized channels.

The phase of the ionization amplitude

$$\delta_j(\boldsymbol{k}) = \arg\langle \Psi_j^{(-)}(\boldsymbol{k})\,|\mathcal{D}|\,\Psi_0\rangle\;,\;\tau_j(\boldsymbol{k}) = \partial\delta_j(\boldsymbol{k})/\partial E \tag{9}$$

is used to calculate the photoemission time delay $\tau_j(\boldsymbol{k})$ in the given emission direction as the phase derivative with respect to the photoelectron energy $E = k^2/2$ (see Equation (S10) of [37]).

Our numerical results for photodetachment of H^- and Li^- are displayed in Figures 4 and 5, respectively. We select these two targets because the threshold laws manifest themselves particularly clearly in negative ions, which have a very simple spectrum and can support only a limited number of discrete excited states in comparison to an infinite number in their neutral atomic counterparts.

The photodetachment cross section (8) as well as the phase and the time delay (9) of H^- near the $n = 2$ threshold are presented in Figure 4. The top panel displays the partial photodetachment cross sections in various channels leaving the H atom in the ground $1s$ and excited $2s$ or $2p$ states. The total cross section is compared with the experiment [50]. The sharp resonance below the $n = 2$ threshold is due to autodetachment of a two-electron bound state into the ground state of the H atom (Feshbach resonance). We see that all four autodetachment channels have very similar cross sections. The time delay in the $n = 2$ channels is strongly angular-dependent. Near the opening of the $n = 2$ channels, it reaches very large negative values.

Figure 5 displays the set of Li^- results analogous to that of H^- shown in Figure 4. The photodetachment cross section exhibited in the top panel of Figure 5 displays a very clear threshold cusp prescribed by the Wigner threshold law [51], which suppresses all the partial waves in the photoelectron wave packet in the newly opened channel except the s-wave.

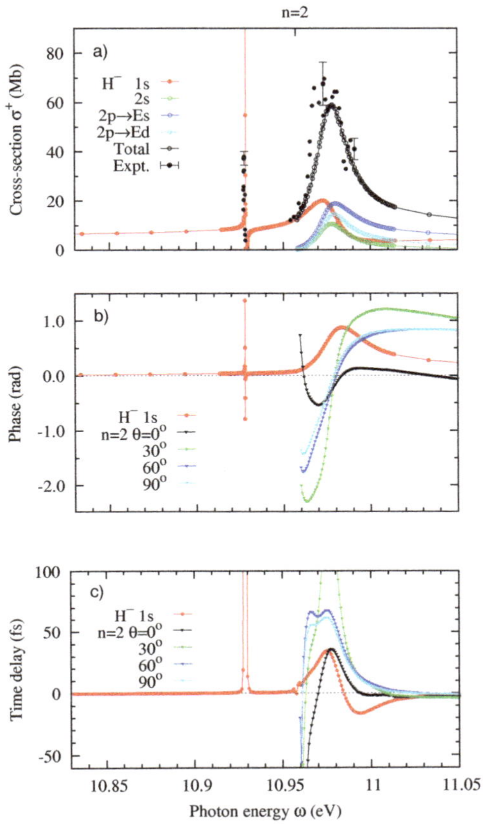

Figure 4. (Color online) (**a**) The cross section (8), (**b**) the phase of the transition amplitude, and (**c**) the time delay (9) of H^- in various photodetachment channels near the $n = 2$ threshold. The experiment [50] is displayed in the top panel.

In the Wigner theory, the cross section near the opening of a new channel can be parameterized as

$$\sigma(E) = \sigma_0 - 2A|E - E_0|^{1/2} \times \begin{cases} \sin^2 \delta_0 & \text{for} \quad E > E_0 \\ \sin \delta_0 \cos \delta_0 & \text{for} \quad E < E_0 \end{cases} . \tag{10}$$

The scattering phase δ_0 is rather small in H^-, as the inelastic threshold is 3/4 Ry above the photodetachment threshold. In contrast, in Li^- , the inelastic $2p$ threshold is only 2 eV above the photodetachment threshold, and the phase δ_0 is rather large. The Wigner Formula (10) predicts the falling cross section above the threshold, while it is always rising below the threshold, hence the cusp formation and the strong dominance of the s-wave in the inelastic channel of Li^- . This dominance reduces dramatically the angular dependence of the phase and time delay. The latter is only weakly angular dependent in comparison with the H^- ion.

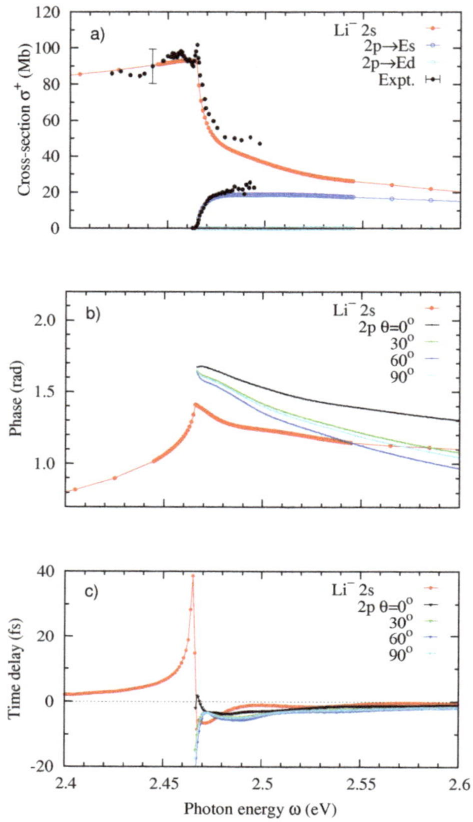

Figure 5. (Color online) (**a**) The cross section (8), (**b**) the phase of the transition amplitude, and (**c**) the time delay (9) of H^- in various photodetachment channels near the $n = 2$ threshold. Experimental cross section results [52,53] are displayed in the top panel.

3.3. *Atomic, Molecular and Optical Science Gateway*

The CCC calculations presented here may be reproduced in a matter of minutes using the Atomic, Molecular and Optical Science (AMOS) Gateway https://amosgateway.org (accessed on 4 February 2022), which is accessible to anyone who is able to authenticate via their own institution. The gateway is a sustainable community-oriented platform for enabling AMOS applications as a service for the AMOS community with intuitive interfaces. Several other AMOS codes are either available, or being made available, via this interface [42].

The gateway-accessible CCC code has been installed and compiled for efficient execution on the machines of the Extreme Science and Engineering Discovery Environment (XSEDE). The code has inbuilt hybrid MPI and OpenMP parallelization, as well as GPU acceleration. The latter has only been implemented recently [54] and continues to be a focus of development. Sample e-Li scattering inputs for the presented calculations are available. The gateway is constantly under development and we expect that more user-friendly interfaces will be built in due course.

4. Conclusions

We are grateful for our interactions with Miron Amusia and their group. Here, we have shown how the incorporation of the Self-Consistent Field and Frozen Core Hartree-Fock routines of the group have played a very important role in the development of the CCC method beyond atomic hydrogen and helium. Not only did this simply broaden the utility of the method but also helped resolve what previously seemed as intractable discrepancies between theory and experiment.

Author Contributions: Software, B.I.S., M.C. and S.P.; formal analysis, D.V.F. and A.S.K (Alisher S. Kadyrov); investigation, A.S.K. (Anatoli S. Kheifets); data curation, X.W.; writing—original draft preparation, I.B. All authors have read and agreed to the published version of the manuscript.

Funding: This research received no external funding.

Institutional Review Board Statement: Not applicable.

Informed Consent Statement: Not applicable.

Data Availability Statement: The numerical data reported in the present work are available on request from the author.

Acknowledgments: The authors gratefully acknowledge the Resources of National Computational Infrastructure facility (NCI Australia), The Pawsey Supercomputer Centre and the Extreme Science and Engineering Discovery Environment (XSEDE), which is supported by National Science Foundation, USA. The Curtin authors are supported by the Australian Research Council.

Conflicts of Interest: The authors declare no known conflict of interest.

References

1. Weigold, E.; Frost, L.; Nygaard, K.J. Large-angle electron-photon coincidence experiment in atomic hydrogen. *Phys. Rev. A* **1980**, *21*, 1950–1954. [CrossRef]
2. Williams, J.F. Electron-photon angular correlations from the electron impact excitation of the 2s and 2p electronic configurations of atomic hydrogen. *J. Phys. B At. Mol. Opt. Phys.* **1981**, *14*, 1197–1217. [CrossRef]
3. Bray, I.; Stelbovics, A.T. Explicit demonstration of the convergence of the close-coupling method for a Coulomb three-body problem. *Phys. Rev. Lett.* **1992**, *69*, 53–56. [CrossRef]
4. Bray, I.; Stelbovics, A.T. Convergent close-coupling calculations of electron-hydrogen scattering. *Phys. Rev. A* **1992**, *46*, 6995–7011. [CrossRef] [PubMed]
5. Bray, I.; Stelbovics, A.T. Calculation of the total ionization cross section and spin asymmetry in electron-hydrogen scattering. *Phys. Rev. Lett.* **1993**, *70*, 746–749. [CrossRef] [PubMed]
6. Karaganov, V.; Bray, I.; Teubner, P.J.O.; Farrell, P. Super-elastic electron scattering on lithium. *Phys. Rev. A* **1996**, *54*, R9–R12. [CrossRef]
7. Karaganov, V.; Bray, I.; Teubner, P.J.O. Super-elastic electron-lithium scattering at 7 and 14 eV. *J. Phys. B At. Mol. Opt. Phys.* **1998**, *31*, L187–L191. [CrossRef]
8. Karaganov, V.; Bray, I.; Teubner, P.J.O. Electron scattering from optically pumped lithium atoms. *Phys. Rev. A* **1999**, *59*, 4407–4417. [CrossRef]
9. McClelland, J.J.; Kelley, M.H.; Celotta, R.J. Superelastic scattering of spin-polarized electrons from sodium. *Phys. Rev. A* **1989**, *40*, 2321–2329. [CrossRef]
10. Scholten, R.E.; Lorentz, S.R.; McClelland, J.J.; Kelley, M.H.; Celotta, R.J. Spin-resolved superelastic scattering from sodium at 10 and 40 eV. *J. Phys. B At. Mol. Opt. Phys.* **1991**, *24*, L653–L659. [CrossRef]
11. Lorentz, S.R.; Scholten, R.E.; McClelland, J.J.; Kelley, M.H.; Celotta, R.J. Spin-resolved elastic scattering of electrons from sodium below the inelastic threshold. *Phys. Rev. Lett.* **1991**, *67*, 3761. [CrossRef] [PubMed]
12. Scholten, R.E.; Shen, G.F.; Teubner, P.J.O. Superelastic electron scattering from sodium: Alignment and orientation parameters. *J. Phys. B At. Mol. Opt. Phys.* **1993**, *26*, 987. [CrossRef]
13. Stockman, K.A.; Karaganov, V.; Bray, I.; Teubner, P.J.O. Superelastic Electron Scattering from Potassium. *J. Phys. B At. Mol. Opt. Phys.* **1998**, *31*, L867–L872. [CrossRef]
14. Stockman, K.A.; Karaganov, V.; Bray, I.; Teubner, P.J.O. Electron impact excitation of the 4^2P state in potassium at 54.4 eV: Differential cross section, alignment and orientation parameters. *J. Phys. B At. Mol. Opt. Phys.* **1999**, *32*, 3003–3013. [CrossRef]
15. Stockman, K.A.; Karaganov, V.; Bray, I.; Teubner, P.J.O. Electron scattering from laser excited states of potassium from 3 to 100 eV. *J. Phys. B At. Mol. Opt. Phys.* **2001**, *34*, 1105–1114. [CrossRef]
16. Chernysheva, L.V.; Cherepkov, N.A.; Radojevic, V. Self-consistent field Hartree-Fock program for atoms. *Comp. Phys. Comm.* **1976**, *11*, 57. [CrossRef]
17. Chernysheva, L.V.; Cherepkov, N.A.; Radojevic, V. Frozen core Hartree-Fock program for atomic discrete and continuous states. *Comp. Phys. Comm.* **1979**, *18*, 87–100. [CrossRef]
18. Bray, I. Convergent close-coupling method for the calculation of electron scattering on hydrogen-like targets. *Phys. Rev. A* **1994**, *49*, 1066–1082. [CrossRef]
19. Fursa, D.V.; Bray, I. Calculation of electron-helium scattering. *Phys. Rev. A* **1995**, *52*, 1279–1298. [CrossRef]
20. Fursa, D.V.; Bray, I. Convergent close-coupling calculations of electron-beryllium scattering. *J. Phys. B At. Mol. Opt. Phys.* **1997**, *30*, L273–L277. [CrossRef]

21. Fursa, D.V.; Bray, I. Convergent close-coupling calculations of electron scattering on helium-like atoms and ions; electron-beryllium scattering. *J. Phys. B At. Mol. Opt. Phys.* **1997**, *30*, 5895–5913. [CrossRef]
22. Bray, I.; Stelbovics, A.T. Calculation of Electron Scattering on Hydrogenic Targets. In *Advances In Atomic, Molecular, and Optical Physics*; Academic Press: Cambridge, MA, USA, 1995; Volume 35, pp. 209–254. [CrossRef]
23. Fursa, D.V.; Bray, I. TOPICAL REVIEW: Convergent close-coupling calculations of electron-helium scattering. *J. Phys. B At. Mol. Opt. Phys.* **1997**, *30*, 757–785. [CrossRef]
24. Bray, I.; Fursa, D.V.; Kadyrov, A.S.; Stelbovics, A.T.; Kheifets, A.S.; Mukhamedzhanov, A.M. Electron- and photon-impact atomic ionisation. *Phys. Rep.* **2012**, *520*, 135–174. [CrossRef]
25. Bray, I.; Abdurakhmanov, I.B.; Bailey, J.J.; Bray, A.W.; Fursa, D.V.; Kadyrov, A.S.; Rawlins, C.M.; Savage, J.S.; Stelbovics, A.T.; Zammit, M.C. Convergent close-coupling approach to light and heavy projectile scattering on atomic and molecular hydrogen. *J. Phys. B At. Mol. Opt. Phys.* **2017**, *50*, 202001. [CrossRef]
26. Yalim, H.; Cvejanovic, D.; Crowe, A. 1s-2p excitation of atomic hydrogen by electron impact studied using the angular correlation technique. *Phys. Rev. Lett.* **1997**, *79*, 2951–2954. [CrossRef]
27. O'Neill, R.W.; van der Burgt, P.J.M.; Dziczek, D.; Bowe, P.; Chwirot, S.; Slevin, J.A. Polarization correlation measurements of electron impact excitation of H(2p) at 54.4 eV. *Phys. Rev. Lett.* **1998**, *80*, 1630–1633. [CrossRef]
28. Williams, J.F.; Mikosza, A.G. Atomic hydrogen revisited: Electron impact excitation to H(2p) at 54 eV. *J. Phys. B At. Mol. Opt. Phys.* **2006**, *39*, 4113–4122. [CrossRef]
29. Mergel, V.; Achler, M.; Dörner, R.; Khayyat, K.; Kambara, T.; Awaya, Y.; Zoran, V.; Nyström, B.; Spielberger, L.; McGuire, J.H.; et al. Helicity Dependence of the Photon-Induced Three-Body Coulomb Fragmentation of Helium Investigated by Cold Target Recoil Ion Momentum Spectroscopy. *Phys. Rev. Lett.* **1998**, *80*, 5301–5304. [CrossRef]
30. Kheifets, A.S.; Bray, I. Calculation of circular dichroism in helium double photoionization. *Phys. Rev. Lett.* **1998**, *81*, 4588–4591. [CrossRef]
31. Achler, M.; Mergel, V.; Spielberger, L.; Dorner, R.; Azuma, Y.; Schmidt-Böcking, H. Photo double ionization of He by circular and linear polarized single-photon absorption. *J. Phys. B At. Mol. Opt. Phys.* **2001**, *34*, 965–981. [CrossRef]
32. Lahmam-Bennani, A.; Duguet, A.; Gaboriaud, M.N.; Taouil, I.; Lecas, M.; Kheifets, A.S.; Berakdar, J.; Cappello, C.D. Complete experiments for the double ionisation of He: (e,3e) cross sections at 1 keV impact energy and small momentum transfer. *J. Phys. B At. Mol. Opt. Phys.* **2001**, *34*, 3073–3087. [CrossRef]
33. Sakhelashvili, G.; Dorn, A.; Höhr, C.; Ullrich, J.; Kheifets, A.S.; Lower, J.; Bartschat, K. Triple Coincidence (e,γ2e) Experiment for Simultaneous Electron Impact Ionization Excitation of Helium. *Phys. Rev. Lett.* **2005**, *95*, 033201. [CrossRef] [PubMed]
34. Kheifets, A.S.; Ivanov, I.A. Convergent close-coupling calculations of two-photon double ionization of helium. *J. Phys. B At. Mol. Opt. Phys.* **2006**, *39*, 1731–1742. [CrossRef]
35. Kheifets, A.S.; Bray, I. Valence-shell double photoionization of alkaline-earth-metal atoms. *Phys. Rev. A* **2007**, *75*, 042703. [CrossRef]
36. Wehlitz, R. Simultaneous emission of multiple electrons from atoms and molecules using synchrotron radiation. *Adv. At. Mol. Opt. Phys.* **2010**, *58*, 1–76. [CrossRef]
37. Schultze, M.; Fiess, M.; Karpowicz, N.; Gagnon, J.; Korbman, M.; Hofstetter, M.; Neppl, S.; Cavalieri, A.L.; Komninos, Y.; Mercouris, T.; et al. Delay in Photoemission. *Science* **2010**, *328*, 1658–1662. [CrossRef]
38. Klünder, K.; Dahlström, J.M.; Gisselbrecht, M.; Fordell, T.; Swoboda, M.; Guénot, D.; Johnsson, P.; Caillat, J.; Mauritsson, J.; Maquet, A.; et al. Probing Single-Photon Ionization on the Attosecond Time Scale. *Phys. Rev. Lett.* **2011**, *106*, 143002. [CrossRef]
39. Kheifets, A.S.; Ivanov, I.A.; Bray, I. Timing analysis of two-electron photoemission. *J. Phys. B At. Mol. Opt. Phys.* **2011**, *44*, 101003. [CrossRef]
40. Kheifets, A.S.; Bray, A.W.; Bray, I. Attosecond Time Delay in Photoemission and Electron Scattering near Threshold. *Phys. Rev. Lett.* **2016**, *117*, 143202. [CrossRef]
41. Kheifets, A.S.; Bray, I. Time delay in two-electron photodetachment and tests of fundamental threshold laws. *Phys. Rev. Res.* **2021**, *3*, 043017. [CrossRef]
42. Schneider, B.I.; Bartschat, K.; Zatsarinny, O.; Hamilton, K.R.; Bray, I.; Scrinzi, A.; Martin, F.; Gonzalez Vasquez, J.; Tennyson, J.; Gorfinkiel, J.D.; et al. Atomic and Molecular Scattering Applications in an Apache Airavata Science Gateway. In *Practice and Experience in Advanced Research Computing*; PEARC'20, Association for Computing Machinery: New York, NY, USA, 2020; pp. 270–277. [CrossRef]
43. Bederson, B. Perfect scattering experiment I. *Comments At. Mol. Phys.* **1969**, *1*, 41.
44. Bederson, B. Perfect scattering experiment II. *Comments At. Mol. Phys.* **1969**, *1*, 65.
45. Andersen, N.; Gallagher, J.W.; Hertel, I.V. Collisional alignment and orientation of atomic outer shells. *Phys. Rep.* **1988**, *165*, 1–188. [CrossRef]
46. Andersen, N.; Bartschat, K.; Broad, J.T.; Hertel, I.V. Collisional alignment and orientation of atomic outer shells. III. Spin-resolved excitation. *Phys. Rep.* **1997**, *279*, 251–396. [CrossRef]
47. Kheifets, A.S.; Bray, I. Calculation of double photoionization of helium using the convergent close-coupling method. *Phys. Rev. A* **1996**, *54*, R995–R997. [CrossRef]
48. Dyall, K.G.; Grant, I.P.; Johnson, C.T.; Parpia, F.P.; Plummer, E.P. GRASP: A general-purpose relativistic atomic structure program. *Comp. Phys. Comm.* **1989**, *55*, 425–456. [CrossRef]
49. Amusia, M.Y. *Atomic Photoeffect*; Plenum Press: New York, NY, USA, 1990.
50. Andersen, H.H.; Balling, P.; Kristensen, P.; Pedersen, U.V.; Aseyev, S.A.; Petrunin, V.V.; Andersen, T. Positions and Isotope Shifts of the H^- ($^1P^0$) Dipole Resonances below the H ($n = 2$) Threshold. *Phys. Rev. Lett.* **1997**, *79*, 4770–4773. [CrossRef]

51. Wigner, E.P. On the Behavior of Cross Sections Near Thresholds. *Phys. Rev.* **1948**, *73*, 1002–1009. [CrossRef]
52. Dellwo, J.; Liu, Y.; Pegg, D.J.; Alton, G.D. Near-threshold photodetachment of the Li^- ion. *Phys. Rev. A* **1992**, *45*, 1544–1547. [CrossRef]
53. Dellwo, J.; Liu, Y.; Tang, C.Y.; Pegg, D.J.; Alton, G.D. Photodetachment cross sections for Li^-. *Phys. Rev. A* **1992**, *46*, 3924–3928. [CrossRef]
54. Bray, I.; Hayat, H.; Fursa, D.V.; Kadyrov, A.S.; Bray, A.W.; Cytowski, M. Calculations of electron scattering on H-like ions. *Phys. Rev. A* **2020**, *101*, 022703. [CrossRef]

atoms

MDPI

Review

Peculiar Physics of Heavy-Fermion Metals: Theory versus Experiment

Vasily R. Shaginyan [1,2,*], Alfred Z. Msezane [2] and George S. Japaridze [2]

1 Petersburg Nuclear Physics Institute of NRC "Kurchatov Institute", 188300 Gatchina, Russia
2 Department of Physics, Clark Atlanta University, Atlanta, GA 30314, USA; amsezane@cau.edu (A.Z.M.); george.japaridze@gmail.com (G.S.J.)
* Correspondence: vrshag@thd.pnpi.spb.ru

Abstract: This review considers the topological fermion condensation quantum phase transition (FCQPT) that leads to flat bands and allows the elucidation of the special behavior of heavy-fermion (HF) metals that is not exhibited by common metals described within the framework of the Landau Fermi liquid (LFL) theory. We bring together theoretical consideration within the framework of the fermion condensation theory based on the FCQPT with experimental data collected on HF metals. We show that very different HF metals demonstrate universal behavior induced by the FCQPT and demonstrate that Fermi systems near the FCQPT are controlled by the Fermi quasiparticles with the effective mass M^* strongly depending on temperature T, magnetic field B, pressure P, etc. Within the framework of our analysis, the experimental data regarding the thermodynamic, transport and relaxation properties of HF metal are naturally described. Based on the theory, we explain a number of experimental data and show that the considered HF metals exhibit peculiar properties such as: (1) the universal T/B scaling behavior; (2) the linear dependence of the resistivity on T, $\rho(T) \propto A_1 T$ (with A_1 is a temperature-independent coefficient), and the negative magnetoresistance; (3) asymmetrical dependence of the tunneling differential conductivity (resistivity) on the bias voltage; (4) in the case of a flat band, the superconducting critical temperature $T_c \propto g$ with g being the coupling constant, while the M^* becomes finite; (5) we show that the so called Planckian limit exhibited by HF metals with $\rho(T) \propto T$ is defined by the presence of flat bands.

Keywords: topology; quantum phase transition; flat bands; fermion condensation; HF metals; thermodynamic; transport properties

PACS: 64.70.Tg; 75.40.Gb; 78.20.-e; 71.10.Hf

Citation: Shaginyan, V.R.; Msezane, A.Z.; Japaridze, G.S. Peculiar Physics of Heavy-Fermion Metals: Theory versus Experiment. *Atoms* **2022**, *10*, 67. https://doi.org/10.3390/atoms10030067

Academic Editors: Anatoli Kheifets, Gleb Gribakin and Vadim Ivanov

Received: 17 May 2022
Accepted: 21 June 2022
Published: 23 June 2022

1. Introduction

Strongly correlated Fermi systems such as heavy-fermion metals, graphene, and high-T_c superconductors exhibit the non-Fermi-liquid (NFL) behavior. Theoretical predictions [1–4] and experimental data collected on many of these systems show that at low temperatures a portion of their excitation spectrum becomes approximately dispersionless, giving rise to so-called flat bands and high-T_c superconductivity, see, e.g., [1,5–12]. The emergence of flat bands at low T indicates that the system is close to a special quantum critical point, namely a topological fermion condensation quantum phase transition (FCQPT), leading to the formation of flat bands dubbed the fermion condensation (FC). The flat bands are formed by the Landau interaction between quasiparticles, while a frustration and van-Hove singularities can facilitate the process. Flat bands have notable features, e.g., raising temperatures, and the superconducting phase transition makes them upward tilted [3,4,13–17]. These observations have been predicted [3,4,14,15,17] and are in accordance with experimental data, see, e.g., [13,16,18]. Moreover, the FC theory allows one to qualitatively and quantitatively evaluate the NFL and Landau Fermi liquid (LFL) behaviors of strongly correlated

Fermi systems, and explain the crossover from one another [1,2,4,15,19,20]. We note that in our review we analyze strongly correlated Fermi systems formed by and located near their topological FCQPT and consider experimental observations that are collected on such systems. Consideration of systems located relatively far from their topological FCQPT is possible within the framework of the FC theory as well, see, e.g., [15,19,20]. We review and explain recent prominent experimental results that to our best knowledge have not found alternative explanations and that strongly suggest that the topological FCQPT is a generic feature of many strongly correlated Fermi systems, being the universal cause of their non-Fermi-liquid behavior, and the fermion condensation theory is able to explain the extraordinary behavior of strongly correlated Fermi systems.

In our review we consider exciting experimental facts such as:

(1) Recent experimental findings of linear dependence on temperature T of the resistivity $\rho(T) \propto T$, collected on high T_c superconductors (HTSC), graphene, heavy fermion (HF) and common metals reveal that the scattering rate $1/\tau$ of charge carriers reaches the Planckian limit $1/(T\tau) = k_B/\hbar$, with $1/\tau$ being the scattering rate and k_B and $\hbar$ being the Boltzmann and Plank constants, respectively [21–24]. Within the framework of the FC theory, we show that the quasi-classical physics is still applicable for describing the linear T-dependence of resistivity of strongly correlated metals at their quantum criticality since flat bands, forming the quantum criticality, generate transverse zero-sound mode with the Debye temperature T_D [25]. At $T \geq T_D$, the mechanism of the linear T-dependence is the same in both ordinary metals and strongly correlated ones and is represented by the electron–phonon scattering. Therefore, it is the electron–phonon scattering at $T \geq T_D$ that leads to the near material-independence of the lifetime τ that is expressed as $1/(\tau T) \sim k_B/\hbar$. As a result, we describe and explain recent exciting experimental observations of universal scattering rate related to the linear T-dependent resistivity of a large number of both strongly correlated Fermi systems and common metals [21–24]. We show that the observed scattering rate is explained by the emergence of flat bands formed by the topological FQCPT rather than by the so-called Planckian limit at which the assumed Planckian scattering rate occurs [25,26]. The Planckian limit then has to occur in common metals. Moreover, in magnetic fields, HF metals transit from the NFL to LFL behavior and $\rho(T) \propto T$ vanishes, being replaced by the LFL behavior $\rho(T) \propto A_2T^2$, with A_2 as the temperature-independent coefficient.

(2) Recent observations of the linear T-dependence, $\rho(T) \propto T$, at low temperatures, $T \to 0$, relate the slope of the linear T-dependent resistivity ρ to the London penetration depth λ_0, indicating a universal scaling property

$$\frac{d\rho}{dT} \propto \lambda_0^2 \tag{1}$$

for a large number of strongly correlated high-temperature superconductors [27]. This scaling relation spans several orders of magnitude in λ_0, attesting to the robustness of the empirical law (1) [28].

(3) We also analyze recent challenging experimental findings of tunneling differential conductivity $dI/dV = \sigma_d(V)$ as a function of the applied bias voltage V, collected under the application of magnetic field B on the twisted graphene and the archetypical heavy-fermion metals $YbRh_2Si_2$ and $CeCoIn_5$ [5,29,30]. We explain the emergence of the asymmetrical part $\Delta\sigma_d = \sigma_d(V) - \sigma_d(-V)$ and demonstrate that $\Delta\sigma_d$ vanishes in magnetic fields as predicted [31].

(4) We consider the recent outstanding experimental observation of the density n_s of superconducting electrons that turns out to be much less than the total density n_ρ of electrons at $T \to 0$ [32] as predicted [33].

(5) We show that the transition temperature T_c is proportional to the superconducting coupling constant g,

$$T_c \propto g. \tag{2}$$

This fact, see Equation (2), leads to creating high-T_c superconductors [1,5–12]. This observation is supported by special features of high-T_c superconductivity based on flat bands, namely that T_c is proportional to the Fermi velocity $V_F \propto 1/N_s(0)$ $V_F \propto T_c$, rather than $N_s(0) \propto 1/V_F \propto T_c$ as stated in standard BCS-like theories [13,16] as predicted [17].

Our results are in good agreement with experimental data and demonstrate that the topological FCQPT is an intrinsic feature of strongly correlated Fermi systems, and the FC theory can be viewed as the universal agent explaining the physics of strongly correlated Fermi systems.

2. Fermion Condensation

The theory of FC has been described several times, see, e.g., [4,15,19,20]; nonetheless, for the readers' convenience, we briefly present this methodology. The usual approach to describe the ensembles of itinerant Fermi particles is the well-known Landau Fermi liquid theory [34,35]. This theory represents the real properties of a solid with itinerant electrons in terms of a Fermi gas of so-called quasiparticles with weak interaction. In this case, the quasiparticles represent the excited states of a solid or liquid states and are responsible for the low temperature thermodynamic, transport and relaxation properties of common metals. These quasiparticle excitations are characterized by the effective mass M^*, that is of the order of the bare mass of electron, M, and depends weakly on external parameters such as temperature T, magnetic field B, external pressure P, etc. [34,35]. However, the LFL theory cannot explain why the effective mass M^* begins to depend strongly on the stimuli above and, for example, can even be a divergent function of magnetic field B or temperature T, see, e.g., [4,15,19,36]. Such a dependence is called the NFL behavior and is connected to the growth of the effective mass that occurs when the system approaches the topological fermion condensation quantum phase transition (FCQPT) leading to an FC state with flat bands [1,4,15,19]. Beyond the FCQPT, the system develops a flat band, formed by FC, and characterized by the topological charge that is different from both the topological charges of the Landau Fermi liquid (LFL) and marginal Fermi liquid, representing a new type of Fermi liquid [2,4,15,19,37]. Thus, the stability of FC is ensured by its topological charge, and it can be destroyed only by the first order phase transition, since the topological charge cannot acquire continuous values [2,15,19,37]. As a result of these unique properties of the FC state, a new state of matter is generated, represented by QSL, HF metals, quasicrystals, 2D liquids such as ^{3}He and high-T_c superconductors, so that 1D, 2D and 3D strongly correlated Fermi systems exhibit universal scaling behavior irrespective of their microscopic structure [15,19,20,38,39].

The main feature of FC theory is the existence of one more instability channel (additional to those of Pomeranchuk) that cannot be described within the framework of the Landau theory of Fermi liquid [35]. Indeed, under some conditions, the effective mass M^* of LFL quasiparticle diverges, see, e.g., [15,19]. As a result, to keep the finite and positive effective mass at zero and finite temperatures, the Fermi surface changes its topology: the Fermi surface transforms into a Fermi layer, as seen in Figure 1. This topological phase transition generates the effective mass dependence on temperature, magnetic field, etc. We assume, without loss of generality [15,19], that the Fermi liquid is homogeneous. That is, in our model we account for the most important and common features only, neglecting marginal effects related to the crystalline anisotropy of solids [15,19,20]. The Landau equation for the quasiparticle effective mass M^* reads [15,34,35]

$$\begin{aligned} \frac{1}{M^*_\sigma(B,T)} &= \frac{1}{M} + \sum_{\sigma_1} \int \frac{\mathbf{p}_F \mathbf{p}}{p_F^3} F_{\sigma,\sigma_1}(p_F, p) \\ &\times \frac{\partial n_{\sigma_1}(\mathbf{p},T,B)}{\partial p} \frac{dp}{(2\pi)^3}, \end{aligned} \tag{3}$$

where $F_{\sigma,\sigma_1}(p_F, p)$ is the interaction function, introduced by Landau. The function $F_{\sigma,\sigma_1}(p_F, p)$, depending on momentum p, Fermi momentum p_F and spin indices σ, σ_1, has the form of spherical harmonics with coefficients taken from the best fit to experiment. The fermion occupation number n in the Fermi–Dirac statistics reads

$$n_\sigma(p,T) = \left\{1 + \exp\left[\frac{(\varepsilon_\sigma(\mathbf{p},T) - \mu_\sigma)}{T}\right]\right\}^{-1}, \tag{4}$$

where $\varepsilon_\sigma(p,T)$ is the single-particle spectrum, and μ_σ is a spin-dependent chemical potential: $\mu_\sigma = \mu \pm \mu_B B$ where μ_B is the Bohr magneton. The magnetic field dependence occurs due to the Zeeman splitting shifting the system from its topological FCQPT [15].

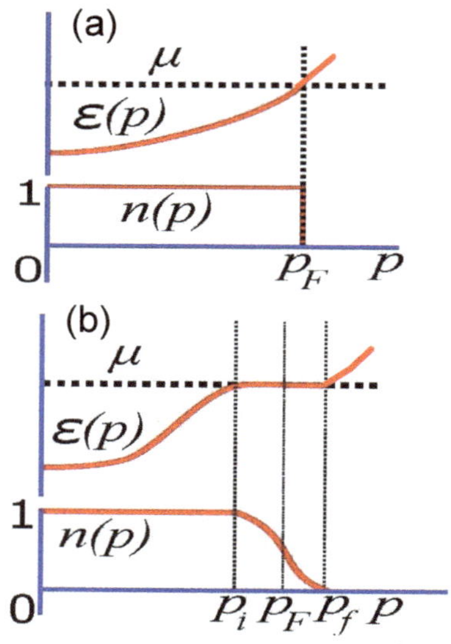

Figure 1. Diagram of flat bands near the FCQPT at zero temperature, $T = 0$. Panel (**a**) shows normal Fermi sphere and corresponding quasiparticles spectrum $\varepsilon(p) \simeq p^2/(2M)$ and occupation number $n(p)$ being a step function. Panel (**b**) displays the system in the FC state after the topological FCQPT. The Fermi sphere alters its topology, which is shown schematically as an emergence of a spherical layer of the thickness $p_f - p_i$. In this case, the Fermi momentum p_F is hidden inside the flat band, defined by the condition $\varepsilon(p) = \mu$ (7). This condition defines the flat band, shown as a dispersionless part of the spectrum $\varepsilon(p) = \mu$, with μ being the chemical potential. The function $n(p)$ decreases gradually from $n(p_i) = 1$ to $n(p_f) = 0$ without violating the Pauli exclusion principle.

The standard procedure for obtaining the single-particle spectrum $\varepsilon_\sigma(p,T)$ in the Landau theory is to vary the system energy $E[n_\sigma(p,T)]$ with regard to the occupation number n

$$\varepsilon_\sigma(p,T) = \frac{\delta E[n(p)]}{\delta n_\sigma(p)}. \tag{5}$$

We note that the Landau interaction entering Equation (3) is not of a special form since it is fixed by the simple condition that the system is in the FCQPT point [15,19]. The explicit form of the variational Equation (5) reads

$$\frac{\partial \varepsilon_\sigma(p,T)}{\partial p} = \frac{p}{M} - \sum_{\sigma_1} \int \frac{\partial F_{\sigma,\sigma_1}(p,p_1)}{\partial \mathbf{p}} n_{\sigma_1}(p_1,T) \frac{d^3 p_1}{(2\pi)^3}, \tag{6}$$

Later on for simplicity, we omit the spin indexes σ. In the FC phase (i.e., beyond the FCQPT) at $T = 0$, Equation (5) takes the form [1]

$$\varepsilon(p, T = 0) = \mu, \ p_i \leq p \leq p_f; \ 0 \leq n(p) \leq 1. \tag{7}$$

where $p_{i,f}$ stands for initial and final momenta (not to be confused with Fermi momentum p_F), where the flat band resides, see Figure 1. Condition (7) defines the flat band since in this case the quasiparticles have no dispersion. By this virtue, quasiparticles have the Fermi velocity $V_F = 0$ and at $T = 0$ are condensed with the same energy $\varepsilon(p, T = 0) = \mu$, representing the superconducting state with the finite order parameter κ, while the superconducting gap $\Delta = 0$, see Section 7. As this resembles the case of Bose condensation, the corresponding phenomenon is called fermion condensation, being separated from LFL by the first order phase transition [1,2,37]. The system with FC acquires properties, being very different from those of ordinary Fermi liquids, since the Fermi liquid with FC forms a new, topologically-protected (and thus "extremely stable") state of matter. This means that if FC is formed in a substance, it will define its properties at $T = 0$ and at elevated temperatures as well. Figure 1 visualizes (at $T = 0$) the consequences of the FCQPT on the Fermi surface, spectrum and occupation number of a Fermi liquid. The transformation from panel (a) (normal Fermi liquid) to panel (b) is represented by altering the Fermi surface topology so that in the normal Fermi liquid the layer of finite length $p_f - p_i$ appears instead of the Fermi surface located at Fermi momentum $p = p_F$. This immediately implies the emergence of the flat part of the spectrum defined by Equation (7), where all the condensed fermions are located. This, in turn, generates the gradual (instead of abrupt on the panel (a) decay of the occupation numbers $n(p)$ from $n = 1$ at $p < p_i$ to $n = 0$ at $p > p_f$.

Equations (3) and (7) allow one to determine the energy spectrum $\varepsilon_\sigma(p, T)$ and occupation numbers $n_\sigma(p, T)$ in a self-consistent way. These quantities, in turn, permit the calculation of the effective mass, $p_F/M^* = \partial\varepsilon(p)/\partial p|_{p=p_F} = V_F$. We emphasize that both magnetic field and temperature dependences of the effective mass $M^*(B, T)$ in the FC phase come from Equation (3) and from the T, B-dependence of $\varepsilon_\sigma(p)$ and $n_\sigma(p)$. Calculated (by Equations (3) and (7)) spectrum and occupation numbers [15] in the FC phase are reported in Figure 2. At (almost) zero temperature, the flat portion of the spectrum is clearly seen at $p_i < p < p_f$. This shape of the spectrum defines $n(p)$ (Figure 2, panel (b)) in the form of "two steps", gradually decaying from one to zero. Simultaneously, at relatively high temperatures (equal to $T/E_F = 0.01$, which at $E_F \sim 1$eV implies $T \simeq 100$ K) this part is rather strongly upward tilted. This shows that finite temperatures erode the FC state, making the effective mass M^* finite, while the system acquires features similar to ordinary Fermi liquid [4,15].

To gain more insights into the physical properties of the FC state, it is helpful to explore the system behavior at $T \to 0$. It was shown earlier [1,15,19] that the ground state of a system with FC is highly degenerate. In this case, the occupation numbers $n_0(p)$ of the FC state quasiparticles (i.e., having dispersionless spectrum or belonging to the flat band) change gradually from $n = 1$ to $n = 0$ at $T = 0$. This variation occurs at $p_i \leq p \leq p_f$. It is clear that such a property of the occupation numbers drastically differs from the property of the usual Fermi–Dirac function property at $T = 0$. Indeed, in that case, the Fermi–Dirac function is represented by the step function between $n = 1$ and $n = 0$ at $p = p_F$, where p_F stands for Fermi momentum, see Figure 1.

At $T = 0$, the infinite degeneracy of the ground state with FC leads to a T-independent entropy term [4,15], remaining finite at $T = 0$ in violation of the Nernst theorem

$$S_0 = -\sum_p [n_0(p) \ln n_0(p) + (1 - n_0(\mathbf{p})) \ln(1 - n_0(p))]. \tag{8}$$

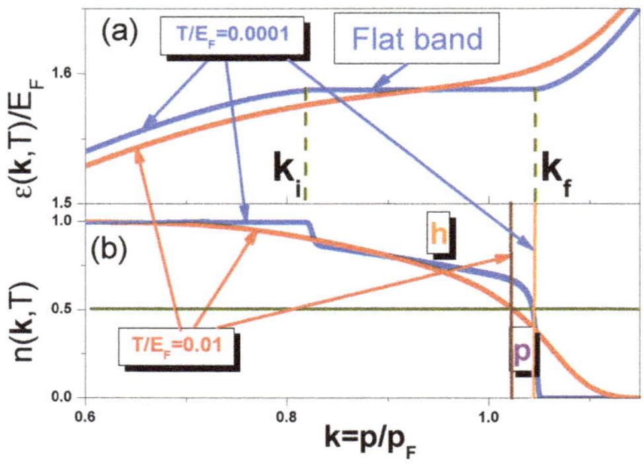

Figure 2. Flat band induced by FC. The calculated single-particle spectrum (**a**) and the quasiparticle occupation number (**b**) at small but finite temperatures versus the dimensionless momentum $k = p/p_F$, where p_F is the Fermi momentum [15]. Temperature is measured in the units of E_F. At $T = 0.01E_F$ and $T = 0.0001E_F$, the vertical lines show the position of the Fermi level E_F at which $n(k,T) = 0.5$ (see the horizontal line in panel (b)). At $T = 0.0001E_F$ (blue curve), the single-particle spectrum $\varepsilon(k,T)$ is almost flat (marked "Flat band") in the range $k_f - k_i$ (with $k_i = p_i/p_F$ and $k_f = k_f/p_F$ denoting, respectively, the initial and final momenta for FC realization, and $k = p/p_F$). Thus, in the range $k_f - k_i$ the density of states $N_0 \to \infty$, and outside the range N_0 is finite. The distribution function $n(k,T)$ becomes more asymmetric with respect to the Fermi level E_F, generating the NFL behavior, and C invariance is broken. To illuminate the asymmetry, the area occupied by holes in panel (b) is labeled h (red) and that occupied by quasiparticles by p (maroon).

Thus, the infinite degeneracy of the FC ground state generated by flat bands, see Refs. [19,20] for a comprehensive discussion. We note that for systems where the Nernst theorem is violated due to the ground state degeneracy is a spin glass [40,41]. It is well known that in normal Fermi liquid the function $n(p)$ at finite temperatures loses its step-like feature at $p = p_F$, becoming continuous around this point. The same is valid for a Fermi liquid with flat bands; this conclusion follows from Equation (4). This means that at small but finite temperatures $T \neq 0$ the degeneracy of the above ground state is lifted, consequently the single-particle energy $\varepsilon(p, T \neq 0)$ acquires a small dispersion [4]

$$\varepsilon(p, T \to 0) = T \ln \frac{1 - n_0(p)}{n_0(\mathbf{p})}. \tag{9}$$

From Equation (9), we see that the dispersion is proportional to T since the occupation numbers n_0 approximately remain the same as at $T = 0$. This means that the entropy S in this case still remains $S(T) \geq S_0$. This situation also jeopardizes the Nernst theorem. To avoid this unphysical situation, the nearly flat bands representing the FC state should acquire dispersion in a way that the excess entropy S_0 should "dissolve" as $T \to 0$. This occurs by virtue of some additions to the FCQPT phase transition such as a ferromagnetic and/or a superconductive one, etc. [4,15,19]. Thus, at low temperatures the FC state has to be consumed by a number of phase transitions. This "consumption" can be viewed as a complicated phase diagram of an HF metal at its quantum critical point. In fact, at $T = 0$ the FC state is represented by the superconducting state with the superconducting order parameter $\kappa = \sqrt{n(p)(1 - n(p))}$ that is finite in the region $(p_i - p_f)$ [15,33,42], for in the region $n(p) < 1$, as shown in Figure 2. Nonetheless, the superconducting gap , $\Delta = 0$, can be absent provided that the superconducting coupling constant $g = 0$. In case of finite g, the gap exhibits very specific non-BCS behavior [43] $\Delta \propto g$, see, e.g., [1,4,44,45] and Section 7.

3. Scaling of Physical Properties

Experimental manifestations of the FC phenomenon correspond to the universal behavior of the physical properties of HF metals [15,19,20,46]. The physical properties of HF metals are formed due to flat bands and are widespread compounds [6]. To reveal the scaling, consider now the approximate solutions of Equation (3) [15,19,20]. At $B = 0$, Equation (3) becomes strongly temperature dependent, which is a typical NFL feature and can be solved analytically [15,19,20,46]:

$$M^*(T) \simeq a_T T^{-2/3}. \tag{10}$$

At $T = 0$, the analytical solution is

$$M^*(B) \simeq a_B B^{-2/3}. \tag{11}$$

Here, a_T and a_B are constants. Under the application of a magnetic field, the system transits to the LFL state with the effective mass becoming almost temperature independent and strongly dependent on B, as seen from Equation (11).

3.1. Internal Variables Revealing the Scaling Behavior

Equations (10) and (11) allow us to construct the approximate solution of Equation (3) in the form $M^*(B,T) = M^*(T/B)$. The introduction of "internal" scales simplifies the problem of constructing the universal scaling of the effective mass M^*, since in that case we eliminate the microscopic structure of the compound in question [15,19,20]. From the Figure 3a, we see that the effective mass $M^*(B,T)$ reaches a maximum M^*_M at a certain temperature $T_M \propto B$ [15]. Accordingly, to measure the effective mass and temperature, it is convenient to introduce the scales M^*_M and T_M. In this case, we have new variables $M^*_N = M^*/M^*_M$ that we call normalized effective mass and $T_N = T/T_M$ that we call normalized temperature. As a result, M^*_N becomes a function of the only variable $T_N \propto T/B$, as seen from Figure 3b.

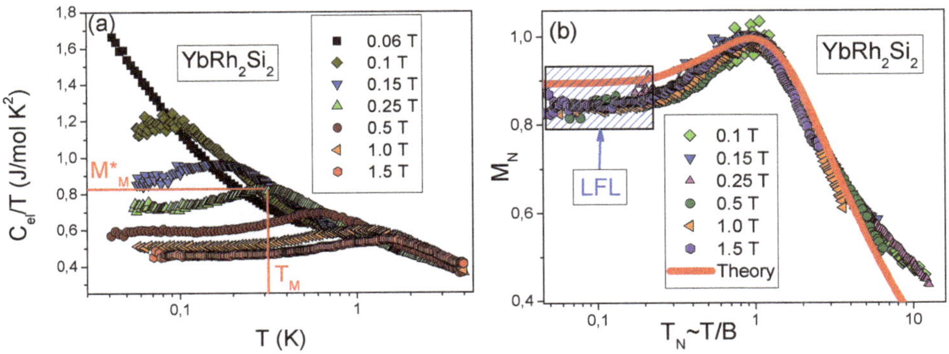

Figure 3. Electronic specific heat of $YbRh_2Si_2$. Panel (**a**): Specific heat C/T, versus temperature T as a function of magnetic field B [36] shown in the legend. Panel (**b**): The normalized effective mass M^*_N as a function of normalized temperature $T_N \propto T/B$. M^*_N is extracted from the measurements of the specific heat C/T on $YbRh_2Si_2$ in magnetic field B [36], see panel (**a**), listed in the legend. Approximate constant effective mass M^* at $T_N < 1$ is typical for the normal Landau Fermi liquids, and is shown by the arrow.

In the vicinity of the FCQPT, the normalized effective mass $M_N^*(T_N)$ can be well approximated by a certain universal function [15,19], interpolating the solutions of Equation (3) between the LFL state, given by Equation (11), and the NFL one, given by Equation (10) [15]

$$M_N^*(T_N) \approx c_0 \frac{1 + c_1 T_N^2}{1 + c_2 T_N^{8/3}}. \quad (12)$$

Here, $T_N = T/T_M$, $c_0 = (1 + c_2)/(1 + c_1)$, where c_1 and c_2 are free parameters. Since the magnetic field B enters Equation (3) as $\mu_B B/T$, the maximum temperature $T_M \sim \mu_B B$. Consequently, from Equation (12),

$$T_M \simeq a_1 \mu_B B; \;\; T_N = \frac{T}{T_M} = \frac{T}{a_1 \mu_B B} \propto \frac{T}{B}, \quad (13)$$

where a_1 is a dimensionless parameter, and μ_B is the Bohr magneton. Equation (13) shows that Equation (12) determines the effective mass as a function of the single variable $T_N \propto T/B$. That is, the curves $M_N^*(T,B)$ merge into a single one $M_N^*(T_N)$, $T_N = T/T_M$, as shown in Figure 4. Since $T_M \propto B$, from Equation (13) we conclude that the curves $M_N^*(T,B)$ coalesce into a single one $M_N^*(T_N = T/B)$, $T_N = T/T_M = T/B$, demonstrating the universal scaling in HF metals [15,19,20]. This universal scaling exhibited by M_N is also shown in Figure 4. We note that Equations (12) and (13) allow one to describe the universal scaling behavior of HF metals, see, e.g., [15,19,20].

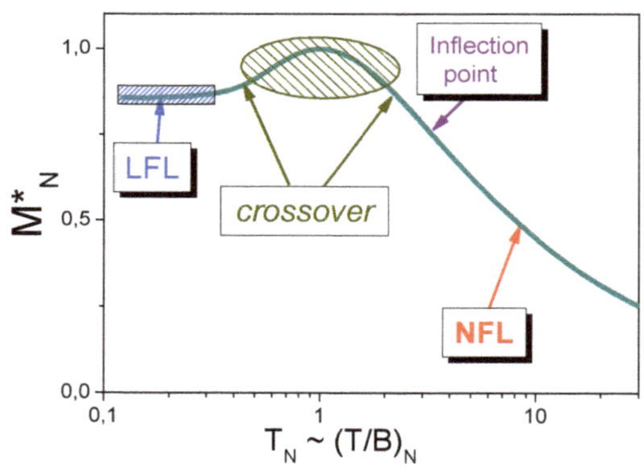

Figure 4. Scaling of the thermodynamic properties governed by the normalized effective mass M_N^* in the case of the application of a magnetic field $T_N \propto T/B$, as follows from Equation (13). The solid curve depicts M_N^* versus normalized temperature T_N. It is clearly seen that at finite $T_N < 1$, the normal Fermi liquid properties take place. At $T_N \sim 1$, M_N^* enters the crossover state, and at growing temperatures it exhibits the NFL behavior.

One more important feature of the FC state is that apart from the fact that the Landau quasiparticle effective mass starts to depend strongly on external stimuli such as T and B, all relations, inherent in the LFL theory, formally remain the same. Namely, the famous LFL relation [35],

$$M^*(B,T) \propto \chi(B,T) \propto \frac{C(B,T)}{T} \propto \gamma_0. \quad (14)$$

still holds. Here, γ_0 is the Sommerfeld coefficient. Expression (14) has been related to the FC case, where the specific heat C, magnetic susceptibility χ and effective mass M^* depend on T and B. Taking Equation (14) into account, we obtain that the normalized values of C/T and χ are of the form [15,19]

$$M^*_N(B,T) = \chi_N(B,T) = \left(\frac{C(B,T)}{T}\right)_N. \tag{15}$$

From Equation (15) we see that the above thermodynamic properties have the same scaling displayed in Figure 4. As a result, we shall see that the observed scaling allows us to construct a general schematic phase diagram, see Section 3.3.

3.2. Magnetoresistance

In the LFL state, the resistivity $\rho(T,B) \propto A_2(B)T^2$. In the case of common metals, it is well known that $\rho(T,B)$ increases with the increasing applied magnetic field B and is described by the Kohler's rule, see, e.g., [47]. In contrast, HF metals exhibit decreasing resistivity in magnetic fields when the metal in question transits from the NFL behavior to the LFL one, see, e.g., [48,49]. The $A(B)$ coefficient, being proportional to the quasiparticle Â—quasiparticle scattering cross section is found to be $A \propto (M^*(B))^2$, as follows from Equation (11) [15,48]. Taking into account Equation (11), we obtain

$$A(B) \simeq A_0 + \frac{D}{B - B_{c0}}, \tag{16}$$

where A_0 and D are fitting parameters. Figure 5 displays experimental data for $A(B)$ collected on two HF metals: $YbRh_2Si_2$ [48] and $Tl_2Ba_2CuO_{6+x}$ [49]. The solid curves represent our calculations, and the inset demonstrates that the well-known Kadowaki–Woods ratio [50] is conserved [48]. This experimental result is in good agreement with Equations (15) and (16).

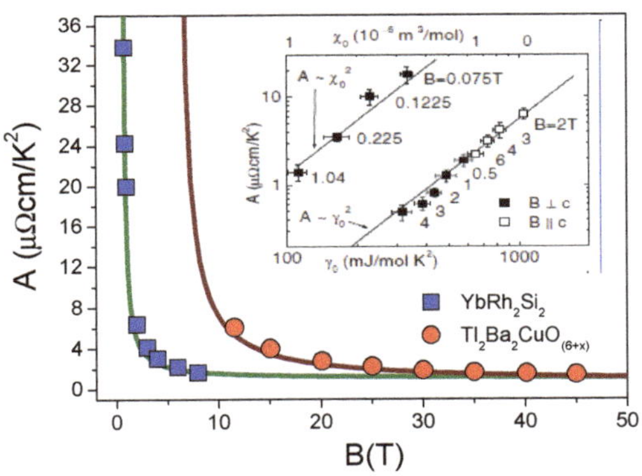

Figure 5. The charge transport coefficient $A(B)$ as a function of magnetic field B obtained in measurements on $YbRh_2Si_2$ [48] and $Tl_2Ba_2CuO_{6+x}$ [49]. The different field scales are clearly indicated. The solid curves represent our calculations based on Equation (16) [15]. The inset (adapted from [51]) shows that $A(B) \propto \chi(B)^2 \propto \gamma_0^2 \propto (C/T)^2$.

To further elucidate the scaling of $A(B)$, we rewrite Equation (16) in the re-scaled variables A/A_0 and B/B_{c0}. Such a recasting immediately reveals the scaling nature of the behavior of these two substances. Both of them are driven to common QCP related to the FCQPT and induced by the application of a magnetic field. As a result, Equation (16) takes the form

$$\frac{A(B)}{A_0} \simeq 1 + \frac{D_N}{B/B_{c0} - 1}, \tag{17}$$

where $D_N = D/(A_0 B_{c0})$ is a constant. From Equation (17), it is seen that upon applying the scaling to both coefficients $A(B)$ for $Tl_2Ba_2CuO_{6+x}$ and $A(B)$ for $YbRh_2Si_2$, they are reduced to a function depending on the single variable B/B_{c0}, thus demonstrating the universal behavior. To support Equation (17), we plot both dependencies in the reduced variables A/A_0 and B/B_{c0} in Figure 6; the universal scaling nature of the coefficients $A(B)$ of these two substances is immediately revealed. We note that the negative magnetoresistance of both $Tl_2Ba_2CuO_{6+x}$ and $YbRh_2Si_2$ results from diminishing $A(B)$ under the application of a magnetic field as follows from Equation (11).

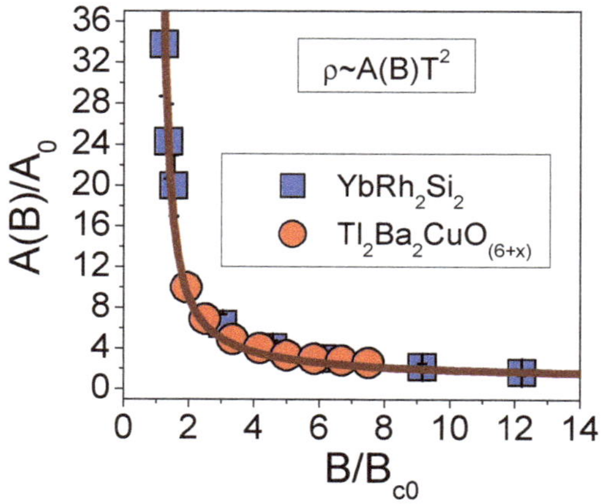

Figure 6. Normalized coefficient $A(B)/A_0 \simeq 1 + D_N/(y-1)$ given by Equation (17) as a function of a normalized magnetic field $y = B/B_{c0}$ shown by squares for $YbRh_2Si_2$ and by circles for high-T_c $Tl_2Ba_2CuO_{6+x}$. D_N is the only fitting parameter.

The scaling behavior of the longitudinal magnetoresistance (LMR) collected on $YbRh_2Si_2$ [48] confirms our above conclusions. This scaling behavior is displayed in Figure 7. Clearly, our calculations are in good agreement with the experimental data. Thus, the fermion condensation theory explains both the negative magnetoresistance and the crossover from the NFL behavior to the LFL one under the application of magnetic fields.

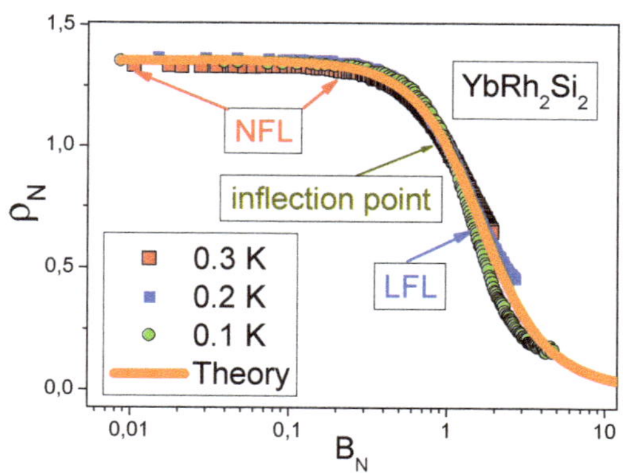

Figure 7. Magnetic field dependence of the longitudinal normalized magnetoresistance LMR versus a normalized magnetic field. The LMR ρ_N was extracted from the LMR of $YbRh_2Si_2$ at different temperatures [48] listed in the legend. The solid line represents our calculations [15]. The arrows show the NFL behavior at $B \ll T$, the inflection point and the LFL behavior at $B \gg T$.

3.3. Schematic Phase Diagram

Based on Equation (12) and Figures 3 and 4, we can construct the schematic $T - B$ phase diagram of HF metals [52], reported in Figure 8. We assume here that at $T = 0$

and $B = B_{c0}$ the system is approximately located at the FCQPT. In the case of $B_{c0} = 0$, the system is located at the FCQPT without tuning. At fixed temperatures, the system is driven by the magnetic field B along the horizontal arrow (from the NFL to the LFL parts of the phase diagram). In turn, at fixed B and increasing T, the system moves from the LFL to the NFL regime along the vertical arrow. The hatched area indicating the crossover between the LFL and the NFL phases separates the NFL state from the slightly polarized paramagnetic LFL state. The crossover temperature $T_M(B)$ is given by Equation (13).

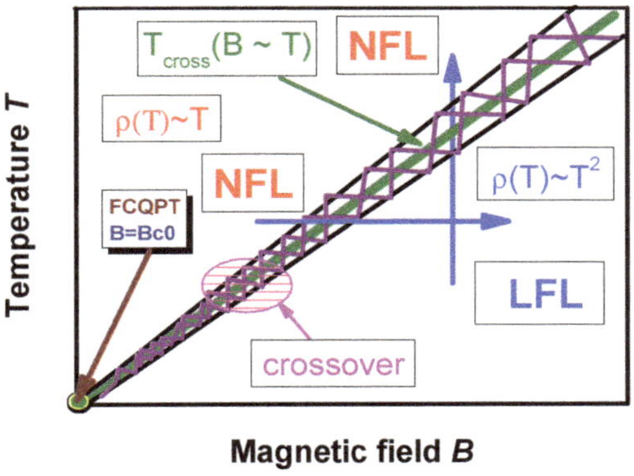

Figure 8. Schematic $T - B$ phase diagram of a strongly correlated Fermi system. The vertical and horizontal arrows crossing the transition region marked by the thick lines depict the LFL–NFL and NFL–LFL transitions at fixed B and T, respectively. At $B < B_{c0}$, the system is in its possible antiferromagnetic (AF) state, with B_{c0} shown by the arrow as denoting a critical magnetic field destroying the AF state. Both the hatched areas shown by the arrow and by the solid curve $T_{\rm cross}(B \sim T)$ represent the crossover separating the domain of NFL behavior from the LFL domain. A part of the crossover is hidden in the possible AF state.

4. The Linear T-Dependent Resistivity and the Planckian Limit

For very different metals such as HF metals, high T_c superconductors and common metals, $\rho(T) \propto T$, the linear dependence of resistivity on temperature and the universality of their fundamental physical properties have been explained within the framework of the FC theory [15,19,25]. On one hand, at low T, the linear T-resistivity

$$\rho(T) = \rho_0 + A_1 T, \tag{18}$$

is experimentally observed in many strongly correlated compounds such as high-temperature superconductors and heavy-fermion metals located near their quantum critical points and therefore exhibiting quantum criticality and a new state of matter, see, e.g., [21,32]. Here, ρ_0 is the residual resistivity and A is a T-independent coefficient. Explanations based on quantum criticality for the T-linear resistivity have been given in the literature, see, e.g., [53–59] and Refs. therein. At room temperatures the T-linear resistivity is exhibited by conventional metals such as Al, Ag or Cu. In the case of a simple metal, the resistivity reads $e^2 n\rho = p_F/(\tau v_F)$ [60], where e is the electronic charge, τ is the lifetime, n is the carrier concentration and p_F and v_F are the Fermi momentum and the Fermi velocity, respectively. Writing the lifetime τ (or inverse scattering rate) of quasiparticles in the form [58,61]

$$\frac{\hbar}{\tau} \simeq a_1 + \frac{k_B T}{a_2}, \tag{19}$$

we obtain [25]

$$a_2 \frac{e^2 n\hbar}{p_F k_B} \frac{\partial \rho}{\partial T} = \frac{1}{v_F}, \tag{20}$$

where $\hbar$ is Planck's constant, k_B is Boltzmanns constant, and a_1 and a_2 are T-independent parameters. Challenging problems for a theory dealing with strongly correlated Fermi systems are:

(1) Experimental data corroborate Equation (20) in the case of both strongly correlated metals and ordinary ones, provided that these demonstrate the linear T-dependence of their resistivity [21], see Figure 9;

(2) Under the application of a magnetic field, HF metals and high-T_c superconductors exhibit the LFL behavior, see Figure 8, and the Planckian limit dissolves in magnetic fields.

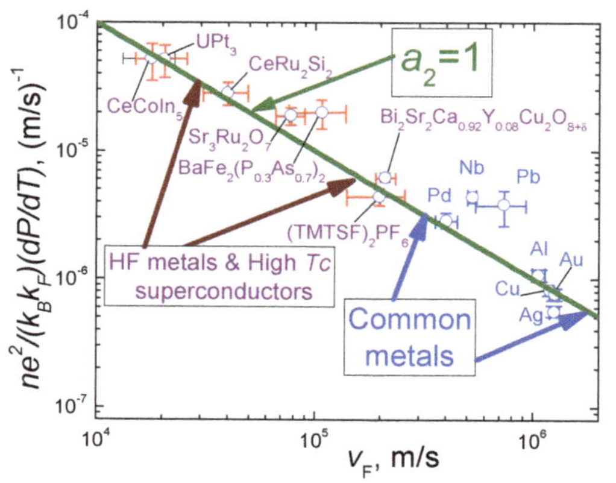

Figure 9. Scattering rates of different strongly correlated metals such as HF metals, high-T_c superconductors, organic metals and conventional metals [21]. All these metals exhibit $\rho(T) \propto T$, and their Fermi velocities V_F vary by two orders of magnitude. The parameter $a_2 \simeq 1$ gives the best fit shown by the solid green line, see Equation (20). The region occupied by the common metals is displayed by the two blue arrows, and the two maroon arrows show the region of strongly correlated metals.

Moreover, the analysis of data in the literature for various compounds and ordinary metals with the linear dependence of $\rho(T)$ shows that the coefficient a_2 is always $0.7 \leq a_2 \leq 2.7$, notwithstanding the large differences in the absolute values of ρ, T and Fermi velocities v_F, varying by two orders of magnitude [21]. As a result, from Equation (19), the T-linear scattering rate is of the universal form, $1/(\tau T) \sim k_B/\hbar$, regardless of different systems displaying the T-linear dependence [19,21,25]. Indeed, this dependence is demonstrated by ordinary metals at temperatures higher than the Debye temperature, $T \geq T_D$, with an electron–phonon mechanism and by strongly correlated metals that are assumed to be fundamentally different from the ordinary ones since the linear T-dependence of their resistivity at temperatures of a few Kelvin is assumed to originate from excitations of electronic origin rather than from phonons [21]. We note that in some cuprates, the scattering rate has a momentum and doping x dependence omitted in Equation (20) [62–64]. Nonetheless, the fundamental picture outlined by Equation (20) is strongly supported by measurements of the resistivity on $Sr_3Ru_2O_7$ for a wide range of temperatures: At $T \geq 100$ K, the resistivity again becomes linearly T-dependent at all applied magnetic fields, as it does at low temperatures and at the critical field $B_c \simeq 7.9$ T but with the coefficient A lower than that seen at low temperatures [21,25]. The same strongly correlated compound exhibits the similar behavior of the resistivity at both quantum critical regime and high temperatures. These facts allow us to expect that the same physics governs the Planckian limit in the case of strongly correlated and ordinary metals. As we will see, the physics here is explained within the fermion condensation theory, and is related to flat bands, the existence of which has been predicted many years ago [1,2,4,15,26,37].

As seen from Figure 9, the scaling relation spans two orders of magnitude in V_F, attesting to the robustness of the observed empirical law [21]. This behavior is explained within the framework of the FC theory since in both cases of common metals and strongly correlated ones, the scattering rate is defined by phonons [25]. In the case of common metals at $T > T_D$, it is well known fact that phonons make a main contribution to the linear dependence of the resistivity, see, e.g., [60]. It has been shown that quasi-classical physics describes the T-linear dependence of the resistivity of strongly correlated metals at $T > T_D$, since flat bands, forming the quantum criticality, generate transverse zero-sound mode with the Debye temperature T_D located within the quantum criticality area [25,57,58]. Therefore, the linear T- dependence is formed by electron–phonon scattering in both ordinary metals and strongly correlated ones. As a result, it is electron–phonon scattering that leads to the near material independence of the lifetime τ that is expressed as

$$\frac{1}{\tau T} \sim \frac{k_B}{\hbar}. \tag{21}$$

We note that there can be another mechanism supporting the linear T-dependence even at $T < T_D$ that fails to warrant a constant τ regardless of the presence of the linear T-dependence of resistivity [25,58]. The mechanism comes from flat bands that are formed by the FC state and contribute to both the linear dependence of the resistivity and to the residual resistivity ρ_0, see Equation (18). Notably, these observations are in good agreement with the experimental data [25,58]. The important point here is that under the application of a magnetic field, the system in question transits from its NFL behavior to an LFL one, and both the flat bands and the FC state are destroyed [15,19], see the $T-B$ phase diagram depicted in Figure 8. Therefore, with resistivity $\rho(T) \propto T^2$, magnetoresistance becomes negative, while the residual resistivity ρ_0 jumps down by a step [19,24,25,58]. Such a behavior is in accordance with experimental data, see, e.g., the case of the HF metal $CeCoIn_5$ [65] that also demonstrates the universal scattering rate at its NFL region, see Figure 9.

5. Asymmetrical Conductivity (Resistivity) of Strongly Correlated Conductors

Direct experimental studies of quantum phase transitions in HTSC and HF metals are of great importance for understanding the underlying physical mechanisms responsible for their anomalous properties. However, such studies of HF metals and HTSC are difficult because the corresponding critical points are usually concealed by their proximity to other phase transitions, commonly antiferromagnetic (AF) and/or superconducting (SC).

Furthermore, extraordinary properties of tunneling conductivity in the presence of a magnetic field were recently observed in a graphene preparation having a flat band [5], as well as in HTSCs and the HF metal $YbRh_2Si_2$ [29,30]. Measuring and analyzing these properties will shed light on the nature of the quantum phase transitions occurring in these substances. Very recently, the scattering rate has been measured in graphene, and it is located near the universal value [23] given by Equation (21), being in accordance with data shown in Figure 9. All these experimental observations qualify graphene as a very interesting material for revealing the physics of strongly correlated Fermi systems.

Most of the experiments on HF metals and HTSCs explore their thermodynamic properties. However, it is equally important to determine other properties of these strongly correlated systems, notably quasiparticle occupation numbers $n(p,T)$ as a function of momentum p and temperature T. These quantities are not linked directly to the density of states (DOS) $N_s(\varepsilon = 0)$ determined by the quasiparticle energy ε or to the behavior of the effective mass M^*. Scanning tunneling microscopy [66–68] and point contact spectroscopy [28,69,70], being sensitive to both the density of states and quasiparticle occupation numbers, are ideal tools for exploring the effects of $\mathcal{C}$ and $\mathcal{T}$ symmetry violation. When $\mathcal{C}$ and $\mathcal{T}$ symmetries are not conserved, the differential tunneling conductivity and dynamic conductance are no longer symmetric functions of the applied voltage V.

Indeed, if under the application of bias voltage V, the current of electrons with the charge $-e$, traveling from HF to a common (i.e., "non-HF") metal changes the sign of a charge carrier to $+e$, then current character and direction alter. Namely, now the carriers are holes with the charge $+e$ traveling from the common to the HF metal. Turning this around, one can obtain the same current of electrons provided that V is changed to $-V$. The resulting asymmetric differential conductivity $\Delta\sigma_d(V) = \sigma_d(V) - \sigma_d(-V)$ becomes nonzero, as seen from Figure 10. On the other hand, if time t is changed to $-t$ (but charge is kept intact), the current changes its direction only. The same result can be achieved by $V \to -V$, and we conclude that $\mathcal{T}$ symmetry is broken, provided that $\Delta\sigma_d(V) \neq 0$. Thus, the presence of $\Delta\sigma_d(V) \neq 0$ signals violation of both $\mathcal{C}$ and $\mathcal{T}$ symmetries. Simultaneously, the change of both $e \to -e$ and $t \to -t$ returns the system to its initial state so that CT symmetry is conserved bearing in mind that the same consideration is true when analyzing $\rho_d(V)$. Note that the parity symmetry P is conserved, and the well-known CPT symmetry is not broken in the considered case. However, the time-reversal invariance and particle-hole symmetry remain intact in normal Fermi systems; the differential tunneling conductivity and dynamic conductance are symmetric functions of V. Therefore, conductivity asymmetry is not observed in conventional metals at low temperatures [28].

To determine the tunneling conductivity, we first calculate the tunneling current $I(V)$ through the contact point between the two metals. This is performed using the method of Harrison [66–68], based on the observation that $I(V)$ is proportional to the particle transition probability introduced by Bardeen [43]. Bardeen considered the probability P_{12} of a particle (say an electron) making a transition from a State 1 on one side of the tunneling layer to a State 2 on the other side. Probability behaves as $P_{12} \sim |t_{12}|^2 N_2(0) n_1(1-n_2)$ where $N_2(0)$ (at $\varepsilon = 0$) is the density of states in State 2, $n_{1,2}$ is the the electron occupation numbers in these states and t_{12} is the transition matrix element. The total tunneling current I is then proportional to the difference between the currents from one to two and that from two to one, and is as follows.

$$\begin{aligned} I \sim P_{12} - P_{21} \sim |t_{12}|^2 N_1(0) N_2(0) \times \\ [n_1(1-n_2) - n_2(1-n_1)] = \\ |t_{12}|^2 N_1(0) N_2(0)(n_1 - n_2). \end{aligned} \tag{22}$$

Harrison applied the WKB approximation to calculate the matrix element [66–68], $t_{12} = t(N_1(0)N_2(0))^{-1/2}$, where t denotes the resulting transition amplitude. Multiplication of expression (22) by two to account for the electron spin and integration over the energy ε leads to the expression for total (or net) tunneling current [66–68]:

$$I(V) = 2|t|^2 \int [n_F(\varepsilon - \mu - V) - n_F(\varepsilon - \mu)] d\varepsilon. \tag{23}$$

Here $n_F(\varepsilon)$ is the electron occupation number for a metal in the absence of a FC, and we have adopted atomic units $e = m = \hbar = 1$, where e and m are the electron charge and mass, respectively. Since temperature is low, $n_F(\varepsilon)$ can be approximated by the step function $\theta(\varepsilon - \mu)$, where μ is the chemical potential.

From Equation (23), it follows that quasiparticles with single-particle energies ε in the range $\mu \leq \varepsilon \leq \mu + V$ contribute to the current, $I(V) = c_1 V$ and $\sigma_d(V) \equiv dI/dV = c_1$, with $c_1 = \text{const}$. Thus, wthin the framework of LFL theory, the differential tunneling conductivity $\sigma_d(V)$, being a constant, is a symmetric function of the voltage V, i.e., $\sigma_d(V) = \sigma_d(-V)$. In fact, the symmetry of $\sigma_d(V)$ holds provided $\mathcal{C}$ and $\mathcal{T}$ symmetries are observed, as is customary for LFL theory. Therefore, $\sigma_d(V)$ is symmetric, and this is common in the case of contact of two ordinary metals (without FC), regardless of whether they are in a normal or superconducting state. Note that a more rigorous consideration of the densities of states N_1 and N_2 entering Equation (22) for $\varepsilon \simeq \mu$ requires their inclusion in the integrand of Equation (23) [71–73]. For example, see Equation (7) of Ref. [73], where this refinement has been carried out for the system of a magnetic adatom and scanning tunneling microscope tip. However, this complication does not break the $\mathcal{C}$ symmetry in the LFL case. Nonetheless, it will be seen below that if the system hosts FC, the presence of the density-of-states factors in the integrand of Equation (23) initiates the asymmetry of the tunneling spectra, since the density of states strongly depends on $\varepsilon \simeq \mu$, see Figure 2. Indeed, the situation becomes quite different in the case of a strongly correlated Fermi system in the vicinity of the FCQPT that causes a flat band [1,2] and violates the $\mathcal{C}$ symmetry [15,19,74]. We note that as we have seen above, the violation of the $\mathcal{C}$ symmetry entails the violation of the $\mathcal{T}$ symmetry. Panel (a) of Figure 2 illustrates the resulting low-temperature single-particle energy spectrum $\varepsilon(k,T)$. Panel (b), which displays the momentum dependence of the occupation numbers $n(k,T)$ in such a system, shows that the flat band induced by the FCQPT, as we have seen above, in fact, violates $\mathcal{T}$ symmetry as well. The broken $\mathcal{C}$ symmetry is reflected in the asymmetry of the regions occupied by particles (labeled p) and holes (labeled h) [15]. We note that the system in its superconducting state and located near the FCQPT exhibits asymmetrical tunneling conductivity, since the $\mathcal{C}$ symmetry remains broken in both the superconducting and the normal states. This observation conforms with the experimental facts [15,70], as seen from Figure 8.

We see from Figure 2 that at low temperatures the electronic liquid of the system has two components. One is an exotic component comprised of heavy electrons occupying momentum range $p_i < p < p_f$ surrounding the Fermi volume near the Fermi surface $p = p_F$. This component is characterized by the superconducting order parameter $\kappa(p) = \sqrt{n(p)(1-n(p))}$. The other component is made up of normal electrons occupying the momentum range $0 \leq p \leq p_i$ [15,33]. In particular, the density of paired charge carriers that form the superfluid density is no longer equal to the total particle density n_{el} represented by paired and unpaired charge carriers. This violation of Leggett's theorem is to be expected since both $\mathcal{C}$ and $\mathcal{T}$ invariants are violated in the NFL state of some HF metals and compounds [15,19,31,74].

We are proposing that for the strongly correlated many-fermion systems in question, the approximate equality $n_s \simeq n_{\rm el}$ that would normally be expected for a real system approximating BCS behavior must be replaced by the inequality $n_s = n_{FC} \ll n_{\rm el}$, where $n_{\rm FC}$ is the density of particles in the FC state [42]. This implies that the main contribution to n_s comes from the FC state. Indeed, the wave function Ξ describing the state of the Cooper pairs as a whole concentrates its associated probability density in the momentum domain of the flat band such that $|\Xi|^2 \propto n_s$, with $|\Xi|^2 \simeq 0$ outside this range. Being defined by the properties of FC, n_s can be very small. Nor does it depend on n_{el}, so it can be expected that $n_s \ll n_{el}$ [33,42].

It is worth noting that the first studies of the overdoped copper oxides suggested that $n_s \ll n_{el}$, but this was attributed to pair-breaking and disorder [75–77], while recent studies with the measurements on ultra-clean samples of $La_{2-x}Sr_xCuO_4$ authenticate the result that $n_s \ll n_{el}$ [32]. It is also relevant that the observed high values of T_c together with the linear dependence of $\rho_{s0} \propto T_c$ [32] of the resistivity are not easily reconciled with the pair-breaking mechanism proposed for dirty superconductors, see, e.g., [53] and Section 7. One cannot expect that such a mechanism would be consistent with high values of T_c and the increase of T_c with doping x. It is worth noting that experimental observation shows that $A_1(x)/T_c(x) \simeq const$ [32,78]. This observation supports the theory of the FC condensation that demonstrates the same result $A_1(x_c - x)/T_c(x_c - x) = const$ [79,80]. Here, x_c is the doping concentration at which the superconductivity sets in, and $(x_c - x) \propto n_s$ [42]. As a result, these evidences support the fermion condensation theory, suggesting the topological FCQPT as the underlying physical mechanism of both the unusual properties of overdoped copper oxides and the asymmetry of tunneling conductivity [1,2,15,19,81].

In case of a strongly correlated Fermi system with FC, the tunneling current becomes [15,31,82,83]

$$I(V) = 2|t|^2 \int [n(\varepsilon - \mu - V, T) - n_F(\varepsilon - \mu, T)] d\varepsilon. \tag{24}$$

Here one of the distribution functions of ordinary metal n_F on the right-hand side of Equation (23) is replaced by $n(\varepsilon, T)$, shown in Figure 2b. As a result, the asymmetric part of the differential conductivity $\Delta\sigma_d(V) = \sigma_d(V) - \sigma_d(-V)$ becomes finite, and we obtain [15,19,31,70,82]

$$\Delta\sigma_d(V) \simeq c\left(\frac{V}{2T}\right)\frac{p_f - p_i}{p_F}, \tag{25}$$

where p_f and p_i define the location of FC, see Figure 2, p_F is the Fermi momentum and c is a constant of order unity.

It is worth noting that Equation (25) is also valid even if the density of states N_1 and N_2 are taken into account, since all this does is change c. Note that the conductivity $\Delta\sigma_d(V)$ remains asymmetric in the superconducting phase of both HTSC and HF metals as well. In such cases, it is again the occupation number $n(p)$ that is responsible for the asymmetric part of $\Delta\sigma_d(V)$, since this function is not appreciably disturbed by the superconductive pairing. This is because usually, in forming the function $n(p)$, the Landau interaction contribution is stronger than that of the superconductive pairing [15]. As a result, $\Delta\sigma_d(V)$ remains approximately the same below the superconducting T_c [15,31]. It is seen from Equation (25) and Figure 10 that with rising temperatures, the asymmetry diminishes and finally vanishes at $T \geq 40$ K. Such a behavior has been observed in measurements on the HF metal $CeCoIn_5$ [84,85], displayed in Figure 10.

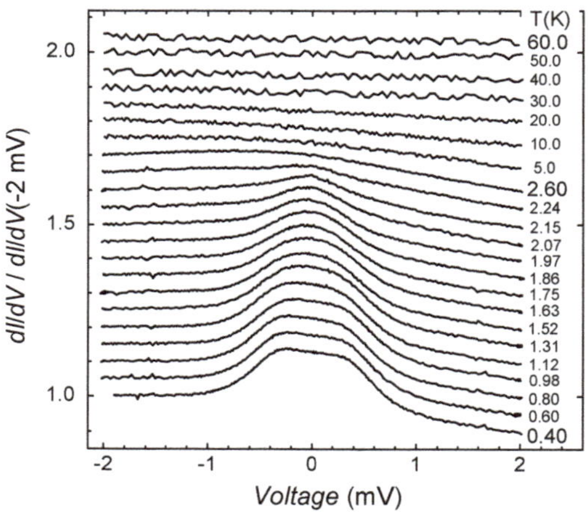

Figure 10. Conductivity spectra $\sigma_d(V) = dI/dV$ measured on the HF metal $CeCoIn_5$ with point contacts ($Au/CeCoIn_5$) over a wide temperature range [84]. Curves $\sigma_d(V)$ are shifted vertically by 0.05 for clarity and normalized by the conductance at -2 mV. The asymmetry develops at $T \simeq 40$ K, becoming stronger at decreasing temperature and persisting below $T < T_c \simeq 2.3$ K in the superconducting state [84].

Under the application of a magnetic field B at sufficiently low temperatures $k_BT \lesssim \mu_BB$, where k_B and μ_B are the Boltzmann constant and the Bohr magneton, the strongly correlated Fermi system transits from the NFL to the LFL regime [15,86]. As we have seen above, the asymmetry of the tunneling conductivity vanishes in the LFL state [15,31,70,82]. It is seen from Figure 11, that $\Delta\sigma_d(V)$, displayed in Figure 10 and extracted from experimental data [85], vanishes in the normal state at sufficiently high magnetic fields applied along the easy axis and low temperatures $k_BT << \mu_B(B - B_c)$ with the critical field $B_c \simeq 5$ T in agreement with the prediction, see, e.g., [15,31,87]. Under this condition, the system transits from the NFL to the LFL behavior, with the resistance ρ becoming a quadratic function of temperature, $\rho(T) \propto T^2$ [15]. The examples of suppression of the asymmetric parts of differential conductivity and resistance under the application of a magnetic field are shown in Figure 11, Figure 12 and Figure 13, respectively.

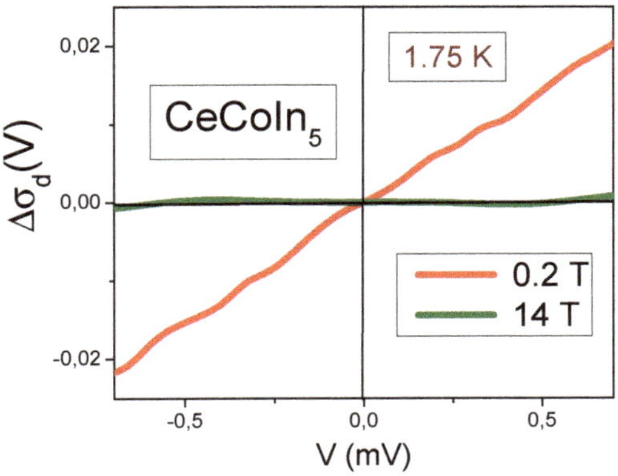

Figure 11. Asymmetric part $\Delta\sigma_d(V)$ of the tunneling differential conductivity measured on $CeCoIn_5$ and extracted from the experimental data [85]. The asymmetric part vanishes at $B = 14$ T and $T = 1.75$ K, with $B_{c0} \simeq 5$ T.

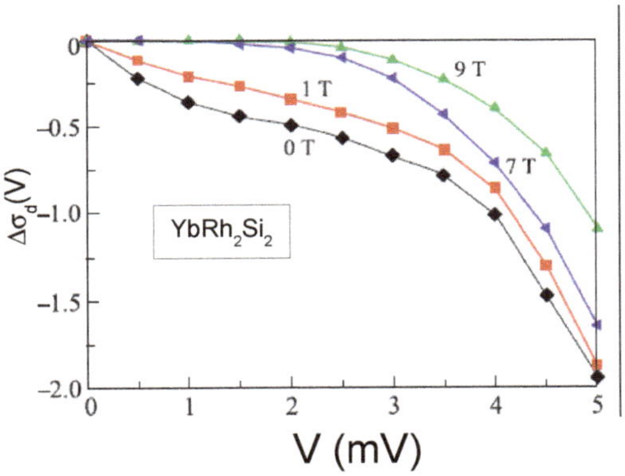

Figure 12. Asymmetric parts $\Delta\sigma_d(V)$ of the tunneling differential conductivity measured on $YbRh_2Si_2$ and extracted from the data shown in Figure 14.

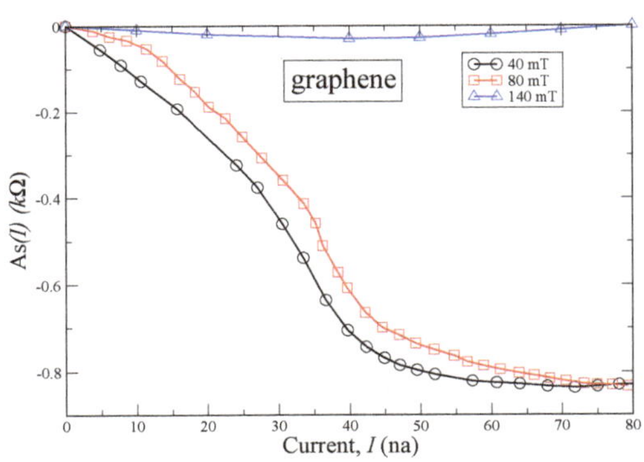

Figure 13. Magnetic field (legend) dependence of the asymmetric part $As(I) = dV/dI(I) - dV/dI(-I)$ versus the current I, extracted from the data of Figure 15 for graphene.

Figure 14 shows the differential conductivity σ_d observed in measurements on $YbRh_2Si_2$ [29,30]. It is seen that asymmetry diminishes with increasing magnetic field B, as the minima of the curves shift to the point $V = 0$, see also Figure 12 for details. The magnetic field is applied along the hard magnetization direction, $B \parallel c$, with $B_c \simeq 0.7$ T [30], where B_c is the critical field suppressing the AF order [51]. The asymmetric part of the tunneling differential conductivity, $\Delta\sigma_d(V)$, extracted from the measurements shown in Figure 14, is displayed in Figure 12. It is seen that $\Delta\sigma_d(V)$ decreases as B increases. We predict that application of the magnetic field in the easy magnetization plane, $B \perp c$ with $B_c \simeq 0.06$ T, leads to a stronger suppression of the asymmetric part of the conductivity, observing that in this case the magnetic field effectively suppresses the antiferromagnetic order and the NFL behavior. Indeed, the experimental data show that low-temperature electrical resistivity $\rho(T)$ of the HF metal $YbRh_2Si_2$, measured at $T \simeq 20$ mK, under the application of the magnetic field $B \geq 75$ mT along an easy magnetization plane, exhibits the LFL behavior $\rho(T) \propto T^2$, while at $B \simeq 60$ mT it demonstrates the NFL behavior, $\rho(T) \propto T$. At the same time, under the application of a magnetic field B along the hard magnetization direction, resistivity shows the LFL behavior at much higher $B \geq 0.8$ T [51]. The same transition from the NFL behavior to the LFL one is observed in measurements of the thermodynamic, transport and relaxation properties, see, e.g., [15,19,51]. We surmise that the asymmetric part $\Delta\sigma_d(V)$ vanishes as soon as $YbRh_2Si_2$ enters its AF state, exhibiting the LFL behavior $\rho(T) \propto T^2$ at $B = 0$ and $T < 70$ mK.

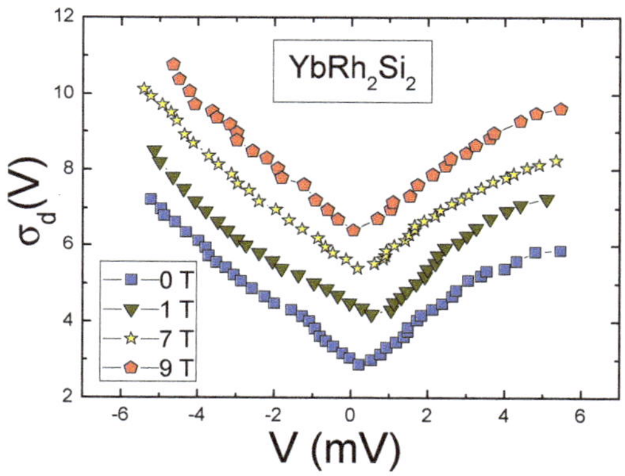

Figure 14. Differential conductivity $\sigma_d(V) = dI/dV$ measured on $YbRh_2Si_2$ under the application of a magnetic field (legend) along the hard magnetization direction [30].

Measuring the differential resistance $\rho_d(V) = dV/dI$ as a function of current I, one finds that the its symmetry properties are the same as those of $\sigma_d(V)$. Namely, under the application of a magnetic field, the asymmetry of the differential resistance vanishes as the system transits into the LFL state. The differential resistance $\rho_d(V)$ of graphene as a function of a direct current I for different magnetic fields B is reported in Figure 15 [5]. The asymmetric part of the differential resistance $As(I) = \rho_d(V) - \rho_d(-V)$ diminishes with an increasing magnetic field, vanishing near $B \simeq 140$ mT. Such a behavior corroborates our conclusion, since the strongly correlated graphene sample has a perfect flat band, implying that the FC effects should be clearly manifested in this material [5].

Thus, in accordance with prediction [15,31,70,82], the asymmetric part tends to zero at tiny magnetic fields of 140 mT, as seen from Figure 13. Note that suppression of the asymmetric part under the application of a magnetic field has been observed in the HF metal $YbCu_{5-x}Al_x$ [81]. The asymmetry persists in the superconducting state of graphene [5] and is suppressed at $B \simeq 80$ mT. Disappearance of the asymmetric part of the differential conductivity in Figure 13 indicates that as the magnetic field increases, graphene transits from the NFL to the LFL state. We remark that the disappearance of the asymmetric part of the differential conductivity was predicted many years before the experimental observations [31,70,82]. It is worth noting that the decrease of the asymmetric part under the application of a magnetic field is an important feature, since the presence of the asymmetric part can be observed by a simple device, e.g., by a diode, since the asymmetric part does not vanish in a magnetic field. Moreover, at $B = 0$, the asymmetric part observed in HF metals and HTSC can be explained in many ways, see, e.g., [88].

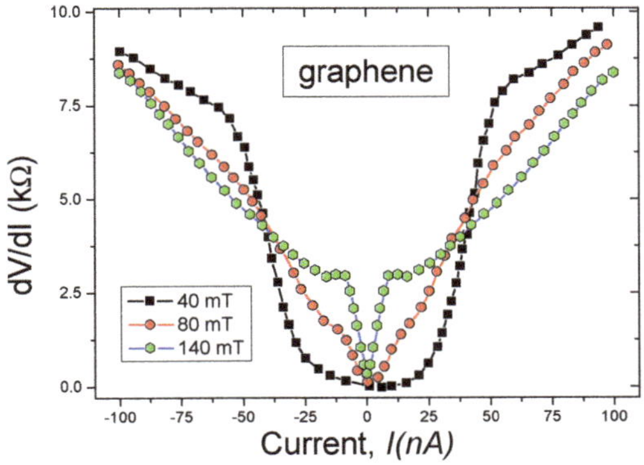

Figure 15. Differential resistance dV/dI of graphene versus current I at different magnetic fields B shown in the legend [5]. Weak asymmetry is observed at small magnetic fields.

To support the statement that the NFL behavior of graphene vanishes in magnetic fields, we surmise that the resistance $\rho(T)$ should exhibit linear dependence $\rho(T) \propto A_1 T$ in the normal state at zero magnetic field, as is generally the case in other strongly correlated Fermi systems. Indeed, at elevated magnetic fields and low temperatures $k_B T << \mu_B B$, the system transits from the NFL behavior to the LFL behavior, causing the resistance to become a quadratic function of temperature $\rho(T) \propto T^2$ that confirms the LFL behavior [15,19,58].

6. Heavy-Fermion Metals and High-Temperature Superconductors: Scaling Relations

It has been shown that the behavior $\rho(T) \propto T$ as $T \to 0$ is an intrinsic property of cuprates associated with a universal scattering rate as well as the property of HF metals [21,22,24], see Section 4. It is stated that the behavior $\rho(T) \propto T$ is achieved when the scattering rate hits the Planckian limit, given by Equation (21), irrespective of the origin of the scattering process [22,24]. However, it is hardly possible that the linear T-dependence of resistivity of common metals is formed by the Planckian limit, as observed in Ref. [21], see Figure 9 and explanation in Ref. [25]. Moreover, HF metals and high-T_c superconductors demonstrate scaling behavior under the application of a magnetic field, pressure, etc., see Figure 3a,b. In magnetic fields, these compounds are shifted from the NFL to the LFL behavior, see, e.g., [15,24]. All these extraordinary features are explained within the framework of the FC theory [1,15,19]. As a result, we can safely suggest that the main reason for the behavior given by Equation (21) is defined by phonons, taking place at $T \geq T_D$ in both strongly correlated Fermi systems and common metals [25].

Another experimental result [27] providing insight into the NFL behavior of strongly correlated Fermi systems is the universal scaling, which can also be explained using the flat band concept. The authors of Ref. [27] measured the temperature dependence $d\rho/dT$ of the resistivity ρ for a large number of HTSC substances for $T > T_c$. Among these were LSCO and the well-known HF compound $CeCoIn_5$; see Table I of Ref. [27]. They discovered quite remarkable behavior: for all substances considered, $d\rho/dT$ shows a linear dependence on the London penetration depth λ_0^2. All of the superconductors considered belong to the London type for which $\lambda_0 >> \xi_0$, where ξ_0 is the zero-temperature coherence length, see, e.g., [42].

It has been shown that the scaling relation [27]

$$\frac{d\rho}{dT} \propto \frac{k_B}{\hbar}\lambda_0^2 \tag{26}$$

remains valid over several orders of magnitude of λ_0, signifying its robustness. At the phase transition point $T = T_c$, the relation (26) yields the well-known Holmes law [27], see also [89] for its theoretical derivation:

$$\sigma T_c \propto \lambda_0^{-2}, \tag{27}$$

in which $\sigma = \rho^{-1}$ is the normal state *dc* conductivity. It has been shown by Kogan [89] that Holms law applies even for the oversimplified model of an isotropic BCS superconductor. Within the same model of a simple metal, one can express the resistivity ρ in terms of microscopic substance parameters [60]: $e^2 n\rho \simeq p_F/(\tau v_F)$, where τ is the quasiparticle lifetime, n is the carrier density, and v_F is the Fermi velocity. Taking into account that $p_F/v_F = M^*$, we arrive at the equation [28]

$$\rho = \frac{M^*}{ne^2\tau}. \tag{28}$$

Note that Equation (28) formally agrees with the well-known Drude formula. It has been shown in Ref. [42] that good agreement with experimental results [32] is achieved when the effective mass and the superfluid density are attributed to the carriers in the FC state only, i.e., $M^* \equiv M_{FC}$ and $n \equiv n_{FC}$. Keeping this in mind and utilizing the relation $1/\tau = k_B T/\hbar$ [19,25,87], we obtain

$$\rho = \frac{M_{FC}}{e^2 n_{FC}}\frac{k_B T}{\hbar} \equiv 4\pi\lambda_0^2\frac{k_B T}{\hbar}, \tag{29}$$

i.e., $d\rho/dT$ is indeed given by the expression (26). Equation (29) demonstrates that fermion condensation can explain all the above experimentally observed universal scaling relations. It is important to note that the FC approach presented here is not sensitive to and transcends the microscopic, non-universal features of the substances under study. This is attributed to the fact that the FC state is protected by its topological structure and therefore represents a new class of Fermi liquids [2,19]. In particular, consideration of the specific crystalline structure of a compound, its anisotropy, its defect composition, etc., do not change our predictions qualitatively. This strongly suggests that the FC approach provides a viable theoretical framework for explaining universal scaling relations similar to those discovered in experiments [27,32]. In other words, condensation of the charge carrier quasiparticles in the considered strongly correlated HTSCs, engendered by a quantum phase transition, is indeed the primary physical mechanism responsible for their observable universal scaling properties. This mechanism can be extended to a broad set of substances with very different microscopic characteristics, as discussed in detail in Refs. [15,19,20].

7. Influence of Superconducting State on Flat Bands

We continue to study Fermi systems with FC at $T = 0$, employing weak BCS-like interaction with the coupling constant g [43]. We analyze the behavior of both the superconducting gap Δ and the superconducting order parameter $\kappa(p)$ as $g \to 0$. In case of BCS-like theories, one obtains the well-know result. Both $\kappa \to 0$ and $\Delta \to 0$, while the FC theory yields $\Delta \propto g$ [1,4,45,90,91]. To study the latter case, we start from the usual pair of equations for the Green's functions $F^+(p,\omega)$ and $G(p,\omega)$ [60]

$$F^+ = \frac{-g\Xi^*}{(\omega - E(p) + i0)(\omega + E(p) - i0)}, \tag{30}$$

$$G = \frac{u^2(p)}{\omega - E(p) + i0} + \frac{v^2(\mathbf{p})}{\omega + E(p) - i0}, \tag{31}$$

where $E^2(p) = \xi^2(p) + \Delta^2$, where $\xi(\mathbf{p}) = \varepsilon(p) - \mu$. Here, $\varepsilon(p)$ is the single particle energy, and μ is the chemical potential. The gap Δ and the function Ξ are given by

$$\Delta = g|\Xi|, \quad i\Xi = \int\int_{-\infty}^{\infty} F^+(\mathbf{p},\omega)\frac{d\omega dp}{(2\pi)^4}. \tag{32}$$

Denoting $v^2(p) = (1 - \xi(p)/E(p))/2$, $v^2(\mathbf{p}) + u^2(p) = 1$, simple algebra yields

$$\xi(p) = \Delta\frac{1 - 2v^2(p)}{2\kappa(p)}. \tag{33}$$

Here $\kappa(p) = u(p)v(p)$ is the superconducting order parameter. It follows from Equation (33) that $\xi \to 0$ when $\Delta \to 0$, provided that $\kappa(p) \neq 0$ in some region $p_i < p < p_f$; thus, the band becomes flat in the region, since $\varepsilon(p) = \mu$ [15,17]. Note that in this case the BCS-like theory gives the standard result implying that both $\Delta = 0$ and $\kappa = 0$ since it is assumed that $\xi(p)$ is fixed. Then, we derive from Equations (32) and (33) that

$$i\Xi = \int_{-\infty}^{\infty} F^+(p,\omega)\frac{d\omega dp}{(2\pi)^4} = i\int \kappa(p)\frac{dp}{(2\pi)^3}. \tag{34}$$

From Equations (32)–(34), we readily see that as $g \to 0$ the superconducting gap $\Delta \to 0$, while the density n_s of the superconducting electrons defined by $\Xi = n_s$ is finite, and the dispersion $\varepsilon(p)$ becomes flat, $\xi = 0$. While $\kappa(p)$ is finite in the region $p_i \leq p \leq p_f$, making Ξ finite. As a result, in systems with FC, the gap Δ vanishes when $g \to 0$, but both the order parameter $\kappa(p)$ and n_s are finite. When the coupling constant g increases, the gap Δ is given by Equation (2), and the superconducting temperature $T_c \propto g\Xi = gn_s$ [1,15]. As a result, one obtains the possibility to construct the room-T_c superconductors [5–12]. At the same time $n_s \ll n_\rho$, where n_ρ is the density of electrons [33,42]. Thus, in case of overdoped superconductors $n_s \ll n_\rho$ rather than $n_s = n_\rho$, as should be in BSC like theories [32,33,42]. Employing Equations (32) and (33), we deduce from Equations (30) and (31) that

$$F^+ = -\frac{\kappa(p)}{\omega - E(\mathbf{p}) + i0} + \frac{\kappa(p)}{\omega + E(p) - i0} \tag{35}$$

$$G = \frac{u^2(p)}{\omega - E(p) + i0} + \frac{v^2(\mathbf{p})}{\omega + E(p) - i0}. \tag{36}$$

In the region occupied by FC, the coefficients $v^2(p)$, $u^2(p) = 1 - v^2(p)$, $v(p)u(p) = \kappa(\mathbf{p}) \neq 0$ are given by $\varepsilon(p) = \mu$, while $E(\mathbf{p}) \to 0$ [1,4,15]. From Equations (35) and (36), it is seen that when $g \to 0$, the equations for $F^+(p,\omega)$ and $G(p,\omega)$ are transformed in the FC region to [90]

$$F^+(p,\omega) = -\kappa(p)\left[\frac{1}{\omega + i0} - \frac{1}{\omega - i0}\right] \tag{37}$$

$$G(p,\omega) = \frac{u^2(p)}{\omega + i0} + \frac{v^2(\mathbf{p})}{\omega - i0}. \tag{38}$$

Integrating $G(p,\omega)$ over ω, one obtains $v^2(\mathbf{p}) = n(p)$. From Equation (32), it follows that Δ is a linear function of g [1,33,45,91]. Since the transition temperature $T_c \sim \Delta \propto g \to 0$, $\kappa(\mathbf{p})$ vanishes at $T \to 0$ via the first order phase transition [2,15]. Thus, on one hand, the FC state with its flat band represents a special solution of the BSC equations. On the other hand, representing a contrast to BSC-like theories, Equation (33) gives the dependence of the spectrum ξ on $\Delta \propto g$, thus, leading to $V_F \propto T_c$ [13,15–17].

Now we use Equation (33) to calculate the effective mass M^* by differentiating both sides of this equation with respect to the momentum p at $p = p_F$ [15,17] and obtain

$$M^* \simeq p_F \frac{p_f - p_i}{2\Delta}. \quad (39)$$

From Equation (39), we obtain that $V_F \propto T_c \propto \Delta$ and conclude

$$V_F \simeq \frac{2\Delta}{p_f - p_i} \propto T_c. \quad (40)$$

From Equations (33) and (40), we see that as $T_c \propto \Delta \to 0$, the Fermi velocity $V_F \to 0$ and the band becomes exactly flat [13,17]. When $T_c \simeq g\Delta$ becomes finite at g increasing, the plateau starts to slightly tilt and is rounded at its end points, as seen from Figure 16. At increasing $\Delta \propto T_c$, both M^* and the density of states $N_s(0)$ are diminished, causing increasing V_F. As seen from Figure 16, the plateau of the flat band of the superconducting system with FC is slightly upward tilted, and M^* is diminished. It follows from Equation (9) that at $T > T_c$ the slope of the flat band is proportional to T, and this dependence can be measured by using ARPES. It is also seen from Figure 2 that both the particle - hole symmetry $\mathcal{C}$ and the time invariance $\mathcal{T}$ are violated generating the asymmetrical differential tunneling conductivity at the NFL behavior, and the NFL behavior is suppressed under the application of a magnetic field that drives the system to its Landau Fermi liquid state, see Section 5.

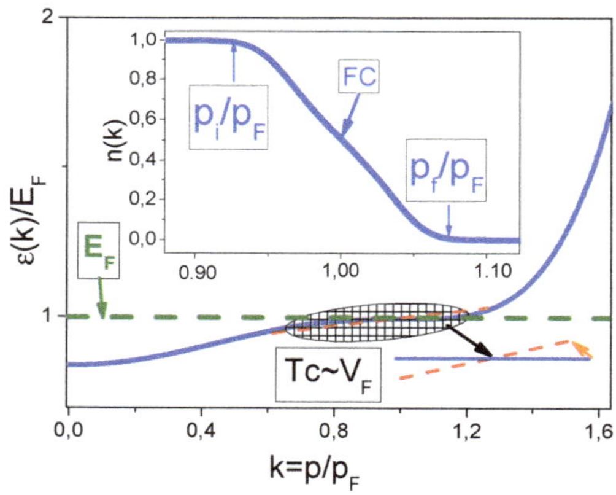

Figure 16. Flat band versus superconducting (SC) state. At $T = 0$, the flat single particle spectrum with $V_F = 0$ is depicted by the solid curve. The transformed flat band by SC with finite V_F is displayed by the red dashed line, see Equation (40). This change is shown by the arrow and by the blue solid and red dashed lines. The dashed area shows the flat band deformation by the SC state. Inset: the occupation numbers $n(k)$ at $T = 0$ as a function of the dimensionless momentum $k = p/p_F$. FC location is displayed by the arrow, with labels p_i/p_F and p_f/p_F revealing the area where $0 < n(p) < 1$, see Equation (7).

Measurements of V_F as a function of T_c [16] are depicted in Figure 17. The inset in Figure 17 shows experimental data collected on the high-T_c superconductor $Bi_2Sr_2CaCu_2O_{8+x}$ in measurements using scanning tunneling microscopy and spectroscopy; here, x is oxygen doping concentration [92]. The integrated local density of states is shown in arbitrary units (au). The straight line depicts the local density of states that is inversely proportional to Δ. Note that the tunneling current is proportional to the integrated local density of states [92]. From the inset, it is clear that the data taken at the position with the highest integrated local density of states has the smallest gap value Δ [92]. These observations are in good agreement with Equations (39) and (40). Thus, our theoretical prediction [15,17] agrees very well with the experimental results [16,92,93]. We note that $V_F \to 0$ as $T_c \to 0$, as seen from Figure 17. This result shows that the flat band is disturbed by the finite value of Δ, and possesses a finite slope that makes $V_F \propto T_c$, as seen from Figure 16. Indeed, from Figure 17, the experimental critical temperatures T_c do not correspond to the minima of the Fermi velocity V_F as they would in any theory wherein pairing is mediated by phonons (bosons) that are insensitive to V_F as they would in any theory wherein pairing is mediated by phonons, or any other bosons, that are insensitive to V_F [16].

Thus, such a behavior is in stark contrast to that expected within the framework of the common BSC-like theories that do not assume that the single particle spectra strongly depends on T_c [15,16,43]. This extraordinary behavior is explained within the framework of the FC theory based on the topological FCQPT, forming flat bands [15,17,19,20].

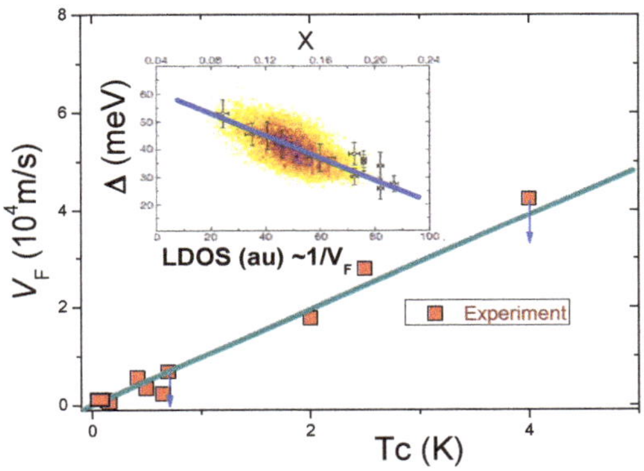

Figure 17. Experimental results (shown by the squares) for the average Fermi velocity V_F versus the critical temperature T_c for graphene (MATBG) [16]. The downward arrows depict that $V_F \leq V_0$, with V_0 the maximal value shown by the red square. Theory is displayed by the solid straight line. Inset is adapted from [92] and shows experimental dependence of the superconducting gap Δ versus the integrated local density of states collected on the high-T_c superconductor $Bi_2Sr_2CaCu_2O_{8+x}$. Here x is oxygen doping concentration. The darker color represents more data points with the same integrated local density of states and the same size gap Δ [92]. The straight blue line shows average value Δ versus the integrated local density of states.

8. Discussion and Conclusions

The central message of the present review article is that if the electronic spectrum of a substance happens to feature a dispersionless part, or flat bands, it is invariably this aspect that is responsible for the measured properties that depart radically from those of the familiar condensed-matter systems described by the Landau Fermi liquid theory. This is the case regardless of the diverse microscopic details characterizing these substances, such as crystal symmetry and structure defects. The explanation of this finding rests on the fact that the fermion condensation most readily occurs in substances hosting flat bands, see, e.g., [1,5–12]. Experimental manifestations of the fermion condensation phenomena are varied, implying that different experimental techniques are most suitable for detecting and analyzing them.

To support the above statements, we have also considered recent challenging experimental observations within the framework of the fermion condensation theory. In summary, we have:

Explained the universal T/B scaling behavior of the thermodynamic and transport properties, including the negative magnetoresistance of the HF metals;

Analyzed the recent challenging experimental facts regarding the tunneling differential conductivity $dI/dV = \sigma_d(V)$ as a function of the applied bias voltage V collected under the application of a magnetic field B on the twisted graphene and the archetypical heavy-fermion metals $YbRh_2Si_2$ and $CeCoIn_5$ [5,29,30];

Explained the emergence of the asymmetrical part $\Delta\sigma_d = \sigma_d(V) - \sigma_d(-V)$ as well as that $\Delta\sigma_d$ vanishes in magnetic fields as was predicted [31];

We further examined the linear dependence on temperature of the resistivity $\rho(T) \propto A_1 T$, demonstrated that $A_1(x_c - x)/T_c(x_c - x) = const$ and explained the data collected on high T_c superconductors, graphene, heavy fermion (HF) and common metals, revealing that the scattering rate $1/\tau$ of charge carriers reaches the Planckian limit;

Elucidated empirical observations of scaling properties [27] within the fermion condensation theory;

Investigated the recent extraordinary experimental observations of the density of superconducting electrons that turns out to be much less than the total density of electrons at $T \to 0$;

Shown that the transition temperature T_c is proportional to the Fermi velocity V_F, $V_F \propto T_c$, rather than $N_s(0) \propto 1/V_F \propto T_c$;

Demonstrated that flat bands make $T_c \propto g$, with g being the coupling constant. It is of crucial importance to note that the flat band superconductivity has already been observed in twisted bilayer graphene, where due to the flat band, the transition temperature T_c highly exceeds the limit dictated by the conventional BCS theory [5–12]. Thus, the basic task now is to attract more experimental groups to search for the room-T_c superconductivity in graphite and other perspective materials.

Indeed, the physics here has been explained within the fermion condensation theory [33] and related to flat bands whose existence was predicted many years ago [1,2,4,15,26,33,37] and paved the way for high-T_c superconductors [5–12]. In conclusion, this is a review of the recent outstanding experimental results that strongly suggest that the topological FCQPT is an intrinsic feature of many strongly correlated Fermi systems and can be viewed as the universal agent defining their non-Fermi liquid behavior. In addition, the fermion condensation theory is able to explain challenging features exhibited by strongly correlated Fermi systems.

Author Contributions: V.R.S., A.Z.M. and G.S.J. designed the project and directed it. All authors have read and agreed to the published version of the manuscript.

Funding: This research received no external funding.

Institutional Review Board Statement: Not applicable.

Informed Consent Statement: Not applicable.

Data Availability Statement: Not applicable.

Acknowledgments: This work was partly supported by U.S. DOE, Division of Chemical Sciences, Office of Basic Energy Sciences, Office of Energy Research.

Conflicts of Interest: The authors declare that they have no known competing financial interests or personal relationships that could have appeared to influence the work reported in this review paper. The authors declare no conflict of interest.

References

1. Khodel, A.V.; Shaginyan, V.R. Superfluidity in systems with fermion condensate. *JETP Lett.* **1990**, *51*, 553.
2. Volovik, G.E. A new class of normal Fermi liquids. *JETP Lett.* **1991**, *53*, 222.
3. Nozières, P. Properties of Fermi liquids with a finite range interaction. *J. Phys. I* **1992**, *2*, 443. [CrossRef]
4. Khodel, V.A.; Shaginyan, V.R.; Khodel, V.V. New approach in the microscopic Fermi systems theory. *Phys. Rep.* **1994**, *249*, 1. [CrossRef]
5. Cao, Y.; Fatemi, V.; Fang, S.; Watanabe, K.; Taniguchi, T.; Kaxiras, E.; Jarillo-Herrero, P. Unconventional superconductivity in magic-angle graphene superlattices. *Nature* **2018**, *556*, 43. [CrossRef]
6. Regnault, N.; Xu, Y.; Li, M.-R.; Ma, D.-S.; Jovanovic, M.; Yazdani, A.; Parkin, S.S.P.; Felser, C.; Schoop, L.M.; Ong, N.P.; et al. Catalogue of flat-band stoichiometric materials. *Nature* **2022**, *603*, 824. [CrossRef]

7. Esquinazi, P.; Heikkilä, T.T.; Lysogorskiy, Y.V.; Tayurskii, D.A.; Volovik, G.E. On the superconductivity of graphite interfaces. *JETP Lett.* **2014**, *100*, 336. [CrossRef]
8. Volovik, G.E. From standard model of particle physics to room-temperature superconductivity. *Phys. Scr.* **2015**, *T164*, 014014. [CrossRef]
9. Peri, V.; Song, Z.-D.; Bernevig, B.A.; Huber, A.S.D. Fragile Topology and Flat-Band Superconductivity in the Strong-Coupling Regime. *Phys. Rev. Lett.* **2021**, *126*, 027002. [CrossRef]
10. Törmä, P.; Peotta, S.; Bernevig, B.A. Superconductivity, superfluidity and quantum geometry in twisted multilayer systems. *Nat. Rev. Phys.* **2022**. [CrossRef]
11. Layek, S.; Monteverde, M.; Garbarino, G.; Mèasson, M.-A.; Sulpice, A.; Bendiab, N.; Rodire, P.; Cazali, R.; Hadj-Azzem, A.; Nassif, V.; et al. Possible high temperature superconducting transitions in disordered graphite obtained from room temperature deintercalated KC8. *arXiv 2022*, arXiv:2205.09358.
12. Ariskina, R.; Stiller, M.; Precker, C.E.; Böhlmann, W.; Esquinazi, P.D. On the Localization of Persistent Currents Due to Trapped Magnetic Flux at the Stacking Faults of Graphite at Room Temperature. *Materials* **2022**, *15*, 3422. [CrossRef] [PubMed]
13. Shaginyan, V.R.; Msezane, A.Z.; Amusia, M.Y.; Japaridze, G.S. Effect of superconductivity on the shape of flat bands. *Europhys. Lett.* **2022**, *138*, 16004. [CrossRef]
14. Khodel, V.A.; Clark, J.W.; Shaginyan, V.R. Rearrangement of the electron Fermi surface in layered compounds. *Solid State Commun.* **1995**, *96*, 353. [CrossRef]
15. Shaginyan, V.R.; Amusia, M.Y.; Msezane, A.Z.; Popov, K.G. Scaling behavior of heavy fermion metals. *Phys. Rep.* **2010**, *492*, 31. [CrossRef]
16. Qin, W.; Zou, B.; MacDonald, A.H. Critical magnetic fields and electron-pairing in magic-angle twisted bilayer graphene. *arXiv 2021*, arXiv:2102.10504.
17. Amusia, M.Y.; Shaginyan, V.R. Quasiparticle dispersion and lineshape in a strongly correlated liquid with the fermion condensate. *Phys. Lett. A* **2000**, *275*, 124. [CrossRef]
18. Melnikov, M.Y.; Shashkin, A.A.; Dolgopolov, V.T.; Huang, S.-H.; Liu, C.W.; Kravchenko, S.V. Indication of band flattening at the Fermi level in a strongly correlated electron system. *Sci. Rep.* **2017**, *7*, 14539. [CrossRef]
19. Amusia, M.Y.; Popov, K.G.; Shaginyan, V.R.; Stephanovich, V.A. Theory of Heavy-Fermion Compounds. In *Springer Series in Solid-State Sciences*; Springer: Berlin/Heidelberg, Germany; New York, NY, USA; Dordrecht, The Netherlands; London, UK, 2015; Volume 182, pp. 1–359
20. Amusya, M.Y.; Shaginyan, V.R. Strongly Correlated Fermi Systems: A New State of Matter. In *Springer Tracts in Modern Physics*; Springer Nature: Cham, Switzerland, 2020; Volume 283.
21. Bruin, J.A.N.; Sakai, H.; Perry, R.S.; Mackenzie, A.P. Similarity of Scattering Rates in Metals Showing T-Linear Resistivity. *Science* **2013**, *339*, 804. [CrossRef]
22. Legros, A.; Benhabib, S.; Tabis, W.; Laliberté, F.; Dion, M.; Lizaire, M.; Vignolle, B.; Vignolles, D.; Raffy, H.; Li, Z.Z.; et al. Universal T-linear resistivity and Planckian dissipation in overdoped cuprates. *Nat. Phys.* **2019**, *15*, 142. [CrossRef]
23. Cao, Y.; Chowdhury, D.; Rodan-Legrain, D.; Rubies-Bigordá, O.; Watanabe, K.; Taniguchi, T.; Senthil, T.; Jarillo-Herrero, P. Strange metal in magic-angle graphene with near Planckian dissipation. *Phys. Rev. Lett.* **2020**, *124*, 076801. [CrossRef] [PubMed]
24. Nakajima, Y.; Metz, T.; Eckberg, C.; Kirshenbaum, K.; Hughes, A.; Wang, R.; Wang, L.; Saha, S.R.; Liu, I.; Butch, N.P.; et al. Quantum-critical scale invariance in a transition metal alloy. *Commun. Phys.* **2020**, *3*, 181. [CrossRef]
25. Shaginyan, V.R.; Popov, K.G.; Khodel, V.A. Quasiclassical physics and T-linear resistivity in both strongly correlated and ordinary metals. *Phys. Rev. B* **2013**, *88*, 115103. [CrossRef]
26. Volovik, G.E. Flat band and Planckian metal. *JETP Lett.* **2019**, *110*, 352. [CrossRef]
27. Hu, T.; Liu, Y.; Xiao, H.; Mu, G.; Yang, Y. Universal linear-temperature resistivity: Possible quantum diffusion transport in strongly correlated superconductors. *Sci. Rep.* **2017**, *7*, 9469. [CrossRef]
28. Shaginyan, V.R.; Msezane, A.Z.; Stephanovich, V.A.; Japaridze, G.S.; Kirichenko, E. Flat bands and strongly correlated Fermi systems. *Phys. Scr.* **2019**, *94*, 065801. [CrossRef]
29. Ernst, S.; Kirchner, S.; Krellner, C.; Geibel, C.; Zwicknagl, C.G.; Steglich, F.; Wirth, S. Emerging local Kondo screening and spatial coherence in the heavy-fermion metal YbRh2Si2. *Nature* **2011**, *474*, 362. [CrossRef]
30. Seiro, S.; Jiao, L.; Kirchner, S.; Hartmann, S.; Friedemann, S.; Krellner, C.; Geibel, C.; Si, Q.; Steglich, F.; Wirth, S. Evolution of the Kondo lattice and non-Fermi liquid excitations in a heavy-fermion metal. *Nat. Commun.* **2018**, *9*, 3324. [CrossRef]
31. Shaginyan, V.R. Dissymetrical Tunneling in Heavy-Fermion Metals. *JETP Lett.* **2005**, *81*, 222. [CrossRef]
32. Božović, J.I.; He, X.; Wu, J.; Bollinger, A.T. Dependence of the critical temperature in overdoped copper oxides on superfluid density. *Nature* **2016**, *536*, 309. [CrossRef]
33. Dukelsky, J.; Khodel, V.A.; Schuck, P.; Shaginyan, V.R. Fermion condensation and non Fermi liquid behavior in a model with long range forces. *Z. Phys. Condens. Matter* **1997**, *102*, 245. [CrossRef]
34. Landau, L.D. The theory of a Fermi liquid. *Sov. Phys. JETP* **1956**, *3*, 920.
35. Lifshits, E.M.; Pitaevsky, L.P. *Statistical Physics, Part 2*; Butterworth-Heinemann: Oxford, UK, 2014.
36. Oeschler, N.; Hartmann, S.; Pikul, A.P.; Krellner, C.; Geibel, C.; Steglich, F. Low-temperature specific heat of $YbRh_2Si_2$. *Phys. B* **2008**, *403*, 1254. [CrossRef]
37. Volovik, G.E. Quantum Phase Transitions from Topology in Momentum Space. *Lect. Notes Phys.* **2007**, *718*, 31.

38. Shaginyan, V.R.; Popov, K.G.; Khodel, V.A. Conventional BCS, unconventional BCS, and non-BCS hidden dineutron phases in neutron matter. *Phys. At. Nucl.* **2014**, *77*, 1063. [CrossRef]
39. Shaginyan, V.R.; Stephanovich, V.A.; Msezane, A.Z.; Schuck, P.; Clark, J.W.; Amusia, M.Y.; Japaridze, G.S.; Popov, K.G.; Kirichenko, E.V. New State of Matter: Heavy Fermion Systems, Quantum Spin Liquids, Quasicrystals, Cold Gases, and High-Temperature Superconductors. *J. Low Temp. Phys.* **2017**, *189*, 410. [CrossRef]
40. Binder, K.; Young, A.P. Spin glasses: Experimental facts, theoretical concepts, and open questions. *Rev. Mod. Phys.* **1986**, *58*, 801. [CrossRef]
41. Mézard, M.; Parisi, G.; Virasoro, M.A. *Spin glass theory and beyond*; World Scientific Lecture Notes in Physics; World Scientific: Singapore, 2004; p. 476. [CrossRef]
42. Shaginyan, V.R.; Stephanovich, V.A.; Msezane, A.Z.; Japaridze, G.S.; Popov, K.G. The influence of topological phase transition on the superfluid density of overdoped copper oxides. *Phys. Chem. Chem. Phys.* **2017**, *19*, 21964. [CrossRef]
43. Bardeen, J. Tunnelling from a Many-Particle Point of View. *Phys. Rev. Lett.* **1961**, *6*, 57. [CrossRef]
44. Törmä, P.; Liang, L.; Peotta, S. Quantum metric and effective mass of a two-body bound state in a flat band. *Phys. Rev. B* **2018**, *98*, 220511(R). [CrossRef]
45. Julku, A.; Peltonen, T.J.; Liang, L.; Heikkilä, T.T.; Törmä, P. Superfluid weight and Berezinskii-Kosterlitz-Thouless transition temperature of twisted bilayer graphene. *Phys. Rev. B* **2020**, *101*, 060505(R). [CrossRef]
46. Shaginyan, V.R. Universal Behavior of Heavy-Fermion Metals Near a Quantum Critical Point. *JETP Lett.* **2004**, *79*, 286. [CrossRef]
47. Xu, J.; Han, F.; Wang, T.-T.; Thoutam, L.R.; Pate, S.E.; Li, M.; Zhang, X.; Wang, Y.-L.; Fotovat, R.; Zhou, X.; et al. Extended Kohler's Rule of Magnetoresistance. *Phys. Rev. X* **2021**, *11*, 041029. [CrossRef]
48. Gegenwart, P.; Westerkamp, T.; Krellner, C.; Tokiwa, Y.; Paschen, S.; Geibel, C.; Steglich, F.; Abrahams, E.; Si, A.Q. Multiple Energy Scales at a Quantum Critical Point. *Science* **2007**, *315*, 969. [CrossRef]
49. Shibauchi, T.; Krusin-Elbaum, L.; Hasegawa, M.; Kasahara, Y.; Okazaki, R.; Matsuda, Y. Field-induced quantum critical route to a Fermi liquid in high-temperature superconductors. *Proc. Natl. Acad. Sci. USA* **2008**, *105*, 7120. [CrossRef]
50. Kadowaki, K.; Woods, S.B. Universal relationship of the resistivity and specific heat in heavy-Fermion compounds. *Solid State Commun.* **1986**, *58*, 507. [CrossRef]
51. Gegenwart, P.; Custers, J.; Geibel, C.; Neumaier, K.; Tayama, T.; Tenya, K.; Trovarelli, O.; Steglich, F. Magnetic-Field Induced Quantum Critical Point in $YbRh_2Si_2$. *Phys. Rev. Lett.* **2002**, *89*, 056402. [CrossRef]
52. Shaginyan, V.R.; Msezane, A.Z.; Popov, K.G.; Japaridze, G.S.; Khodel, V.A. General properties of phase diagrams of heavy-fermion metals. *Europhys. Lett.* **2014**, *106*, 37001. [CrossRef]
53. Lee-Hone, N.R.; Dodge, J.S.; Broun, D.M. Disorder and superfluid density in overdoped cuprate superconductors. *Phys. Rev. B* **2017**, *96*, 024501. [CrossRef]
54. Varma, C.M.; Littlewood, P.B.; Schmitt-Rink, S.; Abrahams, E.; Ruckenstein, A.E. Phenomenology of the normal state of Cu-O high-temperature superconductors. *Phys. Rev. Lett.* **1989**, *63*, 1996. [CrossRef]
55. Simon, M.E.; Varma, C.M. Detection and Implications of a Time-Reversal Breaking State in Underdoped Cuprates. *Phys. Rev. Lett.* **2002**, *89*, 247003. [CrossRef] [PubMed]
56. Phillips, P. Normal state of the copper oxide high-temperature superconductors. *Phil. Trans. R. Soc. A* **2011**, *369*, 1572. [CrossRef] [PubMed]
57. Khodel, V.A.; Clark, J.W.; Shaginyan, V.R.; Zverev, M.V. Second wind of the Dulong-Petit Law at a quantum critical point. *JETP Lett.* **2010**, *92*, 532. [CrossRef]
58. Shaginyan, V.R.; Msezane, A.Z.; Popov, K.G.; Clark, J.W.; Zverev, M.V.; Khodel, V.A. Magnetic field dependence of the residual resistivity of the heavy-fermion metal $CeCoIn_5$. *Phys. Rev. B* **2012**, *86*, 085147. [CrossRef]
59. Taupin, M.; Paschen, S. Are Heavy Fermion Strange Metals Planckian? *Crystals* **2021**, *12*, 251. [CrossRef]
60. Abrikosov, A.A.; Gor'kov, L.P.; Dzyaloshinski, I.E. *Methods of Quantum Field Theory in Statistical Physics*; Prentice-Hall: London, UK, 1963.
61. Aynajian, P.; Neto, E.; Gyenis, A.; Baumbach, R.E.; Thompson, J.D.; Fisk, Z.; Bauer, E.D.; Yazdani, A. Visualizing heavy fermions emerging in a quantum critical Kondo lattice. *Nature* **2012**, *486*, 201. [CrossRef]
62. Peets, D.C.; Hawthorn, D.G.; Shen, K.M.; Kim, Y.-J.; Ellis, D.S.; Zhang, H.; Komiya, S.; Ando, Y.; Sawatzky, G.A.; Liang, R.; et al. X-Ray Absorption Spectra Reveal the Inapplicability of the Single-Band Hubbard Model to Overdoped Cuprate Superconductors. *Phys. Rev. Lett.* **2009**, *103*, 087402. [CrossRef]
63. French, M.M.J.; Analytis, J.G.; Carrington, A.; Balicas, L.; Hussey, N.E. Tracking anisotropic scattering in overdoped $Tl2Ba2CuO_{6+}\delta$ above 100 K. *New J. Phys.* **2009**, *11*, 055057. [CrossRef]
64. Alldredge, J.W.; Lee, J.; McElroy, K.; Wang, M.; Fujita, K.; Kohsaka, Y.; Taylor, C.; Eisaki, H.; Uchida, S.; Hirschfeld, P.; et al. Evolution of the electronic excitation spectrum with strongly diminishing hole density in superconducting $Bi_2Sr_2CaCu_2O_8+\delta$. *Nat. Phys.* **2008**, *4*, 319. [CrossRef]
65. Paglione, J.; Tanatar, M.A.; Hawthorn, D.G.; Boaknin, E.; Hill, R.W.; Ronning, F.; Sutherland, M.; Taillefer, L.; Petrovic, C.; Canfield, P.C. Field-Induced Quantum Critical Point in $CeCoIn_5$. *Phys. Rev. Lett.* **2003**, *91*, 246405. [CrossRef]
66. Harrison, W.A. Tunneling from an Independent-Particle Point of View. *Phys. Rev.* **1961**, *123*, 85. [CrossRef]
67. Zagoskin, A.M. *Quantum Theory of Many-Body Systems*; Springer: New York, NY, USA, 1998.
68. Deutscher, G. Andreev-Saint-James reflections: A probe of cuprate superconductors. *Rev. Mod. Phys.* **2005**, *77*, 109. [CrossRef]

69. Andreev, A.F. The Thermal Conductivity of the Intermediate State in Superconductors. *Sov. Phys. JETP* **1964**, *19*, 1228.
70. Shaginyan, V.R.; Popov, K.G. Asymmetric tunneling, Andreev reflection and dynamic conductance spectra in strongly correlated metals. *Phys. Lett. A* **2007**, *361*, 406. [CrossRef]
71. Giaever, I. Energy Gap in Superconductors Measured by Electron Tunneling. *Phys. Rev. Lett.* **1960**, *5*, 147. [CrossRef]
72. Nicol, J.; Shapiro, S.; Smith, P.H. Direct Measurement of the Superconducting Energy Gap. *Phys. Rev. Lett.* **1960**, *5*, 461. [CrossRef]
73. Schiller, A.; Hershfield, S. Theory of scanning tunneling spectroscopy of a magnetic adatom on a metallic surface. *Phys. Rev. B* **2000**, *61*, 9036. [CrossRef]
74. Shaginyan, V.R.; Msezane, A.Z.; Japaridze, G.S.; Stephanovich, V.A. Violation of the Time-Reversal and Particle-Hole Symmetries in Strongly Correlated Fermi Systems: A Review. *Symmetry* **2020**, *12*, 1596. [CrossRef]
75. Uemura, Y.J.; Luke, G.M.; Sternlieb, B.J.; Brewer, J.H.; Carolan, J.F.; Hardy, W.N.; Kadono, R.; Kempton, J.R.; Kiefl, R.F.; Kreitzman, S.R.; et al. Universal Correlations between Tc and n_s/m^* (Carrier Density over Effective Mass) in High-Tc Cuprate Superconductors. *Phys. Rev. Lett.* **1989**, *62*, 2317. [CrossRef]
76. Uemura, Y.J.; Keren, A.; Le, L.; Luke, G.M.; Wu, W.D.; Kubo, Y.; Manako, T.; Shimakawa, Y.; Subramanian, M.; Cobb, J.L.; et al. Magnetic-field penetration depth in $Tl_2Ba_2CuO_{6+}\delta$ in the overdoped regime. *Nature* **1993**, *364*, 605. [CrossRef]
77. Bernhard, C.; Niedermayer, C.; Binninger, U.; Hofer, A.; Wenger, C.; Tallon, J.L.; Williams, G.V.M.; Ansaldo, E.J.; Budnick, J.I.; Stronach, C.E.; et al. Magnetic penetration depth and condensate density of cuprate high-T_c superconductors determined by muon-spin-rotation experiments. *Phys. Rev. B* **1995**, *52*, 10488. [CrossRef]
78. Putzke, C.; Benhabib, S.; Tabis, W.; Ayres, J.; Wang, Z.; Malone, L.; Licciardello, S.; Lu, J.; Kondo, T.; Takeuchi, T.; et al. Reduced Hall carrier density in the overdoped strange metal regime of cuprate superconductors. *Nat. Phys.* **2021**, *17*, 826. [CrossRef]
79. Khodel, V.A.; Clark, J.W.; Popov, K.G.; Shaginyan, V.R. Occurrence of Flat Bands in Strongly Correlated Fermi Systems and High-T_c Superconductivity of Electron-Doped Compounds. *JETP Lett.* **2015**, *101*, 413. [CrossRef]
80. Khodel, V.A.; Clark, J.W.; Zverev, M.V. Topological disorder triggered by interaction-induced flattening of electron spectra in solids. *Phys. Rev. B* **2020**, *102*, 201108. [CrossRef]
81. Pristàš, G.; Reiffers, M.; Bauer, E.; Jansen, A.G.M.; Maude, A.D.K. Suppression of asymmetric differential resistance in the non-Fermi-liquid system $YbCu_{5-x}Al_x$ ((x=1.3 - 1.75) in high magnetic fields. *Phys. Rev. B* **2008**, *78*, 235108. [CrossRef]
82. Shaginyan, V.R.; Popov, K.G.; Stephanovich, V.A.; Kirichenko, E.V. Asymmetrical tunneling in heavy fermion metals as a possible probe for their non-Fermi liquid peculiarities. *J. Alloy. Compd.* **2007**, *442*, 29. [CrossRef]
83. Shaginyan, V.R.; Amusia, M.Y.; Msezane, A.Z. Quasiparticles and order parameter near quantum phase transition in heavy fermion metals. *Phys. Lett. A* **2005**, *338*, 393. [CrossRef]
84. Park, W.K.; Greene, L.H.; Sarrao, J.L.; Thompson, J.D. Andreev reflection at the normal-metal/heavy-fermion superconductor $CeCoIn_5$ interface. *Phys. Rev. B* **2005**, *72*, 052509. [CrossRef]
85. Shrestha, K.; Zhang, S.; Greene, L.H.; Lai, Y.; Baumbach, R.E.; Sasmal, K.; Maple, M.B.; Park, W.K. Spectroscopic evidence for the direct involvement of local moments in the pairing process of the heavy-fermion superconductor $CeCoIn_5$. *Phys. Rev. B* **2021**, *103*, 224515. [CrossRef]
86. Pogorelov, Y.G.; Shaginyan, V.R. Transition from Non-Fermi Liquid Behavior to Landau-Fermi Liquid Behavior Induced by magnetic Fields. *JETP Lett.* **2002**, *76*, 532. [CrossRef]
87. Shaginyan, V.R.; Amusia, M.Y.; Msezane, A.Z.; Stephanovich, V.A.; Japaridze, G.S.; Kirichenko, E.V. Flat Bands and Salient Experimental Features Supporting the Fermion Condensation Theory of Strongly Correlated Fermi Systems. *Phys. Atom. Nucl.* **2020**, *83*, 132. [CrossRef]
88. Yang, Y. Two-fluid model for heavy electron physics. *Rep. Prog. Phys.* **2016**, *79*, 074501. [CrossRef] [PubMed]
89. Kogan, V.G. Homes scaling and BCS. *Phys. Rev. B* **2013**, *87*, 220507(R). [CrossRef]
90. Shaginyan, V.R.; Msezane, A.Z.; Stephanovich, V.A.; Kirichenko, E.V. Quasiparticles and quantum phase transition in universal low-temperature properties of heavy-fermion metals. *Europhys. Lett.* **2006**, *76*, 898. [CrossRef]
91. Kopnin, N.B.; Heikkilä, T.T.; Volovik, G.E. High-temperature surface superconductivity in topological flat-band systems. *Phys. Rev. B* **2011**, *83*, 220503(R). [CrossRef]
92. Pan, S.H.; O'Neal, J.P.; Badzey, R.L.; Chamon, C.; Ding, H.; Engelbrecht, J.R.; Wang, Z.; Eisaki, H.; Uchida, S.; Guptak, A.K.; et al. Microscopic electronic inhomogeneity in the high-Tc superconductor $Bi_2Sr_2CaCu_2O_{8+x}$. *Nature* **2001**, *413*, 282. [CrossRef]
93. Saito, Y.; Ge, J.; Watanabe, K.; Taniguchi, T.; Young, A.F. Independent superconductors and correlated insulators in twisted bilayer graphene. *Nat. Phys.* **2020**, *16*, 926. [CrossRef]

atoms

MDPI

Review

Recent Progress in Low-Energy Electron Elastic-Collisions with Multi-Electron Atoms and Fullerene Molecules †

Alfred Z. Msezane * and Zineb Felfli

Department of Physics and Center for Theoretical Studies of Physical Systems, Clark Atlanta University, Atlanta, GA 30314, USA; zfelfli@cau.edu

* Correspondence: amsezane@cau.edu

† Professor M. Ya. *Amusia taught us not to aim at merely reproducing experimental results but at understanding the fundamental physics.*

Abstract: We briefly review recent applications of the Regge pole analysis to low-energy $0.0 \leq E \leq 10.0$ eV electron elastic collisions with large multi-electron atoms and fullerene molecules. We then conclude with a demonstration of the sensitivity of the Regge pole-calculated Ramsauer–Townsend minima and shape resonances to the electronic structure and dynamics of the Bk and Cf actinide atoms, and their first time ever use as novel and rigorous validation of the recent experimental observation that identified Cf as a transitional element in the actinide series.

Keywords: Regge poles; generalized bound states; actinide atoms; lanthanide atoms; fullerene molecules; elastic cross sections; binding energies; Ramsauer–Townsend minima; shape resonances; electron affinities

Citation: Msezane, A.Z.; Felfli, Z. Recent Progress in Low-Energy Electron Elastic-Collisions with Multi-Electron Atoms and Fullerene Molecules. *Atoms* **2022**, *10*, 79. https://doi.org/10.3390/atoms10030079

Academic Editor: Anatoli Kheifets

Received: 19 May 2022
Accepted: 12 July 2022
Published: 29 July 2022

1. Introduction

Progress towards the theoretical understanding of the fundamental mechanism underlying stable negative-ion formation in low-energy electron collisions with complex heavy multi-electron atoms and fullerene molecules has been very slow. This physical mechanism is of fundamental importance in physics and chemistry. More specifically, it has important implications for a wide range of applications, from catalysis to drug delivery and water purification. Unfortunately, the complexity of the interactions among electrons in heavy multi-electron atoms and fullerene molecules has, for a long time, made it virtually impossible to reliably predict the energetics of the electron binding and the properties of the resulting negative ions. A theoretical breakthrough was achieved in low-energy electron scattering from complex heavy multi-electron systems through our rigorous Regge pole method, wherein is embedded the electron–electron correlation effects and the core-polarization interaction, identified as the two crucial physical effects responsible for electron attachment resulting in stable negative-ion formation.

Consequently, the robust Regge pole method has allowed us to reliably explore, for the first time ever, negative-ion formation in complex heavy multi-electron systems such as the lanthanide and actinide atoms, as well as the fullerene molecules through the electron elastic total cross sections (TCSs) calculation. Importantly, these directly yield the anionic binding energies (BEs), the shape resonances (SRs) and the Ramsauer–Townsend (R-T) minima. From the TCSs unambiguous and reliable ground, metastable and excited state negative-ion BEs of the formed anions during the collisions are extracted and compared with the measured and/or calculated electron affinities (EAs) of the atoms and fullerene molecules. The novelty and generality of the Regge pole approach is in the extraction of rigorous negative-ion BEs from the TCSs, without any assistance whatsoever from either experiments or any other theories. Whether the measured EAs are identified with the ground state anionic BEs or with the excited states anionic BEs of the formed negative ions during the collisions, the rigorous Regge pole-calculated BEs are available to guide measurements.

Essential to the understanding of chemical reactions involving negative ions are accurate and reliable atomic and molecular affinities [1]. Moreover, low-energy electron collisions, resulting in negative-ion formation, provide a special insight into quantum dynamics [2]. Consequently, the careful determination of the EAs is needed. The Ramsauer–Townsend (R–T) effect is an important *inter alia* for understanding sympathetic cooling and the production of cold molecules using natural fermions and SRs. Additionally, the EAs provides a stringent test of theoretical calculations when their results are compared with those from reliable measurements. For ground state collisions, the Regge pole-calculated negative-ion BEs correspond to the challenging to calculate theoretically EAs, yielding outstanding agreement with the standard measured EAs for Au, Pt and the highly radioactive At atoms as well as for the C_{60} and the C_{70} fullerene molecules. In general, for the fullerenes C_{20} through $C_{92,}$ our Regge pole-calculated ground-state anionic BEs have been found to excellently match with the measured EAs. These results give great credence to the power and ability of the Regge pole method to produce unambiguous and reliable ground state anionic BEs of complex heavy systems through the TCSs calculation. Significantly, the Regge pole method achieves the remarkable feat without the assistance from experiments or any other theories.

Unfortunately, for most of the lanthanide atoms, producing sufficient anions that can be used in photodetachment experiments is challenging [3]. Due to their radioactive nature the actinide atoms are difficult to handle experimentally. Thus, there is a great need for reliable theoretical EAs to guide measurements. The EAs of atomic Au, Pt, and At have been measured [4–9], including those of the C_{60} and C_{70} fullerene molecules [10–14]. For the highly radioactive At atom, various sophisticated theoretical calculations, including the Multiconfiguration Dirac Hartree–Fock (MCDHF) value [15] agree excellently with the measured EAs [9]. Reference [9] employed the Coupled-Cluster method, while [15] used the MCDHF method. Furthermore, in [15] an extensive comparison among various sophisticated theoretical EAs has been carried out. For all these atoms, the measured EAs matched excellently the Regge pole-calculated BEs of the anionic ground states of the formed negative ions during the collisions, see Table 1 for comparisons. Moreover, the measured EAs of the fullerenes C_{20} through to C_{92} agree excellently with the Regge pole-calculated anionic ground states BEs [16,17]. This gives great credence to our interpretation of the EAs of these complex systems, viz. as corresponding to the ground state BEs of the formed negative ions during the collisions.

Recently, the EAs of the highly radioactive actinide atoms Th [18] and U [19,20] were measured as well. The experimentalists concluded that the EAs of both Th and U corresponded to the BEs of the weakly bound electron to the neutral atoms. For the Ti atom, two measurements obtained the EAs as 0.377 eV [21] and 0.075 eV [22]. The former value is close to various theoretical calculations [23,24], including the Regge pole-calculated BE of the second excited state, namely 0.281 eV [25]. However, the value of 0.075 eV [22] is close to the Regge pole BE of the highest excited state of the formed Ti; anion, 0.0664 eV; its ground state BE is 2.42 eV [25]. The measured EA of Hf is 0.178 eV [26]. It is close to the Regge pole SR at 0.232 eV, the RCI EA of 0.114 eV [27] and the Regge pole second excited state anionic BE of 0.113 eV [28]. The Hf highest excited state BE is at 0.017 eV [29]. Indeed, here we are faced with the problem of interpretation of what is meant by the EA.

For the lanthanide atoms problems regarding what is meant by the EA have already been discussed [29,30]. Briefly, for the Nd atom, there are two measured EA values, viz. 1.916 eV [31] and 0.0975 eV [32]. The value of [31] is close to the Regge pole ground state anionic BE value of 1.88 eV [33], while the EA of [32] is close to the RCI EA [34] and the Regge pole anionic BE of the highest excited state [33]. Similarly, the measured EAs for the Eu atom are 0.116 eV [3] and 1.053 eV [35]. The former value agrees excellently with the Regge pole BE of the highest excited state, viz. 0.116 eV [33], and with the RCI EA of 0.117 eV [36]. The EA of [35] agrees very well with the Regge pole-metastable anionic BE value of 1.08 eV [33]. For the large Tm atom, the measured EA [37] is close to the Regge pole-metastable BE (1.02 eV) [33]. Clearly, the results here demonstrate the need for an unambiguous meaning of the EA. The crucial question also considered here is: does the EAs of heavy multi-electron systems (atoms and fullerene molecules) correspond to the BE of the attached electron in the ground, metastable or excited state of the formed negative ion during the collision? Indeed, the meaning of the measured EAs of multi-electron atoms and fullerene molecules is also discussed here within the context of two prevailing viewpoints:

(1) The first considers the EA to correspond to the electron BE in the ground state of the formed negative ion during collision; it is exemplified by the measured EAs of Au, Pt and At atoms and the fullerene molecules from C_{20} through C_{92}.
(2) The second view identifies the measured EA with the BE of electron attachment in an excited state of the formed anion. The measured EAs of Ti, Hf, lanthanide and actinide atoms provide representative examples of this viewpoint.

We conclude the paper with a demonstration of the first ever use of the Regge pole-calculated TCSs as probes of the electronic structures of the actinide atoms Bk and Cf to identify the transitional element in the actinide series [38].

2. Method of Calculation

In this paper we have used the rigorous Regge pole method to calculate the electron elastic TCSs. Regge poles, singularities of the S-matrix, rigorously define resonances [39,40] and in the physical sheets of the complex plane they correspond to bound states [41]. In [42], it was confirmed that the Regge poles formed during low-energy electron elastic scattering become stable bound states. In the Regge pole method, also known as complex angular momentum (CAM), the important and revealing energy-dependent Regge Trajectories are also calculated. Their effective use in low-energy electron scattering has been demonstrated in [33,43], for example. The near-threshold electron–atom/fullerene collision TCS resulting in negative-ion formation as resonances is calculated using the Mulholland formula [44]. In the form below, the TCS fully embeds the essential electron–electron correlation effects [45,46] (atomic units are used throughout):

$$\sigma_{tot}(E) = 4\pi k^{-2}\int_0^\infty \mathrm{Re}[1 - S(\lambda)]\lambda d\lambda - 8\pi^2 k^{-2}\sum_n \mathrm{Im}\frac{\lambda_n \rho_n}{1+\exp(-2\pi i\lambda_n)} + I(E) \quad (1)$$

In Equation (1), $S(\lambda)$ is the S-matrix, $k = \sqrt{2mE}$, m being the mass and E the impact energy, ρ_n is the residue of the S-matrix at the nth pole, $\lambda_n$and $I(E)$ contains the contributions from the integrals along the imaginary λ-axis (λ is the complex angular momentum); its contribution has been demonstrated to be negligible [33].

As in [47], here we consider the incident electron to interact with the complex heavy system without consideration of the complicated details of the electronic structure of the system itself. Therefore, within the Thomas–Fermi theory, Felfli et al. [48] generated the robust Avdonina-Belov-Felfli (ABF) potential which embeds the vital core-polarization interaction.

$$U(r) = -\frac{Z}{r(1+\alpha Z^{1/3}r)(1+\beta Z^{2/3}r^2)} \quad (2)$$

In Equation (2), Z is the nuclear charge, α and β are variation parameters. For small r, the potential describes Coulomb attraction between an electron and a nucleus, $U(r) \sim -Z/r$, while at large distances it has the appropriate asymptotic behavior, *viz.* $\sim -1/(\alpha\beta r^4)$ and accounts properly for the polarization interaction at low energies. For an electron, the source of the bound states giving rise to Regge Trajectories is the attractive Coulomb well it experiences near the nucleus. The addition of the centrifugal term to the well 'squeezes' these states into the continuum [49]. For larger complex angular momentum (CAM) λ, the effective potential develops a barrier. Consequently, a bound state crossing the threshold energy E = 0 in this region may become a long-lived metastable state or an excited state. As a result, the highest "bound state" formed during the collision is identified with the highest excited state, here labeled as EXT-1. As E increases from zero, the second excited state may form with the anionic BE labeled, EXT-2. For the metastable states, similar labeling is used as MS-1, MS-2, etc. However, it should be noted here that the metastable states are labeled relative to the anionic ground state. Regge poles are generalized bound-states, namely solutions of the Schrödinger equation where the energy (E) is real and positive, and λ is complex. The CAM methods have the advantage in that the calculations are based on a rigorous definition of resonances, viz. as singularities of the S-matrix, see [49,50].

The strength of this extensively studied potential [51,52] lies in that it has five turning points and four poles connected by four cuts in the complex plane. The presence of the powers of Z as coefficients of r and r^2 in Equation (2) ensures that spherical and non-spherical atoms and fullerenes are correctly treated. Moreover, small and large systems are appropriately treated. The effective potential $V(r) = U(r) + \lambda(\lambda+1)/2r^2$ is considered here as a continuous function of the variables r and complex λ. The details of the numerical evaluations of the TCSs have been described in [46] and references therein; see also [53]. In the solution of the Schrödinger equation as described in [46], the parameters "α" and "β" of the potential, Equation (2) are varied. With the optimal value of $\alpha = 0.2$ the β-parameter is then varied carefully and when the dramatically sharp resonance appears in the TCS, it is indicative of negative ion formation; this energy position matches the measured EAs of the atom/fullerene, for example Au or C_{60} fullerene. This has been found to be the case in all the atoms and fullerenes we have investigated thus far.

3. Results

3.1. Cross Sections for the Representative Atom Au and Fullerene Molecule C_{60}

Figure 1, taken from Ref. [54] presents TCSs for atomic Au and fullerene molecule C_{60}. They typify the TCSs of complex heavy multi-electron atoms and fullerene molecules, respectively. Importantly, they are characterized by dramatically sharp resonances representing negative-ion formation in ground, metastable and excited anionic states, R–T minima and SRs. In both Figs. the red curves represent ground states electron TCSs, while the green curves denote excited state TCSs. Here the ground states anionic BEs in both Au and C_{60} appearing at the absolute R–T minima matched excellently with the measured EAs, see Figure 1 and Table 1 for comparisons with various measurements. In both systems, the ground states anionic BEs determine their EAs and not the excited anionic BEs (green curves). The data in Table 1 for Pt, At, Ti and Hf atoms as well as for the C_{70} fullerene were extracted from similar curves, as in Figure 1. Notably, the TCSs for the atoms and fullerene molecules become more complicated as the systems considered become larger as exemplified by the actinide atoms in Ref. [54] and the fullerene molecules in [16].

The availability of excellent measured EAs for the Au and Pt atoms [4–8] and the C_{60} fullerene molecule [10–12] allowed us to implement the rigorous Regge pole method to complex multi-electron atoms and fullerene molecules. Excellent agreement with the measurements were obtained for the Au, Pt and At atoms and the fullerene molecules C_{60} and C_{70}, as demonstrated in Table 1. Thus, the Regge pole-calculated ground state anionic BEs were benchmarked on the measured EAs of both Au and C_{60}. Subsequently, the Regge pole method was implemented in the calculation of low-energy electron elastic TCSs for various complex multi-electron atoms, including atomic Hf and Ti. The calculated ground states anionic BEs for the Au, Pt and At atoms matched excellently with the measured EAs of these atoms. For the fullerene molecules C_{20} through to C_{92}, the obtained anionic ground states BEs [16,17] generally agreed very well with the measured EAs. Indeed, the Regge pole method accomplished an unprecedented feat in the calculation of the challenging to theoretically calculate EAs of both multi-electron atoms and the fullerene molecules, from C_{20} through to C_{92}.

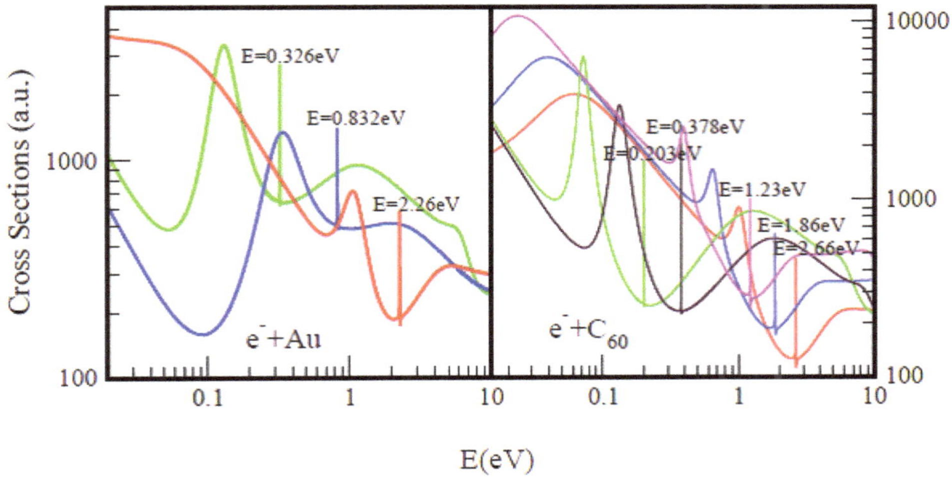

Figure 1. Total cross sections (a.u.) for electron elastic scattering from atomic Au (**left panel**) and the fullerene molecule C_{60} (**right panel**) are contrasted. For atomic Au the red, blue and green curves represent TCSs for the ground, metastable and excited states, respectively. For the C_{60} fullerene the red, blue and pink curves represent TCSs for the ground and the metastable states, respectively, while the green and brown curves denote TCSs for the excited states. The very sharp resonances in both figures Correspond to the Au^- and $C_{60}{}^-$ anionic formation. The anionic Bes in the Figures are intended to guide the eye; the complete values are also presented in Table 1 for better comparisons.

Table 1 already demonstrates the ambiguous meaning of the measured EAs in the large atoms Ti and Hf versus the meaning in the Au, Pt and At atoms. The interpretation of the EAs in the former atoms is that they correspond to the anionic BEs of excited states, while in the latter atoms the EAs are identified with the Regge pole BEs of the formed anions in their ground states.

Table 1. Negative-ion binding energies (BEs) and ground states Ramsauer–Townsend (R-T) minima, all in eV extracted from TCSs of the atoms Au, Pt, At, Ti and Hf and the fullerene molecules C_{60} and C_{70}. They are compared with the measured electron affinities (Eas) in eV. GRS, MS-n and EXT-n (n = 1, 2) refer, respectively, to ground, metastable and excited states. Experimental EAs, EXPT and theoretical EAs are also included. The numbers in the square brackets are the references.

System Z	BEs GRS	BEs MS-1	BEs MS-2	EAs EXPT	BEs EXT-1	BEs EXT-2	R-T GRS	BEs/EAs Theory
Au 79	2.26	0.832	-	2.309 [4] 2.301 [5] 2.306 [6]	0.326	-	2.24	2.262 [25]
Pt 78	2.16	1.197	-	2.128 [4] 2.125 [7] 2.123 [8]	0.136	-	2.15	2.163 [25]
At 85	2.41	0.918	-	2.416 [9]	0.115	0.292	2.43	2.38 [15] 2.42 [55] 2.51 [56] 2..80 [57]
C_{60}	2.66	1.86	1.23	2.684 [10] 2.666 [11] 2.689 [12]	0.203	0.378	2.67	2.663 [17] 2.63 [58] 2.57 [59]

Table 1. *Cont.*

System Z	BEs GRS	BEs MS-1	BEs MS-2	EAs EXPT	BEs EXT-1	BEs EXT-2	R-T GRS	BEs/EAs Theory
C_{70}	2.70	1.77	1.27	2.676 [11] 2.72 [13] 2.74 [14]	0.230	0.384	2.72	3.35 [60] 2.83 [60]
Ti 81	2.42	-	-	0.377 [21] 0.075 [22]	0.066	0. 281	2.40	0.27 [23] 0.291 [24]
Hf 72	1.68	0.525	-	0.178 [26]	0.017	0.113	1.67	0.114 [27] 0.113 [28]

3.2. Ground State Fullerene Cross Sections

The main reason for following up with the fullerene molecules immediately after Section 3.1 is that excellent measured EAs from C_{20} through to C_{92} are available in the literature. Benchmarked on the measured EAs of C_{60} and C_{70}, as indicated under Section 3.1, the Regge pole method was used to calculate the ground state electron elastic TCSs of the fullerene molecules from C_{20} through to C_{240} [16,17]. It is noted here that in the paper [17] only the ground state anionic BEs were calculated and some of the fullerenes ground state BEs can be found there. The novelty and generality of the Regge pole approach is in the extraction of the anionic BEs from the calculated TCSs of the fullerenes, for ground state collisions these BEs yield the unambiguous and definitive challenge to calculate theoretically EAs. In [16,17], the ground state anionic BEs of the fullerenes C_{20} through to C_{92} were found to generally match excellently with the measured EAs. Indeed, these results provided great credence to the ability of the Regge pole method to extract from the calculated TCSs reliable EAs of the fullerene molecules for the first time. The obtained agreement represented an unprecedented accomplishment by the Regge pole method, requiring no assistance whatsoever from either experiment or any other theory for the feat. This allowed the interpretation of the EAs of fullerenes as corresponding to the ground state anionic BEs calculated by the Regge pole method. For the fullerenes, other theories continue to struggle to go beyond the theoretically simple C_{20} and C_{60} fullerenes.

The focus in this Section is on the ground state anionic BEs of the fullerenes from C_{20} through to C_{240} [16,17]. These ground states TCSs are typified by the red curve of the C_{60} TCSs of Figure 1. In addition to the ground state curve, the revealing metastable and excited state TCSs curves demonstrate the richness in structure of the fullerene TCSs (the larger fullerenes reveal more metastable and excited state TCSs than shown in Figure 1). For the C_{20}, C_{24}, C_{26}, C_{28} and C_{44} we have used the data of [16] to demonstrate the reliability of the Regge pole-calculated anionic BEs, since their measured EAs are available. The BE values in Table 2 cover a wider range of the fullerene anionic BEs than those shown in the papers [16,17]. They include values where the experimental EAs are unavailable. In [16], the smaller fullerenes C_{20}, C_{24}, C_{26}, and C_{28} as well as the larger fullerenes C_{92} and C_{112} were studied to assess the extent to which fullerenes behave like "big atoms", as suggested by Amusia [61]. These TCSs were found to be characterized generally by ground, metastable and excited states negative-ion formation, R–T minima and SRs. The ground states anionic BEs correspond to the measured EAs of the fullerenes, see Table 2 as well as Figure 1 for the C_{60} TCSs. The Regge pole method does not determine the orbital angular momentum of the attached electron. This is particularly important for the C_{60} fullerene since there is an uncertainty in the literature regarding whether the C_{60} EA of 2.66 eV corresponds to an s-state or a *p*-state of the attached electron. Since the ground state anionic BE (EA) of C_{60} is determined here, we believe that a structure-type calculation could use our calculated ground state anionic BE (EA) of C_{60} to determine the ground-state fine-structure energies, since the metastable energies are also available. Thus, the lingering question could be answered.

The Regge pole-calculated low-energy electron elastic TCSs for the ground and the first (highest) excited states of fullerenes are robust. For C_{20} (smallest fullerene), the first excited state TCS (highest TCS) resembles that of atomic Au, see Figure 1. Defining R as the value of the ratio of the second to the first R–T minima in the first excited state TCS of C_{20}, in [16] we explored the variation of R from C_{20} through to C_{70}. We found that for C_{20} R (~1.4), greater than unity was close to that for Au (~2.) or Th (~1.9), indicative of atomic behavior, while for C_{24} R was about 1.0. For C_{70}, R was less than 0.5 demonstrating strong departure from atomic behavior due to the significant polarization interaction in C_{70}; which also induces long-lived metastable anions in the C_{70} TCSs. When probed with low-energy electrons, the results for C_{20} exhibited fullerene behavior consistent with the view that fullerenes behave like "big atoms" [61]. The atomic behavior quickly disappears with the increase in the fullerene size. As seen from the Figures of [16], the behavior in C_{28} is no longer atomic because R is less than unity. By C_{92}, the departure from atomic behavior has become significant, due to the increase in the polarization interaction in these larger systems.

For C_{20}, the excited state TCS [16] exhibits a deeper R–T minimum near threshold in comparison with the second R–T minimum, while the ground state TCS ends with a deep R–T minimum, wherein appears the dramatically sharp resonance representing the stable negative ion formed in the ground state during the collision, see also Figure 1. These characteristic R–T minima, also observed in the Dirac R-matrix low-energy electron elastic scattering cross sections calculations for the heavy, alkali-metal atoms Rb, Cs and Fr [62], manifest that the important core-polarization interaction has been accounted for adequately in our calculation, consistent with the conclusion in [63]. The vital importance of the core-polarization interaction in low-energy electron collisions with atoms and molecules was recognized and demonstrated long ago, see [64] for examples and references therein. In C_{20}, the TCSs are characterized by a ground, metastable and excited states TCSs. However, the C_{24}, C_{26} and C_{28} TCSs consist of more metastable and excited states TCSs. Suffice to state that the increased energy space determined mainly by the ground states BEs is conducive to the appearance of the polarization-induced metastable TCS in general. Indeed, these results reveal the complicated interplay between the R–T minima and the shape resonances.

Notably, in all the fullerene molecules investigated here, the ground states anionic BEs occur at the absolute R–T minima of the TCSs, see Figure 1 for example. This facilitates considerably the determination of unambiguous and reliable EAs of the fullerene molecules. Noted here also is that generally the sharp resonances of the metastable TCSs lie between the ground states SRs and the dramatically sharp resonances of the ground states. In Table 2, we have presented various fullerene anionic BEs, but mainly ground states BEs and compared them with the measured EAs where these are available. The results demonstrate the power of the Regge pole method to reliably calculate the anionic BEs of fullerene molecules.

Clearly, the Regge pole approach, entirely new in the field of electron-cluster/fullerene collisions, implemented here represents a theoretical breakthrough in low-energy electron scattering investigations of fullerenes/clusters and complex heavy atoms. Its implementation should speed up the long overdue fundamental theoretical understanding of the mechanism underlying low-energy electron scattering from fullerenes, including heavy and complex atoms, leading to negative ion formation. These results should also help in the construction of the popular square-well potentials for the investigated fullerenes. Most importantly, its great strength is in the ability to produce reliable data without assistance from experiments and/or other theories.

Table 2. Fullerene ground (GR-S), metastable (MS-*n*, *n* = 1, 2, 3) and first excited (EXT-1), second excited (EXT-2) and third excited (EXT-3) anionic states binding energies (BEs). The measured EAs are represented as EXPT and the other theoretical values are denoted as Theory. All the energies are in eV and the numbers in the square brackets are the references. For most of the fullerenes the parameters "α" and "β" of the potential, Equation (2) are tabulated in [17].

System	EA EXPT.	BE(Ours) GR-S	BEs MS-1	BEs MS-2	BEs MS-3	BEs EXT-1	BEs EXT-2	BEs EXT-3	EA Theory
C_{20}	2.44 [65] 2.60 [66] 2.70 [67]	2.72	1.48	-	-	0.466	-	-	-
C_{24}	3.750 [66] 2.90 [67]	3.79	2.29	1.41	-	0.428	0.801	-	-
C_{26}	3.100 [66] 2.95 [67]	2.67	1.59	-	-	0.464	-	-	-
C_{28}	2.80 [66] 3.00 [67]	3.10	1.80	-	-	0.305	0.505	-	-
C_{44}	3.30 [68]	3.15	1.89	1.47	-	0.319	0.492	-	
C_{74}	3.28 [69]	4.03	2.83	2.01	1.48	0.251	0.407	0.643	
C_{76}	2.89 [69]	2.79							
C_{78}	3.10 [69]	2.98							
C_{80}	3.17 [69]	3.28							
C_{82}	3.14 [69]	3.15							3.37 [59]
C_{84}	3.05 [13]	2.94							
C_{86}	≥3.0 [13]	2.92							
C_{90}	≥3.0 [13]	3.06							
C_{92}	≥3.0 [13]	3.09	2.35	1.58	-	0.266	-	-	
C_{100}	-	3.67	2.70	2.04	-	0.242	0.379	0.531	
C_{112}	-	3.31	2.53	1.73	-	0.243	0. 315	0.519	
C_{120}	-	3.74	2.97	2.04	1.58	0.244	0.372	0.576	
C_{124}	-	3.06	2.30	1.71	-	0.289	0.393	0.569	
C_{132}	-	3.59	2.60	1.93	-	0.251	0.338	-	
C_{136}	-	3.75	2.64	2.19	1.67	0.260	0.345	0.488	
C_{140}	-	3.94	3.06	2.23	1.75	0.360	0.562	0.716	
C_{180}	-	3.75	2.64	2.19	1.67	0.260	0.345	0.488	2.61 [70]
C_{240}	-	4.18	-	-	-	-			3.81 [71] 2.32 [72]

3.3. Cross Sections for the Large Atoms Hf, Pt, Au, Ti and At

In the context of the viewpoints (1) and (2) of the Introduction, it is appropriate to discuss the measured EAs of the large atoms Hf(72), Pt(78), Au(79), Ti(81) and the radioactive At (85) in an attempt to understand the meaning of the measured EAs of the Hf and Ti atoms (the numbers within the brackets are the Zs). That is, do their EAs correspond to electron BEs in the ground, the metastable or the excited states of the formed anions during the collision? As seen from Table 1, the meaning of the EAs of Au, Pt and At is clear, namely it corresponds to the ground state anionic BEs of the formed anions during the collisions. However, for Hf and Ti the meaning lacks definitiveness.

For clarity, Figure 2 shows the TCSs for the Hf atom; a similar Figure was obtained for the Ti atom. As seen from the Figure, it is difficult to understand any selection of the anionic BEs other than the ground state anionic BE as the EA of Hf. A similar argument applies to the Ti atom. The measured EA of Hf at 0.178 eV [26], the Regge pole-calculated SR of 0.232 eV, the RCI EA of 0.114 eV [27] and the Regge pole-calculated second excited state anionic BE of 0.113 eV [28] are reasonably close together. The highest excited state BE of Hf is at 0.017 eV [29]. The TCSs for Hf presented in Figure 2 demonstrate the additional presence to the above discussed anionic BEs, a metastable TCS (green curve) and a ground state TCS (pink curve) with anionic BEs of 0.525 eV and 1.68 eV, respectively. Indeed, here we are faced with the problem of interpretation of what is meant by the EA.

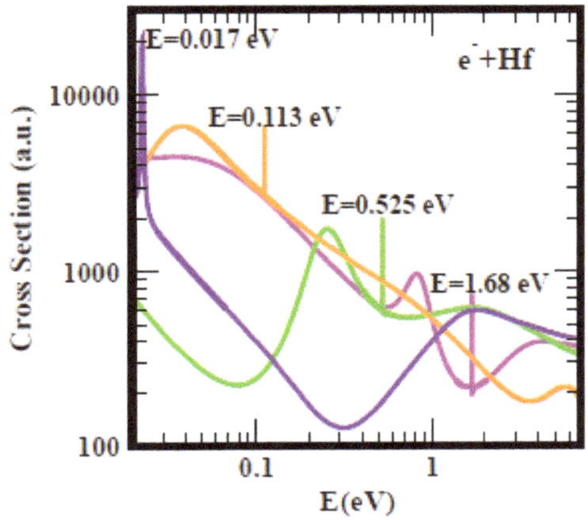

Figure 2. Total cross sections (a.u.) for electron elastic scattering from Hf. The pink, green, orange and purple curves represent the TCSs for the ground, metastable and the two excited states, respectively. The dramatically sharp resonances correspond to the Hf; anionic formation during the collisions.

As indicated in the Introduction, for the Ti atom two measurements obtained its EAs as 0.377 eV [21] and 0.075 eV [22]. The former value is close to various theoretical calculations [23,24], including the Regge pole-calculated BE of the second excited anionic state, namely 0.281 eV [56], see comparisons in Table 1. However, the measured value of 0.075 eV [22] and the Regge pole-calculated BE of the highest excited state of the formed Ti; anion, 0.0664 eV are close together. We note that the Regge pole-calculated ground state anionic BE of Ti is 2.42 eV [56], very close to that of the At atom. Clearly, the results of the Hf and Ti atoms are difficult to interpret without a rigorous theoretical data, as discussed under Section 3.4 dealing with the lanthanide atoms.

3.4. *Cross Sections for the Lanthanide Atoms*

The EA provides a stringent test of theory when the theoretical EAs are compared with those from reliable measurements. This statement holds strongly in the case of the lanthanide atoms. The general problem of interpretation of the measured EAs of the lanthanide atoms has been exposed in [29], as well as elucidated through the rigorous Regge pole method [73]. Appropriately, we begin this section by placing in perspective the existing measurements/calculations of the EAs of the lanthanide atoms. Low-energy electron elastic collision cross sections for the lanthanide atoms, La through Lu were first investigated using the CAM (Regge pole) method [33]. Unfortunately, the investigation was limited to the near threshold energy region, $0.0 \leq E \leq 1.0$ eV and focused upon the comparison with the existing measured and theoretical EAs and never questioned the meaning of the EAs. The CAM calculated TCSs were found to be characterized generally by dramatically sharp resonances whose energy positions were identified with the measured/calculated EAs of the lanthanide atoms. The extracted EAs from the TCSs varied from a low value of 0.016 eV for Tm to a high value of 0.631 eV for atomic Pr. In that paper [33], one sees the effective use of the Regge Trajectories and the Im L (L is the complex angular momentum) in analyzing and interpreting the results. Moreover, the lanthanide parameters "α" and "β" of the potential, Equation (2) are tabulated in that paper [33].

Subsequently, when the energy range was increased from 1.0 eV to 10.0 eV, ground, metastable and excited states anionic BEs were clearly revealed and delineated. Then, the question persisted: do the measured EAs of the lanthanide atoms correspond to the anionic BEs of electron attachment in the ground, metastable or excited states of the formed anions during the collision? For atomic Eu, the resonance at E = 0.116 eV with Im L = 7.6×10^{-6} of the Figure 6 of Ref. [33] should be compared with the results of Figure 2 of Ref. [30] reproduced here for convenience as Figure 3 (left panel). In Figure 13 of Ref. [33] depicting the TCSs of Tm the dramatically sharp resonance at E = 0.016 eV with Im L = 3.4×10^{-5} should be viewed in the context of the recent Figure 2 of Ref. [30], also presented here as Figure 3 (right panel).

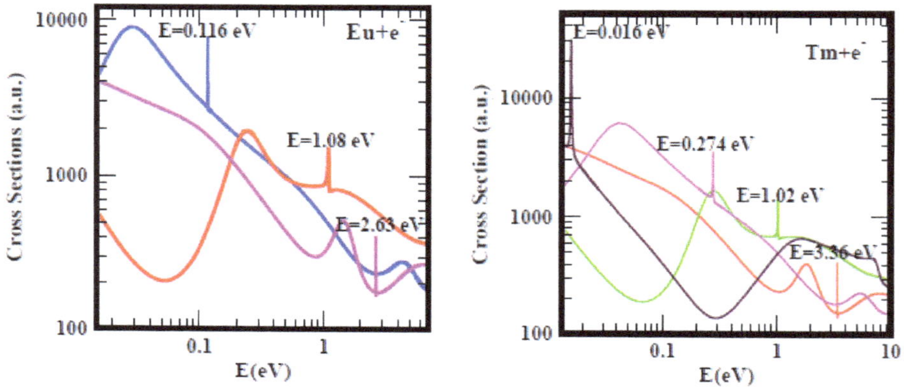

Figure 3. Total cross sections (a.u.) for electron elastic scattering from atomic Eu (**left panel**) and Tm (**right panel**). For Eu the pink, red and blue curves represent the TCSs for the ground, metastable and the excited states, respectively. For Tm atom the red, green, pink and black curves represent the TCSs for the ground, metastable and the two excited states, respectively. The dramatically sharp resonances in both figures correspond to the Eu; and Tm; negative-ions formed during the collision.

The lanthanide and the Hf atoms provide clear cases of the ambiguous and confusing measured and/or calculated EA values. As examples, for Eu we focus on the ground state, pink curve with the BE value of 2.63 eV and the blue curve with the BE of 0.116 eV, corresponding to an excited state TCS. The measured EA (0.116 eV) [3] is in outstanding agreement with the excited state BE value above and the RCI calculated EA (0.117 eV) [36], see Table 1 of [30]. The metastable BE value of 1.08 eV, red curve in Figure 3 (left panel) agrees excellently with the measured EA (1.053 eV) [35]. This clearly demonstrates the ambiguous and confusing meaning of the measured EA of Eu by Refs. [3,35]. Does the EA of Eu correspond to the BE of electron attachment in the metastable state or in the excited state of the formed anion during the collision? Similarly with the case of the Tm atom, the Regge pole calculated ground and excited states BEs are, respectively, 3.36 eV and 0.274 eV. The measured EA of Tm is 1.029 eV [37] and agrees excellently with the Regge pole calculated metastable state BE value of 1.02 eV, green curve in the Figure 3 (right panel). Accordingly, here the meaning of the measured EA of Tm corresponds to the BE of the metastable state. In both Eu and Tm atoms, the meaning of the measured EAs is ambiguous and confusing as well.

A comment is appropriate here regarding the importance or unimportance of Relativistic effects in the calculation of the EAs using the Regge pole method. With a relatively high Z of 63, but a small measured EA of 0.116 eV [3], the Eu atom provides a stringent test of the nonrelativistic CAM method when its prediction (EA = 0.116 eV) [33] is compared with that calculated using the MCDF-RCI (EA = 0.117 eV) [36]. The interpretation aside, the results demonstrate the unimportance of relativistic effects in the calculation of the EA of Eu. Indeed, the EAs calculated using structure-based theoretical methods tend to be riddled with uncertainty and lack definitiveness for complex multi-electron systems and fullerene molecules. For instance, relativistic effects in gold chemistry were investigated by Wesendrup et al. [74] who performed large-scale fully Relativistic Dirac–Hartree–Fock and MP2 calculations as well as nonrelativistic pseudopotential calculations and obtained the EAs of 2.19 eV and 1.17 eV, respectively. These values should be contrasted with the CAM calculated value of 2.263 eV [25] and compared with the measured EAs of Au in Table 1. Of importance, also, is the review on relativistic effects in homogeneous gold catalysis [75].

3.5. Cross Sections for the Actinide Atoms

In [54], we investigated the low-energy electron scattering from the radioactive actinide atoms Th, Pa, U, Np and Pu through the elastic TCSs calculations. The objective was to delineate and identify the characteristic resonance structures, as well as to understand and assess the reliability of the existing theoretical EAs. The recent measurement of the EA value of Th warrants some remark. There is no reason whatsoever for the selective comparison of data by the experiment; there are calculated EAs in the literature [36,76]. Particularly interesting in the study above [54], is the finding for the first time that the TCSs for atomic Pu exhibited fullerene molecular behavior near threshold through the TCS of the highest excited state, while maintaining the atomic character through the ground state TCS. Also, the first appearance of the near threshold deep R–T minimum in the actinide TCSs was first identified in the TCSs of atomic Pu, see Figure 5 of [54].

Figure 4, taken from Ref. [54] with a slight modification due to recalculation presents the TCSs for atomic Th (top figure) and U (bottom figure). They typify the TCSs of the complex multi-electron actinide atoms. Importantly, they are characterized by dramatically sharp resonances representing negative-ion formation in the ground, metastable and excited anionic states, R–T minima and SRs. In both Figures, the red curves represent electron attachment in the ground states while the pink curves denote the highest excited states. For Th, Figure 4 (top) the measured and the calculated EA values are 0.608 eV and 0.599 eV [18], respectively. These values are close to the Regge pole-calculated anionic BE of the second excited state, pink curve (0.549 eV). Close to this value there is a SR at 0.61 eV defined by the blue curve; the ground state anionic BE is at 3.09 eV. Not shown is the highest excited state curve with anionic BE value of 0.149 eV. Importantly, here we note the clear atomic behavior exhibited by the Th TCSs (pink curve) as expected [16]. However, the pink curve in the U TCSs shows strong fullerene behavior [16]. The EAs of U have been measured very recently to be 0.315 eV [19] and 0.309 eV [20] as well as calculated to be 0.232 eV [20]. These values are close to the Regge pole anionic BE value of 0.220 eV for the highest excited state, see Table 3 and Ref. [54] for additional comparisons. Here we do not understand the inconsistency in the meaning of the EAs in Figures 1 and 4, namely as corresponding to the BEs of electron attachment in the ground and the excited anionic states, respectively.

Of particular interest and importance here are the contrasted TCSs for atomic Am and Lr in [77]. In that paper, the polarization-induced TCS (brown curve) of Am exhibited a deep R–T minimum near the threshold. This brown curve behaves similarly to that in the TCSs of Pu, while in Lr it has already flipped over to a SR. Where does the actual flipping take place? This is the subject of Section 3.7. To understand the measurement [38], we need the previously unavailable data for the Cf, Fm and Md atoms to determine where the actual flipping takes place. There are no measured EAs for the actinide atoms beyond U to compare our BEs. However, theoretical EAs are available [36,76,78,79] and these have been compared with our data for the actinide atoms [54,77,80], see also Table 3. For atomic Lr the EA values of 0.310 eV [78] and 0.295 eV [36] are very close to the Regge pole BE of the highest excited state, namely 0.321 eV. These values should guide the reliable measurement of the EA of Lr. We have brought together the data for most of the actinide atoms in Table 3 mainly for convenience of analysis. Hitherto fore, they were scattered all over the literature. Importantly, the recent data from [80] are crucial for understanding Section 3.7.

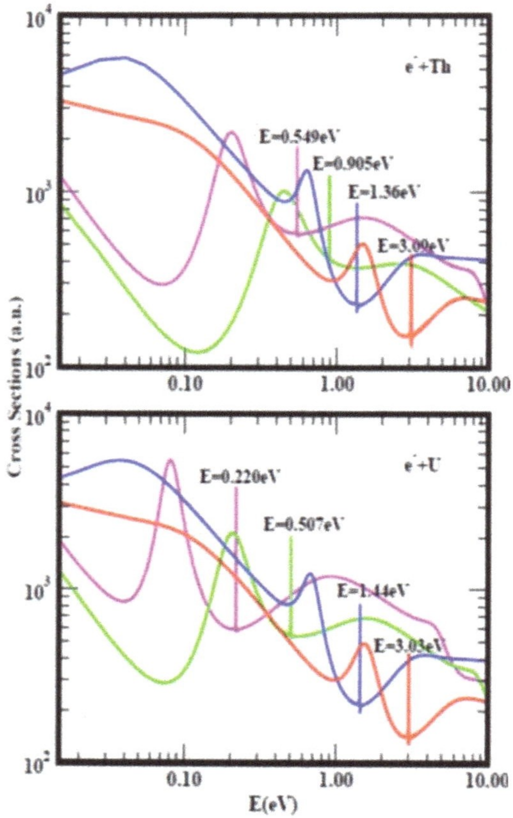

Figure 4. Total cross sections (TCSs) for atomic Th (**top figure**) and U (**bottom figure**) Figures. In this paper the relevant curves in both TCSs are the ground states (red curves) and the excited states (pink curves). The dramatically sharp resonances in both Figures with attendant BEs represent electron attachment. These BE values are intended to guide the eye; the complete values are presented in Table 3. There are also shape resonances next to these sharp peaks.

Table 3. Negative ion binding energies (BEs) in eV and energy positions of ground-state Ramsauer–Townsend (R-T) minima, in eV obtained from the TCSs for the actinide atoms from Pu through Lr. Additionally, included for comparison are the data for Au, Th and U. GRS, MS-*n* and EXT-*n* (*n* = 1, 2) represent ground, metastable and excited states, respectively. The experimental EAs, EXPT and the theoretical EAs, including RCI [36] and GW [76] are also presented. The numbers in the square brackets are the references.

System/ Z	BEs GRS	BEs MS-1	BEs MS-2	EAs EXPT	BEs EXT-1	BEs EXT-2	R-T GRS	BEs/EAs Theory	EAs [36]	EAs [76]
Au 79	2.26	0.832	-	2.309 [4]	0.326	-	2.24	2.262 [25]	-	-
Th 90	3.09	1.36	0.905	0.608 [18]	0.149	0.549	3.10	0.599 [18]	0.368	1.17

Table 3. *Cont.*

System/ Z	BEs GRS	BEs MS-1	BEs MS-2	EAs EXPT	BEs EXT-1	BEs EXT-2	R-T GRS	BEs/EAs Theory	EAs [36]	EAs [76]
U 92	3.03	1.44	-	0.315 [19] 0.309 [20]	0.220	0.507	3.01	0.232 [20] 0.175 [36]	0.373	0.53
Pu 94	3.25	1.57	1.22	N/A	0.225	0.527	3.26	-	0.085	−0.503 −0.276
Am 95	3.25	1.58	0.968	N/A	0.243	0.619	3.27	-	0.076	0.103 0.142
Bk 97	3.55	1.73	0.997	N/A	0.267	0.505	3.53	-	0.031	−0.503 −0.276
Cf 98	3.32	1.70	0.955	n/A	0.272	0.577	3.34	-	0.018 0.010	−0.777 −1.013
Es 99	3.42	1.66	0.948	N/A	0.272	0.642	3.44	-	0.002	0.103 0.142
Fm 100	3.47	1.79	1.02	N/A	0.268	0.623	3.49	-	-	0.597 0.354
Md 101	3.77	1.81	0.996	N/A	0.259	0.700	3.78		-	1.224 0.978
No 102	3.83	1.92	1.03	N/A	0.292	0.705	3.85	-	-	−2.302 −2.325
Lr 103	3.88	1.92	1.10	N/A	0.321	0.649	3.90	0.310 [78] 0.160 [78] 0.476 [79]	0.465 0.295	−0.212 −0.313

It is now clear why many existing experimental measurements and sophisticated theoretical calculations have considered the anionic BEs of the stable metastable and/or excited negative ion formation to correspond to the EAs of the considered lanthanide and actinide atoms. This is contrary to the usual meaning of the EAs found in the standard measurement of the EAs of such complex systems as atomic Au, Pt and the radioactive At, as well as of the fullerene molecules. In these systems, the EAs correspond to the ground state BEs of the formed negative ions. The negative ions obtained here are also important in catalysis.

3.6. Fullerene Negative-Ion Catalysis

3.6.1. Overview

The extensive and crucial applications of fullerenes in science, nanotechnology and industrial research, as well as in astrophysics have motivated this study. The acceptor material used particularly in modern organic solar cells is usually a fullerene derivative [81]. Understanding the stability and degradation mechanism of organic solar cells is essential before their commercialization. Toward this end, designing polymers and fullerenes with larger electron affinity (EA) has been proposed [82]. This motivated our first ever study of the large fullerenes [83], as well as the present study to search for fullerenes with larger EAs. The rich long-lived metastable resonances that characterize the Regge pole-calculated large fullerenes TCSs, presented for the first time in [83], support the important conclusion that the experimentally detected fullerene isomers correspond to the metastable states [84] and further confirm the need to identify and delineate the resonance structures in gentle electron scattering.

The fundamental mechanism underlying atomic negative-ion catalysis was proposed by our group in the context of muon catalyzed nuclear fusion [85,86]. The mechanism involves anionic molecular complex formation in the transition state (TS), with the atomic negative-ion breaking the hydrogen bond strength. The mechanism has been demonstrated in the synthesis of H_2O_2 from H_2O catalyzed using the Au; and Pd; anions to understand the experiments of Hutchings and collaborators [87–89], in the catalysis of light, intermediate and heavy water to the corresponding peroxides [90] and in the oxidation of methane to methanol without the CO_2 emission [91]. More recently, the experiment [89] has used the less expensive atomic Sn for possible water purification in the developing world. In this context, we explored [92] the effectiveness of the fullerene anions C_{20}; to C_{136}; in the catalysis of water oxidation to peroxide and water synthesis from H_2 and O_2 hoping to find inexpensive effective negative-ion catalysts.

3.6.2. Results

The electron elastic TCSs for the typical large fullerenes C_{100}, C_{120} and C_{140} demonstrate negative-ion formation [92–94] with significant differences among their EAs, namely 3.67 eV, 3.74 eV and 3.94 eV, respectively, see also Table 2. It is now clear that the ground state anionic BEs located at the absolute R–T minima of the ground state TCSs yield the challenging to calculate theoretically EAs. Indeed, the R–T minimum provides an excellent environment that is conducive to negative-ion catalysis and the creation of new molecules. The underlying physics in the fullerene TCSs has already been explained in [93,94]. The obtained results are consistent with the observation that low-energy electron-fullerene interactions are characterized by rich resonance structures [95,96] and that the experimentally detected fullerene isomers correspond to the metastable TCSs [84]. They also support the conclusion that the EAs of fullerene molecules are relatively large. The results of [83] including those presented in Table 2 should satisfy part of the requirement to increase fullerene acceptor resistance to degradation by the photo-oxidation mechanism through the use in organic solar cells of fullerenes with high EAs [82]. The extracted EAs from the TCSs could also be used to construct the widely used simple model potentials for the fullerene shells, including endohedral fullerenes [97–105], as well as in the study of the stability of An@C_{40}(An = Th, —–, Md) [106]. Notably, the EAs are at the hearts of many of the model potentials. Indeed, the rich resonance structures in the fullerenes TCSs and their large Eas explain the tendency of fullerenes to form compounds with electron-donor anions and their vast applications as well.

The utility of the generated fullerene anions has been demonstrated in the catalysis of water oxidation to peroxide and water synthesis from H_2 and O_2 using the anionic fullerene catalysts C_{20}^- to C_{136}^- [92]. Figure 5 taken from [92], demonstrates the Density Functional Theory (DFT) calculated TS energy barriers for both processes. DFT and dispersion corrected DFT approaches have been employed for the TS evaluations. Geometry optimization of the structural molecular conformation utilized the gradient-corrected Perdew-Burke-Ernzerhof parameterizations [107] of exchange-correlation as implemented in DMol3 [108]. DFT calculated energy barriers reduction in the oxidation of H_2O to H_2O_2 catalyzed using the anionic fullerene catalysts C_{20}^- to C_{136}^- are shown in the Figure 5 (left panel). The results in Figure 5 (right panel), also from [92] are for the water synthesis from H_2 and O_2 catalyzed using the anionic fullerene catalysts C_{20}^- to C_{136}^- as well. For both water oxidation and water synthesis DFT TS calculations found the $C_{52;}$ and C_{60}^- anions to be numerically stable and the C_{36}^- and C_{100}^- anions to increase the energy barriers the most in the water oxidation to H_2O_2 and water synthesis using H_2 and O_2, respectively. The C_{136}^- anion has proved to be the most effective in reducing the energy barrier significantly when catalyzing both water oxidation to peroxide and synthesis from H_2 and O_2. Importantly, a single large fullerene such as the C_{136}, or even the C_{74} could replace the Au, Pd and Sn atoms in the catalysis of H_2O_2 from H_2O in the experiments [87–89] acting as a multiple-functionalized catalyst. These fullerenes have their metastable Bes close to the EAs of the atoms used in the experiments. Thus, an inexpensive dynamic water purification system for the developing world could be realized [89].

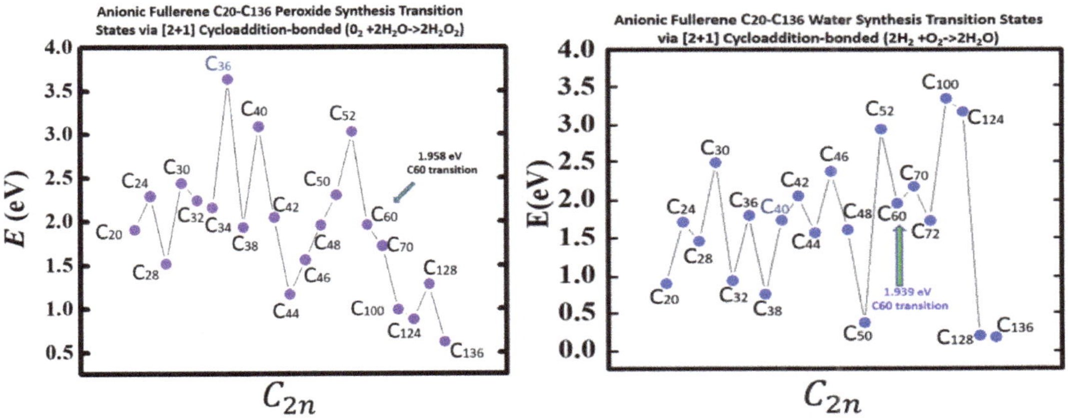

Figure 5. Transition state energy barriers of anionic fullerenes sizes C_{20}^- to C_{136}^- for catalyzing water oxidation to peroxide (**left panel**) and catalyzing hydrogen and oxygen synthesis to water (**right panel**).

Indeed, the utility of the fullerene molecular anions has been demonstrated in the catalysis of water oxidation to peroxide and water synthesis from H_2 and O_2 using the catalysts C_{20}^- to C_{136}^-. DFT TS calculations found C_{52}^- and C_{60}^- anions numerically stable for both. The C_{136}^- anion has proved to be the most effective in reducing the energy barrier significantly when catalyzing both water oxidation to peroxide and synthesis.

3.7. Atomic Structure and Dynamics of Bk and Cf: Experiment Versus Theory

The recent experiment [38] using nanogram material identified a weak spin-orbit-coupling in atomic Bk while a jj coupling scheme described atomic Cf. It concluded that these observations strengthen Cf as a transitional element in the actinide series. Here the Regge pole-calculated low-energy electron elastic TCSs for Bk and Cf atoms are used as novel validation of the experimental observation through the sensitivity of R–T minima and SRs to the electronic structure of these atoms.

In Figure 6, we present the Regge pole-calculated electron elastic TCSs for the Bk (left panel) and Cf (right panel) actinide atoms. As seen the TCSs are characterized generally by dramatically sharp resonances, representing ground, metastable and excited states negative-ion formation, SRs (broad peaks) and R–T minima. Moreover, the highest excited states TCSs (green curves) exhibit fullerene molecular behavior [16]. The energy positions of the sharp resonances, well delineated correspond to the BEs of the formed negative ions during the electron collisions with the Bk and Cf atoms. Each figure consists of BEs of the ground (red curve), metastable (blue and orange curves) and excited (green and brown curves) states TCSs. At first glance these TCSs appear a little complicated. However, they can be readily understood if each curve is discussed separately, see also [93,94] for example. The R–T minima manifest the effects of the polarization interaction [62], while the SRs convey the trapping effect of the centrifugal potential. The underlying physics has already been discussed in [93,94]; it will not be repeated here.

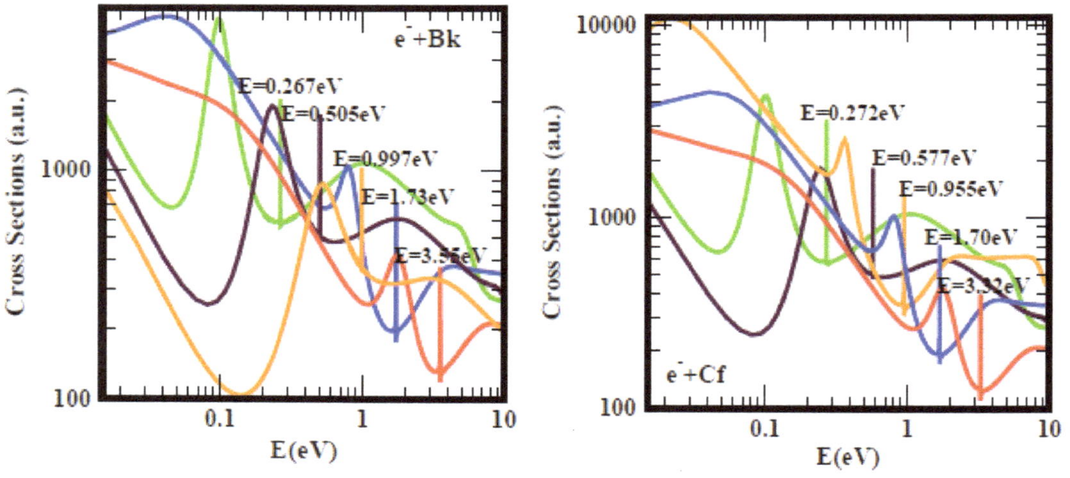

Figure 6. Total cross sections (a.u.) for electron elastic scattering from atomic Bk (**left panel**) and Cf (**right panel**). For both Bk and Cf the red and the blue and orange curves represent TCSs for the ground state and the metastable states, respectively. The green and the brown curves denote TCSs for the excited states. The orange curve with a deep R–T minimum in the Bk TCSs is the polarization-induced TCS due to size. In the Cf TCSs the orange curve has flipped over to a pronounced shape resonance very close to threshold. The dramatically sharp resonances in both figures correspond to the Bk; and Cf; anions formed during the collisions. The labeled BEs are intended to guide the eye; the complete values are presented in Table 3 for clarity.

For our objective here, we focus mainly on the polarization-induced TCSs (orange curves) and the ground state TCSs in both Figs. 3.7. For a better understanding and appreciation of the results, it is appropriate to place in perspective the polarization-induced TCSs that are characterized by a deep R–T minimum near threshold in the TCSs of Bk and a pronounced SR very close to threshold in the Cf TCSs. The polarization-induced TCS with the deep R–T minimum near threshold first appeared in the actinides TCSs through the atomic Pu TCSs [54]. It was attributed to the size effect and the first 6d-orbital collapse impacting the polarization interaction significantly. The first 6d-collapse occurred in the transition Np[Rn]$7s^2 5f^4 6d$ to Pu[Rn]$7s^2 5f^6$. This caused the ground state anionic BE of the Np atom to increase from 3.06 eV to 3.25 eV in Pu. Moreover, the anionic BE of the first metastable state increased from 1.47 eV in Np to 1.57 eV in Pu, see also Table 1 of Ref. [54]. It is the increase in the ground state energy space that facilitated the first appearance of the polarization-induced metastable TCS with the deep R–T minimum near threshold to appear in the Pu TCSs. This R–T minimum in the Pu TCSs continued through the Bk TCSs [77]. Indeed, the increase in the number of polarization-induced TCSs with size has already been clearly demonstrated in the fullerene molecular TCSs [83,92], see also Figure 1 of [54].

The second 6d-orbital collapse occurs in the transition Cm[Rn]$7s^2 5f^7 6d$ to Bk[Rn]$7s^2 5f^9$. To facilitate the discussion, we have included for convenience Table 4; the data have been taken from [54,77,80]. Table 4 shows that the ground state anionic BE increased significantly from 3.32 eV in Cm to 3.55 eV in Bk, thereby widening the energy space for the flip over to take place. Subsequently, the ground state anionic BE dropped from 3.55 eV in Bk to 3.32 eV in Cf. Similarly, for the first metastable states the anionic BEs increased from 1.57 eV in Cm to 1.73 eV in Bk, while it decreased to 1.70 eV in Cf. In Table 4, it is informative to look at mainly the ground state energy values. Briefly, in Bk the ground state anionic BE is 3.55 eV; it decreased to 3.32 eV in Cf after the R–T minimum flipped over to a SR very close to threshold. This is indicative of the smaller energy space required by the SR compared to the R–T minimum. On the other hand, in Pu the ground state anionic BE is 3.25 eV having increased from 3.06 eV in Np to accommodate the first appearance of the polarization-induced TCS with the deep R–T minimum in Pu. Here we also see the appearance of the 1.22 eV BE as MS-2; the Figures are also quite informative.

Table 4. Negative-ion binding energies (BEs), in eV extracted from the TCSs for the actinide atoms Np through Fm are taken from [54,77,80]. GRS, MS-*n* and EX-*n* (*n* = 1, 2) represent respectively ground, metastable and excited anionic states. The experimental EAs, EXPT, denoted by N/A are unavailable.

Z Atom	BEs GRS	EAs EXPT	BEs MS-1	BEs MS-2	BEs EX-2	BEs EX-1
93 Np	3.06	N/A	1.47	-	0.521	0.248
94 Pu	3.25	N/A	1.57	1.22	0.527	0.225
95 Am	3.25	N/A	1.58	0.968	0.619	0.243
96 Cm	3.32	N/A	1.57	1.10	0.519	0.258
97 Bk	3.55	N/A	1.73	0.997	0.505	0.267
98 Cf	3.32	N/A	1.70	0.955	0.577	0.272
99 Es	3.42	N/A	1.66	0.948	0.642	0.272
100 Fm	3.47	N/A	1.79	1.02	0.623	0.268

The flip over of the near threshold Ramsauer–Townsend minimum from the Bk polarization-induced metastable TCS to a pronounced shape resonance very close to threshold in the Cf metastable TCS provides a sensitive probe of the electronic structure and dynamics of these atoms, thereby permitting the first ever use of the R–T minimum and the SR as novel confirmation of Cf as a transitional element in the actinide series, consistent with the experimental observation [38]. Indeed, the rigorous Regge pole method requires no assistance whatsoever from either experiments or any other theory for the remarkable feat, namely of probing reliably the electronic structure of these complicated actinide atoms.

4. Summary and Conclusions

The Regge pole-calculated low-energy electron elastic total cross sections (TCSs) of complex heavy multi-electron systems are characterized generally by dramatically sharp resonances manifesting negative-ion formation. These yield directly the anionic binding energies (BEs), the shape resonances (SRs) and the Ramsauer–Townsend(R-T) minima. From the TCSs unambiguous and reliable ground, metastable and excited states negative-ion BEs of the formed anions during the collisions are extracted and compared with the measured and/or calculated electron affinities (EAs) of the atoms and fullerene molecules. The novelty and generality of the Regge pole approach is in the extraction of rigorous negative-ion BEs from the TCSs, without any assistance whatsoever from either experiment or any other theory. The EA provides a stringent test of theoretical calculations when their results are compared with those from reliable measurements. For ground states collisions, the Regge pole-calculated negative ion BEs correspond to the challenging to calculate theoretically EAs, yielding outstanding agreement with the standard measured EAs for Au, Pt and the highly radioactive At atoms as well as for the C_{60} and C_{70} fullerenes. For C_{20} through to C_{92} fullerenes our Regge pole-calculated ground-state anionic BEs matched in general excellently the measured EAs. These results give great credence to the power and ability of the Regge pole method to produce unambiguous and reliable ground state anionic BEs of complex heavy systems through the TCSs calculation.

The meaning of the measured EAs of multi-electron atoms and fullerene molecules has also been considered here within the context of two prevailing viewpoints:

(1) The first considers the EA to correspond to the electron BE in the ground state of the formed negative ion during collision; it is exemplified by the measured EAs of Au, Pt and At atoms and the fullerene molecules from C_{20} through to C_{92}.

(2) The second view identifies the measured EA with the BE of electron attachment in an excited state of the formed anion. The measured EAs of Ti, Hf, lanthanide and actinide atoms provide representative examples of this viewpoint.

This experimental breakthrough [38], including the recent first ever measurements of the EAs of the highly radioactive element At [9], as well as the Th [18] and U [19,20] atoms represent significant advances in the measurements of the challenging highly radioactive elements. In addition, more such measurements of other radioactive atoms can be expected in the near future. Consequently, reliable theoretical predictions are essential for a fundamental understanding of the underlying physics. Here we have presented an entirely new approach to the validation of the experimental observation in [38], namely through the behavior of the R–T minima and the SRs in the metastable electron elastic TCSs of atomic Bk and Cf. Finally, with the available ground, metastable and excited negative-ion BEs calculated here for the multi-electron atoms and the fullerene molecules, sophisticated theoretical methods such as the Dirac R-matrix, Coupled-Cluster method, MCDHF, MCDF-RCI, etc. can now be used to generate reliable EAs, wave functions and fine-structure energies. Indeed, for unambiguous and definitive meaning of the EAs of multi-electron atoms and the fullerene molecules our anionic BEs can be used in sophisticated theoretical methods to carry out careful investigations such as has been done in [15] for the At atom.

Author Contributions: Z.F. and A.Z.M. are responsible for the conceptualization, methodology, investigation, formal analysis and writing of the original draft, as well as rewriting and editing. A.Z.M. is also responsible for securing the funding for the research. All authors have read and agreed to the published version of the manuscript.

Funding: U.S. DOE, Division of Chemical Sciences, Geosciences and Biosciences, Office of Basic Energy Sciences, Office of Energy Research, Grant: DE-FG02-97ER14743.

Acknowledgments: Research was supported by the U.S. DOE, Division of Chemical Sciences, Geosciences and Biosciences, Office of Basic Energy Sciences, Office of Energy Research, Grant: DE-FG02-97ER14743. The computing facilities of National Energy Research Scientific Computing Center, also funded by U.S. DOE are greatly appreciated.

Conflicts of Interest: The authors declare no conflict of interest or state.

References

1. Kasdan, K.; Lineberger, W.C. Alkali-metal negative ions. II. Laser photoelectron spectrometry. *Phys. Rev. A* **1974**, *10*, 1658. [CrossRef]
2. Field, D.; Jones, N.C.; Ziesel, J.-P. Europhys. *News* **2002**, *33*, 1.
3. Cheng, S.-B.; Castleman, A.W. Direct experimental observation of weakly-bound character of the attached electron in europium anion. *Sci. Rep.* **2015**, *5*, 12414. [CrossRef] [PubMed]
4. Hotop, H.; Lineberger, W.C. Dye-laser photodetachment studies of Au^-, Pt^-, PtN^-, and Ag^-. *J. Chem. Phys.* **2003**, *58*, 2379. [CrossRef]
5. Andersen, T.; Haugen, H.K.; Hotop, H. Binding Energies in Atomic Negative Ions: III. *J. Phys. Chem. Ref. Data* **1999**, *28*, 1511. [CrossRef]
6. Zheng, W.; Li, X.; Eustis, S.; Grubisic, A.; Thomas, O.; De Clercq, H.; Bowen, K. Anion photoelectron spectroscopy of $Au^-(H_2O)$ 1, 2, Au2-(D2O) 1–4, and $AuOH^-$. *Chem. Phys. Lett.* **2007**, *444*, 232–236. [CrossRef]
7. Gibson, D.; Davies, B.J.; Larson, D.J. The electron affinity of platinum. *J. Chem. Phys.* **1993**, *98*, 5104. [CrossRef]
8. Bilodeau, R.C.; Scheer, M.; Haugen, H.K.; Brooks, R.L. Near-threshold laser spectroscopy of iridium and platinum negative ions: Electron affinities and the threshold law. *Phys. Rev. A* **1999**, *61*, 012505. [CrossRef]
9. Leimbach, D.; Karls, J.; Guo, Y.; Ahmed, R.; Ballof, J.; Bengtsson, L.; Pamies, F.B.; Borschevsky, A.; Chrysalidis, K.; Eliav, E.; et al. The electron affinity of astatine. *Nat. Commun.* **2020**, *11*, 3824. [CrossRef]
10. Huang, D.-L.; Dau, P.D.; Liu, H.T.; Wang, L.-S. High-resolution photoelectron imaging of cold C60- anions and accurate determination of the electron affinity of C60. *J. Chem. Phys.* **2014**, *140*, 224315. [CrossRef]
11. Brink, C.; Andersen, L.H.; Hvelplund, P.; Mathur, D.; Voldstad, J.D. Laser photodetachment of C60- and C70- ions cooled in a storage ring. *Chem. Phys. Lett.* **1995**, *233*, 52–56. [CrossRef]
12. Wang, X.-B.; Ding, C.F.; Wang, L.-S. High resolution photoelectron spectroscopy of C_{60}^-. *J. Chem. Phys.* **1999**, *110*, 8217. [CrossRef]
13. Boltalina, O.V.; Sidorov, L.N.; Borshchevsky, A.Y.; Sukhanova, E.V.; Skokan, E.V. Electron affinities of higher fullerenes. *Rapid Commun. Mass Spectrom.* **1993**, *7*, 1009. [CrossRef]

14. Palpant, B.; Otake, A.; Hayakawa, F.; Negishi, Y.; Lee, G.H.; Nakajima, A.; Kaya, K. Photoelectron spectroscopy of sodium-coated C60 and C70 cluster anions. *Phys. Rev. B* **1999**, *60*, 4509. [CrossRef]
15. Si, R.; Froese Fischer, C. Electron affinities of At and its homologous elements Cl, Br, I. *Phys. Rev. A* **2018**, *98*, 052504. [CrossRef]
16. Msezane, A.Z.; Felfli, Z. New insights in low-energy electron-fullerene interactions. *Chem. Phys.* **2018**, *503*, 50. [CrossRef]
17. Felfli, Z.; Msezane, A.Z. Simple method for determining fullerene negative ion formation. *Eur. Phys. J. D* **2018**, *72*, 78. [CrossRef]
18. Tang, R.; Si, R.; Fei, Z.; Fu, X.; Lu, Y.; Brage, T.; Liu, H.; Chen, C.; Ning, C. Candidate for Laser Cooling of a Negative Ion: High-Resolution Photoelectron Imaging of Th−. *Phys. Rev. Lett.* **2019**, *123*, 203002. [CrossRef]
19. Tang, R.; Lu, Y.; Liu, H.; Ning, C. Electron affinity of uranium and bound states of opposite parity in its anion. *Phys. Rev. A* **2021**, *103*, L050801. [CrossRef]
20. Ciborowski, S.M.; Liu, G.; Blankenhorn, M.; Harris, R.M.; Marshall, M.A.; Zhu, Z.; Bowen, K.H.; Peterson, K.A. The electron affinity of the uranium atom. *J. Chem. Phys.* **2021**, *154*, 224307. [CrossRef]
21. Carpenter, D.L.; Covington, A.M.; Thompson, J.S. Laser-photodetachment-electron spectroscopy of Ti-. *Phys. Rev. A* **2000**, *61*, 042501. [CrossRef]
22. Tang, R.; Fu, X.; Ning, C. Accurate electron affinity of Ti and fine structures of its anions. *J. Chem. Phys.* **2018**, *149*, 134304. [CrossRef]
23. Arnau, F.; Mota, F.; Novoa, J.J. Accurate calculation of the electron affinities of the group-13 atoms. *Chem. Phys.* **1992**, *166*, 77. [CrossRef]
24. Wijesundera, W.P. Theoretical study of the negative ions of boron, aluminum, gallium, indium, and thallium. *Phys. Rev. A* **1997**, *55*, 1785. [CrossRef]
25. Felfli, Z.; Msezane, A.Z.; Sokolovski, D. Near-threshold resonances in electron elastic scattering cross sections for Au and Pt atoms: Identification of electron affinities. *J. Phys. B* **2008**, *41*, 105201. [CrossRef]
26. Tang, R.; Chen, X.; Fu, X.; Wang, H.; Ning, C. Electron affinity of the hafnium atom. *Phys. Rev. A* **2018**, *98*, 020501. [CrossRef]
27. Pan, L.; Beck, D.R. Calculations of Hf− electron affinity and photodetachment partial cross sections. *J. Phys. B At. Mol. Opt. Phys.* **2010**, *43*, 025002. [CrossRef]
28. Felfli, Z.; Msezane, A.Z.; Sokolovski, D. Strong resonances in low-energy electron elastic total and differential cross sections for Hf and Lu atoms. *Phys. Rev. A* **2008**, *78*, 030703. [CrossRef]
29. Felfli, Z.; Msezane, A.Z. Conundrum in Measured Electron Affinities of Complex Heavy Atoms. *J. At. Mol. Condens. Nano Phys.* **2018**, *5*, 73. [CrossRef]
30. Felfli, Z.; Msezane, A.Z. Low-Energy Electron Elastic Total Cross Sections for Ho, Er, Tm, Yb, Lu, and Hf Atoms. *Atoms* **2020**, *8*, 17. [CrossRef]
31. Davis, V.T.; Thompson, J.S.; Covington, A. Laser photodetachment electron spectroscopy studies of heavy atomic anions. *NIMB* **2005**, *241*, 118. [CrossRef]
32. Fu, X.; Lu, Y.; Tang, R.; Ning, C. Electron affinity measurements of lanthanide atoms: Pr, Nd, and Tb. *Phys. Rev. A* **2020**, *101*, 022502. [CrossRef]
33. Felfli, Z.; Msezane, A.Z.; Sokolovski, D. Resonances in low-energy electron elastic cross sections for lanthanide atoms. *Phys. Rev. A* **2009**, *79*, 012714. [CrossRef]
34. O'Malley, S.M.; Beck, D.R. Valence calculations of lanthanide anion binding energies: 6*p* attachments to $4f^n6s^2$ thresholds. *Phys. Rev. A* **2008**, *78*, 012510. [CrossRef]
35. Davis, V.T.; Thompson, J.S. An experimental investigation of the atomic europium anion. *J. Phys. B* **2004**, *37*, 1961. [CrossRef]
36. O'Malley, S.M.; Beck, D.R. Valence calculations of actinide anion binding energies: All bound 7p and 7s attachments. *Phys. Rev. A* **2009**, *80*, 032514. [CrossRef]
37. Davis, V.T.; Thompson, J.S. Measurement of the electron affinity of thulium. *Phys. Rev. A* **2001**, *65*, 010501. [CrossRef]
38. Müller, A.; Deblonde, G.J.P.; Ercius, P.; Zeltmann, S.E.; Abergel, R.J.; Minor, A.M. Probing electronic structure in berkelium and californium via an electron microscopy nanosampling approach. *Nat. Commun.* **2021**, *12*, 948. [CrossRef]
39. Frautschi, S.C. *Regge Poles and S-Matrix Theory*; Benjamin: New York, NY, USA, 1963; Chapter X.
40. D'Alfaro, V.; Regge, T.E. *Potential Scattering*; North-Holland: Amsterdam, The Netherlands, 1965.
41. Omnes, R.; Froissart, M. *Mandelstam Theory and Regge Poles: An Introduction for Experimentalists*; Benjamin: New York, NY, USA, 1963; Chapter X.
42. Hiscox, A.; Brown, B.M.; Marletta, M. On the low energy behavior of Regge poles. *J. Math. Phys.* **2010**, *51*, 102104. [CrossRef]
43. Thylwe, K.W. On relativistic shifts of negative-ion resonances. *Eur. Phys. J. D* **2012**, *66*, 7. [CrossRef]
44. Mulholland, H.P. An asymptotic expansion for Σ(2n + 1)exp (Àσ(n + 1/2)2). *Math. Proc. Camb. Philos. Soc.* **1928**, *24*, 280–289. [CrossRef]
45. Macek, J.H.; Krstic, P.S.; Ovchinnikov, S.Y. Regge Oscillations in Integral Cross Sections for Proton Impact on Atomic Hydrogen. *Phys. Rev. Lett.* **2004**, *93*, 183203. [CrossRef]
46. Sokolovski, D.; Felfli, Z.; Ovchinnikov, S.Y.; Macek, J.H.; Msezane, A.Z. Regge oscillations in electron-atom elastic cross sections. *Phys. Rev. A* **2007**, *76*, 012705. [CrossRef]
47. Dolmatov, V.K.; Amusia, M.Y.; Chernysheva, L.V. Electron elastic scattering off A@C60: The role of atomic polarization under confinement. *Phys. Rev. A* **2017**, *95*, 012709. [CrossRef]

48. Felfli, Z.; Belov, S.; Avdonina, N.B.; Marletta, M.; Msezane, A.Z.; Naboko, S.N. Semiclassical approach to Regge poles trajectories calculations for nonsingular potentials: Thomas–Fermi type. In Proceedings of the Third International Workshop on Contemporary Problems in Mathematical Physics, Cotonou, Republic of Benin, 1–7 November 2003; Govaerts, J., Hounkonnou, M.N., Msezane, A.Z., Eds.; World Scientific: Singapore, 2004; pp. 218–232.
49. Sokolovski, D.; Msezane, A.Z.; Felfli, Z.; Ovchinnikov, S.Y.; Macek, J.H. What can one do with Regge poles? *Nucl. Instrum. Methods Phys. Res. Sect. B* **2007**, *261*, 133. [CrossRef]
50. Connor, J.N.L. New theoretical methods for molecular collisions: The complex angular-momentum approach. *J. Chem. Soc. Faraday Trans.* **1990**, *86*, 1627. [CrossRef]
51. Belov, S.; Thylwe, K.-E.; Marletta, M.; Msezane, A.Z.; Naboko, S.N. On Regge pole trajectories for a rational function approximation of Thomas–Fermi potentials. *J. Phys. A* **2010**, *43*, 365301. [CrossRef]
52. Thylwe, K.-E.; McCabe, P. Partial-wave analysis of particular peaks in total scattering cross sections caused by a single partial wave. *Eur. Phys. J. D* **2014**, *68*, 323. [CrossRef]
53. Burke, P.G.; Tate, C. A program for calculating regge trajectories in potential scattering. *Comput. Phys. Commun.* **1969**, *1*, 97. [CrossRef]
54. Felfli, Z.; Msezane, A.Z. Negative Ion Formation in Low-Energy Electron Collisions with the Actinide Atoms Th, Pa, U, Np and Pu. *Appl. Phys. Res.* **2019**, *11*, 52. [CrossRef]
55. Li, J.; Zhao, Z.; Andersson, M.; Zhang, X.; Chen, C. Theoretical study for the electron affinities of negative ions with the MCDHF method. *J. Phys. B* **2012**, *45*, 165004. [CrossRef]
56. Felfli, Z.; Msezane, A.Z.; Sokolovski, D. Slow electron elastic scattering cross sections for In, Tl, Ga and at atoms. *J. Phys. B At. Mol. Opt. Phys.* **2012**, *45*, 045201. [CrossRef]
57. Zollweg, R.J. Electron Affinities of the Heavy Elements. *J. Chem. Phys.* **1969**, *50*, 4251. [CrossRef]
58. Zakrzewski, V.G.; Dolgounitcheva, O.; Ortiz, J.V. Electron propagator calculations on the ground and excited states of C60(-). *J. Phys. Chem. A* **2014**, *118*, 7424. [CrossRef] [PubMed]
59. Nagase, S.; Kabayashi, K. Theoretical study of the lanthanide fullerene CeC82. Comparison with ScC82, YC82 and LaC82. *Chem. Phys. Lett.* **1999**, *228*, 106. [CrossRef]
60. Tiago, M.L.; Kent, P.R.C.; Hood, R.Q.; Reboredo, F. A Neutral and charged excitations in carbon fullerenes from first-principles many-body theories. *J. Chem. Phys.* **2008**, *129*, 084311. [CrossRef] [PubMed]
61. Amusia, M.Y. Fullerenes and endohedrals as "big atoms". *Chem. Phys.* **2013**, *414*, 168. [CrossRef]
62. Bahrim, C.; Thumm, U. Low-lying $^3P^o$ and $^3S^e$ states of Rb−, Cs−, and Fr−. *Phys. Rev. A* **2000**, *61*, 022722. [CrossRef]
63. Johnson, W.R.; Guet, C. Elastic scattering of electrons from Xe, Cs+, and Ba2+. *Phys. Rev. A* **1994**, *49*, 1041. [CrossRef]
64. Brigg, W.J.; Tennyson, J.; Plummer, M. R-matrix calculations of low-energy electron collisions with methane. *J. Phys. B At. Mol. Opt. Phys.* **2014**, *47*, 185203. [CrossRef]
65. Prinzbach, H.A.; Weller, A.; Landenberger, P.; Wahl, F.; Worth, J.; Scott, L.T.; Gelmont, M.; Olevano, D.; von Issendorff, B. C20 Carbon Clusters: Fullerene–Boat–Sheet Generation, Mass Selection, Photoelectron Characterization. *Chem.-A Eur. J.* **2008**, *12*, 6268. [CrossRef]
66. Achiba, Y.; Kohno, M.; Ohara, M.; Suzuki, S.; Shiromaru, H.J. Electron detachment spectroscopic study on carbon and silicon cluster anions. *J. Electron Spectrosc. Relat. Phenom.* **2005**, *142*, 231. [CrossRef]
67. Yang, S.; Taylor, K.J.; Craycraft, M.J.; Conceicao, J.; Pettiette, C.L.; Cherhnovsky, O.; Smalley, R.E. UPS of 2–30-atom carbon clusters: Chains and rings. *Chem. Phys. Lett.* **1988**, *144*, 431. [CrossRef]
68. Kietzmann, H.; Rochow, R.; Gantefor, G.; Eberhardt, W.; Vietze, K.; Seifert, G.; Fowler, P.W. Electronic structure of small fullerenes: Evidence for the high stability of C32. *Phys. Rev. Lett.* **1998**, *81*, 5378. [CrossRef]
69. Boltalina, O.V.; Dashkova, E.V.; Sidorov, L.N. Gibbs energies of gas-phase electron transfer reactions involving the larger fullerene anions. *Chem. Phys. Lett.* **1996**, *256*, 253. [CrossRef]
70. Xu, L.; Cai, W.; Shao, X. Systematic search for energetically favored isomers of large fullerenes C_{122}–C_{130} and C_{162}–C_{180}. *Comput. Mater. Sci.* **2008**, *41*, 522. [CrossRef]
71. Cabrera-Trujillo, J.M.; Alonso, J.A.; Iniguez, M.P.; Lopez, M.J.; Rubio, A. Theoretical study of the binding of Na clusters encapsulated in the C_{240} fullerene. *Phys. Rev. B* **1996**, *53*, 16059. [CrossRef]
72. Tarento, R.J.; Joyes, P.Z. Size dependence of the electronic and magnetic properties of fullerenes (C_{60}-C_{240}). *Eur. Phys. J. D* **1996**, *37*, 165. [CrossRef]
73. Msezane, A.Z. A Rigorous Model of Electron Attachment in Lanthanide Atoms. Research Outreach. 2021. Available online: https://researchoutreach.org/ (accessed on 4 July 2022).
74. Wesendrup, R.; Laerdahl, J.K.; Schwerdtfeger, P. Relativistic effects in gold chemistry. VI. Coupled cluster calculations for the isoelectronic series AuPt−, Au2,AuPt−, Au2, and AuHg+AuHg+. *J. Chem. Phys.* **1999**, *110*, 9457. [CrossRef]
75. Gorin, D.; Toste, F. Relativistic effects in homogeneous gold catalysis. *Nature* **2007**, *446*, 395. [CrossRef]
76. Guo, Y.; Whitehead, M.A. Electron affinities of alkaline-earth and actinide elements calculated with the local-spin-density-functional theory. *Phys. Rev. A* **1989**, *40*, 28. [CrossRef]
77. Msezane, A.Z.; Felfli, Z. Low-Energy Electron Elastic Collisions with Actinide Atoms Am, Cm, Bk, Es, No and Lr: Negative-Ion Formation. *Atoms* **2021**, *9*, 84. [CrossRef]

78. Eliav, E.; Kaldor, U.; Ishikawa, Y. Transition energies of ytterbium, lutetium, and lawrencium by the relativistic coupled-cluster method. *Phys. Rev. A* **1995**, *52*, 291. [CrossRef]
79. Borschevsky, A.; Eliav, E.; Vilkas, M.J.; Ishikawa, Y.; Kaldor, U. Transition energies of atomic lawrencium. *Eur. Phys. J. D* **2007**, *45*, 115–119. [CrossRef]
80. Felfli, Z.; Msezane, A.Z. Low-Energy Electron Elastic Total Cross Sections for the Large Actinide Atoms Cf, Fm and Md. *Appl. Phys. Res.* **2022**, *14*, 15. [CrossRef]
81. Speller, E.M. The significance of fullerene electron acceptors in organic solar cell photo-oxidation. *Mater. Sci. Technol.* **2017**, *33*, 924. [CrossRef]
82. Hoke, E.T.; Sachs-Quintana, I.T.; Lloyd, M.T.; Kauvar, I.; Mateker, W.R.; Nardes, A.M.; Peters, C.H.; Kopidakis, N.; McGehee, M.D. The role of electron affinity in determining whether fullerenes catalyze or inhibit photooxidation of polymers for solar cells. *Adv. Energy Mat.* **2012**, *2*, 13. [CrossRef]
83. Msezane, A.Z.; Felfli, Z.; Shaginyan, V.R.; Amusia, M.Y. Anionic formation in low-energy electron scattering from large fullerenes: Their multiple functionalization. *Int. J. Curr. Adv. Res.* **2017**, *6*, 8503. [CrossRef]
84. Kronik, L.; Fromherz, R.; Ko, E.; Ganteför, G.; Chelikowsky, J.R. Highest electron affinity as a predictor of cluster anion structures. *Nat. Mater.* **2002**, *1*, 49. [CrossRef]
85. Msezane, A.Z.; Felfli, Z.; Sokolovski, D. Novel mechanism for nanoscale catalysis. *J. Phys. B* **2010**, *43*, 201001. [CrossRef]
86. Armour, E.A.G. Muon, positron and antiproton interactions with atoms and molecules. *J. Phys. Conf. Ser.* **2010**, *225*, 012002. [CrossRef]
87. Edwards, J.K.; Carley, A.F.; Herzing, A.A.; Kiely, C.J.; Hutchings, G.J. Direct synthesis of hydrogen peroxide from H_2 and O_2 using supported Au–Pd catalysts. *J. Chem. Soc. Faraday Discuss.* **2008**, *138*, 225. [CrossRef] [PubMed]
88. Edwards, J.K.; Solsona, B.; Landon, P.; Carley, A.F.; Herzing, A.; Watanabe, M.; Kiely, C.J.; Hutchings, G.J. Direct synthesis of hydrogen peroxide from H_2 and O_2 using Au–Pd/Fe_2O_3 catalysts. *J. Mater. Chem.* **2005**, *15*, 4595. [CrossRef]
89. Freakley, S.J.; He, Q.; Harrhy, J.H.; Lu, L.; Crole, D.A.; Morgan, D.J.; Ntainjua, E.N.; Edwards, J.K.; Carley, A.F.; Borisevich, A.Y.; et al. Palladium-tin catalysts for the direct synthesis of H_2O_2 with high selectivity. *Science* **2016**, *351*, 959. [CrossRef]
90. Tesfamichael, A.; Suggs, K.; Felfli, Z.; Wang, X.-Q.; Msezane, A.Z. Atomic Gold and Palladium Negative-Ion Catalysis of Light, Intermediate, and Heavy Water to Corresponding Peroxides. *J. Phys. Chem. C* **2012**, *116*, 18698. [CrossRef]
91. Msezane, A.Z.; Felfli, Z.; Tesfamichael, A.; Suggs, K.; Wang, X.-Q. Gold anion catalysis of methane to methanol. *Gold Bull.* **2012**, *3*, 127. [CrossRef]
92. Felfli, Z.; Suggs, K.; Msezane, A.Z. *To be published*. 2022.
93. Msezane, A.Z. Negative Ion Binding Energies in Complex Heavy Systems. *J. At. Mol. Condens. Nano Phys.* **2018**, *5*, 195. [CrossRef]
94. Msezane, A.Z.; Felfli, Z. Low-energy electron scattering from fullerenes and heavy complex atoms: Negative ions formation. *Eur. Phys. J. D* **2018**, *72*, 173. [CrossRef]
95. Elhamidi, O.; Pommier, J.; Abouaf, R.J. Low-energy electron attachment to fullerenes and in the gas phase. *J. Phys B* **1997**, *30*, 4633. [CrossRef]
96. Huang, J.; Carman, H.S.; Compton, R.N. Low-Energy Electron Attachment to C60. *J. Phys. Chem.* **1995**, *99*, 1719. [CrossRef]
97. Amusia, M.Y.; Baltenkov, A.S.; Chernysheva, L.V. Modification of the Endohedral Potential after Instant Ionization of an Inner Atom. *J. Exp. Theor. Phys. Lett.* **2020**, *111*, 18. [CrossRef]
98. Amusia, M.Y.; Baltenkov, A.S.; Krakov, B.G. Photodetachment of negative C60− ions. *Phys. Lett. A* **1998**, *243*, 99. [CrossRef]
99. Ivanov, V.K.; Kashenock, G.Y.; Polozkov, R.G.; Solov'yov, A.V. Photoionization cross sections of the fullerenes C20 and C60 calculated in a simple spherical model. *J. Phys. B At. Mol. Opt. Phys.* **2001**, *34*, L669. [CrossRef]
100. Phaneuf, R.A.; Kilcoyne, A.L.D.; Aryal, N.B.; Baral, K.K.; Esteves-Macaluso, D.A.; Thomas, C.M.; Hellhund, J.; Lomsadze, R.; Gorczyca, T.W.; Ballance, C.P.; et al. Probing confinement resonances by photoionizing Xe inside a C+60 molecular cage. *Phys. Rev. A* **2013**, *88*, 053402. [CrossRef]
101. Amusia, M.Y.; Chernysheva, L.V.; Dolmatov, V.K. Confinement and correlation effects in the Xe@C60 generalized oscillator strengths. *Phys. Rev. A* **2011**, *84*, 063201. [CrossRef]
102. Lin, C.Y.; Ho, Y. Photoionization of atoms encapsulated by cages using the power-exponential potential. *J. Phys. B At. Mol. Opt. Phys.* **2012**, *45*, 145001. [CrossRef]
103. Dolmatov, V.K.; Connerade, J.-P.; Baltenkov, A.S.; Manson, S.T. Structure and photoionization of confined atoms, Radiat. *Phys. Chem.* **2004**, *70*, 417. [CrossRef]
104. Baltenkov, A.; Manson, S.T.; Msezane, A.Z. Jellium model potentials for the C60 molecule and the photoionization of endohedral atoms, A@C60. *J. Phys. B* **2015**, *48*, 185103. [CrossRef]
105. Madjet, M.E.; Chakraborty, H.S.; Manson, S.T. Giant Enhancement in Low Energy Photoemission of Ar Confined in C60. *Phys. Rev. Lett.* **2007**, *99*, 2430. [CrossRef]
106. Ryzhkov, M.V.; Delley, B. Electronic structure of predicted endohedral fullerenes An@ C40 (An = Th–Md). *Comput. Theor. Chem.* **2013**, *1013*, 70. [CrossRef]
107. Tkatchenko, A.; Scheffler, M. Accurate Molecular Van Der Waals Interactions from Ground-State Electron Density and Free-Atom Reference Data. *Phys. Rev. Lett.* **2009**, *102*, 73005. [CrossRef]
108. *DMol3*; Accelrys Software Inc.: San Diego, CA, USA, 2011.

atoms

MDPI

Perspective

Probing C_{60} Fullerenes from within Using Free Electron Lasers

Nora Berrah

Physics Department, University of Connecticut, Storrs, CT 06268, USA; nora.berrah@uconn.edu

Abstract: Fullerenes, such as C_{60}, are ideal systems to investigate energy redistribution following substantial excitation. Ultra-short and ultra-intense free electron lasers (FELs) have allowed molecular research in a new photon energy regime. FELs have allowed the study of the response of fullerenes to X-rays, which includes femtosecond multi-photon processes, as well as time-resolved ionization and fragmentation dynamics. This perspective: (1) provides a general introduction relevant to C_{60} research using photon sources, (2) reports on two specific X-ray FEL-based photoionization investigations of C_{60}, at two different FEL fluences, one static and one time-resolved, and (3) offers a brief analysis and recommendations for future research.

Keywords: fullerenes; C_{60}; photoionization; fragmentation; free electron laser; femtosecond; dynamics; pump probe

Citation: Berrah, N. Probing C_{60} Fullerenes from within Using Free Electron Lasers. *Atoms* **2022**, *10*, 75. https://doi.org/10.3390/atoms10030075

Academic Editors: Anatoli Kheifets, Gleb Gribakin and Vadim Ivanov

Received: 20 May 2022
Accepted: 13 July 2022
Published: 14 July 2022

1. Introduction

The behavior of C_{60} in relation to ionizing radiation is important because it sheds light on the fundamental many-body problem due to the numerous nuclei–electron responses exhibited in a large molecule. Investigating C_{60} using different light sources may reveal responses, which may impact other research fields. We thus studied C_{60} using short wavelength lasers, and in particular, with free electron lasers (FELs), which are still relatively new light sources compared to table-top lasers or synchrotron facilities. FELs have opened up new research opportunities because they deliver photons in a new energy regime for many scientific fields, from physics to chemistry, as well as to matter under extreme conditions and biology [1]. These VUV/X-ray lasers are accelerator-based tools, which are a hybrid between synchrotron radiation facilities and typical table-top lasers. FELs produce high brightness radiation with typical femtosecond (fs) pulse duration [2] and have been available since 2005, with the first VUV FLASH FEL at DESY in Germany [3]. There are currently several X-ray FELs around the world in addition to XUV FELs. The first X-ray FEL, the Linac Coherent Light Source (LCLS) at SLAC National Accelerator Laboratory [1,2], was commissioned and made available to scientists in 2009. It was used to carry out the static and time-resolved work reported here. The LCLS so far has a repetition rate up to 120 Hz, including a fs time scale where the pulse duration can be as short as 2–3 fs and as long as 500 fs [1,2]. Since 2017, the LCLS FEL also provides pulses as short as ~280 attoseconds (*as*) in the soft X-ray regime, which is an unprecedented technical progress that already impacts current X-ray *as* research [4].

X-ray FELs have opened up the possibility to investigate the ultra-fast response of matter to intense femtosecond X-ray pulses as well as to their pulse duration. The initial research on atoms and small molecules uncovered new aspects of this response, such as rapid sequences of inner-shell photoionization and the Auger–Meitner decay [5,6]. More recently, fullerenes were investigated with FELs because they bridge the gap between molecules and nanoparticles and are model systems for studying the dynamical behavior of large systems when exposed to intense, X-ray short pulses. Fullerenes have displayed molecular [7] and bulk [8] behavior and have proven to be an excellent testing ground for experiments and theories [9]. The behavior of C_{60} in relation to ionizing radiation is intriguing due to the numerous nuclei–electron responses exhibited, since it consists of a cage of 60 atoms with 240 valence electrons [10]. The interaction of such a large system is key to investigating many-body problems induced on the system's electrons by the photon electric field. The photon interaction with the electronic fullerene's degrees of freedom results in electronic dynamics, which leads to nuclei dynamics, since they are inter-connected. Thus, the investigation of the interaction of C_{60} with photons has been carried out extensively, theoretically and experimentally [11], to advance, in some cases, the quantitative understanding of the electronic and nuclear structure of these large molecules.

The photon used by the scientific community to study C_{60} varied from IR to X-rays, and some studies were static, while others were time dependent, in order to understand the ultra-fast dynamics that arises in these systems subsequent to photoabsorption. The time-resolved experiments and calculation of the interaction of C_{60} with photons have the ultimate goal to control ultra-fast molecular dynamics and understand the chemical transformation at the fs timescale. These studies include probing the multi-electron interactions in fullerenes as well as between the electrons and the carbon nuclei [12].

Non-linear physics and strong-field table-top laser research with C_{60} was extensively studied, and the photoionization mechanisms were found to be wavelength and pulse duration dependent [13]. The single-active-electron (SAE) method was used to calculate the ionization of C_{60} in intense, 4×10^{13} W/cm^2 laser pulses with durations between 27 and 70 fs and for a wide range of wavelengths ranging from 395 to 1800 nm [14], which agreed with the measurements. For a long I.R. wavelength of 1800 nm and 70 fs pulse duration, the SAE picture predicts "over the barrier" ionization for a peak intensity of 10^{15} W/cm^2, leading to non-fragmented parent molecule but highly charged $C_{60}{}^{q+}$ (q = 1–12) [12] ions. At a short wavelength of 355 nm, the excitation of C_{60} with 10 ns pulses led to fragmentation by delayed ionization and C_2 emission as well as other fragments, even for small intensities of about 2×10^6 W/cm^2 [15].

We investigated and we report here on two studies regarding the X-ray ionization and fragmentation of C_{60} under high- and mid-fluence X-ray femtosecond pulses from the LCLS. One study is static, and the other one is time resolved. The results of the static study demonstrated that intense X-ray FEL multiply ionizes the parent molecules before breaking into molecular ions as well as into highly charged atomic C states. This work contributed toward understanding the radiation damage, and in particular, electronic damage, due to X-ray radiation, which is essential to understand for the progress of bio-molecular imaging. The result of the time-resolved study gave new insights into the dynamics of the C_{60} fragmentation subsequent to mid-fluence absorption of the X-ray photon. It demonstrated the importance of chemical effects and charge transfer in stabilizing the molecule against fragmentation over several hundred femtoseconds after the X-ray pump pulse. We conclude this article by identifying and providing recommendations for future research opportunities using *as* pump-probe techniques.

2. Materials and Methods

The measurement of the ions resulting from the ionization and fragmentation dynamics of C_{60} under X-ray exposure was achieved by using a magnetic bottle spectrometer [16] at the LCLS AMO hutch. The method is described elsewhere [16]; thus, our description here is brief. The C_{60} sample was produced via a collimated molecular beam using an evaporative oven introducing the gas phase C_{60} into the vacuum chamber. The oven was resistively heated and had a small nozzle and skimmer through which the C_{60} molecules entered the interaction region. The oven was heated to ~700–800 K, and a liquid-nitrogen-cooled dump opposite the skimmer captured the target after it passed through the interaction region. X-ray optics focused the incoming X-ray pulses to a peak focal intensity of ~ 10^{15} W cm^{-2}. The magnetic bottle spectrometer consisted of a 2 m long ion drift path, providing high ion mass-to-charge and KE resolution. The pulse energies of the two beams, centroid photon energy and pulse duration were recorded and used as statistical filters in the analysis. The interaction region, defined by the intersection of the focused X-ray beam and the molecular beam, covered a volume around the focus with an inhomogeneous X-ray spatial fluence distribution. Therefore, the measured data contain contributions from a wide range of fluence (volume integrated signal), with peak X-ray fluence only at the center of the focus. To model the interaction region, the X-ray spatial fluence distribution was calibrated using the ion yields from Ar. This spatial fluence distribution was then applied in the modeling of the X-ray interaction with C_{60} to account for low- and high-fluence regions, allowing quantitative comparison of theoretical predictions to the experimental data.

3. Results

3.1. Photoionization of C_{60} with High-Fluence X-ray FEL Pulses

One of the scientific motivations to study C_{60} with fs X-ray FELs is because these sources target atomic orbitals instead of molecular orbitals, and they allow simple measurements of the response of inner-shell electrons (localized with each atom forming the molecule) compared to the complex response of the molecular orbitals composed of all valence electrons. Probing inner-shell electrons with short wavelengths allows an efficient probing of physical and chemical phenomena from within, since it is an inside-out ionization. Inner-shell photoionization was carried out with synchrotron radiation [17], but the difference between these two light sources is that FELs have a fs time structure and are super intense [18] compared to synchrotrons. A synchrotron pulse on average has about 10^4 photons while FELs have about 10^{12}, allowing the investigation of non-linear processes as well as time-resolved photoionization and fragmentation of the molecules [19].

Another motivation to choose to carry out the photoionization of C_{60} with intense FELs was to understand the radiation damage of large systems because this finding could contribute to the understanding of biomolecular imaging using X-ray scattering techniques, which do not provide detailed spectroscopic information. C_{60} is considered a benchmark molecule because it consists of chemically bonded carbon atoms with representative bond lengths and damage processes to bio-molecules [20]. Although FELs provide the incident brightness needed to achieve diffraction-limited atomic resolution experiments, they, nevertheless, induce possible electronic and structural damage, altering the sample despite the use of short pulse durations [21].

We carried out an experimental and theoretical investigation of C_{60} dynamics with intense fs X-rays to provide a spectroscopic study that offers either a quantitative or qualitative understanding of molecular dynamics. To understand the effects of increased per-atom fluence dose in the photon–molecule interaction, we used the large photoabsorption cross-section of carbon 1 s electrons. We ionized C_{60} with 485 eV photon energy to reach conditions in which each atom in a C_{60} molecule in the X-ray focus absorbs multiple photons. The study was performed with three pulse durations (4 fs, 60 fs, 90 fs) to ascertain the effect of the pulse duration. The core ionization induces the Auger–Meitner process, resulting in many photo- and Auger–Meitner electrons due to the cyclic multiphoton ionization. This process leads to secondary ionization of C_{60}, and it fragments ions by the photo- and Auger–Meitner electrons, which are found to be weak in isolated small molecules and absent in atoms. These effects were found, however, to be very significant for the ionization and fragmentation of C_{60} under high photon dose rate conditions and had to be incorporated in the model calculation to account for the experimental data. The C_{60} molecule charges up to $C_{60}{}^{8+}$, based on our observation, and because of the short C–C bond lengths, it fragments via Coulomb repulsion into molecular and C ion charge states distribution from C^{+} to C^{6+} at 90 fs pulse duration. At a shorter pulse duration, 4 fs, the higher charge state obtained is only C^{5+}, since the pulse is shorter, giving less multi-cyclic photoionization and Auger–Meitner decay. Our investigation focused on the production of charged atomic C states and not on the molecular fragments.

We show in Figure 1 some of the multi-ionized parent C_{60} as well as its fragmentation products, such as the molecular charged carbon chains, compared with the model [22]. Not all fragments are shown or taken into consideration in the calculation here; thus, the sum over all fractional fragmentation ion yields is not unity. Furthermore, we do not measure the formation of neutral fragments in our experiment. As can be seen, there is not an agreement between the experimental molecular data and the model, which was tailored for the formation of charged atomic C states. The calculation is based on a model that follows both electronic and ionic dynamics in space and time, where the atoms/ions are treated as classical particles using Newtonian mechanics, but the rate equations and the cross sections are introduced quantum mechanically. The charges interact via Coulomb forces, and a non-relativistic equation of motion is used [22]. However, the molecular effects were not included, which is the reason for the discrepancy between the measurements and the calculation for the molecular fragment shown in Figure 1. Nonetheless, in the case of atomic carbon charge states, this model, which describes the charged particles behaving as if they were classical particles, agrees well with the experimental data, as shown in Figure 2.

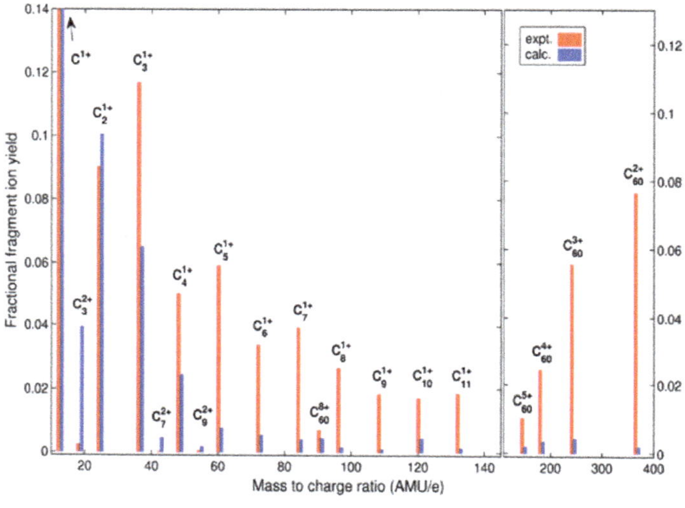

Figure 1. Sequential multi-photon ionization of C_{60} displaying molecular ion fragments. The photon energy was 485 eV, the pulse duration was 90 fs, and the pulse energy was 0.61 mJ.

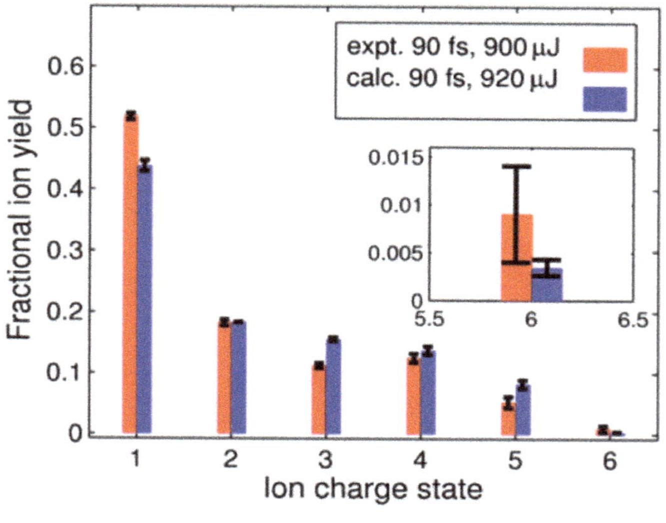

Figure 2. Charged atomic C state distribution measured with pulse duration of 90 fs and 900 μJ.

Figure 2 depicts the comparison between the experimental data and the model of the atomic C^{n+} (n = 1–6) ion charge states generated after the Coulomb explosion of ${C_{60}}^{8+}$. As can be seen, the pulse intensity allowed the formation of fully stripped C ions with 90 fs pulse duration. The model initially predicted more abundant charge states; however, the strong recombination of electrons with the C ions after the pulse ends led to the observed ion yield results [22].

In summary, the static experimental and theoretical investigation of C_{60} with intense pulses resulted in demonstrating that an intense X-ray FEL multiply ionizes the parent molecules before breaking into molecular ions as well as into highly charged atomic C states. We learned that, in the case of charged atomic C, secondary ionization (collisional ionization by trapped electrons) and recombination of electrons with C ion fragments were extremely important to the interaction of C_{60} with intense photons because the interaction created a microplasma that allowed high kinetic energy electrons to also ionize the C_{60} atomic fragments. The calculation for the charged atomic C states will not have agreed with the measurement if it did not include these effects. These effects were not sufficient for the case of the molecular fragment ions, which was not the focus of our investigation. Our work with intense X-ray pulses also contributed a detailed understanding of electronic damage (photoelectron, Auger–Meitner electrons) due to X-ray radiation, which is essential for the progress of bio-molecular imaging [22].

3.2. Time-Resolved Photoionization of C_{60} with Mid-Fluence X-ray FEL Pulses

The instrumental advances that provided the generation of pairs of synchronized femtosecond X-ray FEL pulses [23] has made it possible to carry out time-resolved studies enabling tracking, probing and ultimately understanding the time evolution of X-ray-induced photo processes. We thus extended our static investigation of C_{60} by carrying out a time-resolved experimental and theoretical investigation of its dynamics with mid-fluence fs X-rays. Specifically, we examined the role of chemical effects, such as chemical bonds and charge transfer, on the fragmentation following multiple ionization of the molecules. The X-ray pump-probe investigations enabled probing charge and nuclear dynamics after inner-shell photoabsorption. In this mid-fluence X-ray multiphoton regime, like in the previous intense case, there is a high degree of ionization, challenging the time-resolved theoretical work because of the response of a large number of degrees of freedom and the formation of highly excited states.

We performed X-ray pump/X-ray probe measurements [23] in which the X-ray pump pulse with 640 eV photons from the LCLS enabled K-shell ionization of the carbon atoms and induced a substantial degree of ionization. We then used an X-ray probe pulse to observe the dynamics initiated by the pump pulse, by detecting molecular and atomic fragment ions. The time evolution of the observed fragment ions was interpreted by numerical simulations. The ion fragments were detected with an ion time-of-flight spectrometer [22], which recorded the evolution of C_{60} into fragment ions. The time delays between the pump and probe pulses were between 25 and 925 fs [23]. The pulse duration of the first and second pulses was 20 fs and 10 fs, respectively. The measured total energy was 0.77 +/− 0.01 mJ, while the estimated energy of the two pulses was 45% for the pump (peak intensity of 4×10^{15} W/cm^2) and 55% for the probe pulse. The theoretical modeling employed a molecular-dynamics-based simulation tool with recent additions to include the treatment of chemical bonds via classical force fields [22] and valence-electron charge transfer [24]. Under the current experimental conditions, multiphoton ionization and Auger–Meitner decay lead to multiply charged C_{60} molecules, stable up to ${C_{60}}^{13+}$, which break up into atomic and molecular fragments.

The examination of the charged atomic C fragments demonstrated a time-delay dependence to characterize the molecular time evolution. Figure 3 compares the experimental and theoretical atomic ion yields for C^{1+} to C^{4+} as a function of pulse delay, showing a qualitative good agreement between the experiment and modeling regarding only the dynamical behavior, not the absolute values of the yield of the charged atomic C states. As can be seen, for both the measurements and the calculations, the behavior of the C^+ ion yield changes little as a function of time delay between 25 and 925 fs. However, the ion yield behavior for C^{2+} - C^{4+} increases steadily over a time-delay range of approximately 600 fs; then, a plateau is observed at a longer time delay.

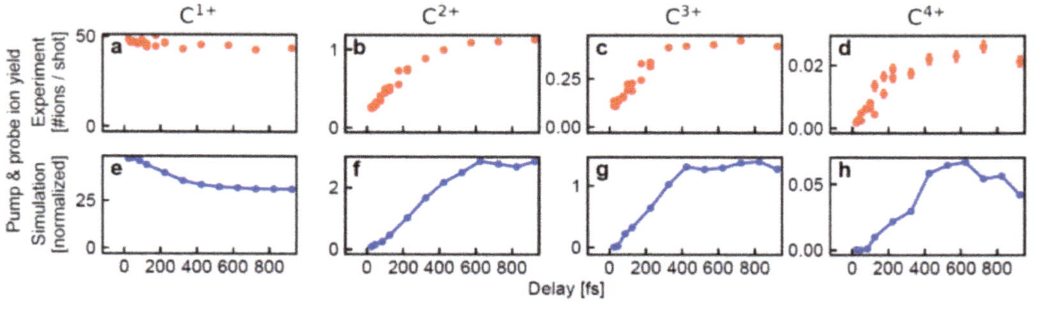

Figure 3. Comparison between experiment and modeling of the time-resolved ionization and fragmentation dynamics of C_{60} displaying charged atomic C fragments. The top panels show the measurements while the bottom panels show the calculations. Panels (**a**,**e**) show the C^{1+}; (**b**,**f**) show the C^{2+}; (**c**,**g**) show the C^{3+}; and (**d**,**h**) show the C^{4+} fragment ions. See text for details.

Our analysis of the data guided by the calculation allowed us to understand the real-time evolution of the parent fragments after the pump pulse. Specifically, we found out that a substantial fraction of the ejected fragments, subsequent to the ionization of C_{60}, are neutral carbon atoms. We calculated that beyond about 300 fs, there is an even larger number of C neutral than C^+ ions. This modeling allowed us to deduce that the probe pulse generates C^{2+} predominantly through the ionization of neutral C by a photoionization and Auger–Meitner decay (P-A) sequence, while C^{3+} is created similarly after ionization of C^{1+} by a P-A sequence. C^{4+} is formed from neutral C via two P-A sequences or from C^{1+} by a P-A sequence and a valence ionization or secondary ionization. Based on the measurement and modeling, we concluded that the production of C^{2+}, C^{3+} and C^{4+} is primarily due to the ionization of C/C^{1+} fragments by the probe pulse.

Our modeling demonstrated that chemical bonds and valence charge transfer are important in the photoionization of C_{60}, since we found no time-delay dependence in the yield of the C^{2+}, C^{3+} and C^{4+} ions without inclusion of such chemical effects. We found that the parent ion ${C_{60}}^{13+}$ does not undergo instantaneous fragmentation because it is delayed, as shown in Figure 3. The fragmentation is delayed relative to the pump pulse due to the still existing chemical bonds because it takes time for the 60-atom system to break up into smaller fragments, most of which exist only transiently for several tens of femtoseconds. In addition, during this structural transformation, it takes additional time for those fragments to eject neutral and singly charged atomic ion fragments. We show the impact of the chemical effects in Figure 4 (top panel) by plotting the maximum atomic displacement for the parent ion ${C_{60}}^{13+}$ based on MD simulations, for two situations: (1) chemical phenomena, such as bonding and charge transfer, are removed; (2) the full model that includes both chemical bonding and charge transfer.

It is clear that chemical effects minimize significantly the maximum displacement of C atoms, demonstrating evidently that C_{60} is structurally resistant, on time scales of tens of femtoseconds, against the impact of X-ray multiphoton ionization. Figure 4 (bottom panel) shows the simulation snapshots of the real-time evolution of ${C_{60}}^{13+}$ for the full model, illustrating the ionization and fragmentation dynamics of C_{60} irradiated by femtosecond X-rays.

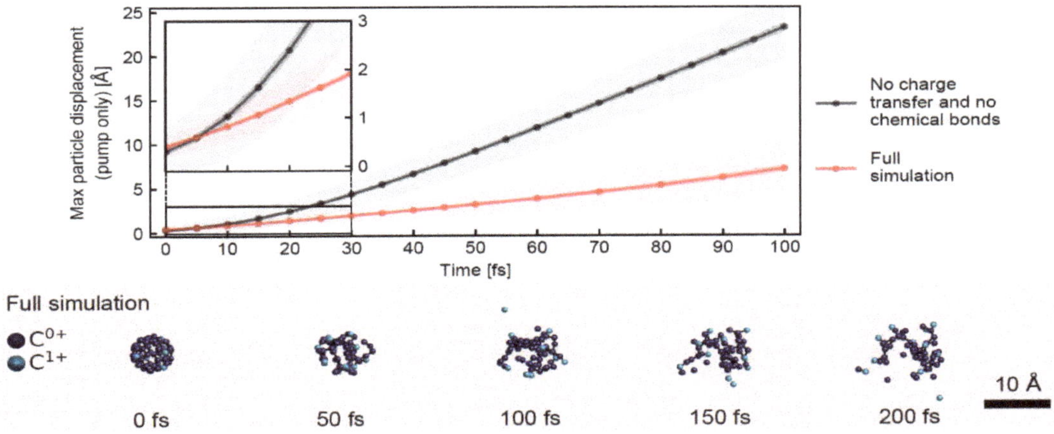

Figure 4. Movie snapshots of the real-space and real-time simulation dynamics of $C_{60}{}^{13+}$ induced by XFEL pulse.

The implications of this work for the field of femtosecond molecular imaging is that charge transfer, nuclear arrangement, chemical bonds, and thus, the chemical structure are resistant to the intense electromagnetic environment created by XFEL irradiation. The impact of this work is that delayed fragmentation will play a critical role in most other intense X-ray multiphoton ionizations of molecules. With the advent of several new FELs all over the world, our results, which lay the foundation for a deeper understanding and quantitative modeling of XFEL-induced radiation damage, will impact biomolecular imaging.

In summary, this time-resolved investigation gave new insight into the dynamics of the C_{60} fragmentation subsequent to mid-fluence absorption of the X-ray photon. It revealed the importance of chemical effects, such as covalent bonding and charge transfer, in stabilizing the molecule against fragmentation over several hundred femtoseconds after the X-ray pump pulse. Such detailed understanding of X-ray-induced ionization dynamics and atomic motions in molecules is crucial for the applications of high-intensity X-ray beams [25].

4. Discussion

Our experimental work in the high- [22] and mid-fluence [25] X-ray regime, investigating the response of C_{60} to fs FEL pulses, revealed new physical and chemical processes that were validated either qualitatively or quantitatively by state-of-the-art simulation methodology. Specifically, the static X-ray high-fluence, 90 fs study showed that the Coulomb explosion subsequent to multiphoton ionization is so violent that it leads to fully stripped C atoms, and the model does not require the inclusion of detailed molecular effects. The most important effect that needed to be included in the calculation was the secondary electron collision with the ions fragments as well as the recapture of electrons by the ion fragments. On the other hand, the time-resolved mid-fluence study demonstrated that chemical bonds and the charge transfer effect were crucial to be included in the calculation to observe dynamics as a function of time delay. These effects had to be included to agree with the measurements that depicted dynamics as a function of time delay. In this situation, the simulations and the experiment revealed that despite significant ionization induced by the ultra-short (20 fs) X-ray pump pulse, the fragmentation of C_{60} was significantly delayed. This work uncovered the persistence of the molecular structure of C_{60}, which hinders fragmentation over a timescale of hundreds of femtoseconds. Furthermore, the calculation demonstrated that a substantial fraction of the ejected fragments is neutral carbon atoms. In fact, it is these neutral fragments' ejection from C_{60}, as the molecule cools off, that prohibits a strong Coulomb explosion. These findings, interpreted by the most advanced modeling and theory, provide insights into X-ray FEL-induced radiation damage in large molecules, including biomolecules [25]. In fact, this work seems to indicate that experiments conducted with up to 20 fs, with mid fluence, will not experience radiation damage due to the delayed ionization. Thus, bio-molecule X-ray diffraction at this fluence will not suffer the "diffract before destroy" scenario. Furthermore, our simulation [25] provides a solid basis for the reliable interpretation of processes in systems even larger than C_{60}. Future XFEL-based experimental research into a wide range of systems will benefit from our results and the theoretical advances needed to interpret the experimental work.

Recently, a time-resolved experiment was conducted with intense XUV photons at the FLASH-II FEL that probed the resonance structure at about 20 eV, and this work might reveal new insight regarding the ionization and fragmentation of C_{60} at lower photon energy [26]. In addition, an experiment was conducted with hard X-rays at LCLS to visualize the light-induced reshaping of C_{60} via X-ray diffraction. Specifically, time-resolved X-ray diffraction images of C_{60} molecules were recorded during and after their interaction with intense near-infrared fields, giving direct access to structural changes of the molecules and their neutral or ionic fragmentation, in real time [27].

The current experiments so far focused on measuring the charged fragments resulting from absorption of X-ray photons by neutral C_{60} to investigate the ionization processes or using X-ray diffraction to visualize the transformation of C_{60}. Future work could investigate the anion, C_{60}^- [28,29] or C_{60}^+ [30]. Additionally, what is missing are experiments that will use photoelectron spectroscopy to measure in detail the ionization and fragmentation dynamics. In particular, the electron-ion coincidence techniques could be used, especially with the now available high repetition rate at FELs sources, to measure the momenta of the ions as well as identify the possible coincidences among the fragments and to visualize the different coincidence channels. Furthermore, the use of ultra-fast electron diffraction (UED) techniques [31] will allow the imaging of the molecular changes as a function of time delay between the pump and probe pulse.

Finally, the exciting advent of *as* X-ray pulses from FELs opens up the opportunity to study the electronic dynamics that precedes the nuclear dynamics [32]. Specifically, inner-shell X-ray pump-probe studies with *as* pulses could allow the probing and understanding of the electronic dynamics on its natural timescale. Currently, *as* pulses from the LCLS-II X-ray FEL are provided to scientists with photon energies tunable across the soft X-ray regime and with spectral brightness six orders of magnitude greater than HHG sources [32]. This *as* capability will soon be reproduced at other FEL facilities around the world, enabling previously impossible as experiments in all research fields.

5. Conclusions

Intense and short X-ray FELs allowed new understanding of the interaction and response of C_{60} fullerenes in a new energy regime. Furthermore, FELs facilities offer the possibility to carry out time-resolved experiments that allow us to extract molecular movies of the system's transformation as it evolves as a function of time. Future work could focus on using shorter X-ray pulses, for example, *as* pulses, as well as using other experimental methodologies, such as electron spectroscopy, coincidence techniques or ultra-fast electron diffraction. The combination of X-ray *as* FEL facilities and differential experimental techniques will enable scientists to understand the first steps in the photoabsorption or scattering investigations, resulting in new breakthrough that will push the frontiers of science and may result in new discoveries.

Funding: This work was funded in part by the Chemical Sciences, Geosciences and Biosciences Division, Office of Basic Energy Sciences, Office of Science, US Department of Energy, grant No. DE-SC0012376.

Institutional Review Board Statement: Not applicable.

Informed Consent Statement: Not applicable.

Data Availability Statement: The data can be made available if requested.

Acknowledgments: We thank A. Sanchez-Gonzalez, Z. Jurek, R. Obaid, H. Xiong, R.J. Squibb, T. Osipov, A. Lutman, L. Fang, T. Barillot, J.D. Bozek, J. Cryan, T.J.A. Wolf, D. Rolles, R. Coffee, K. Schnorr, S. Augustin, H. Fukuzawa, K. Motomura, N. Niebuhr, L.J. Frasinski, R. Feifel, C.P. Schulz, K. Ueda, K. Toyota, S.-K. Son, K. Ueda, T. Pfeifer, J.P. Marangos, B.F. Murphy, L. Avaldi, P. Bolognesi, C. Bostedt, J. Eland, M. Guehr, J. Farrell, J. Glownia, D.T. Ha, K. Hoffmann, E. Kukk, B. McFarland, C. Miron, M. Mucke and R. Santra.

Conflicts of Interest: The authors declare no conflict of interest.

References

1. Berrah, N.; Bucksbaum, P.H. The Ultimate X-ray Machine. *Sci. Am.* **2014**, *310*, 64–71. [CrossRef] [PubMed]
2. Emma, P.; Akre, R.; Arthur, J.; Bionta, R.; Bostedt, C.; Bozek, J.; Brachmann, A.; Bucksbaum, P.; Coffee, R.; Decker, F.-J.; et al. First lasing and operation of an ångstrom-wavelength free-electron laser. *Nat. Photon.* **2010**, *4*, 641–647. [CrossRef]
3. Ackermann, W.; Asova, G.; Ayvazyan, V.; Azima, A.; Baboi, N.; Ahr, J.B.; Balandin, V.; Beutner, B.; Brandt, A.; Bolzmann, A.; et al. Operation of a free-electron laser from the extreme ultraviolet to the water window. *Nat. Photon.* **2007**, *1*, 336–342. [CrossRef]
4. Duris, J.; Li, S.; Driver, T.; Champenois, E.G.; MacArthur, J.P.; Lutman, A.A.; Zhang, Z.; Rosenberger, P.; Aldrich, J.W.; Marinelli, A.; et al. Tunable isolated attosecond X-ray pulses with gigawatt peak power from a free-electron laser. *Nat. Photon.* **2020**, *14*, 30–36. [CrossRef]
5. Berrah, N.; Fang, L.; Murphy, B.; Osipov, T.; Ueda, K.; Kukk, E.; Feifel, R.; van der Meulen, P.; Salen, P.; Schmidt, H.T.; et al. Double Core-Hole Spectroscopy for Chemical Analysis with an Intense X-Ray Femtosecond Laser. *Proc. Natl. Acad. Sci. USA* **2011**, *108*, 16912. [CrossRef]
6. Young, L.; Kanter, E.P.; Krässig, B.; Li, Y.; March, A.M.; Pratt, S.T.; Santra, R.; Southworth, S.H.; Rohringer, N.; DiMauro, L.F.; et al. Femtosecond electronic response of atoms to ultraintense X-rays. *Nature* **2010**, *466*, 56–61. [CrossRef] [PubMed]
7. Johansson, J.O.; Campbell, E.E.B. Probing excited electronic states and ionisation mechanisms of fullerenes. *Chem. Soc. Rev.* **2013**, *42*, 5661–5671. [CrossRef] [PubMed]
8. Feng, M.; Zha, J.; Petek, H. Atomlike, hollow-core-bound molecular orbitals of C_{60}. *Science* **2008**, *320*, 359–362. [CrossRef]
9. Mignolet, B.; Johansson, J.O.; Campbell, E.E.B. Probing rapidly ionizing super-atom molecular orbitals in C_{60}: A computational and femtosecond photoelectron spectroscopy study. *Chem. Phys. Chem.* **2013**, *14*, 3332–3340. [CrossRef]
10. Li, H.; Mignolet, B.; Wachter, G.; Skruszewicz, S.; Zherebtsov, S.; Süßmann, F.; Kessel, A.; Trushin, S.A.; Kling, N.G.; Kübel, M.; et al. Coherent Electronic Wave Packet Motion in C_{60} Controlled by the Waveform and Polarization of Few-Cycle Laser Fields. *Phys. Rev. Lett.* **2015**, *114*, 123004. [CrossRef]
11. Xiong, H.; Mignolet, B.; Fang, L.; Osipov, T.; Wolf, T.; Wolf, J.A.; Sistrunk, E.; Gühr, M.; Remacle, R.; Berrah, N. The Role of Super-Atom Molecular Orbitals in Doped Fullerenes in a Femtosecond Intense Laser Field. *Sci. Rep.* **2017**, *7*, 121. [CrossRef]
12. Bhardwaj, V.R.; Corkum, P.B.; Rayner, D.M. Internal laser-induced dipole force at work in C_{60} molecule. *Phys. Rev. Lett.* **2003**, *91*, 203004. [CrossRef]
13. Li, H.; Mignolet, B.; Wang, Z.; Betsch, K.J.; Carnes, K.D.; Ben-Itzhak, I.; Cocke, C.L.; Remacle, F.; Kling, M.F. Transition from SAMO to Rydberg State Ionization in C_{60} in Femtosecond Laser Fields. *J. Phys. Chem. Lett.* **2016**, *7*, 4677–4682. [CrossRef] [PubMed]
14. Jaroń-Becker, A.; Becker, A.; Faisal, F.H.M. Single-active-electron ionization of C_{60} in intense laser pulses to high charge states. *J. Chem. Phys.* **2007**, *126*, 124310. [CrossRef] [PubMed]
15. Lebeault, M.-A.; Baguenard, B.; Concina, B.; Calvo, F.; Climen, B.; Leépine, F.; Bordas, C. Decay of C_{60} by delayed ionization and C_2 emission: Experiment and statistical modeling of kinetic energy release. *J. Chem. Phys.* **2012**, *137*, 054312. [CrossRef] [PubMed]
16. Berrah, N.; Fang, L.; Osipov, T.; Jurek, Z.; Murphy, B.F.; Santra, R. Emerging photon technologies for probing ultrafast molecular dynamics. *Faraday Disc.* **2014**, *171*, 471–485. [CrossRef]
17. Canton, S.E.; Yencha, A.J.; Kukk, E.; Bozek, J.D.; Lopes, M.C.A.; Snell, G.; Berrah, N. Experimental Evidence of a Dynamic Jahn-Teller Effect in $C_{60}{}^+$. *Phys. Rev. Lett.* **2002**, *89*, 45502. [CrossRef]
18. Duris, J.P.; MacArthur, J.P.; Glownia, J.M.; Li, S.; Vetter, S.; Miahnahri, A.; Coffee, R.; Hering, P.; Fry, A.; Welch, M.E.; et al. Controllable X-Ray Pulse Trains from Enhanced Self-Amplified Spontaneous Emission. *Phys. Rev. Lett.* **2021**, *126*, 104802. [CrossRef]
19. Bucksbaum, P.; Berrah, N. Brighter and faster: The promise and challenge of the X-ray free-electron laser. *Phys. Today* **2015**, *68*, 26. [CrossRef]
20. Luo, Y.R. *Comprehensive Handbook of Chemical Bond Energies*; CRC Press: Boca Raton, FL, USA, 2007.
21. Hoener, M.; Fang, L.; Kornilov, O.; Gessner, O.; Pratt, S.T.; Gühr, M.; Kanter, E.P.; Blaga, C.; Bostedt, C.; Bozek, J.D.; et al. Ultra-intense X-ray Induced Ionization, Dissociation and Frustrated Absorption in Molecular Nitrogen. *Phys. Rev. Lett.* **2010**, *104*, 253002. [CrossRef]
22. Murphy, B.F.; Osipov, T.; Jurek, Z.; Fang, L.; Son, S.-K.; Mucke, M.; Eland, J.; Zhaunerchyk, V.; Feifel, R.; Avaldi, L.; et al. Femtosecond X-ray-induced explosion of C_{60} at extreme intensity. *Nat. Commun.* **2014**, *5*, 4281. [CrossRef] [PubMed]
23. Lutman, I.A.; Maxwell, T.J.; MacArthur, J.P.; Guetg, M.W.; Berrah, N.; Coffee, R.N.; Ding, Y.; Huang, Z.; Marinelli, A.; Moeller, S.; et al. Fresh-slice multicolour X-ray free-electron lasers. *Nat. Photon.* **2016**, *10*, 745. [CrossRef]

24. Kumagai, Y.; Jurek, Z.; Xu, W.; Fukuzawa, H.; Motomura, K.; Iablonskyi, D.; Nagaya, K.; Wada, S.-I.; Mondal, S.; Tachibana, T.; et al. Radiation-induced chemical dynamics in Ar clusters exposed to strong X-ray pulses. *Phys. Rev. Lett.* **2011**, *20*, 223201. [CrossRef] [PubMed]
25. Berrah, N.; Sanchez-Gonzalez, A.; Jurek, Z.; Obaid, R.; Xiong, H.; Squibb, R.J.; Osipov, T.; Lutman, A.; Fang, L.; Barillot, T.; et al. X-ray multiphoton ionization of molecules: Femtosecond-resolved observation of delayed fragmentation and evaporation of neutral atoms. *Nat. Phys.* **2019**, *15*, 1279. [CrossRef]
26. LaForge, A.; Mishra, D.; Obaid, R.; Trost, F.; Meiste, S.; Lindenblatt, H.; Rosenberger, P.; Biswas, S.; Shashank, P.; Moshammer, R.; et al. XUV pump-probe of the shape resonance in C_{60}. *Phys. Rev. Lett.* **2022**. *manuscript in preparation.*
27. Schnorr, K.; Augustin, S.; Saalmann, U.; Schmid, G.; Rouzee, A.; Obaid, R.; AlHaddad, A.; Berrah, N.; Blaga, C.; Bostedt, C.; et al. Visualizing the light-induced reshaping of C_{60} via X-ray diffraction. *Science* **2022**. *submitted, under review.*
28. Amusia, M.Y.; Baltenkov, A.S.; Krakov, B.G. Photodetachment of negative C_{60}^- ions. *J. Phys. Lett. A* **1998**, *243*, 99–105. [CrossRef]
29. Bilodeau, R.C.; Gibson, N.D.; Walter, C.W.; Esteves-Macaluso, D.A.; Schippers, S.; Müller, A.; Phaneuf, R.A.; Aguilar, A.; Hoener, M.; Rost, J.M.; et al. Single-Photon Multiple-Detachment in Fullerene Negative Ions: Absolute Ionization Cross Sections and the Role of the Extra Electro. *Phys. Rev. Lett.* **2013**, *111*, 043003. [CrossRef]
30. Scully, S.W.J.; Emmons, E.D.; Gharaibeh, M.F.; Phaneuf, R.A.; Kilcoyne, A.L.D.; Schlachter, A.S.; Schippers, S.; Müller, A.; Chakraborty, H.S.; Madjet, M.E.; et al. Photoexcitation of a volume plasmon in C_{60} ions. *Phys. Rev. Lett.* **2005**, *94*, 065503. [CrossRef]
31. Wolf, T.J.A.; Sanchez, D.M.; Yang, J.; Parrish, R.M.; Nunes, J.P.F.; Centurion, M.; Coffee, R.; Cryan, J.P.; Gühr, M.; Hegazy, K.; et al. The photochemical ring-opening of 1,3-cyclohexadiene imaged by ultrafast electron diffraction. *Nat. Chem.* **2019**, *11*, 504–509. [CrossRef]
32. Li, S.; Driver, T.; Rosenberger, P.; Champenois, E.G.; Duris, J.; Al-Haddad, A.; Averbukh, V.; Barnard, J.C.T.; Berrah, N.; Bostedt, C.; et al. Real-time Observation of Coherent Electron Motion by Attosecond Streaking of Auger-Meitner Decay. *Science* **2022**, *375*, 285–290. [CrossRef] [PubMed]

atoms

MDPI

Article

Quasifree Photoionization under the Reaction Microscope

Sven Grundmann [1], Florian Trinter [1,2], Yong-Kang Fang [3], Kilian Fehre [1], Nico Strenger [1], Andreas Pier [1], Leon Kaiser [1], Max Kircher [1], Liang-You Peng [3], Till Jahnke [4], Reinhard Dörner [1] and Markus S. Schöffler [1,*]

1 Institut für Kernphysik, Goethe-Universität, Max-von-Laue-Straße 1, 60438 Frankfurt, Germany; grundmann@atom.uni-frankfurt.de (S.G.); trinter@atom.uni-frankfurt.de (F.T.); fehre@atom.uni-frankfurt.de (K.F.); strenger@atom.uni-frankfurt.de (N.S.); pier@atom.uni-frankfurt.de (A.P.); kaiser@atom.uni-frankfurt.de (L.K.); kircher@atom.uni-frankfurt.de (M.K.); doerner@atom.uni-frankfurt.de (R.D.)

2 Molecular Physics, Fritz-Haber-Institut der Max–Planck–Gesellschaft, Faradayweg 4-6, 14195 Berlin, Germany

3 State Key Laboratory for Mesoscopic Physics and Frontiers Science Center for Nano-Optoelectronics, School of Physics, Peking University, Beijing 100871, China; yongkangfang@pku.edu.cn (Y.-K.F.); liangyou.peng@pku.edu.cn (L.-Y.P.)

4 European XFEL, Holzkoppel 4, 22869 Schenefeld, Germany; jahnke@atom.uni-frankfurt.de

* Correspondence: schoeffler@atom.uni-frankfurt.de

Abstract: We experimentally investigated the quasifree mechanism (QFM) in one-photon double ionization of He and H_2 at 800 eV photon energy and circular polarization with a COLTRIMS reaction microscope. Our work provides new insight into this elusive photoionization mechanism that was predicted by Miron Amusia more than four decades ago. We found the distinct four-fold symmetry in the angular emission pattern of QFM electrons from H_2 double ionization that has previously only been observed for He. Furthermore, we provide experimental evidence that the photon momentum is not imparted onto the center of mass in quasifree photoionization, which is in contrast to the situation in single ionization and in double ionization mediated by the shake-off and knock-out mechanisms. This finding is substantiated by numerical results obtained by solving the system's full-dimensional time-dependent Schrödinger equation beyond the dipole approximation.

Keywords: two-electron systems; one-photon double ionization; many-electron correlation

Citation: Grundmann, S.; Trinter, F.; Fang, Y.-K.; Fehre, K.; Strenger, N.; Pier, A.; Kaiser, L.; Kircher, M.; Peng, L.-Y.; Jahnke, T.; Dörner, R.; Schöffler, M.S. Quasifree Photoionization under the Reaction Microscope. *Atoms* **2022**, *10*, 68. https://doi.org/10.3390/atoms10030068

Academic Editor: Yew Kam Ho

Received: 18 May 2022
Accepted: 22 June 2022
Published: 28 June 2022

1. Introduction

The quasifree mechanism (QFM) is a special case of photo-double ionization (PDI) that was predicted by M. Amusia et al. in 1975 [1]. The name of the process originates from the idea that the photon interacts with a quasifree electron pair without involvement of the nucleus. The kinematic profile of QFM is characterized by electrons being emitted back-to-back with equal kinetic energy, which leave the nucleus with close-to-zero recoil momentum [2,3]. The interaction between photons and atoms is generally dominated by electric-dipole contributions, but the QFM profile is forbidden in a dipole transition due to angular momentum and parity conservation [4]. Hence, QFM facilitates double ionization by means of a pure electric-quadrupole transition. As QFM ejects two electrons only from the small part of the initial-state two-electron wave function where both electrons are spatially close together [5,6], its transition amplitude is extremely small and experimental investigations of QFM are challenging [7]. However, nearly four decades after the prediction, the existence of the quasifree mechanism was finally experimentally confirmed by Schöffler et al. in 2013 through the observation of doubly charged helium nuclei with close-to-zero momentum [8]. Note that this signature of QFM is similar to what is found for double ionization by Compton scattering which becomes the dominant double-ionization channel at high photon energies [9–11].

The recently renewed interest in nondipolar photoionization in the one-photon and strong-field ionization regimes (see, e.g., Refs. [12–17]) encouraged further experimental and theoretical investigations of QFM. In 2018, electrons emitted back-to-back with equal energy were observed for helium PDI at 1100 eV photon energy [18]. This work displayed the angular emission pattern of electrons originating from a pure quadrupole transition. Two years later, QFM was confirmed for H_2 PDI at 800 eV photon energy and it was shown how the QFM cross section relates to the initial spatial probability density at the two-electron cusp, which is the point where both electrons coalesce [19]. In the present work, we examine once again the experimental data used in Refs. [18,19] to continue the investigation of QFM. First, we will show that the angular distribution of QFM electrons originating from H_2 PDI displays the same four-fold symmetry that was already observed for He PDI. Furthermore, second, we will provide evidence for the assumption that the photon momentum is not imparted onto the center of mass in quasifree photoionization. The latter finding is supported by numerical results from solving the full-dimensional time-dependent Schrödinger Equation (TDSE) beyond the dipole approximation.

2. Methods

The two experiments reported here have been performed at beamline P04 at the PETRA III synchrotron (DESY, Hamburg, Germany [20]) using circularly polarized light at 800 eV photon energy. We employed a cold target recoil ion momentum spectroscopy (COLTRIMS) reaction microscope [21–23] and intersected a supersonic gas jet of He or H_2 with the photon beam at a right angle. Due to the supersonic expansion, the H_2 were in their vibrational ground state. The charged reaction fragments were projected by an electric field and guided by a magnetic field to two time- and position-sensitive detectors with delay-line position readout [24,25]. The initial momenta after PDI were retrieved from the particles' times-of-flight and positions of impact. The concept of a COLTRIMS reaction microscope is illustrated in Figure 1. The experimental results reported in the present work were obtained from the same experimental runs as Refs. [18,19], where further experimental details can be found.

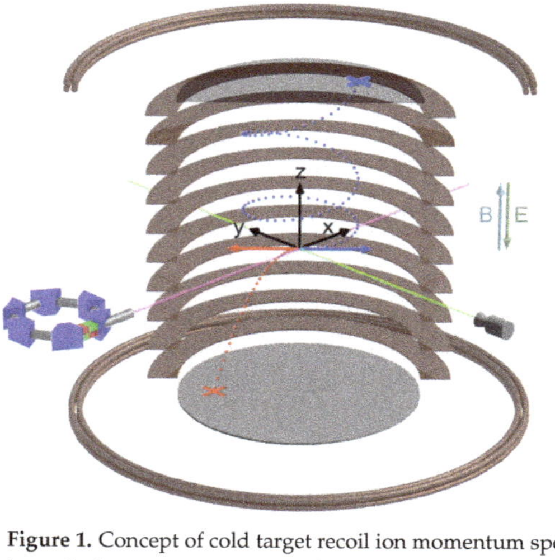

Figure 1. Concept of cold target recoil ion momentum spectroscopy (COLTRIMS). A supersonic jet (green) of a target gas is crossed with synchrotron light (violet) at a right angle. A homogeneous electric field *E*, generated by a spectrometer (copper plates), and a homogeneous magnetic field *B*, created by a Helmholtz pair (copper tubes), guide the charged reaction fragments (red trajectory: ion, blue trajectory: electron) towards time- and position-sensitive detectors. The initial three-dimensional momentum vector of each reaction fragment (blue and red arrows) was calculated from the time-of-flight and the position of impact on the detectors (marked with a blue and a red cross).

The two-electron nondipole TDSE code used here was developed based on our previous dipole code for helium, which has been successfully applied in a series of studies on the two-photon double ionization of helium [26–28]. The form of the Hamiltonian with the nondipole corrections and the chosen parameters in the calculations can be found in Ref. [29]. After the end of the light pulse, the two-electron wave function was further propagated freely for a time of 10 a.u. Then, it was projected onto the uncorrelated symmetrical product of two single-electron scattering states to obtain the joint momentum distribution of the two ejected electrons $P(\boldsymbol{k}_1, \boldsymbol{k}_2)$. The momentum spectrum of the ion $P(\boldsymbol{Q})$ could then be obtained from $P(\boldsymbol{k}_1, \boldsymbol{k}_2)$ by momentum conservation: $P(\boldsymbol{Q}) = \int P(\boldsymbol{k}_1, \boldsymbol{k}_{\text{photon}} - \boldsymbol{k}_1 - \boldsymbol{Q}) d\boldsymbol{k}_1$. For the case of the dipole approximation, $\boldsymbol{k}_{\text{photon}}$ was set to 0. In the calculations, a linearly polarized light pulse along the z axis was adopted, with its propagation direction in the x axis. Therefore, the momentum spectrum of the ion in the propagation direction of the light pulse was given by $P(Q_x) = \iint dQ_z dQ_y P(\boldsymbol{Q})$. For the distribution of $P(\boldsymbol{Q})$ in the PDI of helium, previous studies showed that the majority of the events are located close to the surface of a sphere in the momentum space with a radius of $p_{\text{single}} = \sqrt{\omega - I_p}$, where ω and I_p, respectively, represent the photon energy and ionization potential of helium. Such a behavior is closely related to the shake-off (SO) and the electron knock-out mechanism [8]. In order to show the effect of the QFM mechanism, we used a similar method as that in Ref. [8]. The integral interval in the light polarization direction of Q_z was limited from $-p_{\text{single}}/4$ to $p_{\text{single}}/4$. Furthermore, the integral interval of Q_y was limited by $|Q_y| \leq p_{\text{single}}/2$.

3. Results

3.1. Separating a Pure Quadrupole Contribution to Photo-Double Ionization

In quasifree photoionization, the photon is absorbed by the electron pair and the nucleus is only a spectator that receives no recoil momentum. By means of momentum and energy conservation, the ejection of two electrons back-to-back with equal energy is a strict consequence of a vanishing recoil momentum, if one neglects the photon momentum. For a dipole transition, this kinematic profile is forbidden for PDI of He and H_2, whose ground-state wave functions both have the same ^1S symmetry [4,30]. Hence, the QFM is a pure quadrupole contribution to PDI and it can be isolated particularly clearly in a differential cross section that shows the double-ionization yield as a function of the electron energy sharing $\varepsilon = E_{e1}/(E_{e1} + E_{e2})$ and the angle enclosed by the two electron momentum vectors $\vartheta_{12} = \cos^{-1}[\boldsymbol{p}_{e1} \cdot \boldsymbol{p}_{e2}/(|\boldsymbol{p}_{e1}| \times |\boldsymbol{p}_{e2}|)]$ (electron mutual angle). This is done in Figure 2 where we show the measured yield from the double ionization of H_2 (A,B) and He (D,E) at 800 eV photon energy as a function of ε and ϑ_{12}. Panels A and D show the full range of the two variables, but panels B and E show only the region indicated by the dashed black lines in A and D. A comparison between panels A and D points out a strong resemblance in the electron emission patterns of H_2 and He double ionization, as expected from the similarities in the electronic ground states. A distinctive difference can be seen around equal energy sharing ($\varepsilon = 0.5$) and back-to-back emission ($\cos\vartheta_{12} = -1$) where there appears to be a noticeable signal in panel A, corresponding to QFM, but apparently a node in panel D. Panels C and D highlight this relevant region of the cross section. While the QFM is evident for H_2, a logarithmic-scale display is required in panel D to make the weak relative contribution of QFM to the total double-ionization cross section of helium visible at all (see Ref. [19] for further discussions on this finding).

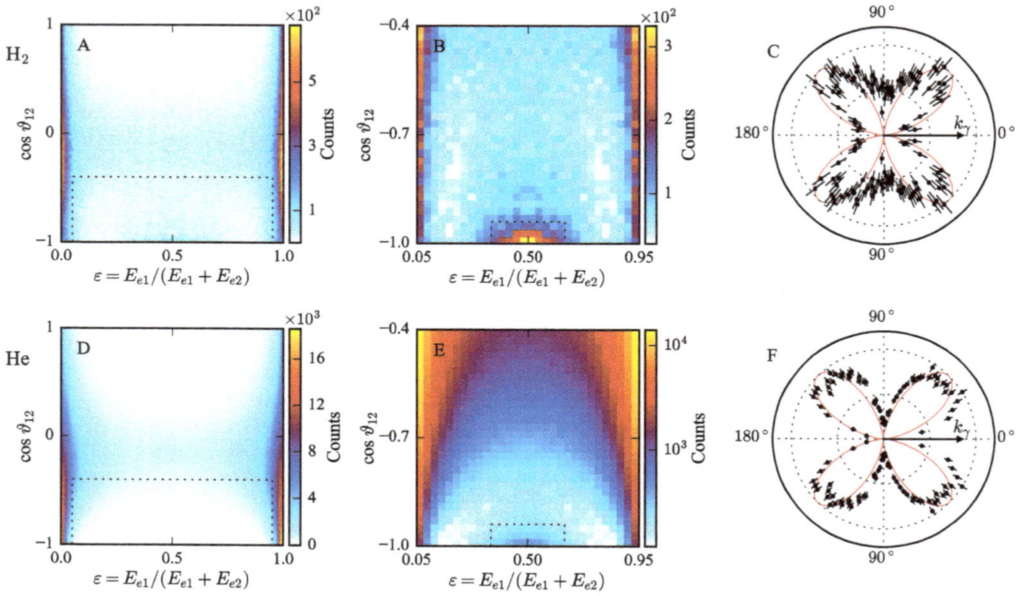

Figure 2. Measured electron distributions of H_2 double ionization (**A–C**) and He double ionization (**D–F**) by a single 800 eV circularly polarized photon. (**A**,**D**) Measured electron yield as a function of the electron energy sharing ε and the electron mutual angle ϑ_{12}. (**B**,**E**) Detailed section from panels A and D as indicated by the dashed black lines therein. Note the logarithmic-scale display in panel E. Contributions around $\varepsilon = 0.5$ and $\cos\vartheta_{12} = -1$ correspond to the QFM electrons. (**C**,**F**) Angular distributions of QFM electrons as a function of the polar angle enclosed by the electron momentum vector and the light propagation direction. The shown data are limited to $0.35 < \varepsilon < 0.65$ and $\vartheta_{12} > 160°$, as indicated by the dashed black lines in panels B and E. For this selection, the dipole contribution to photoionization vanishes. The red lines represent $|Y_{21}|^2$ and are normalized to the data points.

To highlight the quadrupole nature of QFM, we selected a subset of our data limited to $0.35 < \varepsilon < 0.65$ and $\vartheta_{12} > 160°$, as indicated by the dashed black lines in Figures 2B,E, and present in Figures 2C,F—the measured electron yield as a function of the angle ϑ_γ, which is enclosed by the electron momentum vector and the light propagation direction. The angular-momentum transfer is an important physical difference between dipole and quadrupole transitions. By definition, a dipole transition transfers one unit of angular momentum to the two-electron final state due to the photon spin, while two units of angular momentum are available in a quadrupole transition. The angular momentum of the outgoing electron wave becomes observable in the angular distribution of the electron. The red lines in Figures 2C,F represent the square of the spherical harmonic for $\ell = 2$ and $m = 1$, $|Y_{21}|^2 \propto \cos^2\vartheta_\gamma \times \sin^2\vartheta_\gamma$, which describes the final-state angular distribution of electrons that result from a pure electric-quadrupole transition from any initial s-subshell. Here, we have chosen the photon propagation direction $\hat{\boldsymbol{k}}_\gamma$ as the quantization axis. The photon spin vector is (anti-)parallel to $\hat{\boldsymbol{k}}_\gamma$ and we get $\Delta\ell = 1$ and $\Delta m = 1$ through the transfer of the spin angular momentum. The additional unit of orbital angular momentum— $\boldsymbol{k}_\gamma \times \boldsymbol{r}_e = \hbar$— is oriented perpendicularly to the quantization axis. It increases the magnitude of the electron angular momentum but has no effect on its projection m onto $\hat{\boldsymbol{k}}_\gamma$. The strong resemblances between the measured angular emission patterns and $|Y_{21}|^2$, as demonstrated in Figures 2C,F, underline that QFM electrons originate from a pure quadrupole contribution to photoionization. Note that this agreement is better for He, and we suspect this is simply due to larger momentum uncertainty in case of H_2. In our experiment, one of the two QFM electron momentum vectors is reconstructed by means of momentum conservation. This is less accurate for H_2 because the ions' center-of-mass momentum has to be retrieved from two protons instead of being measured via the doubly charged nucleus. Thus, the momentum uncertainty propagating to the calculated electron is larger in the case of H_2 than for He (see Refs. [15,19] for further details). While such a four-fold symmetry in the angular emission pattern of QFM electrons has already been shown for He PDI at 1100 eV photon energy [18], the results shown in Figure 2C for H_2 PDI further support our current understanding of quasifree photoionization.

3.2. Transfer of Photon Momentum

Nondipolar photoionization induces a forward-backward asymmetry in light propagation direction into the momentum distributions of the reaction fragments. This is due to the nonzero linear photon momentum $\boldsymbol{k}_\gamma$ and the interference between electric dipole and higher multipole contributions to the photoionization process. The question of which fragments obtain the photon momentum after the reaction has been investigated since the early days of photoionization studies [31,32]. In the case of photo-single-ionization, momentum and energy conservation dictate that the center of mass—which is essentially the residual photo-ion—obtains the photon momentum [21]. Additional degrees of freedom allow for a more intricate sharing of the photon momentum between the reaction fragments in photo-double ionization. However, a recent experiment–theory collaboration investigated He PDI up to 1100 eV photon energy and showed that the momentum distribution of helium nuclei after double ionization exhibits the same forward-backward asymmetry as helium nuclei from single ionization [29]. In this photon–energy range, He PDI is dominated by the shake-off process while the quasifree mechanism is negligible in absolute terms. Apparently, the shake-off process, where the second electron is emitted due to electron–electron correlation [33,34], treats the photon momentum similarly to how single ionization does, and the photon momentum is imparted onto the doubly charged helium nucleus. However, the quasifree mechanism proceeds without involvement of the nucleus, as the photon couples directly to the two electrons. Hence, one could expect that the photon momentum is not imprinted onto the photo-ion.

In order to test this assumption, we inspect the momentum distributions of photo-ions after He PDI at 800 eV photon energy in Figure 3 for SO (A) and QFM (B) by selecting subsets of our data.

In panel A, the measured photo-ion momenta accumulate on a semicircle with a radius that equals the maximum electron momentum and which is off-centered to the right by the magnitude of the photon momentum ($k_\gamma = 0.215$ au), indicating the transfer of the photon momentum onto the photo-ion. Note that this is the same behavior that has previously been shown for He single ionization [35]. In panel B, however, the measured average momentum of QFM photo-ions in light propagation direction is much closer to zero. This is even more apparent in panel C, where we projected the distribution shown in panel B onto the light propagation direction and determined the center through a Gaussian fit (solid blue line). Our experimental results speak in favor of the assumption that QFM photo-ions do not receive the photon momentum in the double-ionization process.

Further proof are the results of nondipole TDSE calculations for He double ionization at 800 eV photon energy, and linearly polarized light that are shown in Figure 3D. The local maximum of the nondipole curve (red line) that corresponds to QFM is exactly at zero, while the outer local maxima are shifted to the right by the magnitude of the photon momentum.

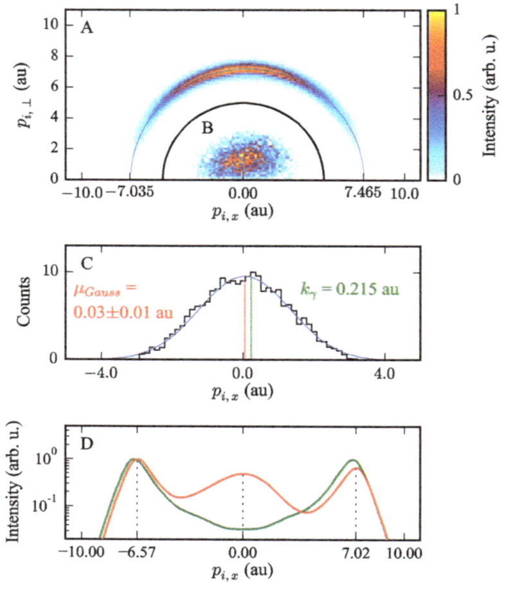

Figure 3. Photo-ion distributions of He double ionization by a single 800 eV photon. (**A**) The experimental data shown are limited to $\varepsilon < 0.005$ or $\varepsilon > 0.995$ and correspond to double ionization through the shake-off process. The blue semicircle is shifted to the right by the photon momentum. The SO photo-ions accumulate on this semicircle. (**B**) The experimental data shown are limited to $0.35 < \varepsilon < 0.65$ and $\vartheta_{12} > 160°$. They resemble double ionization via QFM. (**C**) Projection of data from B onto the x axis (black) and Gaussian fit (blue) to obtain the center of the momentum distribution of QFM photo-ions along the light propagation direction (red, μ_{Gauss}). Note that the indicated error is the standard deviation of the mean value from the Gaussian fit estimated as the square root of the respective diagonal element of the covariance matrix. The green line indicates the photon momentum for comparison. (**D**) Dipole (green) and nondipole (red) TDSE calculations for helium double ionization with 800 eV linearly polarized photons. The polarization vector is parallel to the z axis and the data shown are limited to $p_{i,y} = 0 \pm \frac{p_{\text{single}}}{2}$ & $p_{i,z} = 0 \pm \frac{p_{\text{single}}}{4}$. Experiment and theory suggest that the photon momentum is not imprinted onto QFM photo-ions.

4. Conclusions

We have investigated the quasifree mechanism of one-photon double ionization of He and H_2 at 800 eV photon energy and addressed two open questions concerning this intriguing process. We found the four-fold symmetry in the angular emission pattern of QFM electrons from H_2 double ionization that underlines the quadrupole nature of the process. Furthermore, we showed that the photon momentum is not transferred onto the photo-ion in quasifree photoionization, which is in contrast to single ionization in general and double ionization by means of electron–electron correlation.

Author Contributions: Conceptualization, M.S.S.; Experimental preapration, S.G., F.T., T.J. and M.S.S.; Experiment performed, S.G., F.T., K.F., N.S., A.P., L.K., M.K., T.J. and M.S.S.; Data analysis, S.G.; Validation, F.T., T.J., R.D. and M.S.S.; Supervision, T.J., R.D. and M.S.S.; Theory calculations, Y.-K.F. and L.-Y.P.; Writing—original draft preparation, S.G.; Writing—review and editing, S.G., F.T., Y.-K.F., K.F., N.S., A.P., L.K., M.K., L.-Y.P., T.J., R.D. and M.S.S.; Visualization, S.G.; Funding acquisition, T.J., R.D. and M.S.S. All authors have read and agreed to the published version of the manuscript.

Funding: This project was funded by DFG and BMBF.

Institutional Review Board Statement: Not applicable.

Informed Consent Statement: Not applicable.

Data Availability Statement: Not applicable.

Acknowledgments: The experimental team thank Miron Amusia for their continued encouragement of the experiments that are hunting for signals of QFM. We acknowledge DESY (Hamburg, Germany), a member of the Helmholtz Association HGF, for the provision of experimental facilities. Parts of this research were carried out at PETRA III and we would like to thank Jörn Seltmann and Kai Bagschik for excellent support during the beam time. We acknowledge support by DFG and BMBF.

Conflicts of Interest: The authors declare no conflicts of interest.

References

1. Amusia, M.Y.; Drukarev, E.G.; Gorshkov, V.G.; Kazachkov, M.O. Two-electron photoionization of helium. *J. Phys. B* **1975**, *8*, 1248. [CrossRef]
2. Amusia, M.Y.; Drukarev, E.G.; Liverts, E.Z. Small recoil momenta double ionization of He and two-electron ions by high energy photons. *Eur. Phys. J. D* **2020**, *74*, 173. [CrossRef]
3. Amusia, M.Y.; Drukarev, E.G.; Liverts, E.Z.; Mikhailov, A.I. Effects of small recoil momenta in one-photon two-electron ionization. *Phys. Rev. A* **2013**, *87*, 043423. [CrossRef]
4. Maulbetsch, F.; Briggs, J.S. Selection rules for transitions to two-electron continuum states. *J. Phys. B* **1995**, *28*, 551. [CrossRef]
5. Suric, T.; Drukarev, E.G.; Pratt, R.H. Characterization of high-energy photoionization in terms of the singularities of the atomic potential. I. Photoionization of the ground state of a two-electron atom. *Phys. Rev. A* **2003**, *67*, 022709. [CrossRef]
6. Amusia, M.Y.; Drukarev, E.G.; Mandelzweig, V.B. Quasifree Mechanism in Ionization Processes. *Phys. Scr.* **2005**, *72*, C22. [CrossRef]
7. Drukarev, E.G. Breakdown of asymptotics of the double photoionization of helium at the high-energy limit. *Phys. Rev. A* **1995**, *51*, R2684. [CrossRef]
8. Schöffler, M.S.; Stuck, C.; Waitz, M.; Trinter, F.; Jahnke, T.; Lenz, U.; Jones, M.; Belkacem, A.; Landers, A.L.; Pindzola, M.S.; et al. Ejection of Quasi-Free-Electron Pairs from the Helium-Atom Ground State by Single-Photon Absorption. *Phys. Rev. Lett.* **2013**, *111*, 013003. [CrossRef]
9. Spielberger, L.; Jagutzki, O.; Dörner, R.; Ullrich, J.; Meyer, U.; Mergel, V.; Unverzagt, M.; Damrau, M.; Vogt, T.; Ali, I.; et al. Separation of Photoabsorption and Compton Scattering Contributions to He Single and Double Ionization. *Phys. Rev. Lett.* **1995**, *74*, 4615. [CrossRef]
10. Spielberger, L.; Jagutzki, O.; Krässig, B.; Meyer, U.; Khayyat, K.; Mergel, V.; Tschentscher, T.; Buslaps, T.; Bräuning, H.; Dörner, R.; et al. Double and Single Ionization of Helium by 58-keV X Rays. *Phys. Rev. Lett.* **1996**, *76*, 4685. [CrossRef]
11. Kircher, M.; Trinter, F.; Grundmann, S.; Kastirke, G.; Weller, M.; Vela-Perez, I.; Khan, A.; Janke, C.; Waitz, M.; Zeller, S.; et al. Ion and Electron Momentum Distributions from Single and Double Ionization of Helium Induced by Compton Scattering. *Phys. Rev. Lett.* **2022**, *128*, 053001. [CrossRef] [PubMed]
12. Smeenk, C.T.L.; Arissian, L.; Zhou, B.; Mysyrowicz, A.; Villeneuve, D.M.; Staudte, A.; Corkum, P.B. Partitioning of the Linear Photon Momentum in Multiphoton Ionization. *Phys. Rev. Lett.* **2011**, *106*, 193002. [CrossRef] [PubMed]
13. Chelkowski, S.; Bandrauk, A.D.; Corkum, P.B. Photon Momentum Sharing between an Electron and an Ion in Photoionization: From One-Photon (Photoelectric Effect) to Multiphoton Absorption. *Phys. Rev. Lett.* **2014**, *113*, 263005. [CrossRef] [PubMed]
14. Hartung, A.; Eckart, S.; Brennecke, S.; Rist, J.; Trabert, D.; Fehre, K.; Richter, M.; Sann, H.; Zeller, S.; Henrichs, K.; et al. Magnetic fields alter strong-field ionization. *Nat. Phys.* **2019**, *15*, 1222–1226. [CrossRef]
15. Grundmann, S.; Trabert, D.; Fehre, K.; Strenger, N.; Pier, A.; Kaiser, L.; Kircher, M.; Weller, M.; Eckart, S.; Schmidt, L.P.H.; et al. Zeptosecond birth time delay in molecular photoionization. *Science* **2020**, *370*, 339–341. [CrossRef]
16. Hartung, A.; Brennecke, S.; Lin, K.; Trabert, D.; Fehre, K.; Rist, J.; Schöffler, M.S.; Jahnke, T.; Schmidt, L.P.H.; Kunitski, M.; et al. Electric Nondipole Effect in Strong-Field Ionization. *Phys. Rev. Lett.* **2021**, *126*, 053202. [CrossRef]
17. Lin, K.; Eckart, S.; Hartung, A.; Trabert, D.; Fehre, K.; Rist, J.; Schmidt, L.P.H.; Schöffler, M.S.; Jahnke, T.; Kunitski, M.; et al. Photoelectron energy peaks shift against the radiation pressure in strong-field ionization. *Sci. Adv.* **2022**, *8*, eabn7386. [CrossRef]
18. Grundmann, S.; Trinter, F.; Bray, A.W.; Eckart, S.; Rist, J.; Kastirke, G.; Metz, D.; Klumpp, S.; Viefhaus, J.; Schmidt, L.P.H.; et al. Separating Dipole and Quadrupole Contributions to Single-Photon Double Ionization. *Phys. Rev. Lett.* **2018**, *121*, 173003. [CrossRef]
19. Grundmann, S.; Serov, V.V.; Trinter, F.; Fehre, K.; Strenger, N.; Pier, A.; Kircher, M.; Trabert, D.; Weller, M.; Rist, J.; et al. Revealing the two-electron cusp in the ground states of He and H_2 via quasifree double photoionization. *Phys. Rev. Res.* **2020**, *2*, 033080. [CrossRef]
20. Viefhaus, J.; Scholz, F.; Deinert, S.; Glaser, L.; Ilchen, M.; Seltmann, J.; Walter, P.; Siewert, F. The Variable Polarization XUV Beamline P04 at PETRA III: Optics, mechanics and their performance. *Nucl. Instrum. Methods Phys. Res. A* **2013**, *710*, 151. [CrossRef]

21. Dörner, R.; Mergel, V.; Jagutzki, O.; Spielberger, L.; Ullrich, J.; Moshammer, R.; Schmidt-Böcking, H. Cold Target Recoil Ion Momentum Spectroscopy: A 'momentum microscope' to view atomic collision dynamics. *Phys. Rep.* **2000**, *330*, 95–192. [CrossRef]
22. Ullrich, J.; Moshammer, R.; Dorn, A.; Dörner, R.; Schmidt, L.P.H.; Schmidt-Böcking, H. Recoil-ion and electron momentum spectroscopy: Reaction-microscopes. *Rep. Prog. Phys.* **2003**, *66*, 1463. [CrossRef]
23. Jahnke, T.; Foucar, L.; Titze, J.; Wallauer, R.; Osipov, T.; Benis, E.P.; Alnaser, A.; Jagutzki, O.; Arnold, W.; Semenov, S.K.; et al. Vibrationally Resolved K-shell Photoionization of CO with Circularly Polarized Light. *Phys. Rev. Lett.* **2004**, *93*, 083002. [CrossRef]
24. Jagutzki, O.; Mergel, V.; Ullmann-Pfleger, K.; Spielberger, L.; Spillmann, U.; Dörner, R.; Schmidt-Böcking, H. A broad-application microchannel-plate detector system for advanced particle or photon detection tasks: Large area imaging, precise multi-hit timing information and high detection rate. *Nucl. Instrum. Methods Phys. Res. A* **2002**, *477*, 244–249. [CrossRef]
25. Jagutzki, O.; Lapington, J.S.; Worth, L.B.C.; Spillman, U.; Mergel, V.; Schmidt-Böcking, H. Position sensitive anodes for MCP read-out using induced charge measurement. *Nucl. Instrum. Methods Phys. Res. A* **2002**, *477*, 256–261. [CrossRef]
26. Zhang, Z.; Peng, L.-Y.; Xu, M.-H.; Starace, A.F.; Morishita, T.; Gong, Q. Two-photon double ionization of helium: Evolution of the joint angular distribution with photon energy and two-electron energy sharing. *Phys. Rev. A* **2011**, *84*, 043409. [CrossRef]
27. Jiang, W.-C.; Peng, L.-Y.; Xiong, W.-H.; Gong, Q. Comparison study of electron correlation in one-photon and two-photon double ionization of helium. *Phys. Rev. A* **2013**, *88*, 023410. [CrossRef]
28. Jiang, W.-C.; Shan, J.-Y.; Gong, Q.; Peng, L.-Y. Virtual Sequential Picture for Nonsequential Two-Photon Double Ionization of Helium. *Phys. Rev. Lett.* **2015**, *115*, 153002. [CrossRef]
29. Chen, S.-G.; Jiang, W.-C.; Grundmann, S.; Trinter, F.; Schöffler, M.S.; Jahnke, T.; Dörner, R.; Liang, H.; Wang, M.-X.; Peng, L.-Y.; et al. Photon Momentum Transfer in Single-Photon Double Ionization of Helium. *Phys. Rev. Lett.* **2020**, *124*, 043201. [CrossRef]
30. Weber, T.; Czasch, A.; Jagutzki, O.; Müller, A.; Mergel, V.; Kheifets, A.; Feagin, J.; Rotenberg, E.; Meigs, G.; Prior, M.H.; et al. Fully Differential Cross Sections for Photo-Double-Ionization of D_2. *Phys. Rev. Lett.* **2004**, *92*, 163001. [CrossRef]
31. Auger, P. Etude expérimentale des directions d'émission des photoélectrons. *J. Phys. Radium* **1927**, *8*, 85–92. [CrossRef]
32. Sommerfeld, A.; Schur, G. Über den Photoeffekt in der K-Schale der Atome, insbesondere über die Voreilung der Photoelektronen. *Ann. Phys.* **1930**, *396*, 409–432. [CrossRef]
33. Briggs, J.S.; Schmidt, V. Differential cross sections for photo-double-ionization of the helium atom. *J. Phys. B* **2000**, *33*, R1. [CrossRef]
34. Kheifets, A. On different mechanisms of the two-electron atomic photoionization. *J. Phys. B* **2001**, *34*, L247. [CrossRef]
35. Grundmann, S.; Kircher, M.; Vela-Perez, I.; Nalin, G.; Trabert, D.; Anders, N.; Melzer, N.; Rist, J.; Pier, A.; Strenger, N.; et al. Observation of Photoion Backward Emission in Photoionization of He and N_2. *Phys. Rev. Lett.* **2020**, *124*, 233201. [CrossRef]

atoms

MDPI

Article

On Producing Long-Lived Spin Polarized Metastable Atoms—Feasibility of Storing Electric Energy †

Horst Schmidt-Böcking * and Gernot Gruber

Institut für Kernphysik, Universität Frankfurt, Max von Laue Str. 1, 60438 Frankfurt, Germany
* Correspondence: schmidtb@atom.uni-frankfurt.de
† Dedicated to Miron Amusia.

Abstract: We describe a method of producing long-lived multiply excited spin polarized atoms or ions, the decay of which is strongly delayed or even blocked by intra-ionic magnetic stabilization. Special configurations with huge internal magnetic fields capture only spin polarized electrons in collisions with spin aligned atomic hydrogen gas targets. It is expected that the spin aligned configuration yields an extremely high internal magnetic field which will effectively block spin flip transitions. By this the lifetime of inner shell vacancies is expected to strongly increase.

Keywords: spin-polarized atoms; highly charged ions; energy storage

Citation: Schmidt-Böcking, H.; Gruber, G. On Producing Long-Lived Spin Polarized Metastable Atoms—Feasibility of Storing Electric Energy. *Atoms* **2022**, *10*, 76. https://doi.org/10.3390/atoms10030076

Academic Editor: Yew Kam Ho

Received: 12 June 2022
Accepted: 15 July 2022
Published: 18 July 2022

Over thirty years ago, Miron Amusia was a frequent visitor to our institute at Frankfurt University. The actual reason for this was that at the time his granddaughter was staying in nearby Wiesbaden. Miron made use of the opportunity of seeing her to visit us at our Frankfurt institute and discuss interesting physics problems. One of them was ball lightning [1]. Miron made mention of research that was going on before 1990 in the then Soviet Union on the question of storing energy in light ball matter. Therefore, the question of the nature of light ball matter was intensively discussed.

The onset was somewhat similar to this: ball lightning occurs in thunderstorms in normal air. Characteristic for it is that it is a plasma-like object. Its size is about that of a ball of 50 cm in diameter, it emits light for about 30 s after creation, and it moves rotating (Figure 1). To account for the 30 s half-life of light emission, the emitting states in the corresponding atoms or ions would have to be metastable. The rotation of the light ball means that its angular momentum probably is the sum of the angular momenta of space quantized atoms, similar to what occurs in the Einstein-de Hass effect [2]. Since air consists mainly of N_2, we were speculating that a special metastable spin polarized configuration of N_2^* or N^* produced in a lightening event might be causing the ball-lightning effect. An N^* configuration could possibly be ($1s^2$, $2s^1$, $2p^3$, $3s^1$), where a 2s electron is excited into the 3s shell by the strongly varying magnetic field in the lightning flash. In the process the excited electron makes a spin flip. Thus, all the electrons in the 2s, 2p, and 3s shells have the same spin orientation (Figure 2). The $1s^2$ remains a closed shell and the electrons in 3s can only decay via spin flip into the 2s state, and the transition is strongly delayed.

Excited atoms and ions with inner shell vacancies have been explored in slow ion-atom collisions. Many groups investigated their formation and decay [3–8]. It is well known that inner shell vacancies of multiply excited hollow atoms or ions decay mostly within fractions of a picosecond by Auger or X-ray emission, when the electrons undergo transitions without spin-flip. In some experiments metastable configurations have been observed with a delayed decay in the nanosecond regime. In nature and also under laboratory conditions so far, no evidence has been found that long-lived (millisecond or longer) metastable multiply excited ionic or atomic systems are created in such collisions. To our knowledge, in all these experiments the spin configuration was not known. Only in some instances of highly ionized few-electron ions the spin configuration was known. Here only one electron is excited, and its decay is highly forbidden, i.e., the state can only decay by magnetic dipole or two-photon transition or spin-flip. Such ionic configurations have lifetimes of the order of microseconds or even longer. Ninomiya et al. [9] found evidence for longer living "metastable hollow atoms" that were formed by penetration of highly charged ions through very narrow channels in thin films. Therefore, to create metastable configurations with multiple vacancies and with long lifetimes one must be able to create ionic configurations in a controlled way where the spin orientation of the electrons can be manipulated.

Figure 1. A 1901 illustration of ball lightning (Ball lightning—Wikipedia).

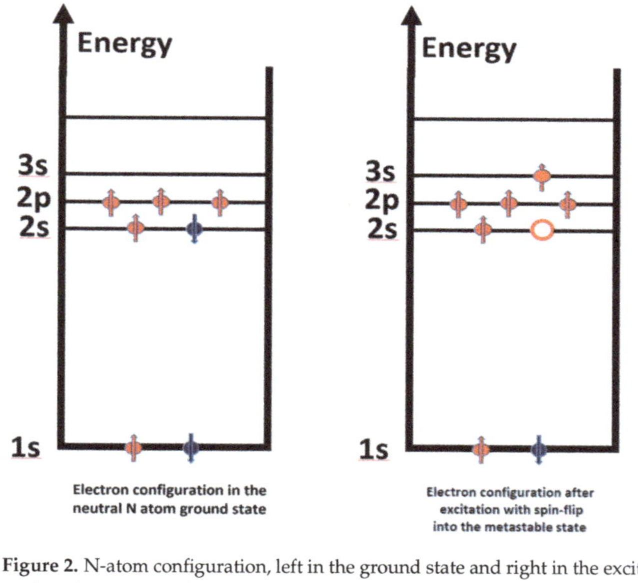

Figure 2. N-atom configuration, left in the ground state and right in the excited state. "Red" electrons are in spin-up state, blue ones in spin-down state. The open circle indicates the vacancy state.

If in a multiple-capture process all captured electrons have parallel spins, the energy barrier for a spin flip should strongly increase. In this case it might be possible to block or delay spin flip transitions due to the very strong internal magnetic field. The multiple-capture process, however, must be fast enough (in the order of a microsecond) that, until the formation of the very high internal magnetic field, the highly excited ion has no time to decay. Due to such a strong internal magnetic field and the blocking of spin-flip transitions caused by it, even in special neutral atoms the lifetime of inner-shell vacancies might be strongly enhanced and may last up to milliseconds or even longer. For such special ionic configurations, an internal magnetic field strength much stronger than achievable up to now in laboratories ($\geq$90 Tesla [10]) may exist.

For obtaining qualitative information on the lifetime of such metastable subjects we contacted experts in calculating the configuration of multiply excited ions or atoms [11]. However, their theoretical approach was not capable of calculating transitions in extremely strong inner atomic magnetic fields. They also were not able to calculate the magnetic field-strength, e.g., for an S = 9 configuration. They told us, however, that no spin-flip should occur when the magnetic field is infinitely large. The theorists thought it is not likely that such configurations would have a lifetime sufficiently long for applications in everyday life, e.g., for efficiently storing energy by use of long living inner shell vacancies in so-called spin-polarized matter. However, nobody excluded that it might be possible nevertheless to produce such matter.

Therefore, this short letter is also written to stimulate the interest of theorists in such spin-polarized configurations. Since to our knowledge no data on the lifetime of such spin-polarized metastable states and the strength of the inner-atom magnetic field is available in literature, we will only outline an experimental route of how to produce such atoms with very high S value in the laboratory. It may be noted in passing that the here presented scheme of producing spin-polarized matter has been patented by the German Patent Office [12] (Figure 3).

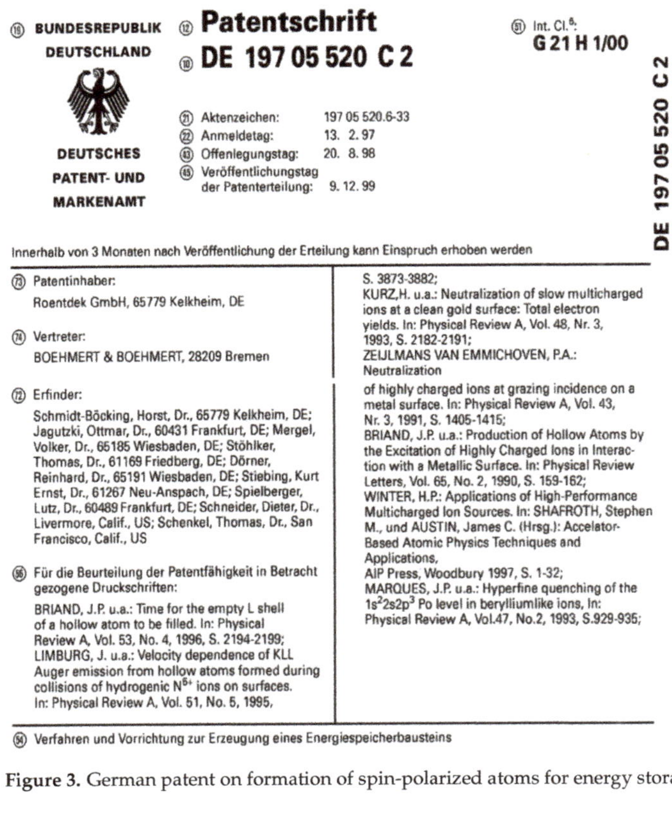

(19) BUNDESREPUBLIK DEUTSCHLAND

DEUTSCHES PATENT- UND MARKENAMT

(12) **Patentschrift**

(10) **DE 197 05 520 C 2**

(51) Int. Cl.⁶:
G 21 H 1/00

(21) Aktenzeichen: 197 05 520.6-33
(22) Anmeldetag: 13. 2. 97
(43) Offenlegungstag: 20. 8. 98
(45) Veröffentlichungstag der Patenterteilung: 9. 12. 99

DE 197 05 520 C 2

Innerhalb von 3 Monaten nach Veröffentlichung der Erteilung kann Einspruch erhoben werden

(73) Patentinhaber:
Roentdek GmbH, 65779 Kelkheim, DE

(74) Vertreter:
BOEHMERT & BOEHMERT, 28209 Bremen

(72) Erfinder:
Schmidt-Böcking, Horst, Dr., 65779 Kelkheim, DE; Jagutzki, Ottmar, Dr., 60431 Frankfurt, DE; Mergel, Volker, Dr., 65185 Wiesbaden, DE; Stöhlker, Thomas, Dr., 61169 Friedberg, DE; Dörner, Reinhard, Dr., 65191 Wiesbaden, DE; Stiebing, Kurt Ernst, Dr., 61267 Neu-Anspach, DE; Spielberger, Lutz, Dr., 60489 Frankfurt, DE; Schneider, Dieter, Dr., Livermore, Calif., US; Schenkel, Thomas, Dr., San Francisco, Calif., US

(56) Für die Beurteilung der Patentfähigkeit in Betracht gezogene Druckschriften:
BRIAND, J.P. u.a.: Time for the empty L shell of a hollow atom to be filled. In: Physical Review A, Vol. 53, No. 4, 1996, S. 2194-2199;
LIMBURG, J. u.a.: Velocity dependence of KLL Auger emission from hollow atoms formed during collisions of hydrogenic N^{6+} ions on surfaces. In: Physical Review A, Vol. 51, No. 5, 1995, S. 3873-3882;
KURZ,H. u.a.: Neutralization of slow multicharged ions at a clean gold surface: Total electron yields. In: Physical Review A, Vol. 48, Nr. 3, 1993, S. 2182-2191;
ZEIJLMANS VAN EMMICHOVEN, P.A.: Neutralization of highly charged ions at grazing incidence on a metal surface. In: Physical Review A, Vol. 43, Nr. 3, 1991, S. 1405-1415;
BRIAND, J.P. u.a.: Production of Hollow Atoms by the Excitation of Highly Charged Ions in Interaction with a Metallic Surface. In: Physical Review Letters, Vol. 65, No. 2, 1990, S. 159-162;
WINTER, H.P.: Applications of High-Performance Multicharged Ion Sources. In: SHAFROTH, Stephen M., und AUSTIN, James C. (Hrsg.): Accelerator-Based Atomic Physics Techniques and Applications, AIP Press, Woodbury 1997, S. 1-32;
MARQUES, J.P. u.a.: Hyperfine quenching of the $1s^22s2p^3$ Po level in berylliumlike ions, In: Physical Review A, Vol.47, No.2, 1993, S.929-935;

(54) Verfahren und Vorrichtung zur Erzeugung eines Energiespeicherbausteins

Figure 3. German patent on formation of spin-polarized atoms for energy storage [12].

Figure 4 schematically shows the experimental arrangement for producing spin-polarized neutral-atom beams. The apparatus consists of three experimental sections: highly charged ions in a special charge state are produced in an Electron Cyclotron Resonance ion source (ECR) [13]. With modern ion source techniques nearly any kind of ions in any charge state can be produced. Present ECR ion sources or EBIS or EBIT devices can produce high intensity low energy beams of highly charged ions.

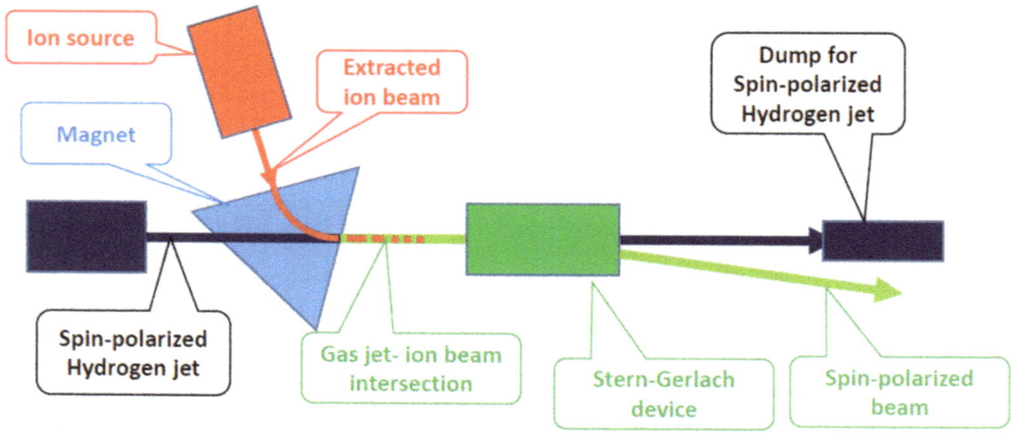

Figure 4. Scheme of experimental set-up to create spin-polarized neutral atoms (see text).

The magnetically selected beam penetrates a nearly completely spin-polarized hydrogen gas-jet and thereby captures within a few microseconds many spin-polarized electrons for getting neutralized. To our knowledge spin-polarized H targets (supersonic jets) with an 80% degree of polarization are extant [14]. Since this kind of target provides a very high target thickness, we believe that the degree of polarization can be enhanced in a second Stern-Gerlach device, with loss of target thickness, however. The non-neutralized beam components are magnetically deflected behind the gas target in the Stern-Gerlach apparatus and removed. The neutralized beam is spin-state selected in the same Stern-Gerlach device [15,16]. In this way a neutral spin-polarized atomic beam is generated: thus, a larger fraction of the highly spin-polarized atoms have possibly the same spin-polarized configuration.

As an example, we consider the production of spin-polarized Kr atoms, starting with the extraction of Kr^{13+} ions from an ECR ion source, having 23 electrons in their ionic shells. This ion charge state is chosen because it has a high internal magnetic field. This is due to the ionic electron-configuration being such that all lower shells ($1s^2$, $2s^2$, $2p^6$, $3s^2$, $3p^6$) are filled and the 3d shell is half filled, $3d^5$. According to Hund's rule [17,18], in this configuration the Kr^{13+} ions should have a total L value of zero, the total S value should be maximum, S = 2.5.

The selected ion beam is injected in a spin-polarized hydrogen-gas target-jet (Figure 4). The intersection of the ion beam and the hydrogen jet is arranged thus that both the ion beam and the atoms in the jet move in parallel (for about 10 cm), improving by their larger intersection length the capture probability. After capturing another 13 spin-polarized electrons, a measurable fraction of the neutral Kr atoms should be in the S state of S = 9.

The neutralized Kr beam is injected in another Stern-Gerlach device which separates the Kr atoms according to their S value and to the orientation of the total spin. Out of the second Stern-Gerlach device several Kr beams emerge, being separated according to their spin polarization.

It may even be possible to produce in such a set-up neutral uranium being in a S = 37 state. One would start with extracting U^{69+} ions from an ion source (ground state configuration: $1s^2$, $2s^2$, $2p^6$, $3s^2$, $3p^6$, $3d^5$) (Figure 5). When these ions capture 69 spin-oriented electrons in a spin-polarized hydrogen-jet all higher states (higher than 3d) can be filled with electrons of the same spin direction in the deepest allowed states. In the most favorable case one might obtain even a ($1s^2$, $2s^2$, $2p^6$, $3s^2$, $3p^6$, $4s^1$, 4p, $4d^5$, $4f^7$, $5s^1$, $5p^3$, $5d^5$, $5f^7$, $5g^9$, $6s^1$, $6p^3$, $6d^5$, $6f^7$, $7s^1$, $7p^3$, $7d^5$, $8s^1$, $8p^2$) configuration. The S value of this configuration would be S = 37.

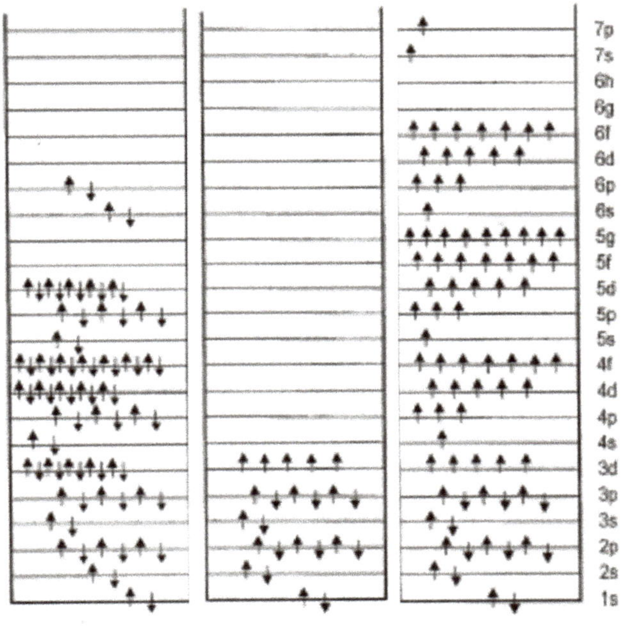

Figure 5. Electron shell configurations. Left, for the neutral uranium atom, middle, for the uranium 69+ ion, right, for a neutral uranium atom after capture of 69 spin-orientated electrons [19] (see text).

Extremely magnetic atoms of this kind may show many new quantum features. The magnetic field might be so strong that, for instance, in Kr atomic bonding proceeds via magnetic forces. This would possibly provide matter with unexpected e-e- correlation features. In addition, the amount of energy stored in the empty states would be huge, because the states of different spin polarization in the atom are empty. For a single uranium atom, it might exceed 30 keV. This is 10,000 times more than a conventional electric battery can store. Whether this would be practically applicable is very unlikely, but who may know?

To conclude: the procedure outlined above of how to produce spin-polarized atoms in an unusual high spin state seems rather exotic. However, we believe that it should be possible, as discussed, to produce sufficiently many atoms of this kind for exploring the physics of such exotic spin-polarized quantum systems in the laboratory. Spin-polarized matter can open up new interesting fields in atomic and solid-state physics.

Without the stimulating discussions with Miron Amusia in Frankfurt more than 30 years ago, we would never have come in contact with the question of what ball- lightning matter might be. We have written this short letter to present the ideas Miron initiated. We think we owe to Miron Amusia the publication of these ideas. When Max Born tried to prevent Otto Stern from performing the Stern-Gerlach experiment because it seemed infeasible to him, Otto Stern replied: no experiment is as dumb as not to perform it.

Author Contributions: All authors have read and agreed to the published version of the manuscript.

Funding: This research received no external funding.

Conflicts of Interest: The authors declare no conflict of interest.

References

1. Wikipedia: Rätselhaftes Gewitterphänomen und Mythos Kugelblitz. Available online: http://en.wikipedia.org/wiki/User:Binadot (accessed on 11 June 2022).
2. Einstein, A.; de Haas, W.J. Experimenteller Nachweis der Ampereschen Molekularströme. *Verhdlgn. DPG* **1915**, *17*, 152.
3. Donets, E.D. The Electron Beam Method of Production of Highly Charged Ions and Its Applications. *Phys. Scr.* **1983**, *T3*, 11. Available online: https://iopscience.iop.org/article/10.1088/0031-8949/1983/T3/003/meta (accessed on 12 June 2022). [CrossRef]
4. Briand, J.P.; De Billy, L.; Charles, P.; Essabaa, S.; Briand, P.; Geller, R.; Desclaux, J.P.; Bliman, S.; Ristori, C. Production of hollow atoms by the excitation of highly charged ions in interactio with a metallic surface. *Phys. Rev. Lett.* **1990**, *65*, 159. Available online: https://journals.aps.org/prl/abstract/10.1103/PhysRevLett.65.159 (accessed on 12 June 2022). [CrossRef] [PubMed]
5. Winter, H. Image charge acceleration of multi-charged argon ions in grazing collisions with an aluminum surface. *EPL Europhys. Lett.* **1992**, *18*, 207. Available online: https://iopscience.iop.org/article/10.1209/0295-5075/18/3/004/meta (accessed on 12 June 2022). [CrossRef]
6. Kurz, H.; Töglhofer, K.; Winter, H.P.; Aumayr, F.; Mann, R. Electron emission from slow hollow atoms at a clean metal surface. *Phys. Rev. Lett.* **1992**, *69*, 1140. Available online: https://journals.aps.org/prl/abstract/10.1103/PhysRevLett.69.1140 (accessed on 12 June 2022). [CrossRef] [PubMed]
7. Das, J.; Morgenstern, R. Measurement of autoionization rates for electron emission above a metal surface. *Phys. Rev. A* **1993**, *47*, R755. Available online: https://journals.aps.org/pra/abstract/10.1103/PhysRevA.47.R755 (accessed on 12 June 2022). [CrossRef] [PubMed]
8. Grether, M.; Spieler, A.; Köhrbrück, R.; Stolterfoht, N. Dynamic K- and L-shell filling of Ne^{9+} projectiles interacting with an Al (111) surface. *Phys. Rev. A* **1995**, *52*, 426. Available online: https://journals.aps.org/pra/abstract/10.1103/PhysRevA.52.426 (accessed on 12 June 2022). [CrossRef] [PubMed]
9. Ninomiya, S.; Yamazaki, Y.; Koike, F.; Masuda, H.; Azuma, T.; Komaki, K.; Kuroki, K.; Sekiguchi, M. Stabilized hollow ions extracted in vacuum. *Phys. Rev. Lett.* **1997**, *78*, 4557. [CrossRef]
10. Research with Highest Electromagnetic Fields—Helmholtz-Zentrum Dresden-Rossendorf, HZDR. Available online: https://www.hzdr.de/db/Topics?pld=60&pNid=0 (accessed on 11 June 2022).
11. Surzhykov, A. (Physikalisch Technische Bundesanstalt, 38116 Braunschweig, Germany); Fritzsche, S. (University of Jena, Jena, Germany). Personal communication, 2012.
12. Schmidt-Böcking, H.; Jagutzki, O.; Mergel, V.; Stöhlker, T.; Dörner, R.; Stiebing, K.E.; Spielberger, L.; Schneider, D. Verfahren und Vorrichtung zur Herstellung eines Energiespeichers. Deutsches Patent 1999 DE 197 05 520 C 2. Available online: https://patents.google.com/patent/DE19705520C2/de (accessed on 12 June 2022).
13. Geller, R. Electron cyclotron resonance sources: Historical review and future prospects. *Rev. Sci. Instrum.* **1998**, *69*, 1302–1310. [CrossRef]
14. Zapfe, K.; Braun, B.; Gaul, H.G.; Grieser, M.; Povh, B.; Rall, M.; Steffens, E.; Stock, F.; Haeberli, W. High density polarized hydrogen gas target for storage rings. *Rev. Sci. Instrum.* **1995**, *66*, 28–31. Available online: https://aip.scitation.org/doi/abs/10.1063/1.1146388 (accessed on 12 June 2022). [CrossRef]
15. Gerlach, W.; Stern, O. Der experimentelle nachweis der richtungsquantelung im magnetfeld. *Z. Phys.* **1922**, *9*, 349–352. Available online: https://link.springer.com/article/10.1007/BF01326983 (accessed on 12 June 2022). [CrossRef]
16. Gerlach, W.; Stern, O. über die richtungsquantelung im magnetfeld. *Ann. Phys.* **1924**, *74*, 673. Available online: https://onlinelibrary.wiley.com/doi/abs/10.1002/andp.19243791602 (accessed on 12 June 2022). [CrossRef]
17. Hund, F. Zur Deutung der Molekelspektren, I. *Z. Phys.* **1927**, *40*, 742–764. [CrossRef]
18. Hund, F. Zur Deutung der Molekelspektren, II. *Z. Phys.* **1927**, *42*, 93–120. [CrossRef]
19. Schmidt-Böcking, H.; Jagutzki, O.; Mergel, V.; Stöhlker, T.; Dörner, R.; Stiebing, K.E.; Spielberger, L.; Schneider, D. On the Formation of "quasi stable" Hollow Atoms towards a High Power Electric Energy Storage. In Proceedings of the 5th International Symposium on Ball Lightnings (ISBL), Niigata, Japan, 26–29 August 1997; p. 133.

atoms

MDPI

Article

Electron Correlations in Sequential Two-Photon Double Ionization of an Ar Atom

Boris M. Lagutin [1], Ivan D. Petrov [1], Victor L. Sukhorukov [2,3,*], Victor A. Kilin [4], Nikolay M. Novikovskiy [2,3], Philipp V. Demekhin [3] and Arno Ehresmann [3]

1 Rostov State Transport University, 344038 Rostov-on-Don, Russia
2 Institute of Physics, Southern Federal University, 344090 Rostov-on-Don, Russia
3 Institute of Physics, Center for Interdisciplinary Nanostructure Science and Technology (CINSaT), University of Kassel, 34132 Kassel, Germany
4 Tomsk Polytechnic University, 634050 Tomsk, Russia
* Correspondence: vlsu16@mail.ru

Abstract: Sequential two-photon ionization is a process that is experimentally accessible due to the use of new free-electron laser sources for excitation. For the prototypical rare Ar gas atoms, a photoelectron spectrum (PES) corresponding to the second step of the sequential two-photon double ionization ($2PDI^{II}$) at a photon excitation energy of 65.3 eV was studied theoretically with a focus on the consequences of electron correlations in the considered process. The calculation predicts many intense lines at low photoelectron energies, which cannot be explained on the basis of a one-electron approximation. The processes that lead to the appearance of these lines include many-electron correlations, either in the first or second step of photoionization. A significant fraction of the intensity of the low-energy part of PES is associated with the Auger decay of the excited states formed at the second step of 2PDI. The shape of the low-energy part of the $2PDI^{II}$ PES is expected to be dependent on both the energy of photon excitations and the flux of the exciting beam.

Keywords: sequential photoionization; double ionization; photoelectron spectra; argon; many-electron correlations; free-electron laser

Citation: Lagutin, B.M.; Petrov, I.D.; Sukhorukov, V.L.; Kilin, V.A.; Novikovskiy, N.M.; Demekhin, P.V.; Ehresmann, A. Electron Correlations in Sequential Two-Photon Double Ionization of an Ar Atom. *Atoms* **2022**, *10*, 139. https://doi.org/10.3390/atoms10040139

Academic Editors: Anatoli Kheifets, Gleb Gribakin and Vadim Ivanov

Received: 14 October 2022
Accepted: 10 November 2022
Published: 13 November 2022

1. Introduction

The first studies of multiphoton photoionization of atoms by free electron lasers (FELs) focused on the charge distributions of ions (see [1–3] and a brief review [4]). Subsequently, it became possible to study photoelectron spectra (PES) due to the multiphoton ionization of atoms with energy and angular resolution [5,6] (see also the review devoted to studies of double- and triple-sequential ionization of atoms [7]).

If the photon excitation energy of an intense photon beam interacting with individual atoms exceeds the ionization threshold of an atom, then the most probable process among the different mechanisms of multiphoton ionization is sequential photoionization [6–8]. The theory to describe sequential two-photon double ionization (2PDI) [9,10] has, in the past, been applied to an experiment performed for Ar when the impinging-photon energy was less than the 3s ionization threshold [4,11]. In this case, the interpretation of the experimental data can be limited to considering the main $3p \dashrightarrow \varepsilon\ell$ channel only, i.e., considering the variety of $3s^2 3p^6 \dashrightarrow 3s^2 3p^4 \varepsilon\ell\varepsilon'\ell'$ channels. At larger photon excitation energies, ω, e.g., when ω exceeds the 3s ionization threshold, it is expected that the influences of many electron correlations will dramatically increase, considerably complicating the interpretation of an experiment. This conjecture is based on past findings, where considerable correlative contributions were identified for conventional single-photon ionization of Ar at these photon excitation energies, namely inter-shell correlations [12–15] and the dipole polarization of electron shells [16].

The main goal of the present paper is, therefore, to study the influences of the most significant single-electron and double-electron excitations on the 2PDI of Ar. We calculated the 2PDI PES of Ar at a photon excitation energy of $\omega = 65.3$ eV for which an experiment was planned in the near future. We limited ourselves to computing the 2PDI^{II} PES obtained for the second step of the sequential two-photon double ionization, which can be extracted from the total signal using its quadratic flux dependence [17].

2. Two-Photon Processes Involving Valence and Sub-Valence Electrons

Possible processes resulting in lines in the 2PDI^{II} PES of Ar predicted for a photon excitation energy of 65.3 eV are presented in Figures 1 and 2.

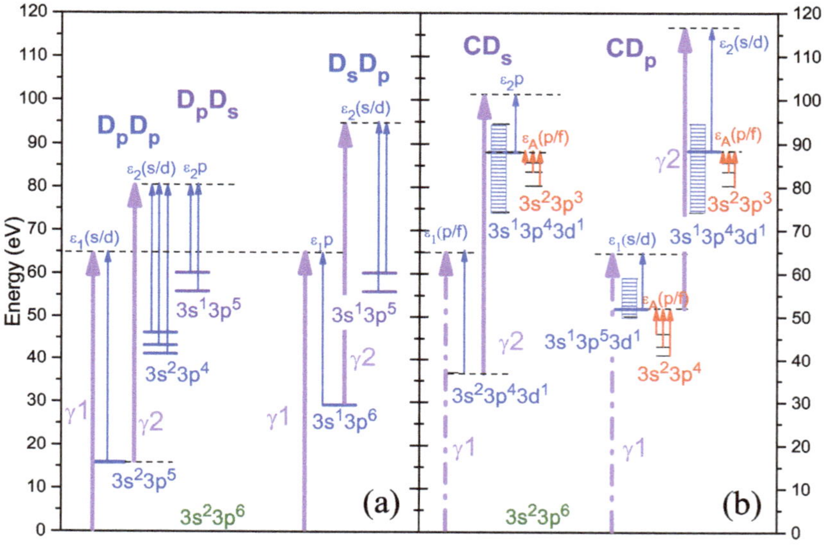

Figure 1. Energy scheme of the sequential two-photon double ionization of argon. (**a**) Direct one-electron ionization at each step. (**b**) Correlational processes relevant to the first ionization step. Designations: bold violet arrows—energies of the exciting photons: solid arrows—direct (*D*) photoionization processes and dash-dotted arrows—correlational (*C*) processes in the photoionization. Thin arrows—kinetic energy of the emitted photoelectrons (blue) and Auger electrons (red). Vertical rectangles represent a set of energy levels of the proper configuration allowed according to the selection rules. Energy levels were computed in the HF approach with DPES. The heights of the rectangles were estimated using the "transition array" technique [18,19].

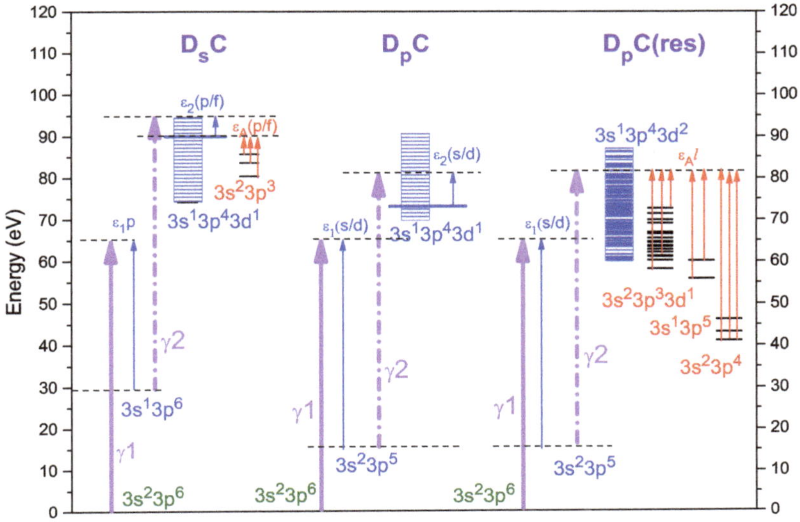

Figure 2. Energy schemes of the sequential two-photon double ionization of argon with a correlational process at the second ionization step. Designations are the same as in Figure 1.

2.1. Direct Processes

In the following, we will call 2PDI^{II} processes, where one electron from a certain shell is removed by a first photon and a second electron is removed from the shell of the ion by a second photon; "direct" processes are abbreviated by the symbol D in the depicted graphs; 2PDI^{II} process, removing two 3p electrons from Ar (D_pD_p), is shown in panel (a), column D_pD_p:

$$3s^2 3p^6 + \gamma_1 \dashrightarrow 3s^2 3p^5 + \varepsilon_1 \ell (\ell = 0, 2); \tag{1}$$

$$3s^2 3p^5 + \gamma_2 \dashrightarrow 3s^2 3p^4 + \varepsilon_2 \ell (\ell = 0, 2). \tag{2}$$

This process was studied in detail both experimentally and theoretically [4,7,11,17,20,21]. It results in three 2PDI^{II} PES lines with the kinetic energy of electrons in the region from 33 to 40 eV. This part of the spectrum is the most intense.

In column D_pD_s of Figure 1a another two-step process is shown in which the first step is as process (1) above, but where in the second step, a 3s electron is removed from the ion

$$3s^2 3p^5 + \gamma_2 \dashrightarrow 3s^1 3p^5 + \varepsilon_2 p \tag{3}$$

In the final stage of the process (3), the strong dipole polarization of the electron shells (DPES) [16] was taken into account. This effect is described by $3p3p - 3sn\ell(\ell = 0, 2)$ excitations in the current approach. As a result, the two lowest levels shown in column D_pD_s of Figure 1a mainly contains the $3s^1 3p^5 (^3P)$ and $3s^1 3p^5 (^1P)$ basis states. Higher levels (not shown in Figure 1a) mainly contain $3s^2 3p^3 n\ell(\ell = 0, 2)(LS)$ basis states and result in low-energy satellite lines (sat).

In the 2PDI^{II} process, a 3s electron is removed from the 3s shell of the atom, and in the second step, a 3p electron is removed from the singly charged ion

$$3s^2 3p^6 + \gamma_1 \dashrightarrow 3s^1 3p^6 + \varepsilon_1 p; \tag{4}$$

$$3s^1 3p^6 + \gamma_2 \dashrightarrow 3s^1 3p^5 + \varepsilon_2 \ell (\ell = 0, 2); \tag{5}$$

is depicted in column D_sD_p of Figure 1a. Processes (4) and (5) result in photoelectron lines for electrons with kinetic energies larger than 30 eV.

2.2. Correlational Processes

The direct (D) processes considered above are transitions that are described by one-electron electric dipole interaction. The other atomic electrons are spectators in the transitions. Below, we consider correlational (C) processes influencing both the first and second steps of the sequential two-photon double ionization.

The D_sC column of Figure 2 (left column) shows a two-step process that leads to the appearance of 2ph(II) lines with electron energy in the range from 0 to 20 eV. The first step of this process is the direct (D) transition (4). Note that in this transition, the important inter-shell correlation between the $3s^13p^6\varepsilon_1 p$ and $3s^13p^5\varepsilon_2 d$ channels [14] in the final state was taken into account. The second step is possible only due to electron correlations (C) either in the initial or the final state, since the direct single-electron transition

$$3s^13p^6 + \gamma_2 \dashrightarrow 3s^13p^43d^1 + \varepsilon_2\ell(\ell = 1,3) \tag{6}$$

is forbidden due to the selection rule.

Possible correlational transitions of the second step of the two-photon D_sC processes in the lowest order of the Coulomb interaction are the following:

$$3s^13p^6 \longrightarrow 3s^13p^43d^1\varepsilon'\ell'(\ell' = 0,2,4) \dashrightarrow 3s^13p^43d^1\varepsilon_2\ell(\ell = 1,3) \tag{7}$$
$$3s^13p^6 \longrightarrow 3s^23p^43d^1 \dashrightarrow 3s^13p^43d^1\varepsilon_2 p \tag{8}$$
$$3s^13p^6 \longrightarrow 3s^13p^4\varepsilon'\ell'(\ell' = 1,3)\varepsilon_2\ell \dashrightarrow 3s^13p^43d^1\varepsilon_2\ell(\ell = 1,3) \tag{9}$$
$$3s^13p^6 \longrightarrow 3s^03p^53d^1\varepsilon_2\ell(\ell = 1,3) \dashrightarrow 3s^13p^43d^1\varepsilon_2\ell(\ell = 1,3) \tag{10}$$
$$3s^13p^6 \dashrightarrow 3s^13p^5\varepsilon'\ell'(\ell' = 0,2) \longrightarrow 3s^13p^43d^1\varepsilon_2\ell(\ell = 1,3) \tag{11}$$
$$3s^13p^6 \dashrightarrow 3s^03p^6\varepsilon_2 p \longrightarrow 3s^13p^43d^1\varepsilon_2 p \tag{12}$$
$$3s^13p^6 \dashrightarrow 3s^13p^53d^1 \longrightarrow 3s^13p^43d^1\varepsilon_2\ell(\ell = 1,3) \tag{13}$$

In all Equations (7)–(13), and following, the dashed arrows denote the electric dipole interaction and the solid arrows denote the Coulomb interaction.

We estimated the amplitudes (7)–(13) using the "transition array" technique [18,19], which was created to solve astrophysical problems in order to avoid cumbersome "level-to-level" calculations in atoms with a large number of open shells. In the case of the problem solved in the present paper, applying the "transition array" technique required calculating discrete and continuum atomic orbitals (AOs) for channels with orbital numbers $\ell = 0 - 4$, in addition to the core AOs. Discrete AOs with an average radius of up to 20 a.u. and continuum AOs with energy $\varepsilon = 800$ Ry were computed in the frozen Hartree–Fock (HF) field of Ar^{2+}. In this configuration space, all single- and double-electron excitations in the initial and final states were taken into account, contributing to the 2PDI process under consideration. Among hundreds of channels contributing to the 2PDI (excluding the direct channels D_pD_p and D_pD_s) the greatest contributions to the 2PDI were given by those that contained a $3d$ electron (see Equations (7)–(13)). At the same time, among all correlational amplitudes (7)–(13), amplitude (7) had the largest value. The main aim of the present work was to identify the main processes contributing to the low-energy part of the photoelectron spectrum rather than a detailed quantitative description of the spectral structure. Therefore, in computing the photoionization cross section (PICS) of the second step of the D_sC process, we restricted our description to the most significant amplitude (7) only and neglected the DPES.

The final states stemming after the D_sC two-photon double-ionization, decay by the Auger process

$$3s^13p^43d^1 \longrightarrow 3s^23p^3\varepsilon_A\ell(\ell = 1,3) \tag{14}$$

The Auger electrons of process (14) also contributed to the $2PDI^{II}$ spectrum at low kinetic energies. This is depicted in column D_sC of Figure 2 by red vertical arrows.

Column D_pC of Figure 2 represents a correlational process similar to D_sC, but creates the $3s^23p^5$ ionic state after the first ionization step.

The $D_pC(res)$ correlational process depicted in the right column of Figure 2 is of particular interest. In this process, the second photon does not knock out the electron, but excites it into states of the $3s^13p^43d^2$ configuration, with a subsequent Auger decay. Since the excitation to the $3s^13p^43d^2$ states is resonant, the shape of the 2PDI^{II} spectrum would strongly depend on the photon excitation energy. The "Auger part" of the photoelectron spectrum should be rich due to a large number of possible Auger decays (see multiple red arrows in the right column of Figure 2):

$$3s^13p^43d^2 \longrightarrow 3s^13p^5\varepsilon_A\ell(\ell=1,3) \tag{15}$$

$$3s^13p^43d^2 \longrightarrow 3s^23p^4\varepsilon_A\ell(\ell=0,2,4) \tag{16}$$

$$3s^13p^43d^2 \longrightarrow 3s^23p^33d^1\varepsilon_A\ell(\ell=1,3) \tag{17}$$

Correlational transitions may also take place in the first step of 2PDI. The CD_s column of Figure 1b shows a two-step process, the first step of which is possible only because of the configuration interaction in the initial or the final state, since the direct transition

$$3s^23p^6+\gamma_1 \dashrightarrow 3s^23p^43d^1+\varepsilon_1\ell(\ell=1,3) \tag{18}$$

is dipole forbidden. One of the possible correlation transitions leading to the appearance of the $3s^23p^43d^1$ ionic state after the first ionization step is the following:

$$3s^23p^6 \dashrightarrow 3s^23p^5\varepsilon'\ell'(\ell'=0,2) \longrightarrow 3s^23p^43d^1\varepsilon_1\ell(\ell=1,3). \tag{19}$$

The second step of the CD_s process is a direct transition

$$3s^23p^43d^1+\gamma_2 \dashrightarrow 3s^13p^43d^1+\varepsilon_2p. \tag{20}$$

The process CD_s result in lines in the same region of the 2PDI^{II} PES as the D_sD_p process, i.e., with an electron kinetic energy larger than 30 eV.

Another possible two-photon process is depicted in the CD_p column of Figure 1b. Similar to the CD_s case, the first step is possible due to the configuration interaction either in the initial or the final configuration only, since the direct transition

$$3s^23p^6+\gamma_1 \dashrightarrow 3s^13p^53d^1+\varepsilon_1\ell(\ell=0,2) \tag{21}$$

is dipole forbidden. One of the possible correlational transitions leading to the appearance of the $3s^13p^53d^1$ ionic state after the first ionization step is the following:

$$3s^23p^6 \dashrightarrow 3s^23p^5\varepsilon'\ell'(\ell'=0,2) \longrightarrow 3s^13p^53d^1\varepsilon_1\ell(\ell=0,2). \tag{22}$$

The second step of the CD_p process is the direct transition

$$3s^13p^53d^1+\gamma_2 \dashrightarrow 3s^13p^43d^1+\varepsilon_2\ell(\ell=0,2). \tag{23}$$

The peculiarity of the CD_p process is the fast Auger decay of the intermediate $3s^13p^53d^1$ states

$$3s^13p^53d^1 \longrightarrow 3s^23p^4\varepsilon_A\ell(\ell=1,3) \tag{24}$$

which reduce the intensities of respective 2PDI^{II} lines (if the intensity of the exciting radiation is small). A similar situation was considered in [22], where the process of the two-photon ionization of the Ne atom was studied.

Our estimates show that the average Auger width of the $3s^13p^53d^1$ states is about 1 eV, whereas the largest photoionization cross section of these states at the energy of 65.3 eV does not exceed 0.5 Mb. Then, at the exciting–radiation flux of 1.5×10^{15} W/cm^2 [4], the probability of process (23) is less than 10^{-9}. When the probability of process (23) reaches a noticeable value at a radiation flux of more than 10^{16} W/cm^2, it will become 5%.

Thereby, at a lower flux of the exciting radiation, the Auger decay (24) will most likely occur after the first ionization step in the CD_p process. Consequently, the ionization of the state of a doubly charged ion will occur in the second step as

$$3s^2 3p^4 + \gamma_2 \dashrightarrow 3s^2 3p^3 + \varepsilon_2 \ell (\ell = 0, 2). \quad (25)$$

or

$$3s^2 3p^4 + \gamma_2 \dashrightarrow 3s^1 3p^4 + \varepsilon_2 p. \quad (26)$$

The kinetic energies of electrons in process (25) exceed 20 eV, and process (26), similar to process (3), with low intensity. Therefore, the CD_p process will not be discussed later.

3. Calculation Details

The atomic orbitals (AOs) used in the calculation of the cross sections for the first ionization step were obtained by solving the non-relativistic Hartree–Fock (HF) equations for the ground $1s^2 2s^2 2p^6 3s^2 3p^6$ configuration of Ar. The AOs of continuum electrons were calculated by solving the HF term-dependent equations for the $3s^2 3p^5 \varepsilon s(^1P)$, $3s^2 3p^5 \varepsilon d(^1P)$ configurations in the case of the $D_p D_p$ process and for the $3s^1 3p^6 \varepsilon p(^1P)$ configuration in the case of the $D_s C$ process with frozen core AOs.

The total experimental cross sections of Ar [23] were compared with the presently computed $\sigma_{3p}(\omega)$ in Figure 3. One can recognize that the HF cross sections calculated in the length form (dash curve) and velocity form (short dash curve) of the dipole transition operator differed by more than 1.5 times at the threshold.

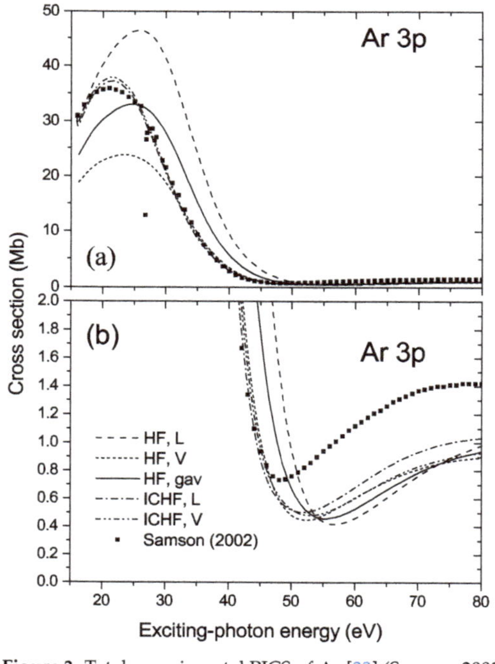

Figure 3. Total experimental PICS of Ar [23] (Samson 2002) and presently computed PICS of the Ar 3p shell in the single-electron HF approximation in length (σ_L^{HF}) and velocity (σ_V^{HF}) forms and the geometric average between the length and velocity forms (σ_{gav}). Results of the calculation, taking into account intra-shell correlation(s) (ICHF) in length and velocity forms are depicted as well. Panel **(b)** represents an enlarged part of panel **(a)** in the region of the Cooper minimum.

We also calculated the $3p$ shell PICS, taking into account the intra-shell correlation(s) (ICHF) via a technique described in [14]. These cross sections (dash–dot curve for the length form and dash–dot-dot curve for the velocity form) are very close to each other and to the experimental cross sections in the photon energy range from the threshold up to 30 eV. At higher photon energies, of more than 30 eV, the theoretical PICS of the $3p$ shell of Ar differ from the experimental ones (see panel (b)), since for these energies, in addition to photoionization of the $3p$ shell, photoionization of the $3s$ shell is possible to the main $3s^13p^6$ and satellite $3s^23p^4n\ell$ ionic states.

Intensities of the photoelectron lines $I_{ij}(\omega)$ were calculated using the formula

$$I_{ij}(\omega) \sim \sigma_{1i}(\omega) \cdot \sigma_{2ij}(\omega), \tag{27}$$

where $\sigma_{1i}(\omega)$ is the cross section to the first 2PDI step—photoionization of the neutral atom leading to the formation of a single-charged ion state i; $\sigma_{2ij}(\omega)$ is the cross section for the second 2PDI step—photoionization of the ionic state i leading to the formation of a double-charged ion state j.

Cross sections $\sigma_{1i}(\omega)$ and $\sigma_{2ij}(\omega)$, calculated within the Hartree–Fock (HF) approximation in the length (σ_L^{HF}) and velocity (σ_V^{HF}) gauges could have significant differences that disappear when taking into account the inter-shell and intra-shell correlations, applying all orders of the perturbation theory [14,15]. In some cases, cumbersome calculations associated with taking into account these correlations can be avoided if the geometric average (gav) value of these cross sections, σ_{gav}, is used:

$$\sigma_{gav} = \left(\sigma_L^{HF} \cdot \sigma_V^{HF}\right)^{1/2} \tag{28}$$

Figure 3 illustrates that, in the case of Ar $3p$-photoionization, using σ_{gav} achieves an inaccuracy of 10% when computing PICS in the photon excitation energy range of interest. Therefore, the σ_{gav}- approximation was used throughout the present paper.

The following set of AOs was used in our calculations of the second step of 2PDI^{II}. In the D_pD_p and the D_pD_s processes, the AOs were obtained by solving the HF equations for the $1s^22s^22p^63s^23p^5$ configuration. The AOs of continuum electrons for the D_pD_p process were calculated by solving the term-depending HF equation for the $3s^23p^4(^1S,\ ^1D,\ ^3P)\varepsilon\ell(LS)$ states with frozen core AOs. The AOs of continuum electrons for the D_pD_s process were calculated similarly but for the $3s^13p^5(^1P,\ ^3P)\varepsilon p(LS)$ states using frozen core AOs. The total cross section was obtained as a sum of partial cross sections over the LS quantum numbers.

In order to take into account the DPES in the second step of the D_pD_s process, the interaction between the $3s^13p^5$ and $3s^23p^3nd$ $(n = 3, 4)$ configurations was taken into account.

The cross sections for the second ionization step in the case of the D_sC process were computed applying the amplitude (7) only, which is described in more detail. The expression for the photoionization cross section is

$$\sigma_2\left(3s^13p^6 \longrightarrow 3s^13p^43d^1, \omega\right) = \frac{4}{3}\pi^2\alpha a_0^2\omega^{\pm 1}\sum_{\ell}\left|A\left(3s^13p^6 \longrightarrow 3s^13p^43d^1\varepsilon_2\ell\right)\right|^2, \tag{29}$$

where the signs (+) and (−) correspond to the length and velocity gauges of the transition dipole operator D, respectively; ω determined by $E(3s^13p^6) + \omega = E(3s^13p^43d^1) + \varepsilon_2$ stands for the photon excitation energy in atomic units; $\alpha = 1/137.036$ is the fine-structure constant; and the square of the Bohr radius $a_0^2 = 28.0028$ Mb converts the atomic units to cross sections in $Mb = 10^{-22}$ m^2.

The transition amplitude (7) is:

$$A\left(3s^13p^6 \longrightarrow 3s^13p^43d^1\varepsilon_2\ell\right) = \sum_{\varepsilon'\ell'} \frac{\langle 3s^13p^43d^1\varepsilon_2\ell|D|3s^13p^43d^1\varepsilon'\ell'\rangle\langle 3s^13p^43d^1\varepsilon'\ell'|H^{ee}|3s^13p^6\rangle}{-IP(3s^13p^43d^1)-\varepsilon'} \quad (30)$$

where D is the dipole transition operator; H^{ee} is the Coulomb interaction operator; $IP(3s^13p^43d^1)$ is the ionization potential for the configuration $3s^13p^43d^1$ in respect with the $3s^13p^6$ state; the summation/integration over ε' includes both the discrete and continuum states.

The calculation was performed using the AOs obtained by solving the HF equations for the $1s^22s^22p^63s^13p^6$ configuration. The $3d$ AO was obtained in the $1s^22s^22p^63s^13p^43d^1$ configuration. The continuum AOs for the intermediate and final $3s^13p^43d^1\varepsilon'\ell'$ and $3s^13p^43d^1\varepsilon_2\ell$ configurations were calculated using frozen core AOs averaged over the configuration. Since the amplitude (7) contained divergent radial integrals $\langle\varepsilon_2\ell|r|\varepsilon'\ell'\rangle$, it was calculated using the correlation function method [24].

When calculating the energy levels of the $1s^22s^22p^63s^13p^43d^1$ configuration, the AOs obtained by solving the HF equations for this configuration were used. The same atomic orbitals were used to calculate the Auger decay probabilities (14).

Intensities of the Auger components for the PES were calculated using the formula

$$I(ij \longrightarrow LS,\omega) = I_{ij}(\omega)\cdot \varkappa(j \longrightarrow LS), \quad (31)$$

where $I(ij \longrightarrow LS,\omega)$ is the intensity of the Auger line corresponding to the decay of the $3s^13p^43d^1(j)$ state to the configuration $3s^23p^3(LS)\varepsilon\ell$; $I_{ij}(\omega)$ is the intensity (27) of the photoelectron line and $\varkappa(j \longrightarrow LS)$ is the Auger yield computed as

$$\varkappa(j \longrightarrow LS) = \Gamma(j \longrightarrow LS)\left[\sum_{L'S'}\Gamma(j \longrightarrow L'S')\right]^{-1}, \quad (32)$$

where $\Gamma(j \longrightarrow LS)$ is the partial width of the $j \longrightarrow LS$ Auger-transition and the summation in the denominator includes all terms of the configuration $3s^23p^3(LS)$, the decays that are energetically allowed. The partial Auger width in the atomic units was calculated as

$$\Gamma(j \longrightarrow LS) = 2\pi\left|\left\langle 3s^23p^3(LS)\varepsilon\ell|H^{ee}|3s^13p^43d^1(j)\right\rangle\right|^2. \quad (33)$$

4. Results and Discussion

The most relevant calculation of the photoelectron spectrum during the sequential two-photon double ionization of the Ar atom, known to date, was the calculation by Kiselev et al. in 2020 [11]. In that work, the photoionization cross sections (at a photon energy of 33.4 eV) were measured and computed using the R-matrix technique with the B-spline R-matrix package [25]. At this photon energy, the D_pD_s direct process described in Section 2.1 and the correlational processes considered in Section 2.2 do not occur due to the insufficient energy of the incident radiation.

The experimental spectrum contains the lines of both the first and second ionization steps. In particular, the electron energies in the $3s^23p^6 + \gamma_1 \dashrightarrow 3s^13p^6\varepsilon_1p$ first-step transition are very close to those for the second-step transition $3s^23p^5 + \gamma_2 \dashrightarrow 3s^23p^4(^1D)\varepsilon_2\ell(\ell = 0,2)$. Therefore, the corresponding lines in the photoelectron spectrum overlap. As a result, it is impossible to determine the intensity of the line corresponding to the $3s^23p^4(^1D)$ state using the experimental spectrum obtained in [11].

In the present work, we calculated the 2PDI^{II} PES of Ar at a photon energy of 33.4 eV. In accordance with the above, we compare the results of both calculations only. From Figure 1 of [11], we estimate that the relative line intensities in the computed 2PDI^{II} PES are $I(^3P) : I(^1D) : I(^1S) = 100 : 45 : 15$. In our calculation, this ratio is 100:32:11. The two calculations yield similar ratios, and we are, thus, confident that our calculation method provides sufficient accuracy to explain the features of a possible new experiment in the "correlation" region of exciting photon energies.

As discussed above, only transitions (1)–(4), and (7) were taken into account in the present calculations. In total, this stage required computing 3 partial PICS to the first 2DPI step (see processes (1) and (4)) and 34 PICS for the second 2DPI step (see processes (2), (3), and (7)). The PICS to the first 2DPI step (1) are depicted in Figure 3. The PICS to the first 2DPI step (4) at an exciting photon energy 65.3 eV are $\sigma_1(\omega = 65.3\ \text{eV}) = 0.22$ Mb. The largest correlational PICS of the second 2DPI step are compared with the PICS of the direct transition (2) in Figure 4. It can be seen that for all three terms $3s^23p^4(^1S,\ ^1D,\ ^3P)$, the shapes of the PICS profiles for direct transitions are qualitatively similar to the usual $3p$ PICS (see Figure 3), which are determined by the Cooper minimum. The correlational cross section has a qualitatively different dependence on the photon excitation energy, which is mainly determined by the energy dependence of the $3d \dashrightarrow \varepsilon f$ transition and, to some extent, is a manifestation of the $3d \dashrightarrow \varepsilon f$ giant resonance. The ratio of partial cross sections for the εp and εf channels in the photon excitation energy range of 50–80 eV varied by almost three times, which suggests a significant energy dependence of the photoelectron angular distribution parameter in the specified energy range and may be of interest for experimental research.

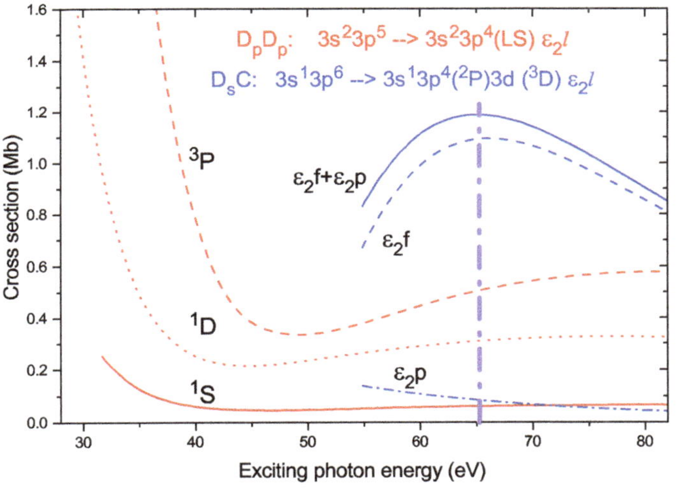

Figure 4. Calculated photoionization cross section(s) (PICS) for the second step of the direct $3s^23p^5 \dashrightarrow 3s^23p^4\varepsilon_2\ell$ and the largest correlational $3s^13p^6 \dashrightarrow 3s^13p^43d^1\varepsilon_2\ell$ 2PDI transitions of Ar. For the correlational transition, partial PICS are also shown. The vertical dash-dotted line shows the photon excitation energy used for computing the photoelectron spectrum in Figure 5.

The photoelectron spectra computed in several approximations are depicted in Figure 5. Figure 5a shows the spectrum due to direct transitions (1)–(3) only. Transitions (1) and (2) lead to the appearance of three 2PDI^{II} lines in the energy range from 32 to 40 eV, corresponding to the $\text{Ar}^{2+}\ 3s^23p^4(^1S,\ ^1D,\ ^3P)$ states, and transitions (1) and (3) lead to the appearance of a structure in the range of electron kinetic energies from 0 to 25 eV. The components of the computed spectrum were convolved using Gaussian functions, FWHM = 0.5 eV. For a better comparison, we present all of the calculated relative intensities of the groups of lines with respect to the total intensity of the $3p^4$ (D_pD_p) multiplet, which is considered 100%. The calculated ratio of the intensities $I(D_pD_p) : I(D_pD_s)$ is $100:25$. Within the D_pD_s structure the ratio of the $I(3s^13p^5)$ group to the $I(3s^23p^3nd^1(sat))$ group is 14:11.

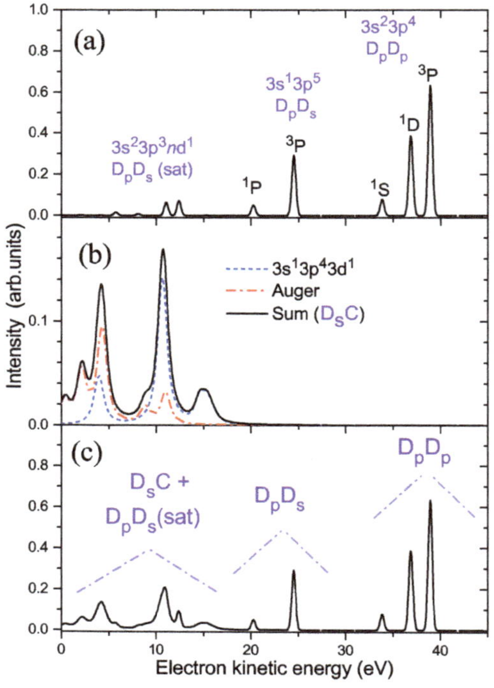

Figure 5. The 2PDI^{II} PES of Ar for photon excitation energy 65.3 eV; (**a**) direct D_pD_p and D_pD_s processes; (**b**) D_sC correlational process: photoelectron lines of processes (4) and (7) ($3s^13p^43d^1$); Auger lines of process (14) (Auger) and total spectrum (Sum); (**c**) the sum of the spectra depicted in panels (**a**,**b**).

The theoretical spectrum caused by the D_sC process is shown in Figure 5b. Accounting for transitions (4) and (7) results in fairly intense structures in the energy range from the threshold to 20 eV (short dash curve). We also took into account the Auger decay channel (14). The lines corresponding to Auger electrons, computed according to Equations (31)–(33), are also shown in Figure 5b by the dash–dot curve. The total spectrum is presented in Figure 5b by the solid curve. For a better perception of the contributions of the photoelectron and Auger-electron parts of the spectrum, the intensity scale was increased by a factor of 5 compared to Figure 5a. Since the Auger width of the transition (14) was 1 eV on average, when constructing this part of the spectrum, its components were convolved using Lorentz functions, FWHM = 1 eV, and additionally convolved by an instrumental broadening using Gaussian functions, FWHM = 0.5 eV. Finally, the total spectrum, taking into account all processes (1)–(3), (4), (7), and (14), is shown in Figure 5c. Both correlation effects have comparable intensities with a ratio $I(3s^13p^43d^1) : I(Auger)$ equaling to 72:64 (in respect with $I(D_pD_p)$).

To summarize, the calculated in the described approximation integral intensity of PES lines for the photoelectrons with low kinetic energies, stemming from the correlational D_sC process, is 136% of the $I(D_pD_p)$ intensity of the direct (non-correlational) two-photon D_pD_p process associated with transitions to the Ar^{2+} $3s^23p^4(^1S,\ ^1D,\ ^3P)$ states.

Further theoretical investigations could be taken into account regarding the DPES, which involves mixing the $3s^13p^6$ configuration with the $3s^23p^4n\ell(\ell = 0,2)$ configurations in the first ionization step (4). In the second ionization step, the amplitude (7), remaining amplitudes (8)–(13), and the DPES in these processes should be taken into account.

5. Conclusions

We computed a theoretical photoelectron spectrum for photoelectrons of low kinetic energies for the second step of sequential two-photon double ionization ($2PDI^{II}$ PES) of Ar at a photon excitation energy of 65.3 eV, exceeding the 3*s* ionization threshold. The calculation takes into account the most significant electron correlations stemming from single- and double-electron excitations.

The calculation predicts that in the low-energy part of the $2PDI^{II}$ PES, correlation satellites should be observed whose intensities exceed the intensity of the main (direct) transition to the Ar^{2+} $3s^2 3p^4(^1S,\ ^1D,\ ^3P)$ states. Correlation satellites consist of lines associated with conventional photoelectron emissions and the Auger decay of the $3s^1 3p^4 n\ell^1$ states; the cross sections of both processes are comparable in magnitude. The lines of the first type should change their energy with changing photon excitation energy, while the energy of the Auger lines should remain unchanged.

The dependence of the intensities of correlation satellites may have resonant characters associated with the excitation of discrete levels at certain energies. In this case, the excitation of resonances at the second step of the two-photon process (e.g., those stemming from the second-step configuration $3s^1 3p^4 3d^2$) will contribute to the $2PDI^{II}$ PES even at small exciting radiation fluxes. At large fluxes, there should be a contribution from resonances excited at the first step (e.g., those stemming from the first-step configuration $3s^1 3p^5 3d^1$). The estimates of the lifetime of the $3s^1 3p^5 3d^1 (LSJ)$ states showed that the contribution of these processes can be significant at fluxes exceeding 10^{16} W/cm^2.

Author Contributions: Conceptualization and methodology V.L.S. and A.E.; data curation, B.M.L., I.D.P., V.A.K. and N.M.N.; formal analysis, validation, visualization, and writing—original draft, B.M.L., I.D.P. and V.L.S.; funding acquisition, project administration and supervision, P.V.D. and A.E.; investigation, B.M.L., I.D.P., V.L.S., V.A.K. and N.M.N.; software, B.M.L., I.D.P., V.L.S. and V.A.K.; writing—review and editing, B.M.L., I.D.P., V.L.S., P.V.D. and A.E. All authors have read and agreed to the published version of the manuscript.

Funding: V.L.S., N.M.N., P.V.D., and A.E. were funded by the Deutsche Forschungsgemeinschaft (DFG)—project no. 328961117-SFB 1319 ELCH (extreme light for sensing and driving molecular chirality).

Data Availability Statement: All the data reported in this work are available from the correspondence author on reasonable request.

Acknowledgments: The authors are grateful to Markus Ilchen for drawing their attention to the problem existing in interpreting the 2PDI of Ar, for his permanent interest in the work, and for discussing the results obtained.

Conflicts of Interest: The authors declare no conflict of interest.

Abbreviations

The following abbreviations are used in this manuscript:

FEL	free electron laser
PES	photoelectron spectrum
$2PDI^{II}$	the second step of the sequential two-photon double ionization
FWHM	Full width on half maximum
PICS	photoionization cross section
DPES	dipole polarization of electron shells
HF	Hartree–Fock

References

1. Moshammer, R.; Jiang, Y.H.; Foucar, L.; Rudenko, A.; Ergler, T.; Schröter, C.D.; Lüdemann, S.; Zrost, K.; Fischer, D.; Titze, J.; et al. Few-Photon Multiple Ionization of Ne and Ar by Strong Free-Electron-Laser Pulses. *Phys. Rev. Lett.* **2007**, *98*, 203001. [CrossRef] [PubMed]
2. Sorokin, A.A.; Bobashev, S.V.; Feigl, T.; Tiedtke, K.; Wabnitz, H.; Richter, M. Photoelectric Effect at Ultrahigh Intensities. *Phys. Rev. Lett.* **2007**, *99*, 213002. [CrossRef] [PubMed]

3. Young, L.; Kanter, E.P.; Krässig, B.; Li, Y.; March, A.M.; Pratt, S.T.; Santra, R.; Southworth, S.H.; Rohringer, N.; DiMauro, L.F.; et al. Femtosecond electronic response of atoms to ultra-intense X-rays. *Nature* **2010**, *466*, 56–62. [CrossRef] [PubMed]
4. Ilchen, M.; Hartmann, G.; Gryzlova, E.V.; Achner, A.; Allaria, E.; Beckmann, A.; Braune, M.; Buck, J.; Callegari, C.; Coffee, R.N.; et al. Symmetry breakdown of electron emission in extreme ultraviolet photoionization of argon. *Nat. Commun.* **2018**, *9*, 4659. [CrossRef] [PubMed]
5. Kurka, M.; Rudenko, A.; Foucar, L.; Kühnel, K.U.; Jiang, Y.H.; Ergler, T.; Havermeier, T.; Smolarski, M.; Schössler, S.; Cole, K.; et al. Two-photon double ionization of Ne by free-electron laser radiation: A kinematically complete experiment. *J. Phys. B At. Mol. Opt. Phys.* **2009**, *42*, 141002. [CrossRef]
6. Rouzée, A.; Johnsson, P.; Gryzlova, E.V.; Fukuzawa, H.; Yamada, A.; Siu, W.; Huismans, Y.; Louis, E.; Bijkerk, F.; Holland, D.M.P.; et al. Angle-resolved photoelectron spectroscopy of sequential three-photon triple ionization of neon at 90.5 eV photon energy. *Phys. Rev. A* **2011**, *83*, 031401. [CrossRef]
7. Grum-Grzhimailo, A.N.; Gryzlova, E.V.; Fritzsche, S.; Kabachnik, N.M. Photoelectron angular distributions and correlations in sequential double and triple atomic ionization by free electron lasers. *J. Mod. Opt.* **2016**, *63*, 334–357. [CrossRef]
8. Bachau, H.; Lambropoulos, P. Theory of the photoelectron spectrum in double ionization through two-photon absorption from He($2s^2$). *Phys. Rev. A* **1991**, *44*, R9–R12. [CrossRef]
9. Fritzsche, S.; Grum-Grzhimailo, A.N.; Gryzlova, E.V.; Kabachnik, N.M. Angular distributions and angular correlations in sequential two-photon double ionization of atoms. *J. Phys. B At. Mol. Opt. Phys.* **2008**, *41*, 165601. [CrossRef]
10. Gryzlova, E.V.; Grum-Grzhimailo, A.N.; Fritzsche, S.; Kabachnik, N.M. Angular correlations between two electrons emitted in the sequential two-photon double ionization of atoms. *J. Phys. B At. Mol. Opt. Phys.* **2010**, *43*, 225602. [CrossRef]
11. Kiselev, M.D.; Carpeggiani, P.A.; Gryzlova, E.V.; Burkov, S.M.; Reduzzi, M.; Dubrouil, A.; Faccialá, D.; Negro, M.; Ueda, K.; Frassetto, F.; et al. Photoelectron spectra and angular distribution in sequential two-photon double ionization in the region of autoionizing resonances of ArII and KrII. *J. Phys. B At. Mol. Opt. Phys.* **2020**, *53*, 244006. [CrossRef]
12. Amusia, M.; Ivanov, V.; Cherepkov, N.; Chernysheva, L. Interference effects in photoionization of noble gas atoms outer s-subshells. *Phys. Lett. A* **1972**, *40*, 361–362. [CrossRef]
13. Amusia, M.Y.; Ivanov, V.; Cherepkov, N.; Chernysheva, L. Intershell and intersubshell effects in photoionization of atoms. *Zh. Eksp. Teor. Fiz.* **1974**, *66*, 1537–1549.
14. Amusia, M.; Cherepkov, N. *Many-Electron Correlations in Scattering Processes;* Case Studies in Atomic Physics; North-Holland: Amsterdam, The Netherlands, 1975; Volume 5, pp. 49–179.
15. Amusia, M.Y. *Atomic Photoeffect*, 1st ed.; Plenum Press: New York, NY, USA, 1990.
16. Sukhorukov, V.; Petrov, I.; Lagutin, B.; Ehresmann, A.; Schartner, K.H.; Schmoranzer, H. Many-electron dynamics of atomic processes studied by photon-induced fluorescence spectroscopy. *Phys. Rep.* **2019**, *786*, 4. [CrossRef]
17. Braune, M.; Hartmann, G.; Ilchen, M.; Knie, A.; Lischke, T.; Reinkster, A.; Meissner, A.; Deinert, S.; Glaser, L.; Al-Dossary, O.; et al. Electron angular distributions of noble gases in sequential two-photon double ionization. *J. Mod. Opt.* **2016**, *63*, 324–333. [CrossRef]
18. Bauche, J.; Bauche-Arnoult, C.; Klapisch, M. Transition Arrays in the Spectra of Ionized Atoms. *Adv. At. Mol. Phys.* **1988**, *23*, 131–195. [CrossRef]
19. Kučas, S.; Karazija, R. Global characteristics of atomic spectra and their use for the analysis of spectra. I. Energy level spectra. *Phys. Scr.* **1993**, *47*, 754–764. [CrossRef]
20. Gryzlova, E.V.; Ma, R.; Fukuzawa, H.; Motomura, K.; Yamada, A.; Ueda, K.; Grum-Grzhimailo, A.N.; Kabachnik, N.M.; Strakhova, S.I.; Rouzée, A.; et al. Doubly resonant three-photon double ionization of Ar atoms induced by an EUV free-electron laser. *Phys. Rev. A* **2011**, *84*, 063405. [CrossRef]
21. Hikosaka, Y.; Fushitani, M.; Matsuda, A.; Endo, T.; Toida, Y.; Shigemasa, E.; Nagasono, M.; Tono, K.; Togashi, T.; Yabashi, M.; et al. Resonances in three-photon double ionization of Ar in intense extreme-ultraviolet free-electron laser fields studied by shot-by-shot photoelectron spectroscopy. *Phys. Rev. A* **2013**, *88*, 023421. [CrossRef]
22. Rohringer, N.; Santra, R. Resonant Auger effect at high x-ray intensity. *Phys. Rev. A* **2008**, *77*, 053404. [CrossRef]
23. Samson, J.; Stolte, W. Precision measurements of the total photoionization cross-sections of He, Ne, Ar, Kr, and Xe. *J. Electron. Spectrosc. Relat. Phenom.* **2002**, *123*, 265–276. [CrossRef]
24. Sukhorukov, V.L.; Lagutin, B.M.; Petrov, I.D.; Schmoranzer, H.; Ehresmann, A.; Schartner, K.H. Photoionization of Kr near 4s threshold. II. Intermediate-coupling theory. *J. Phys. B At. Mol. Opt. Phys.* **1994**, *27*, 241–256. [CrossRef]
25. Zatsarinny, O. BSR: B-spline atomic R-matrix codes. *Comput. Phys. Commun.* **2006**, *174*, 273–356. [CrossRef]

atoms

MDPI

Article

A Glimpse into Photodetachment Spectra of Giant and Nested Fullerene Anions

Valeriy K. Dolmatov [1,*,†,‡] **and Steven T. Manson** [2,‡]

1 Department of Chemistry and Physics, University of North Alabama, Florence, AL 35632, USA
2 Department of Physics and Astronomy, Georgia State University, Atlanta, GA 30030, USA
* Correspondence: vkdolmatov@una.edu
† Retired. Current address: 1057 Dixie Ave., Florence, AL 35630, USA.
‡ These authors contributed equally to this work.

Abstract: We focus on the study of the photodetachment of bare, i.e., single-cage $(C_N)^-$ as well as nested (multi-cage) $(C_N@C_M@\ldots)^-$ singly charged fullerene anions. We calculate the attached electron's wavefunctions, energies, oscillator strengths and photodetachment cross sections of the C_{60}^-, C_{240}^-, C_{540}^-, $(C_{60}@C_{240})^-$, $(C_{60}@C_{540})^-$, $(C_{240}@C_{540})^-$ and $(C_{60}@C_{240}@C_{540})^-$ fullerene anions, where the attached electron is captured into the ground s-state by the resultant external field provided by all fullerene cages in the anion. The goal is to gain insight into the changes in behavior ofphotodetachment of this valence electron as a function of the different geometries and potentials of the various underlying fullerenes or nested fullerenes (fullerene onions) both due to their increasing size and due to "stuffing" of a larger bare fullerene with smaller fullerenes. To meet this goal, we opt for a simple semi-empirical approximation to this problem: we approximate each individual fullerene cage by a rigid potential sphere of a certain inner radius, thickness and potential depth, as in numerous other model studies performed to date. The results reveal a number of rather significant differences in the wavefunctions, oscillator strengths and photodetachment cross sections among these fullerene anions, some of which are completely counter-intuitive. The results obtained can serve as a "zeroth-order-touchstone" for future studies of single-cage and nested fullerene anions by more rigorous theories and/or experiments to build upon this work to assess the importance of interactions omitted in the present study.

Keywords: photodetachment; carbon fullerenes; carbon fullerene anions; carbon fullerene onions

Citation: Dolmatov, V.K.; Manson, S.T. A Glimpse into Photodetachment Spectra of Giant and Nested Fullerene Anions. *Atoms* **2022**, *10*, 99. https://doi.org/10.3390/atoms10040099

Academic Editor: Eugene T. Kennedy

Received: 24 August 2022
Accepted: 19 September 2022
Published: 22 September 2022

1. Introduction

The photoionization/photodetachment of various neutral ($q = 0$) and charged ($q \neq 0$) fullerenes, $C_N^{\pm q}$, and their endohedral counterparts, $A@C_N^{\pm q}$ (where A is the atom encapsulated inside $C_N^{\pm q}$ cage), has been the subject of experimental (see, e.g., [1–7] and references therein) as well as intense systematic theoretical studies for many years now (see, e.g., a recent review paper [8] with an abundance of references therein). In particular, Professor M.Y. Amusia, to the legacy of whom this Special Issue of *Atoms* is devoted to, has contributed vastly to the study of the interaction of particles and light with fullerenes and endo-fullerenes, see, e.g., [9–18], to name a few.

Although the research on the interaction of fullerenes and endo-fullerenes with light has also touched upon the subject of fullerene anions (see, e.g., [1–4,8,9,19–24] and references therein), yet, to the best of the authors' knowledge, the subject of photodetachment of giant fullerenes anions [$(C_N)^-$ with $N \gg 60$] as well as of nested fullerene anions, $(C_n@C_{m>n}@\ldots)^-$, referred to as fullerene onion-anions in the present paper, has not been studied. Given the current strong interest in studying various elementary processes of basic importance involving fullerene formations, it is appealing to fill in this gap in the present state of knowledge. The present paper remedies the situation by presenting a first insight into the phenomenon of photodetachment of both giant fullerene and fullerene onion-anions.

In general, elementary processes involving fullerene formations present a formidable multifaceted problem for theorists, thereby requiring the investment of considerable efforts to comprehensibly address all facets of the problem as well as the interaction(s) between them. Therefore, before investing such efforts in a comprehensive study, a kind of roadmap is needed as a guide to the subsequent comprehensive study of this multifaceted problem. Thus, the main narrow goal of the present study is to gain insight, using the simplest reasonable approximation, for modifications of the photodetachment cross sections of giant and nested fullerene anions owing to changes in their geometry induced by stuffing of a larger bare fullerene with smaller and smaller fullerenes: $(C_{240})^-$, $(C_{540})^-$, $(C_{60}@C_{240})^-$, $(C_{60}@C_{540})^-$, $(C_{240}@C_{540})^-$ and $(C_{60}@C_{240}@C_{540})^-$. To meet this goal, we approximate each individual fullerene cage by a rigid potential sphere of a certain inner radius $r_{\rm in}$, thickness Δ and potential depth U_0, as in many earlier model studies of fullerene-involved processes cited above. Within the framework of this approximation, we detail how the photodetachment of the fullerene onion-anions differs crucially from the photodetachment of the largest bare (i.e., single-cage) fullerene anion owing to the differences in geometries between the fullerene formations.

To label the discrete states occupied by the attached electron in a fullerene anion, we adopt, just as a matter of labeling, the traditional notation used for atoms, i.e., the $n\ell$-notation, where ℓ is the orbital quantum number and $n \geq \ell + 1$. Thus, in our notations, the first s-state of the attached electron is $1s$, the next s-state is a $2s$ state, the first p-state is a $2p$ state, the next p-state is a $3p$ state, and so on.

Finally, atomic units (a.u.) ($|e| = \hbar = m = 1$, where e and m are the electron's charge and mass, respectively, and $\hbar$ is a reduced Planck's constant) are used throughout the paper unless stated otherwise.

2. Review of Theory

We model an individual C_N cage (N being the number of carbon atoms in the cage) by a $U_{C_N}(r)$ spherical annular potential of the inner radius, $r_{\rm in}$, finite thickness, Δ, and depth, U_0:

$$U_{C_N}(r) = \begin{cases} -U_0, & r_{\rm in} \leq r \leq r_{\rm in} + \Delta \\ 0, & \text{otherwise.} \end{cases} \quad (1)$$

Such modeling of a C_N cage was suggested in the early work by Puska and Nieminen [25] and, since then, has found an extensive use in numerous studies to date; the reader is referred to [5,13,16,18,21,22,25–27] and to the review paper [8] for many more references on the subject, as well as, e.g., to references [6,8–10,18,21,22,29,31,46] from [21] (and references therein).

We emphasize that, with regard to C_{60}, such model has been proven [5,18,26] (and references therein) to produce results in a reasonable agreement with the experimental photoionization spectrum of endohedral $Xe@C_{60}^+$ [5] and a qualitative and even semi-quantitative agreement with experimental differential elastic electron scattering off C_{60} [6]. Such modeling was also shown [27] to result in a semi-quantitative agreement with some of the most prominent features of the $e^- - C_{60}$ total elastic electron scattering cross section predicted by a far more sophisticated ab initio molecular-Hartree–Fock approximation [27]. This lays out a supporting background for a reasonable suitability of such modelling of C_{60} for the application to photodetachment of a C_{60} fullerene anion as well. Furthermore, our model replaces the earlier fullerene-anion-photodetachment approximations [9,19], which utilized the idea of an infinitesimally thin fullerene wall, by a more realistic finite-width-wall approximation, which is certainly an improvement to the cited approximations.

A fullerene onion, then, is modeled by a potential which is a linear combination of the corresponding $U_{C_N}(r)$ potentials, as in [28]:

$$U_{C_N}@C_M@\ldots = U_{C_{60}} + U_{C_{240}} + \ldots. \tag{2}$$

The parameters r_{in}, U_0 and Δ of the individual C_{60}, C_{240} and C_{540} fullerene cages in fullerene onion-anions are assumed to be the same as for the corresponding isolated bare (single-cage) fullerenes. In the present paper, we take the values for $r_{\rm in}$, Δ and U_0 for $C_{60}/C_{240}/C_{540}$ from [28]: $r_{\rm in} = 5.8/12.6/18.8, \Delta = 1.9/1.9/1.9$and $U_0 = 8.22/10/12$ eV, respectively.

A fullerene anion, C_N^-, or a fullerene onion-anion, $(C_N@C_M@\ldots)^-$, then, is formed by binding of an external electron into a s-state or a p-state in the field of corresponding $U_{C_N}(r)$ or $U_{C_N@C_M@\ldots}$ potential, respectively. Thus, the bound, $P_{n\ell}$, and continuum, $P_{\epsilon\ell}$, radial wavefunctions for the attached electron in a corresponding fullerene anion satisfy the radial Schrödinger equation:

$$-\frac{1}{2}\frac{d^2P_{n/\epsilon\ell}}{dr^2} + \left[U_C + \frac{\ell(\ell+1)}{2r^2}\right]P_{n/\epsilon\ell}(r) = E_{n/\epsilon\ell}P_{n/\epsilon\ell}(r). \tag{3}$$

Here, n and ℓ are is the principal and orbital quantum numbers, respectively, ϵ is the photoelectron energy and U_C is the fullerene potential determined by Equations (1) or (2), respectively.

This equation is solved with the following boundary conditions for the discrete and continuum states:

$$P_{n\ell}(r)|_{r\to 0,\infty} = 0, \text{ whereas } P_{\epsilon\ell}(r)|_{r\gg 1} \to \sqrt{\frac{2}{k\pi}}\sin\left(kr - \frac{\pi\ell}{2} + \delta_\ell(\epsilon)\right). \tag{4}$$

Here, $\delta_\ell(\epsilon)$ is the phase of the continuum state wavefunction and k is the photoelectron momentum.

Note that such model of fullerene anion photodetachment is similar in spirit to the one suggested earlier [9,19], albeit there is a Dirac-bubble potential, rather than the spherical annular potential, was used to approximate the C_{60} cage.

The photodetachment cross sections, $\sigma_{n\ell\to\epsilon,\ell\pm1}$, as well as the oscillator strengths of the discrete, $f_{n\ell\to n',\ell\pm1}$, and continuum, $f_{n\ell\to\epsilon,\ell\pm1}$, spectra of fullerene anions, were calculated using well-known formulas, see, e.g., [29]:

$$\sigma_{n\ell\to\epsilon,\ell\pm1} = \frac{4}{3}\pi^2\alpha\frac{N_{n\ell}}{2\ell+1}\omega d^2_{\ell\pm1}, \tag{5}$$

$$f_{n\ell\to n',\ell\pm1} = \frac{N_{n\ell}}{3(2\ell+1)}\omega d^2_{\ell\pm1}, \tag{6}$$

$$f_{n\ell\to\epsilon,\ell\pm1} = \frac{1}{2\pi^2\alpha}\int_0^\infty \sigma_{n\ell\to\epsilon,\ell\pm1}d\omega. \tag{7}$$

Here, α is the fine-structure constant, ω is the photon energy, $N_{n\ell}$ is the number of electron in the $n\ell$ state (a single electron in our case), and $d_{\ell\pm1}$ is the reduced radial matrix element for the transition from the $n\ell$ state to a $n'(\epsilon), \ell\pm1$ final state.

3. Results and Discussion

3.1. Single-Cage Fullerene Anions

As the first step, we scrutinize the $1s$ ground-states and $2p$ excited-states in the bare fullerene anions: C_{60}^-, C_{240}^- and C_{540}^-. We note that our calculations revealed no np excited-states with $n > 2$ in any of these anions. The corresponding ground-state $P_{1s}(r)$ and excited-state $P_{2p}(r)$ radial functions and the corresponding E_{1s} and E_{2p} energies of the attached electron in the C_{60}^-, C_{240}^- and C_{540}^- anions are presented in Figure 1.

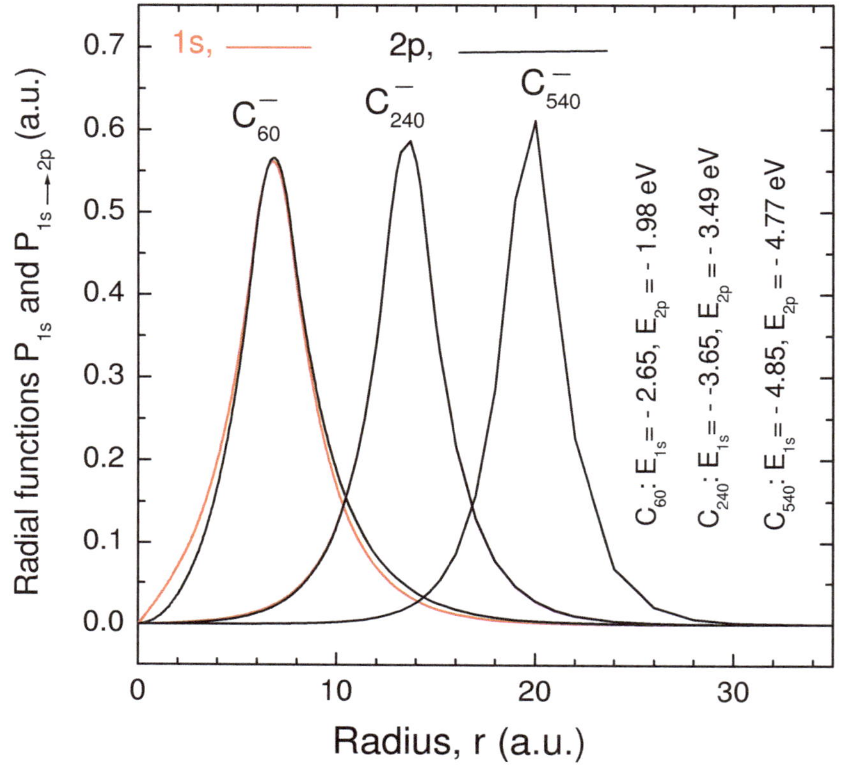

Figure 1. Calculated ground-state $P_{1s}(r)$ and excited- state $P_{2p}(r)$ radial functions and corresponding E_{1s} and E_{2p} binding energies of the attached electron in C_{60}^-, C_{240}^-, and C_{240}^- anions, as designated.

One can see that the P_{1s} and P_{2p} functions reach their maxima within the wall of a corresponding fullerene, i.e., within $5.8 < r < 7.7$ in C_{60}, $12.6 < r < 14.5$ in C_{240} and $18.85 < r < 20.75$ in C_{540}. This, actually, has been expected, for an obvious reason. A strikingly unexpected result (at first glance), though, is that $P_{1s} \approx P_{2p}$ to a high degree of approximation, particularly in C_{240}^- and C_{540}^-. This seems strange, because the Schrödinger equations for a s-state and a p-state differ by the presence of a centrifugal potential $U_{cfg} = \frac{\ell(\ell+1)}{2r^2}$ for a p-state. Correspondingly, the P_{1s} function should have differed from the P_{2p} function. To understand why the situation is opposite to the expected one, we depict, for the case of C_{60}, the cage model potential $U_{C_{60}}(r)$, the centrifugal potential $U_{cfg} = \frac{\ell(\ell+1)}{2r^2} = \frac{1}{r^2}$ for a p-electron, and the $2p$ probability density distribution, $\rho_{2p} = P_{2p}^2$, in Figure 2.

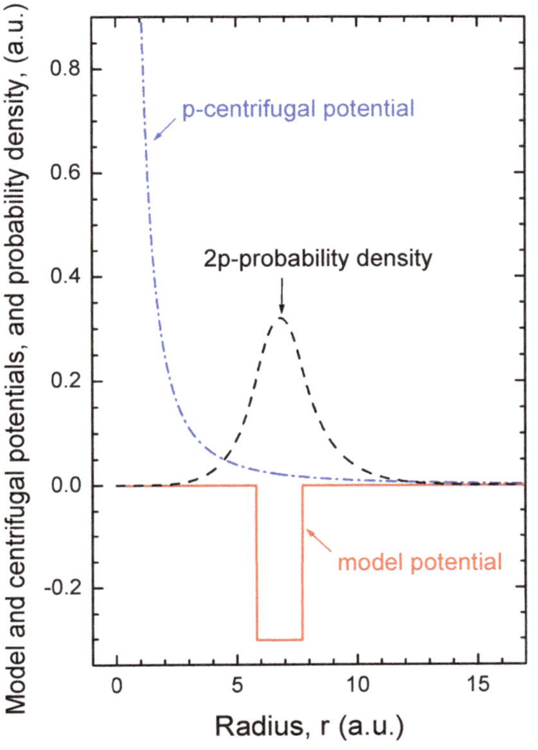

Figure 2. Calculated cage model potential $U_{C_{60}}(r)$, the centrifugal potential $U_{cfg} = \frac{\ell(\ell+1)}{2r^2} = \frac{1}{r^2}$ for a p-electron, and the $2p$ probability density distribution, $\rho_{2p} = P^2_{2p}$, as designated.

One can see that, inside the hollow interior of C_{60} ($U_{C_{60}} = 0$), the ρ_{2p} probability density is practically a zero up to about $r = 3$. Therefore, the presence of the centrifugal potential, however large it is, does not matter in this spatial region. ρ_{2p} starts differ from $\rho_{2p} \approx 0$ between approximately $3 < r < 5.8$. There, however, U_{cfg} is already small and, additionally, only less than 20% of electronic charge is concentrated in this spatial region. Hence, again, a role of the small $U_{cfg} \neq 0$ is largely obliterated in there. Inside the C_{60} wall itself U_{cfg}, on average, is less than 3% of $U_{C_{60}} = 0.302$, whereas outside of the C_{60} wall, $U_{cfg} \ll U_{C_{60}}$ in addition to a rapidly damping probability density distribution. Thus, it now becomes clear that the presence of the centrifugal potential for a p-electron cannot make the solution of the Schrödinger equation to differ any notably from its solution for a s-electron. This discussion explains why P_{1s} differs from P_{2p} only insignificantly, in C^-_{60}. Additionally, we believe that the reader can easily extend this discussion of the behavior of P_{1s} and P_{2p} in C^-_{60} to giant fullerene anions to understand why P_{1s} and P_{2p} become practically identical in each of the C^-_{240} and C^-_{540} anions.

Although the energies are generally more sensitive quantities to parameters in the Schrödinger equation than the wavefunctions, the difference between E_{1s} and E_{2p} binding energies is more noticeable than between the wavefunctions, although still small: it is about 25% of E_{1s} for C^-_{60}, 4% for C^-_{240}, and 1.6% for C^-_{540}. Note that the difference between E_{1s} and E_{2p} is decreasing with increasing size of the fullerene anion. The largest energy difference 25% is in C^-_{60}, as is the largest difference between P_{1s} and P_{2p} (see Figure 1). This is because the $2p$-centrifugal potential energy in C^-_{60} is larger than in other fullerene anions, owing to a significantly smaller size of the C_{60} cage as compared to the other two.

Because $P_{1s} \approx P_{2p}$, the corresponding $f_{1s\to 2p}$ oscillator strengths in the C_{60}^-, C_{240}^- and C_{540}^- anions must be large. Our calculations show that $f_{1s\to 2p} \approx 0.807$, 0.962, and 0.679 in C_{60}^-, C_{240}^- and C_{540}^-, respectively (see Table 1 for more details).

Table 1. Calculated E_{1s} ground-state energies, ω_{2p} and ω_{3p} energies of the $1s \to 2p\&3p$ transitions (all in eV), discrete $f_{1s\to 2p\&3p}$, and continuum, $f_{1s\to \epsilon p}$, oscillator strengths in the single-cage and multi-cage fullerene anions. Note, our calculations showed no existence of the np excited states with $n > 2$ in the single-cage fullerene anions.

Anions	E_{1s}	ω_{2p}	ω_{3p}	$f_{1s\to 2p}$	$f_{1s\to 3p}$	$f_{1s\to(2p+3p)}$	$f_{1s\to \epsilon p}$
C_{60}^-	−2.654	0.6743	-	0.807	-	0.807	0.194
C_{240}^-	−3.646	0.155	-	0.692	-	0.692	0.307
C_{540}^-	−4.855	0.0831	-	0.6795	-	0.6795	0.326
$(C_{60}@C_{240})^-$	−3.691	0.1782	1.8314	0.7345	0.025	0.7595	0.239
$(C_{60}@C_{540})^-$	−4.855	0.0831	2.8902	0.6795	0.000	0.6796	0.324
$(C_{240}@C_{540})^-$	−4.903	0.092	1.514	0.696	0.052	0.748	0.260
$(C_{60}@C_{240}@C_{540})^-$	−4.903	0.093	1.483	0.699	0.052	0.751	0.256

We note that the oscillator strength $f_{1s\to 2p}$ is decreasing with increasing size of the fullerene cage. At first glance this is strange, because the approximate equality $P_{1s} \approx P_{2p}$ is getting only stronger with increasing size of the fullerene cage, as discussed above. Thus, the overlap between P_{1s} and P_{2p} is increasing and so should have been $f_{1s\to 2p}$ as well, with increasing size of the anion. However, the $\omega_{1s\to 2p} \equiv \omega_{2p}$ excitation energy (see Table 1), is decreasing with the increasing size of the fullerene cage. This counterbalances the increase in the overlap between P_{1s} and P_{2p}, thereby resulting in a smaller $f_{1s\to 2p}$ (which is proportional to ω_{np}) in a bigger fullerene anion. This decrease in the $f_{1s\to 2p}$ oscillator strength with increase in the fullerene size leads to an important conclusion. Namely, we conclude there is an increasing transfer of oscillator strength of a fullerene anion from a discrete spectrum to continuum with increasing size of the fullerene cage, as clearly follows from the oscillator strength sum rule: $f_{1s\to \epsilon p} = 1 - f_{1s\to 2p}$. Calculated $f_{1s\to \epsilon p}$'s are presented in Table 1 as well. At this point it is important to emphasize that the continuum oscillator strengths, presented in Table 1, were calculated using Equation (7) rather than as $1 - f_{1s\to 2p}$ from the sum rule. The fact that the independently calculated $f_{1s\to \epsilon p}$ and the $f_{1s\to \epsilon p} = 1 - f_{1s\to 2p}$ are equal to a high degree of approximation speaks about the adequacy of the calculated photodetachment cross sections themselves, discussed later in the paper.

3.2. Fullerene Onion-Anions

We now move to the discussion of the wavefunctions of the valence electron in the fullerene onion-anions $(C_{60}@C_{240})^-$, $(C_{60}@C_{540})^-$, $(C_{240}@C_{540})^-$ and $(C_{60}@C_{240}@C_{540})^-$. We note first that the potentials of these fullerene onions are, obviously, either double-well or triple-well potentials. Correspondingly, one can expect a greater number of bound states available to the attached electron in these fullerene onion-anions. In our case, the calculations predicted the existence of only two discrete p-states—the $2p$ and $3p$ excited states—, in contrast to only the $2p$ excited-state in the bare fullerene anions. The corresponding P_{1s}, P_{2p} and P_{3p} functions are plotted in Figure 3 where a number of new features are exhibited.

The most striking discovery relates to the behavior of the P_{3p} excited-state wave functions. Their highly peculiar behavior is completely different from the behavior of the P_{1s} and P_{2p} functions in any of these fullerene onion-anions. Indeed, we find that a significant part of the P_{3p} function and, thus, the electron density of the attached electron, is packed inside the wall of the inner cage directly adjacent to a larger fullerene cage in each of these double- and triple-cage fullerene onion-anions. Although this is evident from Figure 3, this is also supported by looking at the mean radii, $\bar{r}_{3p}$, of the $3p$ orbitals in these fullerene onion-anions as well. The calculated $\bar{r}_{3p}$'s are: $\bar{r}_{3p} \approx 7$ in both $(C_{60}@C_{540})^-$

and $(C_{60}@C_{240})^-$ (thus, the $3p$ orbital falls into the C_{60} potential well), whereas $\bar{r}_{3p} \approx 14$ in both $(C_{240}@C_{540})^-$ and $(C_{60}@C_{240}@C_{540})^-$ (thus, $\bar{r}_{3p}$ falls into the C_{240} potential well). This is in a sharp contrast to the P_{1s} and P_{2p} functions that are mainly packed in the potential well of the largest fullerene cage (C_{540}, in our case) in corresponding fullerene onion-anions, respectively, as is evident from Figure 3 (also, $\bar{r}_{1s} \approx \bar{r}_{2p} \approx 20$ in all fullerene onion-anions under discussion). Especially surprising is the behavior of the P_{3p} function in $(C_{60}@C_{540})^-$, where its probability density is almost entirely located inside C_{60}, despite the size of C_{60} being significantly smaller than C_{540}, so that the C_{60} potential well should not have affected the attached valence electron at all, as in the case of the P_{1s} and P_{2p} functions (see Figure 3c,d). In any case, the behavior of P_{3p} in these fullerene onion-anions is extraordinary, a complete break with conventional wisdom.

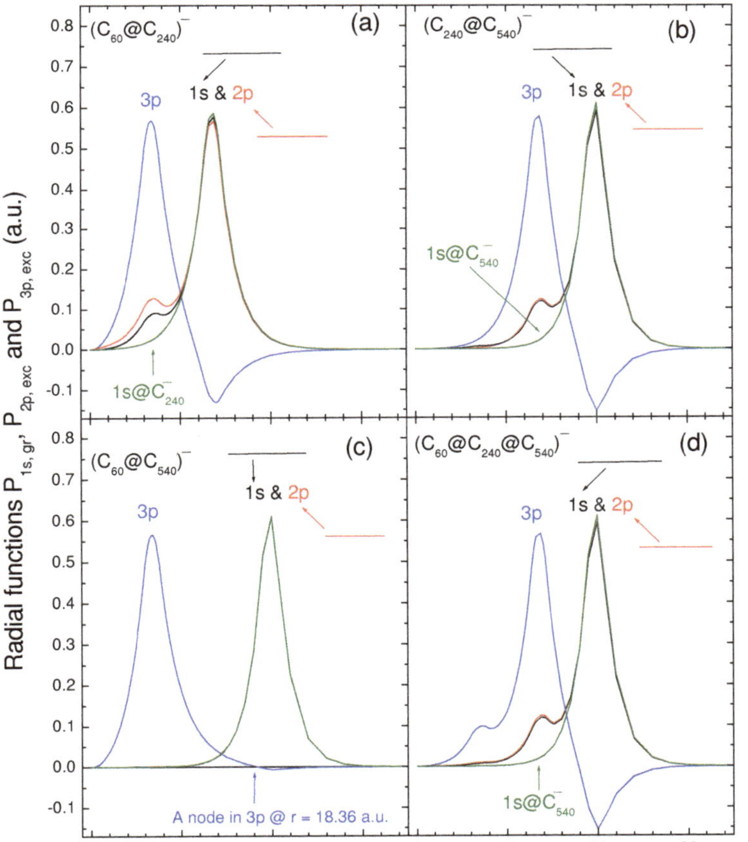

Figure 3. Calculated radial ground-state $P_{1s}(r)$ and excited-state P_{2p} and P_{3p} (due to the $1s \rightarrow np$ transitions, $n = 2, 3$) of the attached electron in fullerene onion-anions: (**a**) $(C_{60}@C_{240})^-$, (**b**) $(C_{240}@C_{540})^-$, (**c**) $(C_{60}@C_{540})^-$, (**d**) $(C_{60}@C_{240}@C_{540})^-$: solid, P_{2p}; dashed, P_{1s}; dash–dot, P_{3p}. Also plotted are the P_{1s} (dash–dot–dot) functions in bare C_{240}^- and C_{540}^-, as designated, for comparison purposes. Note, to avoid any confusion, that the graphs for the P_{1s} and P_{2p} functions in the fullerene onion-anions tightly overlap with each other and are practically indistinguishable from each other with some exception in the case of $(C_{60}@C_{240})^-$.

We interpret the predicted behavior of the P_{3p} excited-state functions in the fullerene onion-anions as being due to both the multi-well nature of the fullerene onion-anion potentials and the fact that, in contrast to the nodeless P_{1s} and P_{2p} functions, the P_{3p} function has one node. That is, the P_{3p} function is distinctly split into an inner part (before the node) and an outer part (beyond the node). It appears that the inner part of the P_{3p} function falls into the potential well associated with a fullerene cage adjacent to the outermost cage in the fullerene onion-anion. Thus, the attached electron partially resides in the inner well.

We note, though, that the behavior of the P_{3p} function in the fullerene onion-anions is somewhat reminiscent of the behavior of the excited P_{3d} and P_{4d} functions, excited from the $3p$ subshell, in endohedral calcium, Ca@C_{60} [30]. There, a significant transfer of the $4d$, but not $3d$, electron density into the inner space of C_{60} was demonstrated. That resulted in a significant increase in the amplitude of the P_{4d} orbital in the inner space of C_{60}. Consequently, the mean radius of the $4d$ orbital was reduced from $\bar{r}_{4d} \approx 14$ in free Ca to only $\bar{r}_{4d} \approx 4.3 < r_c = 5.8$ in Ca@C_{60} [30]. That situation, in turn, was commented on to be somewhat reminiscent of the behavior of the excited $4f$ and $5f$ orbitals in Ba^+ [31,32] that was shown to be due to the double-well nature of the potential of Ba^+ that caused partial orbital collapse of $5f$ into the inner well, thereby causing $5f$, rather than $4f$, to have the greater amplitude near $r = 0$.

3.3. Oscillator Strengths and Photodetachment Cross Sections

Calculated oscillator strengths, $f_{1s\to np}$, of the C_{60}^-, C_{240}^-, and C_{540}^- bare fullerene anions as well as the $(C_{60}@C_{240})^-$, $(C_{60}@C_{540})^-$, $(C_{240}@C_{540})^-$, and $(C_{60}@C_{240}@C_{540})^-$ fullerene onion-anions are/were listed in Table 1 which contains a wealth of information. Since a principal goal of this work is to explore the spectral distribution of oscillator strength, we focus on a comparison among the total oscillator strengths of the discrete spectra, i.e., $f_{1s\to 2p} + f_{1s\to 3p} \equiv f_{(2p+3p)}$, for the single and nested fullerene cages Thus, as we transition from $C_{240}^- \to (C_{60}@C_{240})^-$, the $f_{(2p+3p)}$ oscillator strength is increased. The same change in $f_{(2p+3p)}$ is characteristic along all other transition paths as well: $C_{540}^- \to (C_{240}@C_{540})^-$, $C_{540}^- \to (C_{60}@C_{240}@C_{540})^-$, and, in principle, $C_{540}^- \to (C_{60}@C_{540})^-$, too. Hence, we have unraveled a general tendency: stuffing of a bigger fullerene cage with a smaller fullerene cage, as well as progressively stuffing the biggest fullerene cage with several smaller fullerene cages, results in the transfer of a part of the oscillator strength from a continuum spectrum into a discrete spectrum.

Now, how does the discovered tendency affect the $1s$-photodetachment cross section, σ_{1s}? Obviously, the total area under the graph for σ_{1s} should be decreasing along the discussed fullerene transition paths. This may result in the disappearance of some of the resonance structures in σ_{1s}, or making them narrower, or decreasing their heights, or lowering the values of other parts of σ_{1s}, or all of the above cumulatively. It is, therefore, extremely interesting to study the modifications in σ_{1s}'s on a comparative one-to-one basis for different fullerene anions.

Calculated σ_{1s}'s for C_{240}^- versus $(C_{60}@C_{240})^-$, as well as C_{540}^- versus $(C_{60}@C_{540})^-$, $(C_{240}@C_{540})^-$, and $(C_{60}@C_{240}@C_{540})^-$ are depicted in Figure 4 as functions of the photoelectron momentum κ, in order to eliminate the impact of differences in $1s$ ionization potentials between the fullerene anions on details of σ_{1s}'s, for the adequacy of the comparison between these anions.

We first note that the calculated cross sections exhibit the oscillatory behavior versus the photoelectron momentum, k. Such resonances have been well understood for both photoionization and photodetachment of, as well as electron scattering by, fullerene and endo-fullerene complexes in a large body of research; we refer the reader to the above references, to the review paper [8] for many more references on the subject, as well as, e.g., to [4–8,10–24,29,31,38,39,46] from [21] (and references therein). Following [33], these resonances are commonly referred to as the *confinement resonances*.

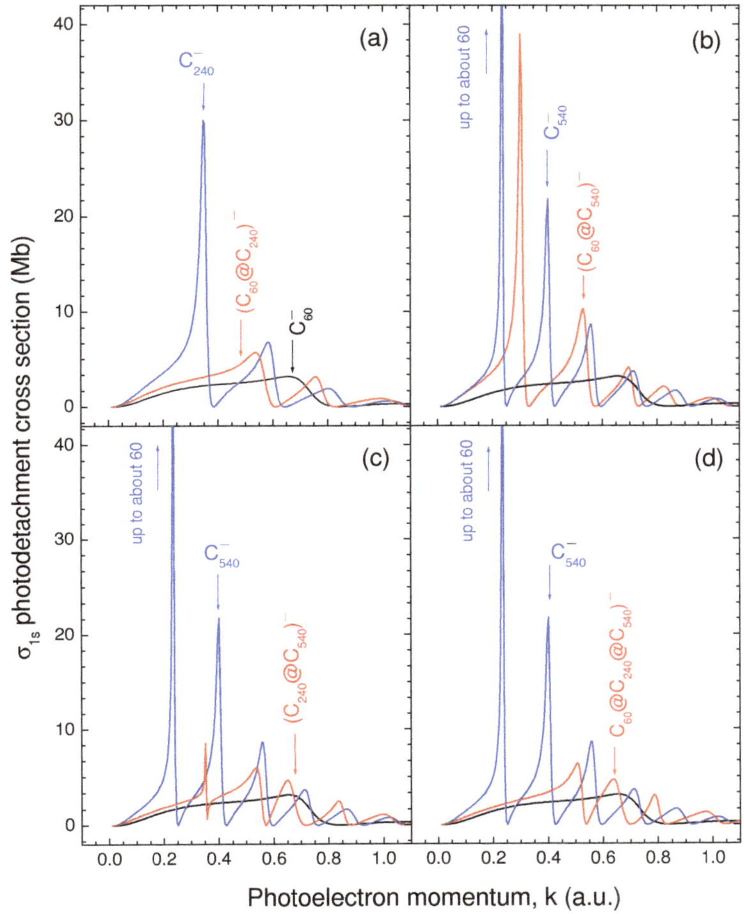

Figure 4. Calculated σ_{1s} photodetachment cross sections of bare fullerene anions and nested fullerene onion-anions, as designated in the figure. Note, on all parts of the figure, the σ_{1s} of C^-_{60} is represented by a dashed-line.

Secondly, note that the prediction mentioned above on the modification of the photodetachment cross section along the path from the bare fullerene anions to the double- and triple-cage fullerene onion-anions is seen to be correct. Indeed, we see the disappearance of one or even two confinement resonance structures (near the lower energy end of the spectrum), and the significant decrease in their amplitudes (except for the case of the $(C_{60}@C_{540})^-$ onion-anion, which is an extraordinary case anyway, as was discussed above). Furthermore, it is interesting that the resonances in σ_{1s} of $(C_{60}@C_{540})^-$ are seen to be shifted towards higher k's, compared to σ_{1s} of the bare $(C_{540})^-$ anion, whereas in all other nested fullerene onion-anions they shift toward lower k's, compared to corresponding bare counterparts.

Thirdly, it is quite interesting that σ_{1s}'s of all fullerene onion-anions, whether double-cage or triple-cage anions, do not differ much in magnitude from σ_{1s} of the smallest C^-_{60} anion in this sequence of fullerene anions (except for σ_{1s} of extraordinary $(C_{60}@C_{540})^-$ at a lower end of the spectrum). To emphasize this, we added σ_{1s} of C^-_{60} to all plots in Figure 4 to facilitate this comparison. One can see that σ_{1s}'s, associated with the nested fullerene onion-anions, oscillate around σ_{1s} of C_{60}^- with average amplitudes that are not much different from σ_{1s} of C_{60}^-.

Lastly, we note that, to check the connection of calculated σ_{1s}'s to calculated oscillator strengths, we calculated the oscillator strength of the continuum spectrum, $f_{1s\to\epsilon p}$, by appropriately integrating σ_{1s}'s in accordance with Equation (7). These calculated $f_{1s\to\epsilon p}$'s are presented in Table 1, and the fact that they have the same values as those obtained from the oscillator sum rule ($f_{1s\to\epsilon p} = 1 - f_{1s\to(2p+3p)}$) speaks to the accuracy of the calculated σ_{1s}'s.

4. Conclusions

In the present paper, we have provided a glimpse into the structure and photodetachment cross sections of bare fullerene anions and nested fullerene onion-anions and uncovered the existence of a number of unusual features. The results were obtained on a zeroth-order basis, so to speak. However, the zeroth-order basis is a valuable and necessarily part of any study of any multielectron atomic and molecular systems and processes. This is because, firstly, it provides a kind of a roadmap where more sophisticated theoretical studies of these systems should be conducted and, secondly, the comparison between the results obtained with a more accurate calculation with those obtained in this zeroth-order study is the only way to understand the importance and strength of the physical interactions that are not accounted in the framework of the zeroth-order approximation. We hope that the results of the present study will serve as an impetus to more complete theoretical studies of the structure and photodetachment spectra of fullerene (onion–)anions, now that we know that they might be quite unusual.

Author Contributions: Conceptualization, V.K.D. and S.T.M.; methodology, V.K.D. and S.T.M.; formal analysis, V.K.D. and S.T.M.; investigation, V.K.D. and S.T.M.; writ-ing—original draft preparation, V.K.D. and S.T.M.; writing—review and editing, V.K.D. and S.T.M.; All authors have read and agreed to the pub-lished version of the manuscript.

Funding: The work of STM was supported by the US Department of Energy, Office of Science, Basic Energy Sciences under Award Number DE-FG02-03ER15428.

Institutional Review Board Statement: Not applicable.

Informed Consent Statement: Not applicable.

Conflicts of Interest: The author declares no conflict of interest.

References

1. Kato, T. Absorption and emission spectra for C_{60}. *Laser Chem.* **1994**, *52*, 155. [CrossRef]
2. Brink, C.; Andersen, L.H.; Hvelplund, P.; Mathur, D.; Voldstad, J.D. Laser photodetachment of C_{60}^- and C_{70}^- ions cooled in a storage ring. *Chem. Phys. Lett.* **1995**, *52*, 233. [CrossRef]
3. Bilodeau, R.C.; Gibson, N.D.; Walter, C.W.; Esteves-Macaluso, D.A.; Schippers, S.; Müller, A.; Phaneuf, R.A.; Aguilar, A.; Hoener, M.; Rost, J.M.; et al. Single-Photon Multiple Detachment in Fullerene Negative Ions: Absolute Ionization Cross Sections and the Role of the Extra Electron. *Phys. Rev. Lett.* **2013**, *111*, 043003. [CrossRef] [PubMed]
4. Müller, A.; Martins, M.; Kilcoyne, A.L.D.; Phaneuf, R.A.; Hellhund, J.; Borovik, A.; Holste K.; Bari, S.; Buhr, T.; Klumpp, S.; et al. Photoionization and photofragmentation of singly charged positive and negative $Sc_3N@C_{80}$ endohedral fullerene ions. *Phys. Rev. A* **2019**, *99*, 063401. [CrossRef]
5. Phaneuf, R.A.; Kilcoyne, A.L.D.; Aryal, N.B.; Baral, K.K.; Esteves-Macaluso, D.A.; Thomas, C.M.; Hellhund, J.; Lomsadze, R.; Gorczyca, T.W.; Ballance, C.P.; et al. Probing confinement resonances by photoionizing Xe inside a C_{60}^+ molecular cage. *Phys. Rev. A* **2013**, *88*, 053402. [CrossRef]
6. Tanaka, H.; Boesten, L.; Onda, K.; Ohashi, O. Cross-beam experiment for the scattering of low-energy electron from gas phase C_{60}. *J. Phys. Soc. Jpn.* **1994**, *63*, 485. [CrossRef]
7. Tanaka, H.; Hoshino, M.; Brunger, M.J. Elastic and inelastic scattering of low-energy electrons from gas-phase C_{60}: Elastic scattering angular distributions and coexisting solid-state features revisited. *Eur. Phys. J. D* **2021**, *75*, 293. [CrossRef]
8. Deshmukh, P.C.; Manson, S.T. Electronic structure and dynamics of confined atoms. *Eur. Phys. J. D* **2021**, *75*, 166. [CrossRef]
9. Amusia, M.Y.; Baltenkov, A.S.; Krakov, B.G. Photodetachment of negative C_{60}^- ions. *Phys. Lett. A* **1998**, *243*, 99. [CrossRef]
10. Amusia, M.Y.; Baltenkov, A.S.; Chernysheva, L.V.; Felfli, Z.; Msezane, A.Z. Dramatic distortion of the 4*d* giant resonance by the C_{60} fullerene shell. *J. Phys. B* **2005**, *38*, L169. [CrossRef]

11. Amusia, M.Y.; Baltenkov, A.S. Effect of plasma oscillations of C_{60} collectivized electrons on photoionization of endohedral noble-gas atoms. *Phys. Rev. A* **2006**, *73*, 062723. [CrossRef]
12. Amusia, M.Y.; Chernysheva, L.V.; Liverts, E.Z. Photoionization of two-shell endohedral atoms. *Phys. Rev. A* **2009**, *80*, 032503. [CrossRef]
13. Amusia, M.Y.; Chernysheva, L.V.; Dolmatov, V.K. Confinement and correlation effects in the Xe@C_{60} generalized oscillator strengths. *Phys. Rev. A* **2011**, *84*, 063201. [CrossRef]
14. Amusia, M.Y. Fullerenes and endohedrals as "big atoms". *Chem. Phys.* **2013**, *414*, 168. 08.007. [CrossRef]
15. Amusia, M.Y.; Chernysheva, L.V. On the Behavior of Scattering Phases in Collisions of Electrons with Multiatomic Objects. *JETP Lett.* **2015**, *101*, 503. [CrossRef]
16. Dolmatov, V.K.; Amusia, M.Y.; Chernysheva, L.V. Effects of target polarization in electron elastic scattering off endohedral A@C_{60}. *Phys. Rev. A* **2017**, *95*, 012709. [CrossRef]
17. Amusia, M.Y.; Baltenkov, A.S. Time delay in electron-C_{60} elastic scattering in a dirac bubble potential model. *J. Phys. B* **2019**, *52*, 015101. [CrossRef]
18. Amusia, M.Y.; Chernysheva, L.V.; Dolmatov, V.K. Angle-differential elastic-electron scattering off C_{60}: A simple semi-empirical theory versus experiment. *J. Phys. B* **2019**, *52*, 085201. [CrossRef]
19. Lohr, L.L.; Blinder, S.M. Electron photodetachment from a Dirac bubble potential. A model for the fullerene negative ion C_{60}^{-}. *Chem. Phys. Lett.* **1992**, *198*, 100. [CrossRef]
20. Baltenkov, A.S.; Msezane, A.Z. Doubly-charged Negative Ion of C_{60}^{-} Molecule. *Proc. Dyn. Syst. Appl.* **2016**, *7*, 239.
21. Dolmatov, V.K. Spin-polarized photoelectron fluxes from fullerene anions. *Atoms* **2020**, *65*, 8040065. 8040065. [CrossRef]
22. Dolmatov, V.K.; Manson, S.T. Photoionization of atoms encapsulated in endohedral ions $A@C_{60}$. *Phys. Rev. A* **2006**, *73*, 013201. [CrossRef]
23. Wang, L.-S.; Conceicao, J.; Jin, C.; Smalley, R.E. Threshold photodetachment of cold C_{60}^{-}. *Chem. Phys. Lett.* **1991**, *182*, 5. [CrossRef]
24. Msezane, A.Z.; Felfli, Z. Recent Progress in Low-Energy Electron Elastic-Collisions with Multi-Electron Atoms and Fullerene Molecules. *Atoms* **2022**, *10*, 79. [CrossRef]
25. Puska, M.J.; Nieminen, R.M. Photoabsorption of atoms inside C_{60}. *Phys. Rev. A* **1993**, *47*, 1181. 47.1181. [CrossRef]
26. Dolmatov, V.K.; Edwards, A. Role of polarizability of a C_N fullerene cage in $A@C_N$ photoionization and $e^- - C_N$ scattering: The size effect. *J. Phys. B* **2019**, *52*, 105001. [CrossRef]
27. Winsted, C.; McKoy, V. Elastic electron scattering by fullerene, C_{60}. *Phys. Rev. A* **2006**, *73*, 012711. RevA.73.012711. [CrossRef]
28. Dolmatov, V.K.; Brewer, P.; Manson, S.T. Photoionization of atoms confined in giant single-walled and multiwalled fullerenes. *Phys. Rev. A* **2008**, *78*, 013415. [CrossRef]
29. Amusia, M.Y.; Chernysheva, L.V. *Computation of Atomic Processes: A Handbook for the ATOM Programs*; IOP Publishing Ltd.: Bristol, UK; Philadelphia, PA, USA, 1997.
30. Connerade, J.-P.; Dolmatov, V.K.; Manson, S.T. A unique situation for an endohedral metallofullerene. *J. Phys. B* **1999**, *32*, L395. [CrossRef]
31. Connerade, J.-P.; Mansfield, M.W.D. Centrifugal barrier perturbation of the nf^2F series in Ba II. *Proc. R. Soc. A* **1975**, *346*, 565. [CrossRef]
32. Connerade, J.-P. *Highly Excited Aroms*; Cambridge: Cambridge, UK, 1998.
33. Connerade, J.-P.; Dolmatov, V.K.; Manson, S.T. On the nature and origin of confinement resonances. *J. Phys. B* **2000**, *33*, 2279. [CrossRef]

atoms

MDPI

Article

Calculation of Low-Energy Positron-Atom Scattering with Square-Integrable Wavefunctions

Sarah Gregg * and Gleb Gribakin

School of Mathematics and Physics, Queen's University Belfast, Belfast BT7 1NN, UK; g.gribakin@qub.ac.uk
* Correspondence: sgregg07@qub.ac.uk

Abstract: The variational method is applied to the low-energy positron scattering and annihilation problem. The ultimate aim of the investigation is to find a computationally economical way of accounting for strong electron–positron correlations, including the effect of virtual positronium formation. The method is applied to the study of elastic s-wave positron scattering from a hydrogen atom. A generalized eigenvalue problem is set up and solved to obtain s-wave positron–hydrogen scattering phase shifts within 8×10^{-3} rad of accepted values. This is achieved using a small number of terms in the variational wavefunction; in particular, only nine terms that depend on the electron–positron distance are included. The annihilation parameter Z_{eff} is also calculated and is found to be in good agreement with benchmark calculations.

Keywords: theoretical atomic and molecular physics; positron; hydrogen; annihilation; phase shift; scattering

Citation: Gregg, S.K.; Gribakin, G.F. Calculation of Low-Energy Positron-Atom Scattering with Square-Integrable Wavefunctions. *Atoms* **2022**, *10*, 97. https://doi.org/10.3390/atoms10040097

Academic Editor: Yew Kam Ho

Received: 1 September 2022
Accepted: 18 September 2022
Published: 22 September 2022

Publisher's Note: MDPI stays neutral with regard to jurisdictional claims in published maps and institutional affiliations.

1. Introduction

The aim of this paper is to explore a numerically frugal method of including important electron–positron correlations in the calculations of positron (e^+) scattering from atoms and molecules. A good understanding of positron interactions with matter is a crucial element in the development of current and future applications of antimatter [1,2]. It is also important for tests of quantum electrodynamics [3] and fundamental experiments with antihydrogen [4].

Since the prediction [5] and discovery [6] of the positron's existence, many experimental and theoretical studies have been focussed on revealing the nature of its interactions with atoms and molecules [7]. Measurements and calculations show that low-energy positron interaction with atoms and molecules is characterized by strong electron–positron correlations. The first of these correlation effects is polarization of the target electron distribution by the positron. It gives rise to the attractive polarization potential with the asymptotic form $-\alpha e^2/2r^4$, where α is the dipole polarizability of the target, e is the charge of the projectile (positron), and r is the distance between the positron and the target. This polarization potential is similar to that which affects electron scattering.

The second correlation effect, which is specific to positrons, is *virtual positronium formation*. Positronium (Ps) is a light hydrogen-like atom that consists of an electron and a positron. Ps has a binding energy of $E_{\text{Ps}} = 6.8$ eV. For positron energies $\varepsilon > E_I - E_{\text{Ps}}$, where E_I is the ionization energy of the target, Ps formation is an important ionization channel in positron collisions. For targets with $E_I > E_{\text{Ps}}$ and $\varepsilon < E_I - E_{\text{Ps}}$, the Ps formation channel is closed. However, atomic electrons can still tunnel from an atom or molecule to the positron to form a Ps-like state temporarily. This effect makes a distinct and sizeable attractive contribution to the interaction of low-energy positrons with atoms and molecular targets. At the same time, this contribution makes positron scattering and annihilation calculations particularly challenging.

Amusia and co-workers [8] were probably the first to recognize the importance of virtual Ps formation. They were able to incorporate this effect and gauge its magnitude using many-body theory calculations for He. (Very recently, this approach was used for positron scattering from atoms with half-filled valence shells [9].) A more accurate approximation for the Ps formation contribution to the positron–atom correlation potential [10] enabled predictions of positron binding to neutral atoms [11] and reliable calculations of positron scattering from noble gas atoms [12]. Ultimately, a consistent *ab initio* method for calculating the Ps formation contribution was developed and tested [13]. It provided a complete and accurate picture of positron scattering and annihilation from noble-gas atoms [14], and has now been generalized to molecular calculations that can yield high-quality predictions of positron–molecule binding energies [15].

Many-body theory allows one to identify the virtual Ps formation contributions with a particular class of diagrams that contribute to the positron–target correlation potential. When other approaches are used, the physical effect of virtual Ps formation is still present, but it manifests itself in a different way. In single-center convergent close-coupling calculations of positron scattering from hydrogen, one observes it as slow convergence with respect to the maximum orbital momentum of the electron and positron states used [16]. This is also seen in configuration–interaction calculations of positron–atom bound states [17]. Such high-angular-momentum states are needed to describe an electron–positron pair (Ps) localized some distance away from the atomic nucleus. This "problem" is immediately removed, however, when a two-center approach is used, in which functions that depend on the electron–positron distance (and hence, describe Ps) are included in the expansion of the wavefunction [18]. It was also seen in Kohn-variational calculations [19,20] that the inclusion of such "virtual Ps" terms in the wavefunction yields significant improvements in the convergence of the scattering phase shifts and a pronounced enhancement of the positron annihilation rate at energies just below the Ps formation threshold. Finally, when the Schwinger multichannel method is used for positron scattering from molecules [21], calculations are significantly improved by adding basis states on extra centers placed away from the atomic nuclei [22]. In this case, such centers help to describe Ps formed virtually outside the molecule. Such "ghost" centers are also used in the most sophisticated many-body theory calculations of positron–molecule binding to enable the accurate description of virtual Ps formation [15].

It can be seen from the above that a well-converged positron scattering calculation should either include a large number of wavefunction terms centered on the nuclei or include terms with explicit dependence on the electron–positron distance. The first approach is more straightforward numerically but may lead to very large basis sizes. The second one is more economical but with an added complexity of dealing with a multicenter problem. It is the latter approach that we want to explore, aiming to include as few correlation terms as strictly necessary to obtain good-quality scattering and annihilation data.

In this paper, the scattering and annihilation of positrons is explored for the positron–hydrogen system through use of the variational method with square-integrable trial wavefunctions. Important correlation effects, including that of virtual Ps formation, are accounted for by including functions which depend on the electron–positron distance. The variational method is set up as a generalized eigenvalue problem. From this, elastic s-wave phase shifts δ and the annihilation parameter Z_{eff} for the e^+-H system are calculated at energies below the Ps formation threshold. Good agreement with benchmark values of both the phase shifts [23–25] and the annihilation parameter [26] is achieved using only a small number of terms in the wavefunction. By providing evidence that this method is a valid approach to the problem, avenues for future research are opened in which more complex matter–antimatter interactions may be explored. It should be added that the positron–hydrogen system has long been used as a testbed for various calculation methods, with many accurate results available at both low and high energies (see, e.g., Refs. [27–29]).

The paper is structured as follows. In Section 2, we set up the generalized eigenvalue problem which is employed to solve the scattering problem and show how to obtain the scattering phase shifts from bound-state calculations. In Section 3, the method is applied to elastic s-wave scattering of a positron from a hydrogen atom. Three sets of phase shift results are presented, beginning with a simple model and progressing toward more detailed descriptions of the system. In Section 4, the annihilation parameter Z_{eff} is calculated using the same trial wavefunctions as in Section 3. Section 5 summarizes the work and indicates its future applications.

2. Scattering as a Bound-State Problem

In this section, we recap how a simple variational method can be used to calculate s-wave elastic scattering phase shifts for scattering from an atomic target.

2.1. Generalized Eigenvalue Problem

The method begins with the choice of a trial wavefunction. Consider a system in the state $|\Psi\rangle$ expanded in terms of a set of linearly independent square-integrable basis functions $\{|\varphi_i\rangle\}_{i=1}^{N}$ which, in general, are neither normalized nor orthogonal:

$$|\Psi\rangle = \sum_{i=1}^{N} c_i \, |\varphi_i\rangle \,. \tag{1}$$

This basis is chosen at the beginning of the problem, and the coefficients c_i are the variational parameters.

Central to the problem is the minimization of the energy functional,

$$\langle E\rangle = \langle\Psi| \, \hat{H} \, |\Psi\rangle \,, \tag{2}$$

with respect to the parameters c_i, whilst holding $\langle\Psi|\Psi\rangle = 1$. The minimum energy calculated using a trial wavefunction provides an upper bound on the exact ground-state energy of the system.

The normalization constraint is imposed during the minimization through use of a Lagrange multiplier E. At the minimum (or a stationary point), we require

$$\frac{\partial}{\partial c_k}\left[\langle\Psi| \, \hat{H} \, |\Psi\rangle - E(\langle\Psi|\Psi\rangle - 1)\right] = 0, \qquad k = 1, \ldots, N. \tag{3}$$

Substituting the expansion of $|\Psi\rangle$ from (1) into (3) gives a system of N linear equations. Assuming that the c_i values are independent of each other, performing partial differentiation with respect to a particular c_k yields the following:

$$\sum_j c_j \, \langle\varphi_k| \, \hat{H} \, |\varphi_j\rangle + \sum_i c_i \, \langle\varphi_i| \, \hat{H} \, |\varphi_k\rangle - E\left(\sum_j c_j \, \langle\varphi_k|\varphi_j\rangle + \sum_i c_i \, \langle\varphi_i|\varphi_k\rangle\right) = 0. \tag{4}$$

Since the matrix elements of the Hamiltonian are real (assuming real basis functions $|\varphi_i\rangle$), the first two sums in (4) are identical and hence may be combined. Similarly, the scalar product of any two basis functions in our problem is real, and the second pair of sums may also be combined. This results in the following equation:

$$\sum_i \underbrace{\langle\varphi_k| \, \hat{H} \, |\varphi_i\rangle}_{H_{ki}} c_i = E \sum_i \underbrace{\langle\varphi_k|\varphi_i\rangle}_{Q_{ki}} c_i, \tag{5}$$

where H_{ki} and Q_{ki} are the elements of matrices $\boldsymbol{H}$ and $\boldsymbol{Q}$, respectively. Hence, (5) takes the form of a matrix equation:

$$\boldsymbol{H}\boldsymbol{c} = E\boldsymbol{Q}\boldsymbol{c}, \tag{6}$$

where the vector $\boldsymbol{c}$ contains the c_i values.

The eigenvalues E_n of the generalized eigenvalue problem (6) are energy eigenvalues of the system with Hamiltonian $\hat{H}$, with the state $|\Psi\rangle$ defined by coefficients c_i, i.e., the elements of the corresponding eigenvector. For a system that has a few bound states or no bound states at all, most of the energy eigenvalues will lie in the continuum. The corresponding states $|\Psi\rangle$, often called *pseudostates*, will not be the true states of the system that represent scattering states. However, it is possible to use the energies and wavefunctions of the pseudostates to determine important properties of the scattering states, e.g., phase shifts or (for positrons) the normalized annihilation rate $Z_{\rm eff}$.

In this work, the generalized eigenvalue problem is solved using Python's `eigh` function [30] which, given matrices $\boldsymbol{H}$ and $\boldsymbol{Q}$, provides the energy eigenvalues and normalized eigenvectors of the system.

2.2. Scattering Phase Shifts

Once the energy eigenvalues have been calculated, they can be used to find the phase shifts δ, e.g., for s-wave scattering. This assumes that the target is spherically symmetric and states $|\Psi\rangle$ have zero total angular momentum. The method used by Gribakin and Swann in [31] is implemented here. Firstly, the eigenvalue problem is solved for a free particle, i.e., using a chosen basis-state expansion but neglecting the interaction between the projectile and the target in $\hat{H}$. The free-particle energy eigenvalues are denoted $E_n^{(0)}$ and increase monotonically with n. Hence, it is possible to introduce an invertible function $f(n)$ such that

$$f(n) = E_n^{(0)} \tag{7}$$

for $n = 1, \ldots, N$. Next, the eigenvalue problem is solved with the full Hamiltonian using the same basis. The corresponding eigenvalues E_n are shifted with respect to those in equation (7), which can be written as

$$E_n = f\left(n - \frac{\delta}{\pi}\right), \tag{8}$$

where δ is the phase shift [31].

Rearranging (8), the phase shifts may be extracted as a function of the energy eigenvalues E_n:

$$\delta = \left[n - f^{-1}(E_n)\right]\pi. \tag{9}$$

With the introduction of a phase shift δ, the function $f(n)$ must now be defined for real values of n, and equivalently, its inverse must be defined for values of energy other than the free-particle energy eigenvalues. The value of $f^{-1}(E)$ for these intermediate values of energy may be found by interpolating between the free-particle energy eigenvalues. However, for particular bases, e.g., those using even-tempered exponents, the function $f(n)$ varies rapidly and is difficult to interpolate accurately. In this case, a new function $g(\ln E)$ can be defined, such that $f^{-1}(E) \equiv g(\ln E)$; this function changes more slowly, making accurate interpolation possible. In the problems that follow, this interpolation is completed using a cubic spline fit to the data for integer n. The phase shift may then be calculated as

$$\delta = [n - g(\ln E_n)]\pi. \tag{10}$$

In the following analysis, the phase shift will be considered as a function of momentum $k = \sqrt{2E}$ rather than energy E, where we use atomic units and assume that the projectile (positron) has a unit mass and set $E = 0$ at the continuum threshold.

3. Calculation of Elastic s-Wave Positron–Hydrogen Phase Shifts

In this section, the variational method is used to calculate elastic scattering phase shifts for a positron scattering from a hydrogen atom. The calculation is restricted to s-wave scattering which dominates at low positron energies.

3.1. One-Particle Problem

To test the method, we consider the simple problem of positron scattering from a "frozen" hydrogen atom. In this case, the wavefunction depends only on the distance between the positron and the nucleus. Figure 1 contains a schematic diagram of the system with the interparticle distances labeled.

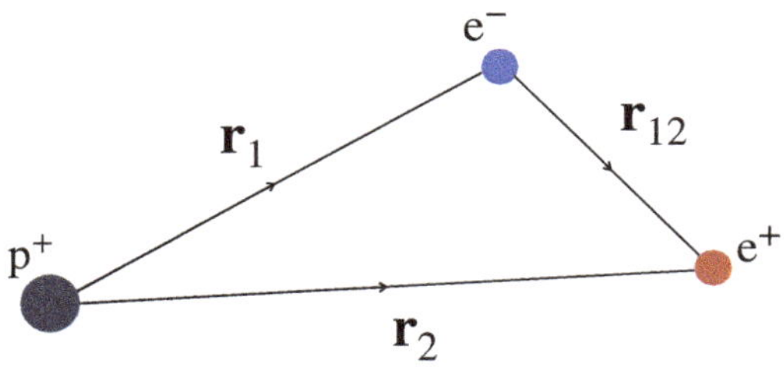

Figure 1. A diagram of the positron–hydrogen system with the interparticle distances labeled.

In the frozen-target approximation, the electron moves in the field of the nucleus (considered infinitely massive) and is "fixed" in the ground (1*s*) state of the atom. The dependence of the wavefunction on r_{12} is neglected. Hence, the total wavefunction becomes a product of the 1*s* electron wavefunction and the unknown positron wavefunction, which we denote $\Psi(r)$, with $r \equiv r_2$. Considering $\Psi(r)$ as the wavefunction of the radial motion, the boundary condition at the origin $\Psi(0) = 0$ is imposed. In this problem, we select a trial wavefunction of the form

$$\Psi(r) = r \sum_{i=1}^{N} c_i \exp(-\beta_i r), \tag{11}$$

where c_i are variational parameters and β_i are chosen real exponents. The corresponding basis functions are

$$\varphi_i(r) = r \exp(-\beta_i r), \qquad i = 1, \dots, N. \tag{12}$$

In the following calculations, the exponents β_i are chosen as

$$\beta_i = \beta_1 \zeta^{i-1}, \qquad i = 1, \dots, N, \tag{13}$$

i.e., forming an even-tempered basis, with $\zeta = 1.5$, $N = 20$ and $\beta_1 = 0.01$.

The elements of matrix $\boldsymbol{Q}$ are calculated as follows:

$$Q_{ij} = \int_0^\infty \varphi_i(r)\varphi_j(r) dr = \frac{2}{(\beta_i + \beta_j)^3}. \tag{14}$$

Next, the Hamiltonian for the system is considered[1]. The electrostatic potential of the ground-state hydrogen atom is (see, e.g., Ref. [32], §36, Problem 2):

$$\hat{U} = \frac{1}{r} + \phi_e(r) = \left(\frac{1}{r} + 1\right) e^{-2r}, \tag{15}$$

where $\phi_e(r)$ is the mean-field potential of the electron cloud and the $1/r$ term accounts for the positron-nucleus interaction.

In addition to $\hat{U}$, the Hamiltonian of the radial motion of the positron contains its kinetic energy, hence

$$\hat{H} = -\frac{1}{2}\frac{d^2}{dr^2} + \left(\frac{1}{r} + 1\right) e^{-2r}. \tag{16}$$

The Hamiltonian matrix elements are calculated as follows:

$$H_{ij} = \int_0^\infty \varphi_i(r)\left(-\frac{1}{2}\frac{d^2\varphi_j(r)}{dr^2}\right)dr + \int_0^\infty \varphi_i(r)\left(\frac{1}{r}+1\right)e^{-2r}\varphi_j(r)dr,$$
$$= -\frac{1}{2}\frac{\beta_i\beta_j}{(\beta_i+\beta_j)^3} + \frac{\beta_i+\beta_j+4}{(\beta_i+\beta_j+2)^3}. \tag{17}$$

The generalized eigenvalue problem (6) for the matrices (14) and (17) was then solved using a simple Python code, and the phase shifts were calculated as described in Section 2.2.

Figure 2 is a plot of n against $\ln E$ for the basis (13). It explains how the phase shifts are found from the pseudostate energy eigenvalues. By construction, for the free-positron eigenvalues $E_n^{(0)}$ (obtained using $\hat{H} = -\frac{1}{2}d^2/dr^2$), the function $g(\ln E_n^{(0)})$ takes integer values, but for the eigenvalues E_n of the positron in the static hydrogen potential, this function takes non-integer values $n - \delta/\pi$, which yield δ for specific positron energies E_n.

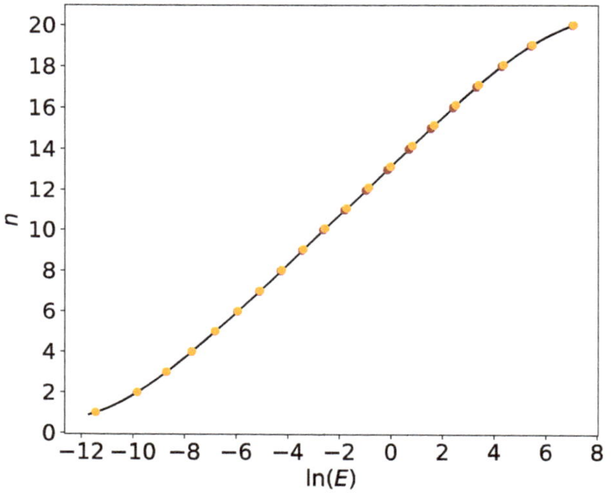

Figure 2. Red circles: values of $n = 1, 2, \ldots$ plotted against $\ln E_n^{(0)}$. Black line: the function $n = g(\ln E)$ obtained using cubic-spline interpolation between the free-particle eigenvalue data. Yellow circles: the points on the interpolated curve for the positron energies E_n in the static hydrogen potential. From this, it can be seen that $\ln E_n$ corresponds to non-integer ordinates $n - \delta/\pi$.

The phase shifts for low-momentum positrons were compared to those obtained from a numerical solution of the radial Schrödinger equation in the static hydrogen potential (obtained using the codes described in [33]). This comparison is displayed in Figure 3. There is a good agreement between these sets of data, especially at low momenta k, providing evidence that the present variational method allows one to extract the scattering phase shifts from a simple bound-state calculation.

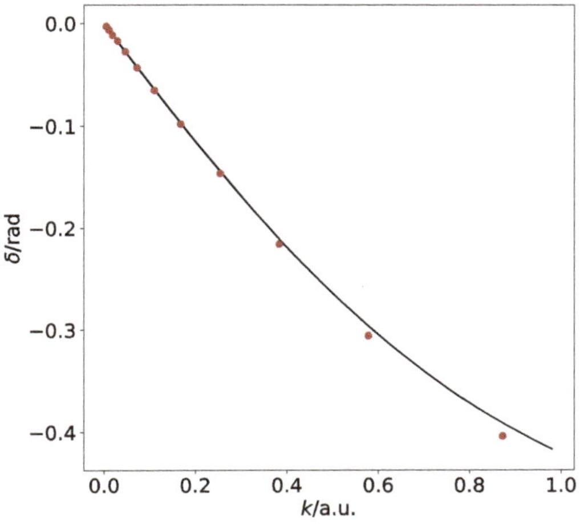

Figure 3. Positron–hydrogen s-wave scattering phase shifts in the static approximation. Red data points: phase shifts obtained using the present variational method. Black line: data obtained by solving the radial Schrödinger equation using the suite of codes described in [33].

3.2. Two-Particle Problem

In this section, a full two-particle dynamics of positron scattering from a hydrogen atom is considered. Here, the electron is no longer fixed in the 1s state of the hydrogen atom and generally, the wavefunction for this system will depend on the distances between all three pairs of particles[2], as labeled in Figure 1. A wavefunction of the following form will be considered:

$$\Psi(r_1, r_2, r_{12}) = \sum_{i=1}^{N} c_i \exp(-\alpha_i r_1 - \beta_i r_2 + \gamma_i r_{12}), \tag{18}$$

where the values of coefficients α_i, β_i and γ_i are chosen and the constants c_i are the variational parameters. The integrals to be evaluated in this section are greatly simplified by using the elliptic (Hylleraas [34]) coordinate system ($s = r_1 + r_2, t = r_1 - r_2, u = r_{12}$), so these coordinates are employed to carry out all of the calculations. The full set of standard integrals used is found in Appendix A. Our calculations begin with a simplified version of (18) using a single value of $\alpha_i = 1$ and $\gamma_i = 0$, i.e., equivalent to the frozen-target approximation of Section 3.1, gradually building toward the more general case. With each added element of flexibility in the wavefunction, a more accurate solution to the scattering problem is obtained.

Labeling the basis functions

$$\varphi_i(r_1, r_2, r_{12}) = \exp(-\alpha_i r_1 - \beta_i r_2 + \gamma_i r_{12}), \tag{19}$$

elements of the overlap matrix $\boldsymbol{Q}$ are calculated as follows:

$$Q_{ij} = \int \varphi_i \varphi_j d\tau, \tag{20}$$

where $d\tau = \pi^2(s^2 - t^2)uds\,dt\,du$ is the volume element, and the integration is over $0 \le s < \infty, 0 \le u \le s, -u \le t \le u$. Substituting φ_i from (19) into (20), we find the overlap integral in the form

$$Q_{ij} = \int \exp\left[2(-A_{ij}s - B_{ij}t + \Gamma_{ij}u)\right] d\tau, \tag{21}$$

where

$$A_{ij} = \frac{1}{4}(\alpha_i + \beta_i + \alpha_j + \beta_j), \tag{22}$$

$$B_{ij} = \frac{1}{4}(\alpha_i - \beta_i + \alpha_j - \beta_j), \tag{23}$$

$$\Gamma_{ij} = \frac{1}{2}(\gamma_i + \gamma_j). \tag{24}$$

This integral shares its structure with the standard integral $\tilde{I}_1$ from Appendix A; hence,

$$Q_{ij} = \tilde{I}_1(A_{ij}, \Gamma_{ij}, B_{ij}). \tag{25}$$

The Hamiltonian operator of the system is given by

$$\hat{H} = -\frac{1}{2}\nabla_1^2 - \frac{1}{2}\nabla_2^2 - \frac{1}{r_1} + \frac{1}{r_2} - \frac{1}{r_{12}}. \tag{26}$$

The first two terms represent the kinetic energy of the electron and positron $\hat{T}$. The third and fourth terms describe the interaction of the electron and positron with the nucleus $\hat{U}$, and the final term represents the electron–positron interaction $\hat{V}$. Hence, the Hamiltonian matrix element H_{ij} is considered as the sum of three contributions:

$$H_{ij} = T_{ij} + U_{ij} + V_{ij}. \tag{27}$$

In the elliptic coordinates, the expectation value of the kinetic energy takes the form

$$\begin{aligned}\langle\Psi|\,\hat{T}\,|\Psi\rangle = \int\Bigg\{&\left(\frac{\partial\Psi}{\partial s}\right)^2 + \left(\frac{\partial\Psi}{\partial t}\right)^2 + \left(\frac{\partial\Psi}{\partial u}\right)^2 \\ &+ \frac{2}{u(s^2-t^2)}\frac{\partial\Psi}{\partial u}\left[s(u^2-t^2)\frac{\partial\Psi}{\partial s} + t(s^2-u^2)\frac{\partial\Psi}{\partial t}\right]\Bigg\}d\tau.\end{aligned} \tag{28}$$

Replacing one of the Ψ by ϕ_i and the other by ϕ_j, and mapping the integrals that arise to the set of standard integrals in Appendix A, one obtains

$$\begin{aligned}T_{ij} = \frac{1}{4}&\left[(\alpha_i+\beta_i)(\alpha_j+\beta_j) + (\alpha_i-\beta_i)(\alpha_j-\beta_j) + 4\gamma_i\gamma_j\right]\tilde{I}_1(A_{ij},\Gamma_{ij},B_{ij}) \\ &-\gamma_i\left[(\alpha_j+\beta_j)\tilde{I}_2(A_{ij},\Gamma_{ij},B_{ij}) + (\alpha_j-\beta_j)\tilde{I}_3(A_{ij},\Gamma_{ij},B_{ij})\right].\end{aligned} \tag{29}$$

The matrix element of the electron and positron interaction with the nucleus is

$$U_{ij} = \int \varphi_i\left[-\frac{2}{s+t} + \frac{2}{s-t}\right]\varphi_j d\tau. \tag{30}$$

This integral is reduced to the standard integrals $\tilde{J}_1$ and $\tilde{J}_3$ (Appendix A), which gives

$$U_{ij} = -2[\tilde{J}_1(A_{ij},\Gamma_{ij},B_{ij}) - \tilde{J}_3(A_{ij},\Gamma_{ij},B_{ij})] + 2[\tilde{J}_1(A_{ij},\Gamma_{ij},B_{ij}) + \tilde{J}_3(A_{ij},\Gamma_{ij},B_{ij})]. \tag{31}$$

Clearly, the $\tilde{J}_1$ terms will cancel here. However, when calculating the free-positron energy eigenvalues, only the first bracketed term on the right-hand side of this equation is required, since the positron–nucleus interaction (second term) is not included in the free-positron Hamiltonian. When both of the Coulomb terms are included, the expression simplifies to

$$U_{ij} = 4\tilde{J}_3(A_{ij},\Gamma_{ij},B_{ij}). \tag{32}$$

Lastly, the matrix element of the electron–positron Coulomb interaction is given by

$$V_{ij} = -\int \varphi_i \frac{1}{u} \varphi_j d\tau. \tag{33}$$

This integral reduces to the standard integral $\tilde{J}_2$ in Appendix A to give

$$V_{ij} = -\tilde{J}_2(A_{ij}, \Gamma_{ij}, B_{ij}). \tag{34}$$

Combining the results in (29), (32) and (34), an expression for the Hamiltonian matrix element is obtained:

$$\begin{aligned} H_{ij} = &\frac{1}{4}\left[(\alpha_i + \beta_i)(\alpha_j + \beta_j) + (\alpha_i - \beta_i)(\alpha_j - \beta_j) + 4\gamma_i\gamma_j\right]\tilde{I}_1(A_{ij}, \Gamma_{ij}, B_{ij}) \\ &- \gamma_i\left[(\alpha_j + \beta_j)\tilde{I}_2(A_{ij}, \Gamma_{ij}, B_{ij}) + (\alpha_j - \beta_j)\tilde{I}_3(A_{ij}, \Gamma_{ij}, B_{ij})\right] \\ &+ 4\tilde{J}_3(A_{ij}, \Gamma_{ij}, B_{ij}) - \tilde{J}_2(A_{ij}, \Gamma_{ij}, B_{ij}). \end{aligned} \tag{35}$$

Note that for $\gamma_i \neq 0$, the matrix element (35) derived using equation (28) is not symmetric, i.e., $H_{ij} \neq H_{ji}$. Hence, H_{ij} must be replaced by the symmetrized combination

$$H'_{ij} = \frac{1}{2}(H_{ij} + H_{ji}), \tag{36}$$

before solving the generalized eigenvalue problem (6), which is consistent with the derivation in Section 2.1.

After setting up the Hamiltonian and overlap matrices, the generalized eigenvalue problem (6) is solved for the energy eigenvalues and eigenvectors. The eigenvalues for the free positron (omitting the positron–nucleus and positron–electron interaction terms) and those for the full Hamiltonian are analyzed to extract the phase shifts, as outlined in Section 2.2. It is noted here that solving (6) provides energy eigenvalues of the whole system. Hence, to obtain the positron energies $E_n^{(0)}$ and E_n, the energy of the ground-state hydrogen atom (-0.5 a.u.) must be subtracted from the eigenvalues.

3.2.1. Reproducing the Frozen-Target Results.

In the first instance, the frozen-target problem is revisited to check that the two-particle code produces the same results as the one-particle code. The dependence of the wavefunction on r_{12} is eliminated by setting $\gamma_i = 0$. In addition, the electron is fixed in the ground state by setting $\alpha_i = 1$. These restrictions will subsequently be lifted to allow the electron to move and to account for the electron–positron correlations. As before, describing the positron requires a wide range of exponents β_i, which are defined as in (13).

Taking all of this into account, we have the following wavefunction:

$$\Psi(r_1, r_2) = \sum_{i=1}^{N} c_i \exp[-r_1 - \beta_i r_2]. \tag{37}$$

Figure 4 shows that the corresponding phase shifts match those from the one-particle problem 3.1. Figure 4 also shows the phase shifts over a larger range of positron momenta. As expected, at large projectile energies, the scattering phase shift tends to zero. Note also that the phase shift is negative at all energies. This is a consequence of the positron–atom interaction being repulsive in the frozen-target approximation.

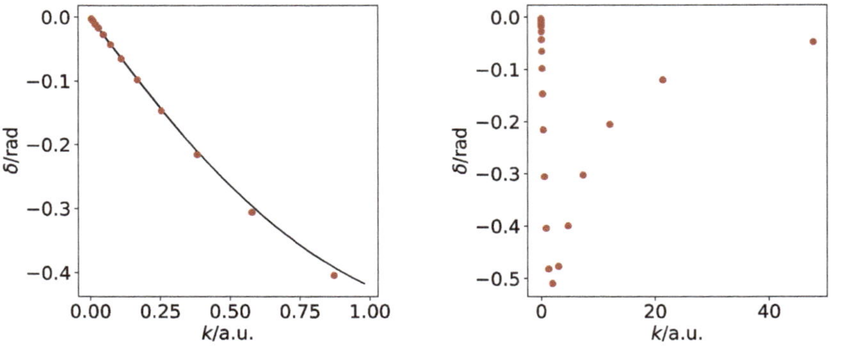

Figure 4. Positron–hydrogen s-wave scattering phase shifts in the static approximation calculated using a two-particle model (red circles), plotted over two contrasting ranges of k. The meaning of the black curve is the same as in Figure 3. Agreement with Figure 3 can be noted.

3.2.2. Variation of α: Radial Correlations.

To probe the effect of electron–positron radial correlations, the wavefunction is augmented by including extra terms with $\alpha_i \neq 1$. Physically, this adjustment allows the incident positron to cause displacement of the atomic electron in the radial direction due to the attraction between the two particles. Inclusion of a term, or several terms, in the wavefunction with $\alpha_i = 0.5$ will facilitate this type of distortion. The wavefunction in this case is written as

$$\Psi(r_1, r_2) = \sum_{i=1}^{N} c_i \exp[-\alpha_i r_1 - \beta_i r_2]. \tag{38}$$

For now, the restriction on the γ_i parameters remains in place, in that $\gamma_i = 0$ and the dependence of the wavefunction on the electron–positron distance is neglected. Note that this approximation corresponds to the so-called Temkin–Poet model that was used earlier to test electron–hydrogen and positron–hydrogen scattering [35].

Firstly, a single term is added with $\alpha_i = 0.5$ and $\beta_i = 0.4$. This value of β_i is chosen because we expect radial correlations to be important when the positron is close to the hydrogen atom. At such distances, it is most likely to attract the electron sufficiently to cause significant distortion. Terms with $\alpha_i = 0.5$ and a full range of β_i values are not immediately introduced because we aim to achieve good accuracy with as few correlation terms as possible. Hence, extra terms are introduced individually to test their importance: if a notable change in the phase shift is seen by adding a particular term, the term is retained and used in the basis. If not, the term is discarded and a different choice is made.

Phase shifts were obtained for various sets of parameters, using up to 20 additional terms. It was found that overall, the effect on the phase shift from adding these terms is small. In Figure 5, a set of results is displayed for a calculation with just three additional terms in the basis, which were found to generate a close-to-maximum shift from the frozen-target results (with $\alpha_i = 0.5$ and $\beta_i = 0.2, 0.4, 1.0$).

A comparison of Figure 5 with Figure 4 shows that the phase shifts have become slightly less negative due to the addition of the extra terms. This means that electron–positron correlations make the positron–atom interaction less repulsive than in the frozen-target case. However, the overall effect of allowing for the radial correlations between the electron and positron remains very small.

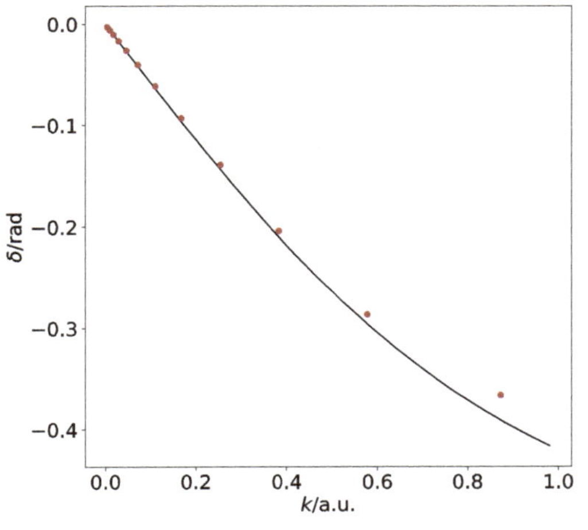

Figure 5. Phase shifts obtained by addition of three terms with $\alpha_i \neq 1$ to the basis. Black line: numerical solution of the radial Schrödinger equation in the static potential of the hydrogen atom included for comparison.

3.2.3. Nonzero γ: Effect of Angular Correlations.

In this section, the flexibility of the wavefunction is increased further by allowing for nonzero values of γ_i, so that the wavefunction takes the most general form

$$\Psi(r_1, r_2, r_{12}) = \sum_{i=1}^{N} c_i \exp(-\alpha_i r_1 - \beta_i r_2 + \gamma_i r_{12}). \tag{39}$$

The addition of r_{12}-dependence to the wavefunction allows for much stronger correlation between the positron and the electron. Physically, these terms account for effects such as virtual Ps formation and polarization of the atom by the positron. In particular, setting $\gamma_i = -0.5$ corresponds to the ground-state Ps wavefunction, allowing the calculation to account for the effect of virtual positronium formation. Note that formation of "real", free Ps is not possible in the chosen energy range, as the incident positron momenta are kept below the Ps formation threshold.

In general, terms with any values of β_i and γ_i may be used, provided that

$$\beta_i - \gamma_i > 0, \tag{40}$$

to ensure that $\Psi(r_1, r_2, r_{12}) \to 0$ for $r_2 \to \infty$.

Taking all of this into consideration, an initial basis was set up identically to that in the frozen-target problem with all values of $\gamma_i = 0$ and $\alpha_i = 1$. The nonzero γ_i terms were added one by one. Quasi-optimal values of the parameters for these extra terms were selected by completing the calculations for different sets of exponents and keeping the term which caused the largest upward change in the phase shifts overall. In the present approach, larger phase shifts are obtained when the energy eigenvalues E_n are lower relative to $E_n^{(0)}$. In a variational calculation, lower energy eigenvalues are obtained when better wavefunctions are used. Hence, it is correct to assume that the best possible choice of terms to add to the basis is that which yields the largest values for the phase shifts. Physically, including electron–positron correlations allows for positron attraction to the atom, increasing the value of the phase shift. Small adjustments are made to all three parameters near the optimum to ensure the best possible value of each parameter, correct to two decimal places.

For a single correlation term, the optimal values of α, β and γ were found to be $\alpha = 0.80$, $\beta = 0.04$ and $\gamma = -0.54$. Once the first term had been optimized, a second was added. The values of the parameters for this term were also selected in the manner described above. With each additional term, an improvement (i.e., increase) in the phase shift values is seen.

This process of adding an individual term may be continued for as many terms as required to reach a desired level of accuracy. However, our aim was to achieve good accuracy using as few terms as possible. Hence, the process was terminated after including a maximum of nine additional terms, yielding a basis with a total of 29 functions.

Figure 6 is an overview of the phase shifts obtained using one, three, five and nine additional terms. These results are also shown in Table 1. Details of the parameters used in these calculations can be found in Appendix B. The results obtained here can be compared to the accurate phase shifts, such as those calculated by Schwartz [23] or, later, by Humberston et al. [24]. In Figure 6, the frozen-target phase shifts are also plotted as a lower bound, while the accurate results from a Kohn variational calculation by Humberston et al. [24] provide an upper bound. It is remarkable that including a single well-chosen correlation term with $\alpha = 0.80$, $\beta = 0.04$ and $\gamma = -0.54$ provides about 80% of the increase in the phase shift with respect to the uncorrelated frozen-target result. Adding the next few correlation terms brings the variational phase shift to within 0.01 rad of the benchmark result.

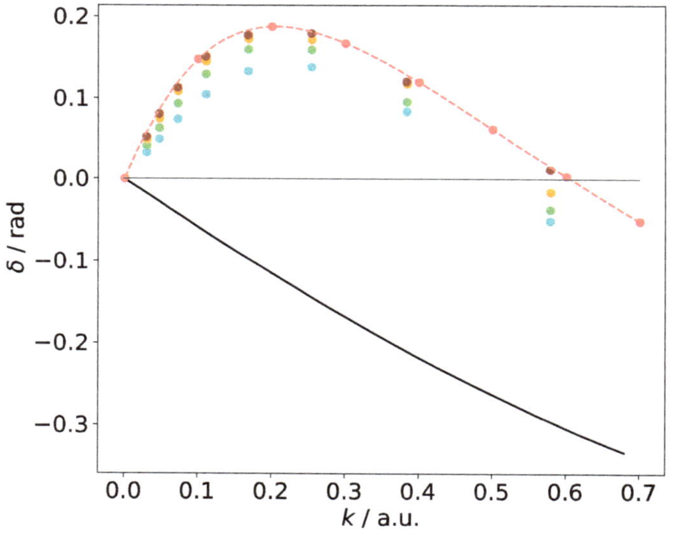

Figure 6. Positron–hydrogen s-wave scattering phase shifts obtained using various numbers of terms with $\gamma_i \neq 0$ in the wavefunction (colored circles, blue: one term, green: three terms, yellow: five terms, red: nine terms). Details of the parameters used in each wavefunction can be found in Appendix B. Pink circles are the Kohn variational calculations of Humberston et al. [24], which are connected by the dashed line to guide the eye. The black solid line is the result of the frozen-target approximation.

In Figure 7, the final phase shifts obtained using all nine correlation (i.e., nonzero γ_i) terms are displayed. Compared with an interpolation of the Kohn variational results of Humberston et al. [24], agreement is seen to within 8×10^{-3} rad. In Table 2, the present results obtained with nine correlation terms are shown alongside the results of Ref. [24] interpolated to the same momentum values. This level of agreement provides evidence that the method employed here is a valid approach to the positron scattering problem, and that it is possible to obtain good-quality scattering data from a bound-state-type calculation that contains only a small number of correlation terms in the wavefunction.

Table 1. Results from positron–hydrogen *s*-wave phase shift calculations with wavefunctions containing $N_E = 1,3,5$ and 9 terms with nonzero γ_i to describe electron–positron correlations (see Appendix B). Phase shift values are shown for the first 11 eigenvalues E_n.

N_E	1		3		5		9	
n	**k/a.u.**	**δ/rad**	**k/a.u.**	**δ/rad**	**k/a.u.**	**δ/rad**	**k/a.u.**	**δ/rad**
1	0.0046	0.0049	0.0046	0.0063	0.0046	0.0075	0.0046	0.0082
2	0.0104	0.0115	0.0104	0.0145	0.0104	0.0176	0.0104	0.0189
3	0.0183	0.0201	0.0183	0.0257	0.0183	0.0307	0.0183	0.0334
4	0.0299	0.0326	0.0299	0.0415	0.0299	0.0494	0.0299	0.0526
5	0.0470	0.0504	0.0470	0.0638	0.0470	0.0752	0.0470	0.0807
6	0.0725	0.0748	0.0725	0.0938	0.0725	0.1086	0.0725	0.1130
7	0.1107	0.1051	0.1107	0.1296	0.1107	0.1449	0.1107	0.1509
8	0.1680	0.1335	0.1680	0.1597	0.1680	0.1725	0.1680	0.1770
9	0.2540	0.1384	0.2540	0.1593	0.2540	0.1713	0.2540	0.1791
10	0.3834	0.0841	0.3834	0.0957	0.3834	0.1176	0.3833	0.1204
11	0.5782	−0.0512	0.5782	−0.0378	0.5782	−0.0163	0.5781	0.0120

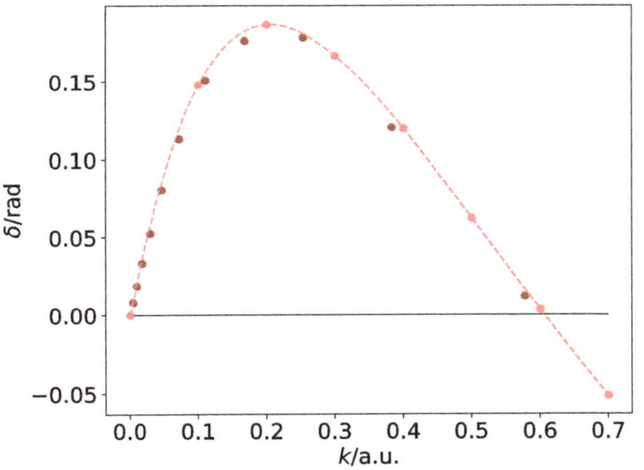

Figure 7. Positron–hydrogen *s*-wave scattering phase shifts obtained using nine nonzero γ_i terms in the wavefunction with various α_i, β_i and γ_i values (red circles). Details of the parameters used in this wavefunction can be found in Appendix B. Pink circles connected by the dashed line are calculations of Humberston et al. [24].

Table 2. Results from positron–hydrogen *s*-wave phase shift calculations with nine $\gamma_i \neq 0$ terms in the wavefunction (see Appendix B) for the first 11 energy eigenvalues E_n. The phase shifts of Humberston et al. [24] δ_{H} interpolated to the same values of k are also shown.

n	**k/a.u.**	**δ/rad**	**δ_{H}/rad**	**Error/rad**
1	0.0046	0.0082	0.0100	0.0018
2	0.0104	0.0189	0.0219	0.0030
3	0.0183	0.0334	0.0375	0.0041
4	0.0299	0.0526	0.0586	0.0059
5	0.0470	0.0807	0.0862	0.0055
6	0.0725	0.1130	0.1202	0.0071
7	0.1107	0.1509	0.1564	0.0055
8	0.1680	0.1770	0.1835	0.0065
9	0.2540	0.1791	0.1815	0.0024
10	0.3833	0.1204	0.1287	0.0083
11	0.5781	0.0120	0.0163	0.0043

4. Calculation of the Annihilation Parameter

In this section, the quality of the variational wavefunctions constructed as described in Section 3 is probed by calculating the normalized annihilation rate, Z_{eff}. Z_{eff} is the effective number of electrons available to the positron for annihilation [36]. For a positron incident on the hydrogen atom, it is given by [13]:

$$Z_{\text{eff}} = \iint \delta(\mathbf{r}_1 - \mathbf{r}_2)|\Psi_{\mathbf{k}}(\mathbf{r}_1, \mathbf{r}_2)|^2 d^3\mathbf{r}_1 d^3\mathbf{r}_2, \tag{41}$$

where the wavefunction is normalized to the incident positron plane wave, i.e., $\Psi_{\mathbf{k}}(\mathbf{r}_1, \mathbf{r}_2) \simeq \psi_{1s}(\mathbf{r}_1)\exp(i\mathbf{k}\cdot\mathbf{r}_2)$, or, for s-wave positron scattering, $\Psi_{\mathbf{k}}(\mathbf{r}_1, \mathbf{r}_2) \simeq \psi_{1s}(\mathbf{r}_1)\sin(kr_2 + \delta)/kr_2$.

Carrying out the integration over $\mathbf{r}_1$ and renaming $\mathbf{r}_2 \equiv \mathbf{r}$ gives

$$Z_{\text{eff}} = \int |\Psi_{\mathbf{k}}(\mathbf{r}, \mathbf{r})|^2 d^3\mathbf{r}. \tag{42}$$

In this integral, $|\Psi_{\mathbf{k}}(\mathbf{r}, \mathbf{r})|^2$ is the electron–positron contact density localized near the atom. Unlike the scattering phase shifts which characterize the wavefunction at large positron distances, the Z_{eff} parameter probes the wavefunction at small positron–atom separations. Here, the bound-state-type variational wavefunction Ψ is proportional to the true continuous spectrum wavefunction $\Psi_{\mathbf{k}}$, i.e., we have

$$\Psi(\mathbf{r}_1, \mathbf{r}_2) = \frac{A}{\sqrt{4\pi}}\Psi_{\mathbf{k}}(\mathbf{r}_1, \mathbf{r}_2), \tag{43}$$

where A is a normalization constant. For s-wave scattering, the normalization constant is obtained from the energy eigenvalue spectrum as (see Ref. [31] for details)

$$A^2 = \frac{2\sqrt{2E}}{\pi}\frac{dE}{dn}. \tag{44}$$

The wavefunctions Ψ generated by solving the generalized eigenvalue problem (as outlined in Section 3) are automatically normalized to unity. Hence, to achieve the correct normalization, these wavefunctions must be divided by A when calculating Z_{eff} from (42). The value of A is calculated for each eigenfunction by substituting the corresponding energy eigenvalue into (44). The derivative dE/dn is evaluated using the function $n = g(\ln E)$ from Section 3 and the fact that

$$\frac{dE}{dn} = E\frac{d\ln E}{dn}. \tag{45}$$

This facilitates a more accurate calculation of the derivative than that obtained by directly calculating dE/dn.

The annihilation parameter is first calculated in the frozen-target approximation, after which the two-particle problem is considered to include electron–positron correlations.

4.1. One-Particle Calculation

In the one-particle frozen-target calculation, the wavefunction only depends on the positron radial coordinate r. The trial wavefunction (11) is employed for this calculation. The c_i values calculated previously are used again here, but each eigenfunction is divided by A to fulfill the normalization condition (43).

In Figure 8, the eigenfunctions obtained using the present method are divided by A and multiplied by $\sqrt{k/\pi}$, and they are compared to the true continuous-spectrum radial function $P_k(r)$ obtained by solving the radial Schrödinger equation [33] and normalized as

$$P_k(r) \simeq \frac{\sin(kr + \delta)}{\sqrt{\pi k}}. \tag{46}$$

For small r, these were found to be in good agreement. As the energy eigenvalues increase, the range of r over which the wavefunctions closely match decreases. However, there is always a very good match at $r \sim 1$ a.u., which dominates in the calculation of Z_{eff}.

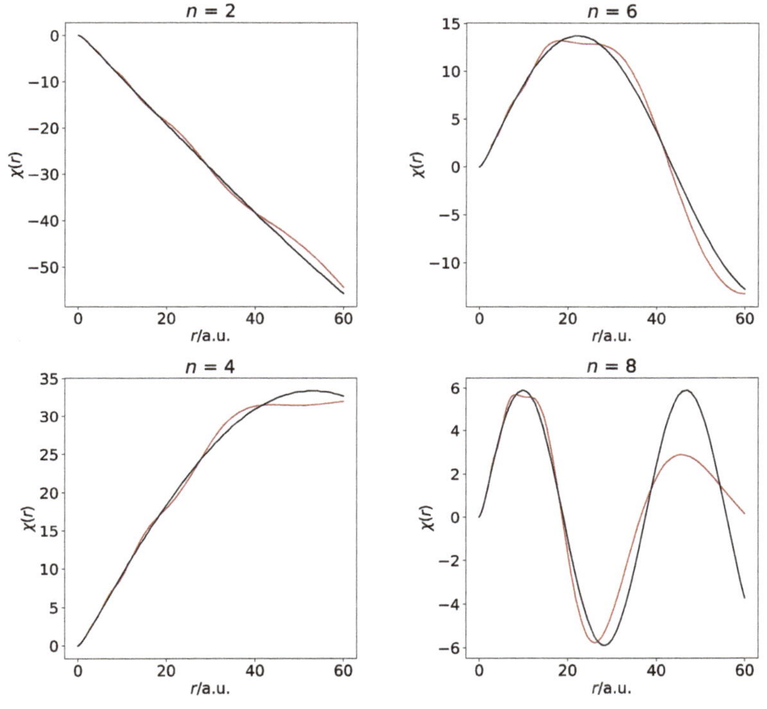

Figure 8. Eigenfunctions obtained from solving the generalized eigenvalue problem (6) in the static approximation (red) and those obtained using the atomic codes [33] (black) for $n = 2, 4, 6$ and 8.

The following integral is used to evaluate Z_{eff} in this problem [13]:

$$Z_{\text{eff}} = \frac{1}{A^2} \int_0^\infty P_{1s}^2(r)\Psi^2(r)r^{-2}dr, \tag{47}$$

where $P_{1s}(r) = 2re^{-r}$ is the ground-state radial wavefunction of the hydrogen atom. Substituting (11) into (47) gives the following expression for the annihilation parameter:

$$Z_{\text{eff}} = \frac{8}{A^2} \sum_{i,j=1}^{N} \frac{c_i c_j}{(2 + \alpha_i + \alpha_j)^3}. \tag{48}$$

Figure 9 shows the corresponding frozen-target Z_{eff} values and compares them with those obtained using true continuous-spectrum positron states in the same approximation [13,33]. Apart from some "noise" related to inaccuracies in the calculation of the normalization constant A at low energies, there is a good general agreement between the two sets of results.

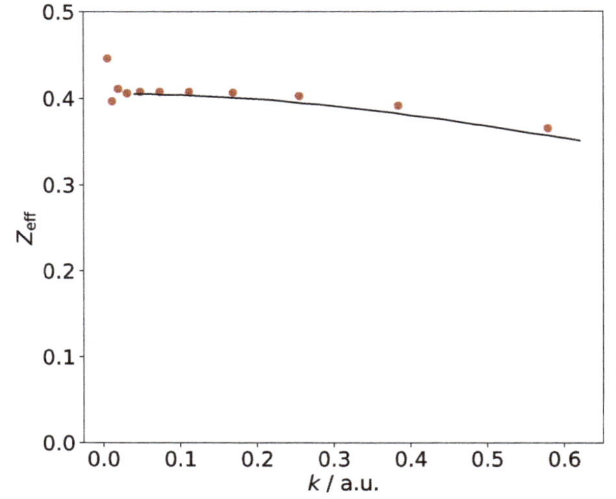

Figure 9. Red circles: values of Z_{eff} obtained in the one-particle model with variational wavefunctions from (48). Black line: Z_{eff} data obtained for the frozen-target model from atomic codes [13,33].

4.2. Two-Particle Calculation

In this section, the annihilation parameter is calculated for the two-particle problem. Firstly, the frozen-target results are reproduced by the two-particle code, using a wavefunction with the form of (39) with all $\alpha_i = 1$ and $\gamma_i = 0$. After verifying that the results match those from the one-particle calculation, these restrictions on the α_i and γ_i values are lifted, and the full correlated wavefunction is used, subject to (40). Here, it is noted that $r_{12} = 0$ in the Z_{eff} calculation, so $\exp(\gamma_i r_{12}) = 1$ for each of the γ_i. In addition, $r_1 = r_2 \equiv r$ is required, and the wavefunction takes the form

$$\Psi(r,r,0) = \sum_{i=1}^{N} c_i \exp[-(\alpha_i + \beta_i)r]. \tag{49}$$

As before, the c_i coefficients are calculated such that the eigenfunctions are normalized to unity. Hence, the normalization must be corrected to satisfy (43) using the value of A for each eigenfunction.

In this case, evaluation of the integral (42) with the wavefunction (49) yields the following expression for Z_{eff}:

$$Z_{eff} = \frac{16\pi^2}{A^2} \int_0^\infty |\Psi(r,r,0)|^2 r^2 dr = \frac{32\pi^2}{A^2} \sum_{i,j=1}^{N} \frac{c_i c_j}{(\alpha_i + \alpha_j + \beta_i + \beta_j)^3}. \tag{50}$$

The coefficients c_i in this calculation differ from those in the one-particle calculation by a factor of $1/2\pi$. Taking this into account, the equivalence of the one- and two-particle results can be verified by setting $\alpha_i = \alpha_j = 1$ in (50) to recover the result from the frozen-target approximation (48). Here, however, the focus is on including terms in the wavefunction to describe electron–positron correlations.

When incorporating correlation terms, the same sets of α_i, β_i and γ_i parameters are used in the basis as for the phase shift calculations in Section 3.2.3 (the values of which are listed in Appendix B). The accuracy of the Z_{eff} calculation does not increase monotonically with the number of correlation terms included the wavefunction, unlike the case of the phase shifts determined by the energy eigenvalues alone. This is shown in Figure 10, where the overall results obtained using one correlation term are more accurate than those obtained using three correlation terms. However, the most accurate set of values was obtained for the wavefunction with the maximum (nine) correlation terms included. The Z_{eff} values from [26] are used as benchmark values to evaluate the accuracy of our calculation. The results from the final calculation are displayed in Table 3 alongside those from [26] interpolated to the same values of momentum.

Table 3. Z_{eff} values obtained using a wavefunction with nine nonzero γ_i terms (Appendix B) to describe electron–positron correlations and the benchmark results from [26] ($Z_{\text{eff,H}}$) interpolated to the same values of momentum.

k/a.u.	Z_{eff}	$Z_{\text{eff,H}}$	Error
0.0046	9.2538	9.6013	0.3475
0.0104	8.1989	9.4644	1.2655
0.0183	8.4351	9.2767	0.8416
0.0299	8.2262	9.0083	0.7821
0.0470	8.0325	8.6212	0.5887
0.0725	7.6456	8.0667	0.4211
0.1107	6.9813	7.2876	0.3063
0.1680	6.0719	6.2319	0.1600
0.2540	4.8017	4.8951	0.0934
0.3833	3.6024	3.4371	−0.1653
0.5781	2.0695	2.3833	0.3138

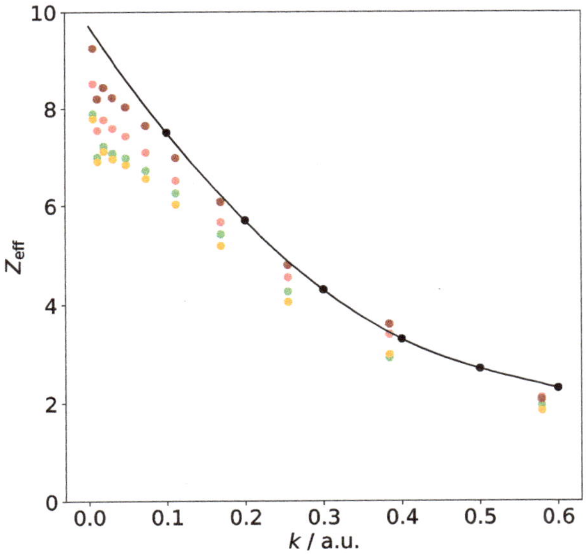

Figure 10. Z_{eff} values obtained in the two-particle model using one (green), three (yellow), five (pink) and nine (red) terms in the wavefunction to describe correlations between the electron and positron. Black circles and line: results from [26] and their interpolation using cubic splines.

Overall, good agreement between our results and the benchmark values is found in the range $k = 0.1$–0.6 a.u. using a trial wavefunction with nine nonzero γ_i terms. Apart from two data points, the accuracy of our variational calculation is better than 10%, which is evidence of the good quality of the wavefunction that incudes only a small number of correlation terms.

5. Conclusions

The variational method was explored as a means of studying positron–hydrogen scattering and annihilation using square-integrable trial wavefunctions. By setting up and solving a generalized eigenvalue problem, s-wave elastic phase shifts and Z_{eff} values for the positron–hydrogen system were obtained in good agreement with benchmark values. Importantly, this was achieved using only a small number of correlation terms in the trial wavefunction, indicating that processes such as virtual positronium formation and polarization of the hydrogen atom can be accounted for using this approach.

Looking forward, this method could facilitate the study of more complex interactions, such as the interaction of positrons with molecules. The key benefit of our approach is the small number of terms required to describe strong electron–positron correlations, meaning that the method is quite economical. To improve upon the current approach, a formal optimization of the trial wavefunction parameters could be performed to increase the accuracy of the calculation. With such a process in place, it would become possible to carry out more complex calculations efficiently.

Author Contributions: Methodology, G.G. and S.G.; software, S.G.; writing, G.G .and S.G.; supervision, G.G. All authors have read and agreed to the published version of the manuscript.

Funding: This research received no external funding.

Data Availability Statement: All data are available from the authors upon request.

Conflicts of Interest: The authors declare no conflict of interest.

Appendix A. Standard Integrals

This appendix contains results for the six standard integrals employed throughout our calculations. These are evaluated using the elliptic coordinate system: $s = r_1 + r_2$, $t = r_1 - r_2$ and $u = r_{12}$:

$$\begin{aligned}\tilde{I}_1(a,b,g) &= \pi^2 \int_0^\infty ds\, e^{-2as} \int_0^s du\, ue^{2bu} \int_{-u}^u dt\, e^{-2gt}(s^2 - t^2) \\ &= \pi^2 \frac{8a^3 - 13a^2b + 6ab^2 - b^3 + bg^2}{8a^3((a-b)^2 - g^2)^3},\end{aligned} \quad \text{(A1)}$$

$$\begin{aligned}\tilde{I}_2(a,b,g) &= \pi^2 \int_0^\infty ds\, se^{-2as} \int_0^s du\, e^{2bu} \int_{-u}^u dt\, e^{-2gt}(u^2 - t^2) \\ &= \pi^2 \frac{5a^2 - 6ab + b^2 - g^2}{8a^2((a-b)^2 - g^2)^3},\end{aligned} \quad \text{(A2)}$$

$$\begin{aligned}\tilde{I}_3(a,b,g) &= \pi^2 \int_0^\infty ds\, e^{-2as} \int_0^s du\, (s^2 - u^2)e^{2bu} \int_{-u}^u dt\, te^{-2gt} \\ &= \pi^2 \frac{g(-5a^2 + 6ab - b^2 + g^2)}{8a^3((a-b)^2 - g^2)^3},\end{aligned} \quad \text{(A3)}$$

$$\begin{aligned}\tilde{J}_1(a,b,g) &= \pi^2 \int_0^\infty ds\, se^{-2as} \int_0^s du\, ue^{2bu} \int_{-u}^u dt\, e^{-2gt} \\ &= \pi^2 \frac{(a-b)^2(4a-b)g + bg^3}{8a^2g((a-b)^2 - g^2)^3},\end{aligned} \quad \text{(A4)}$$

$$\tilde{J}_2(a,b,g) = \pi^2 \int_0^\infty ds\, e^{-2as} \int_0^s du\, e^{2bu} \int_{-u}^u dt\, e^{-2gt}(s^2 - t^2)$$
$$= -\pi^2 \frac{-5a^2 + 4ab - b^2 + g^2}{8a^3((a-b)^2 - g^2)^2}, \quad \text{(A5)}$$

$$\tilde{J}_3(a,b,g) = \pi^2 \int_0^\infty ds\, e^{-2as} \int_0^s du\, ue^{2bu} \int_{-u}^u dt\, te^{-2gt}$$
$$= -\pi^2 \frac{(a-b)g}{2a((a-b)^2 - g^2)^3}. \quad \text{(A6)}$$

Appendix B. Parameters for Positron-Scattering Wavefunction Bases

The positron–hydrogen wavefunctions used in Sections 3.2 and 4.2 all contain the 20 basis functions defined by (13) with $n = 20$, $\zeta = 1.5$ and $\beta_1 = 0.01$ (where all $\alpha_i = 1$ and $\gamma_i = 0$). Additional terms with varying α_i and nonzero γ_i values were then incorporated to account for electron–positron correlations. The table below contains the sets of quasi-optimal values for the parameters obtained using the method described in Section 3.2. These values are displayed for wavefunctions containing one, three, five and nine terms with nonzero γ_i where N_E denotes the number of nonzero γ_i terms.

Table A1. Parameters used for the electron–positron correlation terms, i.e, terms with nonzero γ_i, in positron–hydrogen wavefunctions with $N_E = 1, 3, 5$ and 9 such terms.

N_E	α	β	γ	N_E	α	β	γ
1	0.80	0.04	−0.54	9	0.79	0.06	−0.53
3	0.80	0.05	−0.56		0.99	0.40	0.11
	0.88	0.45	0.10		1.00	0.14	−0.03
	0.97	0.28	−0.46		0.40	0.04	−0.38
5	0.80	0.05	−0.54		0.85	0.88	−0.38
	0.98	0.46	0.06		0.29	0.40	−0.70
	0.99	0.14	−0.11		0.85	0.50	−0.27
	0.45	0.12	−0.67		0.84	−0.20	−0.65
	0.86	0.93	−0.31		0.99	0.67	−0.01

Notes

1 Atomic units are used throughout (in which $e = m = \hbar = 1$, where e is the elementary charge and m is the electron or positron mass).

2 For a state with a zero total angular momentum, the wavefunction is spherically symmetric, so there is no dependence on the directions of $\mathbf{r}_1$ and $\mathbf{r}_2$, except the angle between them, i.e., dependence on r_{12}.

References

1. Chua, S.; Groves, A. *Biomedical Imaging Applications and Advances*; Woodhead Publishing: Cambridge, UK, 2014.
2. Tuomisto, F.; Makkonen, I. Defect identification in semiconductors with positron annihilation: Experiment and theory. *Rev. Mod. Phys.* **2013**, *85*, 1583–1631. [CrossRef]
3. Ishida, A.; Namba, T.; Asai, S.; Kobayashi, T.; Saito, H.; Yoshida, M.; Tanaka, K.; Yamamoto, A. New precision measurement of hyperfine splitting of positronium. *Phys. Lett. B* **2014**, *734*, 338. [CrossRef]
4. Ahmadi, M.; Alves, B.X.R.; Baker, C.J.; Bertsche, W.; Capra, A.; Carruth, C.; Cesar, C.L.; Charlton, M.; Cohen, S.; Collister, R.; et al. Investigation of the fine structure of antihydrogen. *Nature* **2020**, *578*, 375–380. [CrossRef]
5. Dirac, P.A.M. Quantised Singularities in the Electromagnetic Field. *Proc. R. Soc. Lond. Ser. A* **1931**, *133*, 60–72. [CrossRef]
6. Anderson, C.D. The positive electron. *Phys. Rev.* **1932**, *43*, 491–494. [CrossRef]
7. Surko, C.M.; Gribakin, G.F.; Buckman, S.J. Low-energy positron interactions with atoms and molecules. *J. Phys. B* **2005**, *38*, R57–R126. [CrossRef]
8. Amusia, M.Y.; Cherepkov, N.A.; Chernysheva, L.V.; Shapiro, S.G. Elastic scattering of slow positrons by helium. *J. Phys. B* **1976**, *9*, L531–L534. [CrossRef]
9. Amusia, M.Y.; Dolmatov, V.K.; Chernysheva, L.V. Positron elastic scattering by a semifilled-shell atom. *J. Phys. B* **2021**, *54*, 185003. [CrossRef]

10. Gribakin, G.F.; King, W.A. The effect of virtual positronium formation on positron-atom scattering. *J. Phys. B* **1994**, *27*, 2639–2645. [CrossRef]
11. Dzuba, V.A.; Flambaum, V.V.; Gribakin, G.F.; King, W.A. Bound states of positrons and neutral atoms. *Phys. Rev. A* **1995**, *52*, 4541. [CrossRef]
12. Dzuba, V.A.; Flambaum, V.V.; Gribakin, G.F.; King, W.A. Many-body calculations of positron scattering and annihilation from noble-gas atoms. *J. Phys. B* **1996**, *29*, 3151. [CrossRef]
13. Gribakin, G.F.; Ludlow, J. Many-body theory of positron-atom interactions. *Phys. Rev. A* **2004**, *70*, 032720. [CrossRef]
14. Green, D.G.; Ludlow, J.A.; Gribakin, G.F. Positron scattering and annihilation on noble-gas atoms. *Phys. Rev. A* **2014**, *90*, 032712. [CrossRef]
15. Hofierka, J.; Cunningham, B.; Rawlins, C.M.; Patterson, C.H.; Green, D.G. Many-body theory of positron binding to polyatomic molecules. *Nature* **2022**, *606*, 688–693. [CrossRef] [PubMed]
16. Bray, I.; Stelbovics, A.T. Convergent close-coupling calculations of low-energy positron–atomic-hydrogen scattering. *Phys. Rev. A* **1993**, *48*, 4787–4789. [CrossRef]
17. Mitroy, J.; Bromley, M.W.J. Convergence of configuration-interaction single-center calculations of positron-atom interactions. *Phys. Rev. A* **2006**, *73*, 052712. [CrossRef]
18. Kadyrov, A.S.; Bray, I. Two-center convergent close-coupling approach to positron-hydrogen collisions. *Phys. Rev. A* **2002**, *66*, 012710. [CrossRef]
19. Humberston, J.; Van Reeth, P. Annihilation in low energy positron-helium scattering. *Nucl. Instrum. Methods Phys. Res. B* **1998**, *143*, 127–134. [CrossRef]
20. Reeth, P.V.; Humberston, J.W. Elastic scattering and positronium formation in low-energy positron-helium collisions. *J. Phys. B* **1999**, *32*, 3651–3667. [CrossRef]
21. Germano, J.S.E.; Lima, M.A.P. Schwinger multichannel method for positron-molecule scattering. *Phys. Rev. A* **1993**, *47*, 3976–3982. [CrossRef]
22. da Silva, E.P.; Germano, J.S.E.; Lima, M.A.P. Annihilation Dynamics of Positrons in Molecular Environments: Theoretical Study of Low-Energy Positron- C_2H_4 Scattering. *Phys. Rev. Lett.* **1996**, *77*, 1028–1031. [CrossRef] [PubMed]
23. Schwartz, C. Electron Scattering from Hydrogen. *Phys. Rev.* **1961**, *124*, 1468–1471. [CrossRef]
24. Humberston, J.W.; Reeth, P.V.; Watts, M.S.T.; Meyerhof, W.E. Positron-hydrogen scattering in the vicinity of the positronium formation threshold. *J. Phys. B* **1997**, *30*, 2477–2493. [CrossRef]
25. Bhatia, A.K. Positron-Hydrogen Scattering, Annihilation, and Positronium Formation. *Atoms* **2016**, *4*, 27. [CrossRef]
26. Humberston, J.W.; Wallace, J.B.G. The elastic scattering of positrons by atomic hydrogen. *J. Phys. B* **1972**, *5*, 1138–1148. [CrossRef]
27. Ghosh, A.; Sil, N.; Mandal, P. Positron-atom and positron-molecule collisions. *Phys. Rep.* **1982**, *87*, 313–406. [CrossRef]
28. Gien, T.T. Coupled-state calculations of positron-hydrogen scattering. *Phys. Rev. A* **1997**, *56*, 1332–1337. [CrossRef]
29. Ghoshal, A.; Mandal, P. Correlated dipole-polarized basis in Schwinger's principle for elastic positron-hydrogen collisions. *Phys. Rev. A* **2005**, *72*, 042709. [CrossRef]
30. Available online: https://docs.scipy.org/doc/scipy/reference/generated/scipy.linalg.eigh.html (accessed on 3 February 2021).
31. Swann, A.R.; Gribakin, G.F. Model-potential calculations of positron binding, scattering, and annihilation for atoms and small molecules using a Gaussian basis. *Phys. Rev. A* **2020**, *101*, 022702. [CrossRef]
32. Landau, L.D.; Lifshitz, E.M. *Quantum Mechanics*, 3rd ed.; Pergamon: Oxford, UK, 1977.
33. Amusia, M.Y.; Chernysheva, L.V. *Computation of Atomic Processes: A Handbook for the ATOM Programs*; Institute of Physics Publishing: Bristol, UK, 1997.
34. Hylleraas, E.A. Über den Grundzustand des Heliumatoms. *Z. Phys.* **1928**, *48*, 469–494. [CrossRef]
35. Kadyrov, A.S.; Bray, I.; Stelbovics, A.T. Near-Threshold Positron-Impact Ionization of Atomic Hydrogen. *Phys. Rev. Lett.* **2007**, *98*, 263202. [CrossRef] [PubMed]
36. Fraser, P. Positrons and Positronium in Gases. In *Advances in Atomic and Molecular Physics*; Academic Press: New York, NY, USA, 1968; Volume 4, pp. 63–107. [CrossRef]

atoms

MDPI

Article

Shake-Off Process in Non-Sequential Single-Photon Double Ionization of Closed-Shell Atomic Targets

Anatoli S. Kheifets

Research School of Physics, The Australian National University, Canberra, ACT 2601, Australia; a.kheifets@anu.edu.au

Abstract: Amusia and Kheifets in 1984 introduced a Green's function formalism to describe the effect of many-electron correlation on the ionization spectra of atoms. Here, we exploit this formalism to model the shake-off (SO) process, leading to the non-sequential single-photon two-electron ionization (double photoionization—DPI) of closed-shell atomic targets. We separate the SO process from another knock-out (KO) mechanism of DPI and show the SO prevalence away from the DPI threshold. We use this kinematic regime to validate our model by making a comparison with more elaborate techniques, such as convergent and time-dependent close coupling. We also use our model to evaluate the attosecond time delay associated with the SO process. Typically, the SO is very fast, taking only a few attoseconds to complete. However, it can take much longer in the DPI of strongly correlated systems, such as the H^- ion as well as the subvalent shells of the Ar and Xe atoms and Cl^- ion.

Keywords: atomic photoionization; many-electron correlation; non-sequential double ionization

PACS: 32.80.Rm 32.80.Fb 42.50.Hz

Citation: Kheifets, A.S. Shake-Off Process in Non-Sequential Single-Photon Double Ionization of Closed-Shell Atomic Targets. *Atoms* **2022**, *10*, 89. https://doi.org/10.3390/atoms10030089

Academic Editor: Grzegorz Piotr Karwasz

Received: 17 August 2022
Accepted: 5 September 2022
Published: 7 September 2022

1. Introduction

The simultaneous removal of two electrons from an atom following absorption of a single photon is an archetypal process driven entirely by many-electron correlation. Such a non-sequential single-photon two-electron ionization (double photoionization or DPI in short) has been the focus of experimental and theoretical activities for several decades [1–3]. The correlation mechanisms of DPI are well understood and can be described as the shake-off (SO) and knock-out (KO) processes [4–7]. Shake-off is driven by a sudden change of the atomic potential after a fast removal of the primary photoelectron. Conversely, knock-out is a slow process in which the departing electron impinges on the ionic core and ejects the secondary photoelectron. A complementary quasi-free mechanism (QFM) of PDI, in which the nucleus remains a spectator, was predicted theoretically by Amusia and co-workers [8]. Recently, the QFM was studied experimentally [9–11], and it was ascribed to a combination of the SO and KO processes.

The first theoretical description of DPI in atoms invoked the lowest order perturbation theory [12–15]. With a growing computational power, more sophisticated non-perturbative methods were developed. The convergent close-coupling (CCC) [16,17] and the time-dependent close-coupling (TDCC) [18] are among the many predictive and accurate numerical techniques.

A resurged interest in DPI was stimulated by a newly acquired experimental capability to resolve atomic photoionization in time. Laser pulses are shaped in such a way that they can probe atomic ionization on the attosecond (1 as = 10^{-18} s) time scale. Firstly, single photoionization was time resolved [19,20]. Then a DPI process was traced in time [21]. The accompanying theoretical studies have also appeared [22,23].

An alternative theoretical approach to DPI can be provided by the single-hole Green's function (SHGF) formalism introduced to photoionization [24,25]. Originally, this method was utilized to calculate shake-up satellites in atomic photoionization spectra [26–29].

However, by construction, the SHGF contains the DPI continuum, which can be attributed to the SO process. This capability of the SHGF method has been overlooked so far. Here, we rectify this omission.

The present work is structured in the following way. We start with a brief introduction of the SHGF technique, using a diagrammatic expansion of the ionization amplitude. We identify the double ionized continuum in this amplitude and link it with the imaginary part of the SHGF self-energy. This allows us to derive the analytic expressions for the DPI cross section resolved with the photoelectron energy as well as the time delay associated with the SO process. Next, we test our energy resolved DPI cross sections against experiments as well as the earlier CCC and TDCC calculations. This way, we identify the photoelectron energy range where the SO process makes the dominant contribution to DPI. Finally, we evaluate the time that it takes to shake off a bound electron. As expected, the SO process is fast with only a few attoseconds needed to shake off the secondary photoelectron into the two-electron continuum. However, there are few notable exceptions when the SO process takes much longer time to complete. We find this situation in strongly correlated targets such as the negative H^- ion as well as the subvalent shells of the Ar and Xe atoms and the Cl^- ion. The binding of the H^- ion is wholly owed to many-electron correlation, and the photoionization of subvalent *ns* shells in Ar, Xe and Cl^- is affected very strongly by correlation satellites. As the result, the SO process in H^- takes as much as 30 as, whereas the similar process in Ar 3*s* and Xe 5*s* requires nearly 50 as to complete. The same process in Cl^- 3*s* takes in excess of 100 as. We conclude by evaluating other components of the measurable time delay in DPI and thus making the case for the experimental resolution of the SO process in time.

2. Theory

The SO process leading to the ejection of two electrons in the continuum can be exhibited graphically by an infinite sequence of diagrams presented in Figure 1a. Here, we use the following graphical symbols. A straight line with an arrow to the right represents the electron continuum states k, f whereas an arrow pointing to the left exhibits the holes in atomic shells i, l, m. Before photoabsorption, all the atomic shells are presumed to be fully occupied and thus closed. The atomic ground state (the "vacuum" state) contains neither holes nor excited electrons. The wavy line denotes the Coulomb interaction between the electrons. The dashed line represents an absorbed photon. A circle denotes the diagonal matrix element of the self-energy part of the Green's function Σ_i. The SHGF self-energy is expanded graphically in more detail in Figure 1b.

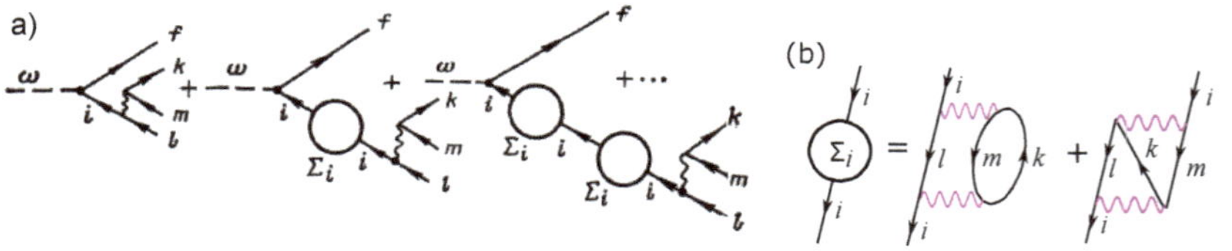

Figure 1. (**a**) Schematic representation of the SO process leading to a two-electron continuum. The circle represents the self-energy part of the SHGF expanded in more detail in (**b**).

The corresponding DPI amplitude can be found by summing a geometric progression of terms in Figure 1a that leads to the following expression:

$$\langle f\|M_{\rm SO}\|i\rangle = \langle f\|r\|i\rangle\left(1-\frac{\Sigma_i(\varepsilon)}{\varepsilon-\epsilon_i-i\delta}\right)^{-1}\frac{\langle ik\|U_L\|lm\rangle}{\varepsilon-\epsilon_i-i\delta}\,. \quad (1)$$

Here, $\langle f\|r\|i\rangle$ is a dipole matrix element of the single-photon absorption and $\langle ik\|U_L\|lm\rangle$ is a Coulomb matrix element, which includes a direct and exchange $l \leftrightarrow m$ terms and involves a transfer of the angular momentum L. Both the dipole and Coulomb matrix elements are reduced to strip their angular momentum projections dependence. The symbol $\varepsilon = \epsilon_f - \omega < 0$ denotes an effective hole energy. In the absence of correlation, $\varepsilon = \epsilon_i$ is the orbital energy. Many-electron correlation adds a discrete spectrum of shake-up satellites and a shake-off continuum, which starts at the DPI threshold $E_\infty = \epsilon_l + \epsilon_m$. The infinitesimally small $\delta \to 0$ in the energy denominator defines the pole bypass.

The imaginary part of Equation (1) gives rise to an additional phase of the DPI amplitude due to the final state correlation:

$$\arg M_{\rm SO} = \arctan \frac{{\rm Im}\Sigma_i(\varepsilon)}{{\rm Re}[\varepsilon - \epsilon_i - \Sigma_i(\varepsilon)]} \approx \arctan \frac{{\rm Im}\Sigma_i(\varepsilon)}{{\rm Re}[\varepsilon - \epsilon_i]} . \tag{2}$$

Here ${\rm Im}\Sigma_i = \pi(2L+1)^{-1}|\langle ik\|U_L\|lm\rangle|^2$. The approximate equality is satisfied under the condition that $|\varepsilon - \epsilon_i| \gg |\Sigma_i(\varepsilon)|$. This condition defines the part of the double ionized continuum sufficiently far from the main photoelectron line.

The energy resolved single-differential cross-section (SDCS) is given by Equation (9) of [25]:

$$\frac{d\sigma_i^{2+}}{d\epsilon_f} = \sigma_i^+ \frac{1}{\pi} \frac{{\rm Im}\Sigma_i(\varepsilon)}{|\varepsilon - \epsilon_i - \Sigma_i(\varepsilon)|^2} \approx \sigma_i^+ \frac{1}{\pi} \frac{{\rm Im}\Sigma_i(\varepsilon)}{|\varepsilon - \epsilon_i|^2} . \tag{3}$$

By solving this equation relative to ${\rm Im}\Sigma_i$, we can express the additional phase of the DPI amplitude due to the final state correlation in the following form:

$$\arg M_{\rm SO} = \arctan \frac{\pi}{\sigma_i^+} \frac{d\sigma_i^{2+}}{d\epsilon_f} |\varepsilon - \epsilon_i| . \tag{4}$$

We note that all the quantities entering this expression are known from the experiment, which are the single photoionization cross-section of the primary photoelectron σ_i^+ and the energy differential DPI cross-section $d\sigma_i^{2+}/d\epsilon_f$.

By integrating the SDCS Equation (3) over the fast photoelectron energy, we can obtain the double-to-single photoionization cross-section ratio:

$$R = \frac{\sigma^{2+}}{\sigma^+} = \frac{1}{\sigma^+} \int_0^\infty \frac{d\sigma^{2+}}{d\epsilon_f} d\epsilon_f . \tag{5}$$

This ratio is known in He, and its isoelectronic sequence of ions [30]. It will serve as a convenient reference in Section 3.2.

Following [25], we can introduce the inverse SHGF

$$F(\varepsilon) = G^{-1}(\varepsilon) = \varepsilon - \epsilon_i - \Sigma_i(\varepsilon) \tag{6}$$

Then the argument of the SO amplitude Equation (2) can be rewritten as

$$\arg M_{\rm SO} = \arg G^{-1}(\varepsilon) = \arg F(\varepsilon) \tag{7}$$

This expression allows to use the integral rule presented by Equation (5) of [25], which relates the energies of the discrete shake-up satellites with the time delay:

$$\sum_{k=0}^{\infty} (\varepsilon_k - E_k) = \frac{1}{\pi} \int_{-\infty}^{E_\infty} \varepsilon \tau(\varepsilon)\, d\varepsilon \,, \text{ where } \tau(\varepsilon) = \frac{\partial}{\partial \varepsilon} \arg M_{\rm SO} = \frac{\partial}{\partial \varepsilon} \arg F(\varepsilon) \tag{8}$$

Here $E_k = \epsilon_l + \epsilon_m - \epsilon_k$ and $E_0 = \epsilon_i$ are non-correlated energies of the shake-up excitations calculated from the HF orbital energies. Meanwhile, ε_k are the corresponding energies shifted by the final-state correlation and found as the poles of the SHGF. The integral time delay rule Equation (8) presents an analytical test for the numerical SO time delay values. This test will be conducted in Section 3.3.

3. Numerical Results

3.1. Computation Details

We use the ATOM suite of programs developed by Miron Amusia and co-workers [31]. The SCFHF and FCHF computer codes calculate the atomic ground and excited states in the self-consistent and frozen-core Hartree–Fock (HF) approximations, respectively. Then the Coulomb matrix elements are evaluated, and the SHGF and its self-energy are found. The latter are used to calculate the SDCS, the R ratio and the time delay associated with the SO process.

We consider the two types of the SO process. In the first type, all the hole states i, l, m are confined to the same ns shell. The fast primary photoelectron is ejected from this shell into the p-continuum, whereas the slow electron is shaken off into the s-continuum. Such an intra-shell SO process takes place in the outer valence shell of the He, Be and Mg atoms as well as the H^- ion. The second inter-shell type of the SO process accompanies ionization of the sub-valent ns shells of noble gas atoms Ne, Ar and Xe. While the primary hole i is made in the ns shell, the secondary holes l, m are made in the outermost np shell. The primary fast photoelectron is ejected in the p-wave, whereas the secondary electron is shaken off primarily into the d-wave.

3.2. Energy Resolved DPI Cross-Sections

In Figure 2, we exhibit the energy resolved DPI cross section of helium at the excess energies above the DPI threshold $E = 100$, 450 and 720 eV (from left to right). The SDCS Equation (3) is symmetrized by adding the two energy distributions of the slow and fast photoelectrons, thus giving it a characteristic U-shape. The present shake-off calculations using Equation (3) are compared with various reference data indicated in the figure caption. The integrated cross-sections of single photoionization σ^+_{1s} are used as tabulated in [32]. We can observe in Figure 2 that the SO mechanism is becoming gradually dominant as the photon energy grows. This is particularly true for a highly asymmetric energy sharing between the photoelectrons. Under this kinematics, the primary photoelectron takes nearly all the photon energy and is ejected in the dipole channel as a p-wave. At the same time, the secondary SO photoelectron is very slow and is ejected almost isotropically as an s-wave in the intra-shell SO process. It is this characteristic photoelectron angular distribution that was observed experimentally in He at $E = 450$ eV [7].

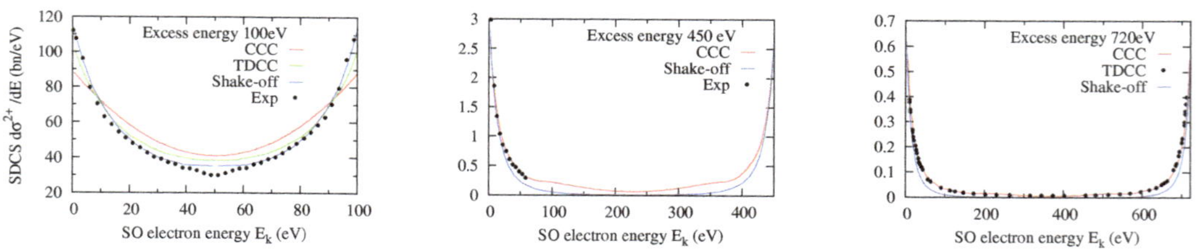

Figure 2. The energy resolved DPI cross-section of helium $d\sigma^{2+}/dE$) (in bn/eV) at the excess energies above the DPI threshold $E = 100$, 450 and 720 eV (from left to right). The present shake-off calculations using Equation (3) are compared with the following reference data. At $E = 100$ eV, the relative measurement [33] is normalized to the TDCC calculation [34] and shown along with an analogous CCC calculation [35]. At $E = 450$ eV, the relative experiment [7] is normalized to the CCC calculation from the same reference. At $E = 720$ eV, the TDCC [36] and the CCC [11] calculations are shown.

Meanwhile, the equal energy sharing between the photoelectrons is affected by a competing KO process and deviates from the present SO predictions. This deviation, which is strongest at $E = 100$ eV, can also be seen at higher photon energies near the mid-point of the photoelectron energy distribution.

The double-to-single ratios as calculated by the SO model using Equation (5) are presented in Table 1. The SO ratio in He is equal to 1.44%, which is very close to the CCC ratio of 1.67% [30]. For other targets, this comparison is less accurate, especially for the H^- ion, which is twice overestimated, while Be and Mg are 50% underestimated.

Table 1. Asymtptoc double-to-single cross-section ratios in various targets as calculated by the SO model using Equation (5) and compared with earlier CCC calculations.

Target	E_0 Ry	Threshold Ry	Raio, % SO	CCC	Ref.
He 1s	1.836	5.807	1.44	1.67	[30]
H^- 1s	1.000	1.055	3.43	1.60	[30]
Be 2s	0.618	1.951	0.26	0.37	[37]
Mg 3s	0.506	1.558	0.16	0.25	[38]

3.3. Time Delay

Results of the time delay calculations by taking the energy derivative of the SO phase Equation (2) are displayed in Figure 3. The horizontal axis in the figure denotes the slow photoelectron energy. It is assumed that the photon energy is very large and nearly all of it is carried away by the second fast photoelectron. The three panels of this figure display the time delay results for the He 1s, Be 2s, Ne 2s and Mg 3s (left), H^- 1s (center) and Ar 3s, Cl^- 3s and Xe 5s (right). We observe a very small SO time delay in He 1s not exceeding a few attoseconds. A similarly small time delay is found in the intra-shell SO process in Be 2s and Mg 3s as well as the inter-shell SO process in Ne 2s. Incidentally, the time delay in the inner 1s shell of Be is much smaller than that in the valence 2s shell, the reason being the Coulomb field of the bare nucleus, which makes many-electron correlation and associated time delay negligible.

The SO time delay in H^- 1s is markedly higher by nearly an order of magnitude. Time delay grows further in Ar 3s and Xe 5s, while it exceeds the 100 as mark in the negative Cl^- ion. We relate this growth of time delay to much bolder shake-up and shake-off processes in these targets, which are strongly affected by many-electron correlation.

Another indication of this effect is presented in Table 2. Here we compare the summary displacement of the main and shake-up satellite lines in the photoelectron spectra of He 1s and Ar 3s with the corresponding time delay integral Equation (8). We observe in this table that the many-electron correlation causes a much stronger line displacement in Ar than in He. This is matched by a much larger SO time delay in Ar in comparison with He.

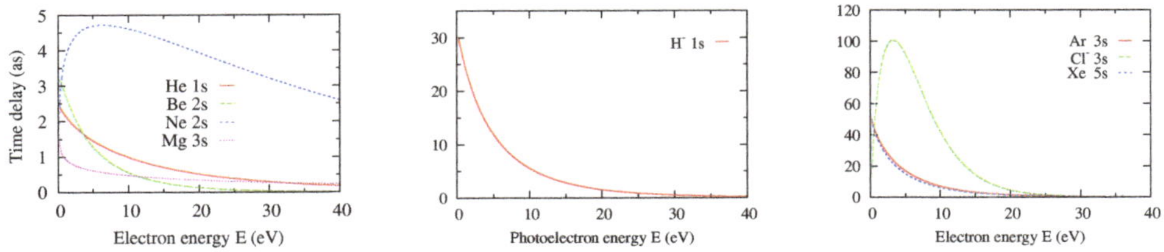

Figure 3. Shake-off time delay (in attoseconds) in He 1s, Be 2s, Ne 2s and Mg 3s (**left**), hydrogen H^- 1s (**center**) and Ar 3s, Cl^- 3s and Xe 5s (**right**), calculated by taking the energy derivative of the SO phase Equation (2).

Table 2. The energies of the shake-up satellites in the photoelectron spectra of He 1*s* and Ar 3*s* as calculated in the HF approximation E_k and as the poles of the SHGF ε_k. The sum $\sum_k (E_k - \varepsilon_k)$ in each target is compared with the corresponding time delay integral Equation (8).

k	E_k	ε_k	$E_k - \varepsilon_k$
	Helium 1*s*		
0	1.8360	1.7287	0.1073
1	4.7360	4.8053	−0.0693
2	5.3540	5.3622	−0.0082
3	5.5540	5.5568	−0.0028
4	5.6460	5.6473	−0.0013
5	5.6950	5.6957	−0.0007
Sum			0.0250
SO integral Equation (8)			0.025
k	E_k	ε_k	$E_k - \varepsilon_k$
	Argon 3*s*		
0	2.5550	2.1508	0.4042
1	2.5920	2.8389	−0.2469
2	2.9630	3.0235	−0.0605
3	3.1070	3.1274	−0.0204
4	3.1780	3.1879	−0.0099
5	3.2180	3.2235	−0.0055
Sum			0.0610
SO integral Equation (8)			0.063

4. Summary and Outlook

In the present work, we demonstrated that the SO process is prevalent in DPI at large photon energies exceeding significantly the double ionization threshold. Under this condition, the two-electron energy sharing is highly asymmetric with the primary photoelectron taking nearly all of the photon energy, while the secondary SO electron is rather slow. The slow electron can be delayed by repeated interaction with the ionic core. In the intra-shell shake-off process, this interaction is confined to the same shell and is typically rather quick. The intra-shell SO processes in He 1*s*, Be 2*s* and Mg 3*s* take no more than several attoseconds to complete. The marked exception is the intra-shell SO in the H^- ion, which may take several tens of attoseconds. We attribute this effect to a strongly correlated nature of H^- which will not bind in the absences of correlation. The inter-shell SO process are more involved and take typically longer time. We observe a considerable delay in the SO of Ar 3*s* and Xe 5*s*. The Cl^- ion, which is iso-electronic to Ar, demonstrates a very significant SO delay exceeding 100 as. All these targets are prone to strong final-state correlation and display intense shake-up satellite spectra with a strong line displacement relative to the corresponding HF energies. The only exception is the inter-shell SO in Ne, which is still rather quick.

It is instructive to compare the SO time delay in DPI with the analogous characteristic of single photon one-electron ionization (single photoionization—SPI). The energy derivative of the SPI amplitude is known as the Wigner time delay. Similarly to electron elastic scattering [39], it characterizes the photoelectron group delay in the dispersive potential of the ionic core. This potential includes an exchange with the core electrons [40]. In addition, the Wigner time delay in SPI is strongly affected by inter-shell correlation [41]. All these characteristics of the Wigner time delay are present in DPI. The SO adds an extra component of the time delay which is specific to DPI.

To resolve the SO process in time, one needs to use various laser-based interferometric techniques which introduce an additional component to the measurable time delay. This component, known commonly as the Coulomb laser coupling (CLC) [42] or the continuum–continuum (CC) correction [43], depends on laser frequency and the asymptotic Coulomb charge Z acting on the departing photoelectron. For the fast primary photoelectron, this charge $Z = 1$ for the neutral targets and $Z = 0$ for the negative ions. In the former case, the fast photoelectron does not experience any CLC correction, while in the latter case, this correction is relatively small because the photoelectron is sufficiently fast. The slow electron sees the asymptotic charge $Z = 2$ for the neutral targets and $Z = 1$ for the negative ions. So while the CLC correction still affects the slow SO elecrtron, it will be relatively weaker.

Funding: This research received no external funding.

Data Availability Statement: The numerical data reported in the present work are available on request from the author.

Acknowledgments: Nearly 40 years on, the legacy of the seminal work initiated by Miron Amusia is still alive and produces fruitful results. Some of these results are reported in the present paper.

Conflicts of Interest: The author declares no known conflict of interest.

References

1. Briggs, J.; Schmidt, V. Differential cross sections for photo-double- ionization of the helium atom. *J. Phys. B At. Mol. Opt. Phys.* **2000**, *33*, R1–R48. [CrossRef]
2. Avaldi, L.; Huetz, A. Photodouble ionization and the dynamics of electron pairs in the continuum. *J. Phys. B* **2005**, *38*, S861. [CrossRef]
3. Wehlitz, R. Simultaneous emission of multiple electrons from atoms and molecules using synchrotron radiation. *Adv. At. Mol. Opt. Phys.* **2010**, *58*, 1–76.
4. McGuire, J.H. *Electron Correlation Dynamics in Atomic Collisions*; Cambridge Monographs on Atomic, Molecular and Chemical Physics; Cambridge University Press: Cambridge, MA, USA , 1997.
5. Samson, J.A. Proportionality of electron-impact ionization to double photoionization. *Phys. Rev. Lett.* **1990**, *65*, 2861–2864. [CrossRef] [PubMed]
6. Kheifets, A.S. On different mechanisms of the two-electron atomic photoionization. *J. Phys. B* **2001**, *34*, L247–L252. [CrossRef]
7. Knapp, A.; Kheifets, A.; Bray, I.; Weber, T.; Landers, A.L.; Schössler, S.; Jahnke, T.; Nickles, J.; Kammer, S.; Jagutzki, O.; et al. Mechanisms of Photo Double Ionization of Helium by 530 eV Photons. *Phys. Rev. Lett.* **2002**, *89*, 033004. [CrossRef]
8. Amusia, M.Y.; Drukarev, E.G.; Gorshkov, V.G.; Kazachkov, M.P. Two-electron photoionization of helium. *J. Phys. B At. Mol. Opt. Phys.* **1975**, *8*, 1248. [CrossRef]
9. Grundmann, S.; Serov, V.V.; Trinter, F.; Fehre, K.; Strenger, N.; Pier, A.; Kircher, M.; Trabert, D.; Weller, M.; Rist, J.; et al. Revealing the two-electron cusp in the ground states of He and H_2 via quasifree double photoionization. *Phys. Rev. Res.* **2020**, *2*, 033080. [CrossRef]
10. Grundmann, S.; Trinter, F.; Bray, A.W.; Eckart, S.; Rist, J.; Kastirke, G.; Metz, D.; Klumpp, S.; Viefhaus, J.; Schmidt, L.P.H.; et al. Separating Dipole and Quadrupole Contributions to Single-Photon Double Ionization. *Phys. Rev. Lett.* **2018**, *121*, 173003. [CrossRef] [PubMed]
11. Schöffler, M.S.; Stuck, C.; Waitz, M.; Trinter, F.; Jahnke, T.; Lenz, U.; Jones, M.; Belkacem, A.; Landers, A.L.; Pindzola, M.S.; et al. Ejection of Quasi-Free-Electron Pairs from the Helium-Atom Ground State by Single-Photon Absorption. *Phys. Rev. Lett.* **2013**, *111*, 013003. [CrossRef]
12. Chang, T.N.; Poe, R.T. Double photoionization of neon. *Phys. Rev. A* **1975**, *12*, 1432–1439. [CrossRef]
13. Carter, S.L.; Kelly, H.P. Double photoionization cross section of argon. *J. Phys. B* **1976**, *9*, L565–L568. [CrossRef]
14. Carter, S.L.; Kelly, H.P. Double photoionization of neon and argon. *Phys. Rev. A* **1977**, *16*, 1525–1534. [CrossRef]
15. Pan, C.; Kelly, H.P. Photoionization cross sections of the Ar atom for production of singly and doubly charged ions near the $2p$ threshold. *Phys. Rev. A* **1989**, *39*, 6232–6240. [CrossRef]
16. Bray, I.; Fursa, D.; Kadyrov, A.; Stelbovics, A.; Kheifets, A.; Mukhamedzhanov, A. Electron- and photon-impact atomic ionisation. *Phys. Rep.* **2012**, *520*, 135–174. [CrossRef]
17. Bray, I.; Fursa, D.V.; Kheifets, A.S.; Stelbovics, A.T. Electrons and photons colliding with atoms: Development and application of the convergent close-coupling method. *J. Phys. B At. Mol. Opt. Phys.* **2002**, *35*, R117–R146. [CrossRef]
18. Pindzola, M.S.; Robicheaux, F.; Loch, S.D.; Berengut, J.C.; Topcu, T.; Colgan, J.; Foster, M.; Griffin, D.C.; Ballance, C.P.; Schultz, D.R.; et al. The time-dependent close-coupling method for atomic and molecular collision processes. *J. Phys. B* **2007**, *40*, R39–R60. [CrossRef]
19. Schultze, M.; Fiess, M.; Karpowicz, N.; Gagnon, J.; Korbman, M.; Hofstetter, M.; Neppl, S.; Cavalieri, A.L.; Komninos, Y.; Mercouris, T.; et al. Delay in Photoemission. *Science* **2010**, *328*, 1658–1662. [CrossRef]
20. Klünder, K.; Dahlström, J.M.; Gisselbrecht, M.; Fordell, T.; Swoboda, M.; Guénot, D.; Johnsson, P.; Caillat, J.; Mauritsson, J.; Maquet, A.; et al. Probing Single-Photon Ionization on the Attosecond Time Scale. *Phys. Rev. Lett.* **2011**, *106*, 143002. [CrossRef] [PubMed]
21. Mansson, E.P.; Guenot, D.; Arnold, C.L.; Kroon, D.; Kasper, S.; Dahlstrom, J.M.; Lindroth, E.; Kheifets, A.S.; L'Huillier, A.; Sorensen, S.L.; et al. Double ionization probed on the attosecond timescale. *Nat. Phys.* **2014**, *292*, 1689. [CrossRef]

22. Kheifets, A.S.; Ivanov, I.A.; Bray, I. Timing analysis of two-electron photoemission. *J. Phys. B At. Mol. Opt. Phys.* **2011**, *44*, 101003. [CrossRef]
23. Kheifets, A.S.; Bray, I. Time delay in two-electron photodetachment and tests of fundamental threshold laws. *Phys. Rev. Res.* **2021**, *3*, 043017. [CrossRef]
24. Wendin, G.; Ohno, M. Strong Dynamical Effects of Many-Electron Interactions in Photoelectron Spectra from 4s and 4p Core Levels. *Phys. Scr.* **1976**, *14*, 148–161. [CrossRef]
25. Amusia, M.Y.; Kheifets, A.S. Effect of correlations on the photoelectron spectrum of atom. *Sov. Phys.-JETP* **1984**, *59*, 710–715.
26. Amusia, M.Y.; Kheifets, A.S. On our ability to measure the singly ionized rare-gas spectroscopic factors using the (γ,e) and (e,2e) reactions. *J. Phys. B* **1985**, *18*, L679–L684. [CrossRef]
27. Kheifets, A.S.; Amusia, M.Y. Relativistic ab initio calculation of the xenon 5s ionization spectra for the (γ,e) and (e,2e) reactions. *Phys. Rev. A* **1992**, *46*, 1261–1269. [CrossRef] [PubMed]
28. Kheifets, A.S. On the two different forms of the spectroscopic factors for the shake-up satellites. *J. Phys. B* **1994**, *27*, L463–L467. [CrossRef]
29. Kheifets, A.S. Green's function calculation of the satellite spectrum of neon. *J. Phys. B* **1995**, *28*, 3791. [CrossRef]
30. Kheifets, A.S.; Bray, I. Photoionization with excitation and double photoionization of helium isoelectronic sequence. *Phys. Rev. A* **1998**, *58*, 4501–4511. [CrossRef]
31. Amusia, M.I.; Chernysheva, L.V. *Computation of Atomic Processes: A Handbook for the ATOM Programs*; Institute of Physics Publishing: Bristol, UK, 1997.
32. Berkowitz, J. *Photoabsorption, Photoionization, and Photoelectron Spectroscopy*; Elsevier Science: Amsterdam, The Netherlands, 2012.
33. Knapp, A.; Walter, M.; Weber, T.; Landers, A.L.; Schössler, S.; Jahnke, T.; Schöffler, M.; Nickles, J.; Kammer, S.; Jagutzki, O.; et al. Energy sharing and asymmetry parameters for photo double ionization of helium 100 eV above threshold in single-particle and Jacobi coordinates. *J. Phys. B* **2002**, *35*, L521–L526. [CrossRef]
34. Colgan, J.; Pindzola, M.S. Double photoionization of helium at high photon energies. *J. Phys. B* **2004**, *37*, 1153–1164. [CrossRef]
35. Kadyrov, A.S.; Bray, I. Convergent close–coupling calculations of positron–hydrogen S–wave model. *Nucl. Instr. Meth. B* **2000**, *171*, 119–125. [CrossRef]
36. Ludlow, J.A.; Colgan, J.; Lee, T.G.; Pindzola, M.S.; Robicheaux, F. Double photoionization of helium including quadrupole radiation effects. *J. Phys. B* **2009**, *42*, 225204. [CrossRef]
37. Kheifets, A.S.; Bray, I. Frozen core model of the double photoionization of beryllium. *Phys. Rev. A* **2002**, *65*, 012710. [CrossRef]
38. Kheifets, A.S.; Bray, I. Valence-shell double photoionization of alkaline-earth-metal atoms. *Phys. Rev. A* **2007**, *75*, 042703. [CrossRef]
39. Wigner, E.P. On the Behavior of Cross Sections Near Thresholds. *Phys. Rev.* **1948**, *73*, 1002–1009. [CrossRef]
40. Rosenberg, L.; Spruch, L. Generalized Levinson theorem: Applications to electron-atom scattering. *Phys. Rev. A* **1996**, *54*, 4985–4991. [CrossRef]
41. Kheifets, A.S. Time delay in valence-shell photoionization of noble-gas atoms. *Phys. Rev. A* **2013**, *87*, 063404. [CrossRef]
42. Pazourek, R.; Nagele, S.; Burgdorfer, J. Time-resolved photoemission on the attosecond scale: Opportunities and challenges. *Faraday Discuss.* **2013**, *163*, 353–376. [CrossRef]
43. Dahlström, J.; Guénot, D.; Klünder, K.; Gisselbrecht, M.; Mauritsson, J.; Huillier, A.L.; Maquet, A.; Taïeb, R. Theory of attosecond delays in laser-assisted photoionization. *Chem. Phys.* **2012**, *414*, 53–64. [CrossRef]

atoms

MDPI

Article

Relativistic Two-Photon Matrix Elements for Attosecond Delays

Jimmy Vinbladh [1,†], Jan Marcus Dahlström [2,*,†] and Eva Lindroth [1,*,†]

1 Department of Physics, Stockholm University, AlbaNova University Center, SE-106 91 Stockholm, Sweden; jimmy.vinbladh@matfys.lth.se
2 Department of Physics, Lund University, Box 118, SE-221 00 Lund, Sweden
* Correspondence: marcus.dahlstrom@matfys.lth.se (J.M.D.); eva.lindroth@fysik.su.se (E.L.)
† These authors contributed equally to this work.

Abstract: The theory of one-photon ionization and two-photon above-threshold ionization is formulated for applications to heavy atoms in attosecond science by using Dirac–Fock formalism. A direct comparison of Wigner–Smith–Eisenbud delays for photoionization is made with delays from the Reconstruction of Attosecond Beating By Interference of Two-photon Transitions (RABBIT) method. Photoionization by an attosecond pulse train, consisting of monochromatic fields in the extreme ultraviolet range, is computed with many-body effects at the level of the relativistic random phase approximation (RRPA). Subsequent absorption and emission processes of infrared laser photons in RABBIT are evaluated by using static ionic potentials as well as asymptotic properties of relativistic Coulomb functions. As expected, light elements, such as argon, show negligible relativistic effects, whereas heavier elements, such a krypton and xenon, exhibit delays that depend on the fine-structure of the ionic target. The relativistic effects are notably close to ionization thresholds and Cooper minima with differences in fine-structure delays predicted to be as large as tens of attoseconds. The separability of relativistic RABBIT delays into a Wigner–Smith–Eisenbud delay and a universal continuum–continuum delay is studied with reasonable separability found for photoelectrons emitted along the laser polarization axis in agreement with prior non-relativistic results.

Keywords: attoscience; attophysics; photoionization; above-threshold ionization; Wigner–Smith–Eisenbud delay; Dirac–Fock; RRPA; RABBIT; krypton; xenon

Citation: Vinbladh, J.; Dahlström, J.M.; Lindroth, E. Relativistic Two-Photon Matrix Elements for Attosecond Delays. *Atoms* **2022**, *10*, 80. https://doi.org/10.3390/atoms10030080

Academic Editors: Anatoli Kheifets, Gleb Gribakin and Vadim Ivanov

Received: 20 June 2022
Accepted: 24 July 2022
Published: 2 August 2022

1. Introduction

The study of attosecond photoionization dynamics has been made possible by coherent light sources in the extreme ultraviolet (XUV) regime based on high-order harmonic generation (HHG) [1]. Experimental techniques that were originally used for pulse characterization, such as the Reconstruction of Attosecond Beating By Interference of Two-photon Transitions (RABBIT) [2] and the attosecond streak-camera [3], have proved useful to gain novel insights into the time it takes for electrons to escape the binding potentials of atoms [4–16], molecules [17–21], and solid-state targets [22–24]. The main observables are delay-dependent modulations in the photoelectron spectra that arise due a phase-locked laser probe field in the infrared (IR) regime [25–30]. For "weak" fields, these modulations can be understood from perturbation theory, where absorption of one XUV photon (Ω) is followed by exchange of one IR photon ($\pm\omega$). It is a rather technical task to evaluate laser-driven continuum–continuum transitions numerically in the presence of the long-range Coulombic potential: $k' \rightarrow k$ [26,31,32]. Thus, analytical continuum–continuum phase shifts $\phi_{cc}(k,k')$, have been derived by using the Wentzel–Kramers–Brillouin (WKB) approximation, in order to interpret the RABBIT delays at sufficiently high kinetic energy of the photoelectrons [33]. Asymptotic theories based on the Eikonal Volkov Approximation (EVA) have also been developed [34]. The main result of these asymptotic theories is that delays observed in RABBIT experiments can be separated into two terms: (i) a finite-difference approximation to the Wigner–Smith–Eisenbud delay of the photoelectron after absorption of one XUV photon: τ_W [35–37], and (ii) a universal continuum–continuum delay: $\tau_{cc}(k;\omega)$, with an analytical expression that only depends on the final momentum of the photoelectron and the frequency of the IR field. In the case of a single angular momentum channel $\lambda = \ell_i + 1$ with ℓ_i being the initial angular momentum, this separation has been successfully implemented to measure the Wigner-like delay of the 2s-orbital in neon atoms [12]. In the more general case, where multiple intermediate angular momenta are populated, $\lambda = \ell_i \pm 1$, the probe process becomes more complicated and care must be taken to account for the weight of all intermediate transitions, which leads to an "effective" Wigner delay [33]. As an example, the RABBIT delay measured close to the 3p-Cooper minimum in argon [38] is much reduced in magnitude when photoelectrons are detected over all emission angles, rather than along the polarization axis of the fields [7]. Nonetheless, the asymptotic theory has been extended to interpret delays from molecules, where contributions of multiple partial waves in the initial orbital and the orientation of the target relative to the laser polarization, adds more complexity to the process [21,39]. Although it has been shown that the separability of delays remains valid at high kinetic energies, by using full two-photon matrix elements from time-independent R-matrix theory [40], the target-specific delay in molecules τ_{PI}, cannot be interpreted as a Wigner-like delay, due to interference effects of multiple partial waves in the two-photon transitions [39] and various channel coupling effects [21,40]. The use of full two-photon R-matrix theory [40] is undoubtedly an important milestone in the field of photoionization delays from molecules, which has allowed for quantitative analysis of many recent experiments [18–21].

In the case of atoms, full two-photon matrix elements have been used for a decade to compute delays in photoionization at various levels of Many-Body Perturbation Theory (MBPT) [41–44]. Although the importance of the random-phase approximation with exchange (RPAE) for attosecond science was first realized by Kheifets [45,46], numerical simulations of the one-photon ionization process, developed by Amusia [47], are inherently insufficient to interpret RABBIT delays. Thus, a two-photon approach was developed, whereby the many-body response of XUV absorption was computed at the level of RPAE, and the IR exchange in the continuum was computed numerically by using an effective one-body ionic potential [41,42]. This two-photon matrix approach has shown good agreement with a range of RABBIT experiments [7,8,12,13,48]. Noble gas atoms consist of multiple valence states, which implies experimentally unresolved ionic states with magnetic quantum numbers $|m| \leq \ell_i$. However, any problem with incoherent final channels can easily be avoided by detecting photoelectrons along the polarization axis $\hat{z}$, where only $m = 0$ contributes. In this configuration, it has been shown that a numerically obtained continuum–continuum delay, τ_{cc}^{MBPT}, can be accurately separated from the one-photon Wigner delay τ_W^{MBPT}, computed for photoelectrons along the polarization axis with the unique ionic state $m = 0$ [41,43]. In this way, a precise separation of delays has been demonstrated down to 5 eV, which is much lower than the high-energy regime predicted by the original asymptotic theory [33]. The two-photon matrix elements have also been used to study effects beyond the asymptotic approximation. Firstly, a strong angle-dependence of RABBIT delays from the isotropic helium atom was evidenced in experiments by Heuser et al. [48]. Secondly, the role of universal asymmetries between absorption and emission processes in the continuum was identified by Busto et al. [49]. Finally, a weak angular-momentum dependence of continuum–continuum phases was measured by Fuchs et al. in helium atoms [50]. The latter discovery was in good agreement with theoretical predictions based on exact two-photon matrix elements for hydrogen, provided by Taïeb [33], as well as full two-photon matrix elements based on MBPT [13]. Thus, several effects that depend on the exact form of continuum states have been identified by using RABBIT delay measurements in recent years [51].

Due to the energy spacing between the odd harmonics from HHG, $\Delta\Omega = 2\omega$, the temporal resolution in traditional RABBIT experiments is limited to probe processes that are much shorter than $2\pi/\Delta\Omega = T_\omega/2 \approx 1.3$ fs (assuming an IR laser system with $\hbar\omega = 1.55$ eV). At a first glance, this seems to preclude any studies of autoionizing dynamics in atoms or molecules, which typically unfold on a time scale of a few femtoseconds, or more [29,52]. However, the subject of combined time–frequency non-linear metrology is quite subtle, and it has been found that a high-energy resolution of photoelectrons in RABBIT sidebands can be used to reconstruct autoionizing processes in time [11]. In this case of resonant excitation, via bound Rydberg states or autoionizating states, it is useful to consider the RABBIT scheme as a combination of one "structured" (resonant) path and another "unstructured" (reference) path [10,11,13,16,53,54]. In this case, the phase variation of the resonant path is typically much stronger than any continuum–continuum (or other non-resonant) phase shift, and the phenomena can be understood by expanding Fano's model for autoionization to laser-assisted photoionization, within the strong-field approximation [55], or by using approximate two-photon two-color matrix elements [56,57]. In the latter works, it was shown that finite pulses, in the time domain, can lead to non-periodic structures in RABBIT experiments due to autoionizing states. The two-photon Fano model has proven essential to disentangle dynamics from multiple autoionizing states measured by the RABBIT technique [14]. Although we find that the theory development for autoionization in RABBIT is another milestone in the field, we will not consider this class of processes in the following work. Rather, we will focus on correlation effects in unstructured continuum, where MBPT is a numerically efficient route to describe correlation effects and RABBIT data can be safely assumed to be periodic.

Despite these many successes, there remained disagreement between experimental and theoretical results for the relative RABBIT delay between the $3p$ and $3s$ orbitals in argon, first measured by Klünder et al. in 2011 [5], which was mostly ascribed to the low signal close to the correlated minimum in the $3s$-partial photoionization cross section [6,41,42,58–60]. The fact that this exceptionally deep minimum from $3s$ arises due to correlation effects, was first showed by Amusia in 1972 by applying the RPAE method to describe photoionization from inner atomic orbitals [61]. By using two-photon matrix elements, it has now been shown that the position, height, and sign of the associated RABBIT delay from $3s$ is similarly sensitive to correlation effects [41,42], which largely stems from the sensitivity of the one-photon Wigner delay peak from the correlated minimum in the photoionization cross section [62]. In order to solve this long-standing problem, a full two-photon two-color RPAE (2P2C-RPAE) method was developed for RABBIT delays [44]. This new method allowed for detailed examination of correlated IR exchange processes. It was found that, apart from a rather minor discrepancy at the correlated $3s$-minimum in argon, the universal separability of the MBPT continuum–continuum delay and Wigner delay was achieved. However, this discrepancy was still not enough to reach agreement with the experimental results [5,6]! It was not until the argon experiment was repeated, with higher photon energies by Alexandridi et al. in 2021 [15], that excellent agreement with 2P2C-RPAE results was found in a broad energy range above the $3s$-minimum in argon. It was also concluded that the long-standing $3p$-$3s$ problem was caused by an "accidental" harmonic satellite, namely the $3s^2 3p^4(^1D)4p(^2P)$ shake-up process, predicted by Wijesundera and Kelly in 1989 by using MBPT [63], which overlapped with the $3s$-RABBIT sidebands. Prior to that, the importance of "two-electron-two-hole" excitations in argon had been found by Amusia and Kheifets by considering effects beyond RPAE in 1981 [64].

The 2P2C-RPAE method also opened up for gauge-invariance tests of the RABBIT theory [44]. It was concluded that the so-called length-gauge formulation of light-matter interaction was much favoured, which is in line with the gauge theory of Kobe [65,66]. In the velocity-gauge formulation of RABBIT, it was found that the interaction with the second photon required a more detailed many-body treatment, beyond the one-body ionic potential, with important contributions from both time-orders of the fields XUV+IR and IR+XUV. Although it was shown that only the *complete* 2P2C-RPAE theory leads to gauge-invariant results, the approximate one-body treatment of the IR-exchange was shown to be an excellent approximation in length gauge. For this reason, we will use the length gauge, with an effective ionic potential to describe IR exchange processes, in our current work, which aims to quantitatively account for relativistic effects in RABBIT experiments.

The study of relativistic effects is quite a recent development in attosecond physics. In our view, Saha et al. have pioneered this field with calculations of relativistic one-photon Wigner delays [62,67,68], based on the relativistic random phase approximation (RRPA). Although RRPA theory was originally developed in the late seventies by Johnson and Cheng to describe one-photon ionization cross sections in heavy elements [69,70], the interest in such phenomena is revived by recent RABBIT experiments that have targeted heavy elements. First, Jordan et al. [71] and Jain et al. [72] have compared photoelectrons from the fine-structure split valance orbitals: $4p_j$ and $5p_j$ with $j = 1/2$ and $3/2$ of krypton and xenon atoms, respectively, and secondly, Jain et al. [73] and Zhong et al. [74], have compared photoelectrons from inner orbitals in xenon, down to the $4d$ orbital. The $4d$ orbital is of special interest because it is known to posses a giant collective resonance in the photoionization cross section, as evidenced by MBPT in the early seventies by Amusia and Wendin [75,76]. Thus, it is now possible to study the role of sizable relativistic effects, such as the spin-orbit effect in xenon, in the time domain with RABBIT. This opens a call for time-dependent methods to solve the Dirac equation for heavy many-electron atoms; as an example we mention the recently developed relativistic time-dependent configuration–interaction singles (RTDCIS) method [77], but also extend the computation of two-color, two-photon matrix elements to the relativistic domain. Concerning the lack of such relativistic theories, we mention that in refs. [71,73], the experiments were accompanied by photoionization delay calculations with one-photon matrix elements at the level of RRPA for XUV absorption, whereas various asymptotic formulas from non-relativistic theory were used to account for IR exchange effects. Our goal here is to treat the whole process within a relativistic framework and below we discuss the different points where the relativistic treatment differs from that of the non-relativistic one with an effective ionic potential for IR exchange [41–43]. We also mention that the method presented here has already been utilized in various projects, such as [49,74], without any detailed description of the theoretical formulation. A full development of the two-photon, two-color relativistic random phase approximation (2P2C-RRPA) is beyond the scope of the present work, but we expect that it would not lead to any major modification of the results presented here, because we base our entire theory on the length gauge formulation of the light–matter interaction, where the one-body ionic potential description of IR exchange processes is a good approximation [44].

In Section 2 below, some basic concepts are introduced, and the relativistic scattering phases, as well as the asymptotic form of the continuum solutions, are discussed in detail. Section 3 discusses photoionization delay in a relativistic framework, and in Section 4 the many-body implementation is outlined, and the technique to calculate the needed two-photon matrix elements is explained. Some results are finally shown in Section 5.

2. Theory

2.1. The Dirac Equation

The starting point for calculations in a relativistic framework is the Dirac equation. We aim here for calculations on many-electron systems, and as a first approximation we let the electron–electron interaction be approximated by an average potential: the relativistic version of the Hartee–Fock (HF) potential, usually called the Dirac–Fock (DF) potential. Each electron is then governed by the one-particle Hamiltonian:

$$h_{DF} = c\boldsymbol{\alpha} \cdot \mathbf{p} + \left(u_{DF} - \frac{e^2}{4\pi\varepsilon_0} \frac{Z}{r} \right) \mathbf{1}_4 + mc^2\beta, \tag{1}$$

with eigenvalues labeled by E, and where $\boldsymbol{\alpha}$ is expressed in Pauli matrices and β has the corresponding form

$$\boldsymbol{\alpha} = \begin{pmatrix} 0 & \sigma \\ \sigma & 0 \end{pmatrix}, \quad \beta = \begin{pmatrix} \mathbf{1}_2 & 0 \\ 0 & -\mathbf{1}_2 \end{pmatrix}. \tag{2}$$

For closed shell atoms, as the rare gases treated here, the Dirac–Fock potential is spherically symmetric, and the two-component radial part of the wave function can be separated out and determined by the radial Hamiltonian

$$\left(h_\kappa^{DF}(r) - mc^2\right)\begin{pmatrix} f_\kappa(r) \\ g_\kappa(r) \end{pmatrix} = \left(E - mc^2\right)\begin{pmatrix} f_\kappa(r) \\ g_\kappa(r) \end{pmatrix} = \epsilon\begin{pmatrix} f_\kappa(r) \\ g_\kappa(r) \end{pmatrix} \tag{3}$$

with

$$\left(h_\kappa^{DF}(r) - mc^2\right) = \begin{pmatrix} u_{DF}(r) - \frac{e^2}{4\pi\varepsilon_0}\frac{Z}{r} & -c\hbar\left(\frac{d}{dr} - \frac{\kappa}{r}\right) \\ c\hbar\left(\frac{d}{dr} + \frac{\kappa}{r}\right) & u_{DF}(r) - \frac{e^2}{4\pi\varepsilon_0}\frac{Z}{r} - 2mc^2 \end{pmatrix}, \tag{4}$$

where f_κ is the upper, typically larger, component, and g_κ the lower, typically smaller, component. The four-component eigenfunction to the one-particle Hamiltonian in Equation (1) can now be written as [78]

$$\psi_{n\ell jm}(r,\theta,\phi) = \begin{pmatrix} \frac{f_{n\ell j}(r)}{r}\chi_{\kappa m}(\theta,\phi) \\ \frac{ig_{n\ell j}(r)}{r}\chi_{-\kappa m}(\theta,\phi) \end{pmatrix}$$

$$= \begin{pmatrix} \frac{f_{nj\ell}(r)}{r}\sum_{\nu,\mu}\langle \ell\mu s\nu \mid jm\rangle \xi_\nu Y_{\ell,\mu}(\theta,\phi) \\ \frac{ig_{nj\ell}(r)}{r}\sum_{\nu,\mu}\langle (2j-\ell)\mu s\nu \mid jm\rangle \xi_\nu Y_{(2j-\ell),\mu}(\theta,\phi) \end{pmatrix}, \tag{5}$$

where $\chi_{\kappa m}(\theta,\phi)$ is a vector coupled function of a spherical harmonic and a spin function ξ_ν. The relativistic quantum number κ is defined by the eigenvalue equation $(\sigma\cdot\boldsymbol{\ell}+1)\chi_{\kappa m} = -\kappa\chi_{\kappa m}$ and takes the value $\kappa = \ell(\ell+1) - j(j+1) - 1/4$. When κ is negative, $(j = \ell + 1/2)$, the spherical harmonic associated with the small component, will be one unit of orbital angular momenta larger than that for the large component, and vice verse for positive κ $(j = \ell - 1/2)$.

The RRPA method, which is also known as the linear response within the time-dependent Dirac–Fock (TDDF) formalism, will be used to describe the atomic response to electromagnetic radiation. It accounts for the interaction with the electromagnetic field in lowest order, including also corrections to the static Dirac–Fock potential by field-perturbed orbitals [47,79]. The method is discussed further in Section 4. In the next section, we will discuss expressions for the radial continuum wave functions at large, but not infinite distances from the ion.

2.2. The Scattering Phase of the Photoelectron

Although the total photoionization cross section is determined by the amplitude of the outgoing electron wave packet, its phase is crucial for its angular dependence as well as its delayed appearance in the continuum. In the following, we discuss the difference of the scattering phase in a relativistic formulation compared to the non-relativistic one.

We consider first an N-electron atom that absorbs a photon and subsequently ejects a photoelectron from orbital b. The radial photoelectron wave function will in the non-relativistic case be described by an outgoing phase-shifted Coulomb wave that asymptotically has the form

$$u^{(1)}_{q,\Omega,b}(r) \approx -\pi M^{(1)}_{\mathrm{nrel}}(q,\Omega,b)\sqrt{\frac{2m}{\pi k\hbar^2}}\, e^{i\left(kr + \frac{Z}{ka_0}\ln 2kr - \ell\frac{\pi}{2} - \sigma_{Z,k,\ell} + \delta_{k,\ell}\right)}. \tag{6}$$

Here energy normalization is assumed, and $M^{(1)}_{\mathrm{nrel}}$ is the non-relativistic electric dipole transition matrix element to the final continuum state q with momenta k, ℓ, and m. Although

$M^{(1)}_{\text{nrel}}$ can be chosen to be real in a one-electron context it will be complex when correlation effects are considered. The Coulomb phase is

$$\sigma_{Z,k,\ell} = \arg\left[\Gamma\left(\ell + 1 + \frac{iZ}{ka_0}\right)\right], \tag{7}$$

for a photoelectron in the field from a point charge of Ze. Note that in Equations (6) and (7), we use the negative Coulomb phase convention, rather than the equivalent positive sign convention that is more commonly used: cf. Equations (1) and (2) in Ref. [44], in order to easily relate the phase expressions to existing relativistic theory in the literature [80]. The additional phase shift $\delta_{k,\ell}$ comes from the short range many-body potential of the final state. The Bohr radius is here denoted with a_0. In the relativistic case, the asymptotic radial wave function will have an upper and a lower component, cf. Equation (5), which will have the form [80]

$$u^{(f,1)}_{q,\Omega,b}(r) \approx -\pi M^{(1)}(q,\Omega,b)\sqrt{\frac{2m}{\pi k\hbar^2}\left(1+\frac{\varepsilon}{2mc^2}\right)} \times e^{i\left(kr+\eta \ln 2kr - \gamma\frac{\pi}{2} - \tilde{\sigma}_{Z,k,\gamma} + \nu + \tilde{\delta}_{Z,k,\gamma}\right)},$$

$$u^{(g,1)}_{q,\Omega,b}(r) \approx -i\zeta\pi M^{(1)}(q,\Omega,b)\sqrt{\frac{2m}{\pi k\hbar^2}\left(1+\frac{\varepsilon}{2mc^2}\right)} \times e^{i\left(kr+\eta \ln 2kr - \gamma\frac{\pi}{2} - \tilde{\sigma}_{Z,k,\gamma} + \nu + \tilde{\delta}_{Z,k,\gamma}\right)}, \tag{8}$$

where the superscripts f and g indicate the large (upper) and small (lower) components respectively and

$$\zeta = \sqrt{\frac{E - mc^2}{E + mc^2}} = \frac{k\hbar}{2mc}\frac{1}{\left(1+\frac{\varepsilon}{2mc^2}\right)} \tag{9}$$

is the relation between the large and small component at infinity. This asymptotic relation is given directly by Equation (4), with $\epsilon = E - mc^2$ being the kinetic energy at infinity. The form of the components in Equation (8) is indeed the same as in the non-relativistic case, but the parameters have slightly changed definition: $M^{(1)}$ is now the relativistic matrix element, and k is calculated from the relativistic kinetic energy as

$$k = \frac{\sqrt{E^2 - m^2c^4}}{\hbar c} = \frac{\sqrt{2\varepsilon m}}{\hbar}\sqrt{1+\frac{\varepsilon}{2mc^2}}. \tag{10}$$

The first factor on the right-hand side of Equation (10) is identical to the non-relativistic expression for k, which is thus only slightly adjusted as long as the kinetic energy of the released electron is modest: $\varepsilon \ll mc^2$. The constant η is given by

$$\eta = Z\alpha E\sqrt{\frac{1}{E^2 - m^2c^4}} = \frac{Z}{a_0 k}\left(\frac{\varepsilon}{mc^2} + 1\right) \tag{11}$$

where α is the fine structure constant, $\alpha = \hbar/(a_0 mc)$. In the non-relativistic limit η will thus tend to $Z/(a_0 k)$ as expected by comparison with Equation (6). The relativistic Coulomb phase is

$$\tilde{\sigma}_{Z,k,\gamma} = \arg[\Gamma(\gamma + i\eta)] \tag{12}$$

with

$$\gamma = \sqrt{\kappa^2 - \alpha^2 Z^2} \tag{13}$$

and

$$\nu = \frac{1}{2}\arg\left[\frac{-\kappa + \frac{iZ}{ka_0}}{\gamma + i\eta}\right]. \tag{14}$$

The phase induced by the short-range part of the many-body potential for the final state is denoted with $\tilde{\delta}_{Z,k,\kappa}$.

2.3. Phase-Shifted Relativistic Coulomb Functions at Large Distances

Calculations on many-body systems have to be done numerically. While the wave function for the escaping photoelectron will differ from the analytically known Coulombic ones at short distances, it will approach a combination of a phase-shifted known regular and irregular Coulomb function outside the core of the remaining ion. Because transition matrix elements between continuum states do not converge on a finite grid, it is convenient to have access to continuum solutions, with a possible phase shift δ, that can be used to continue the integration to infinity. We are here interested to find expressions for the relativistic case, but it is illustrative to compare with the more studied non-relativistic formulation.

The solutions to the hydrogen-like Schrödinger equation with positive energy is given by the Coulomb functions (see e.g., [81]). The regular Coulomb function is in particular

$$F_\ell(\eta_{\text{nrel}}, kr) = \frac{1}{2}e^{\frac{\pi}{2}\eta_{\text{nrel}}}\frac{|\Gamma(\ell+1+i\eta_{\text{nrel}})|}{(2\ell+1)!}e^{-ikr}(2kr)^{\ell+1}M(\ell+1+i\eta_{\text{nrel}}, 2\ell+2, 2ikr), \tag{15}$$

where M is the confluent hypergeometric function, σ is defined in Equation (7) and $\eta_{\text{nrel}} = Z/(a_0k)$. Non-relativistic Coulomb functions expressions, valid for large kr, are provided in Ref. [82]:

$$F_\ell = \bar{g}\cos\Delta_{\text{nrel}} + \bar{f}\sin\Delta_{\text{nrel}} \tag{16}$$

$$G_\ell = \bar{f}\cos\Delta_{\text{nrel}} - \bar{g}\sin\Delta_{\text{nrel}}, \tag{17}$$

for the regular, F_ℓ, and irregular, G_ℓ, Coulomb functions respectively, where

$$\Delta_{\text{nrel}} \equiv kr + \frac{Z}{ka_0}\ln 2kr - \frac{\pi}{2}\ell - \sigma_{Z,k,\ell} + \delta \tag{18}$$

and $\bar{f}$ and $\bar{g}$, which depend on Z, r, k, and ℓ, can be obtained through simple recursive formulas given in ref. [82]. When $r \to \infty$, $\bar{g} \to 0$ and $\bar{f} \to 1$ and thus the regular function approaches a sin-function, and the irregular a cos-function, both with amplitude one. The combination

$$F_\ell(\eta_{\text{nrel}}, kr) - iG_\ell(\eta_{\text{nrel}}, kr) \tag{19}$$

will thus asymptotically approach an outgoing wave, with modulus square equal to unity. Energy normalized continuum functions are obtained by multiplications with $\sqrt{2m/\pi k\hbar^2}$.

It is interesting to note that Equations (16) and (17) imply that the irregular (regular) function can readily be obtained when the regular (irregular) one is at hand. In the former case, the irregular solution is found as

$$G_\ell = \frac{\left(\frac{dF_\ell}{dr} - \frac{F_\ell}{\bar{g}^2+\bar{f}^2}\left(\frac{d\bar{g}}{dr}\bar{g} + \frac{d\bar{f}}{dr}\bar{f}\right)\right)}{k + \eta/r + \frac{1}{\bar{g}^2+\bar{f}^2}\left(\frac{d\bar{g}}{dr}\bar{f} - \frac{d\bar{f}}{dr}\bar{g}\right)}. \tag{20}$$

Turning to the relativistic Coulomb problem, we set out to find the relativistic counterparts to Equations (16) and (17), which to the best of our knowledge, are not available in the literature. The exact two-component relativistic regular, $\tilde{F}_\gamma$, and irregular, $\tilde{G}_\gamma$, solutions are given in a pioneering article by Johnson and Cheng [80]. In particular the regular solution is

$$\tilde{F}_\gamma(\eta, kr) = \sqrt{\frac{E+mc^2}{2E}}\frac{1}{2}e^{\frac{\pi}{2}\eta}\frac{|\Gamma(\gamma+i\eta)|}{\Gamma(2\gamma+1)}(-2ikr)^\gamma e^{ikr} \begin{pmatrix} \left(-\kappa+\frac{iZ}{ka_0}\right)M_\gamma + (\gamma-i\eta)M_{\gamma+1} \\ -i\zeta\left(\left(-\kappa+\frac{iZ}{ka_0}\right)M_\gamma - (\gamma-i\eta)M_{\gamma+1}\right) \end{pmatrix} \tag{21}$$

with ζ, γ, η and k given in Equations (9)–(11) and (13), and the short-hand notation

$$\begin{aligned} M_\gamma &= & M(\gamma-i\eta, 2\gamma+1, -2iz) \\ M_{\gamma+1} &= & M(\gamma+1-i\eta, 2\gamma+1, -2iz) \end{aligned} \tag{22}$$

has been used for the confluent hypergeometric functions.

An asymptotic expansion of the confluent hypergeometric function, M can be found in ref. [83], which indeed can be used to obtain asymptotic expansions for $\tilde{F}_\gamma$ and $\tilde{G}_\gamma$ on forms similar to Equations (16) and (17):

$$\tilde{F}_\gamma = \sqrt{\frac{E+mc^2}{2E}}\begin{pmatrix} \bar{f}_{\text{large}}\cos\Delta - \bar{g}_{\text{large}}\sin\Delta \\ -\zeta(\bar{g}_{\text{small}}\cos\Delta + \bar{f}_{\text{small}}\sin\Delta) \end{pmatrix} \tag{23}$$

and

$$\tilde{G}_\gamma = \sqrt{\frac{E+mc^2}{2E}}\begin{pmatrix} -\left(\bar{g}_{\text{large}}\cos\Delta + \bar{f}_{\text{large}}\sin\Delta\right) \\ -\zeta(\bar{f}_{\text{small}}\cos\Delta - \bar{g}_{\text{small}}\sin\Delta) \end{pmatrix} \tag{24}$$

with

$$\Delta = kr + \eta\ln 2kr - \pi\gamma/2 - \tilde{\sigma}_{Z,k,\gamma} + \nu + \tilde{\delta} \tag{25}$$

with $\tilde{\sigma}$ and ν given in Equations (12) and (14). The possible extra phase shift is denoted by $\tilde{\delta}$. In the non-relativistic limit $\Delta \to \Delta_{\text{nrel}} \pm \pi/2$, for $\kappa > 0$ and $\kappa < 0$ respectively, and thus the sin/cos—functions in Equations (16) and (17) are replaced with $\mp\cos$ / $\pm\sin$ in the upper components of Equation (23) and (24). The relativistic $\bar{f}, \bar{g}$ functions are obtained as

$$\bar{f}_{\text{large/small}} = \text{Re}(\aleph_\pm) \tag{26}$$

$$\bar{g}_{\text{large/small}} = \text{Im}(\aleph_\pm) \tag{27}$$

from

$$\aleph_\pm = \sum_{n=0}\frac{(\gamma-i\eta)_n(-\gamma-i\eta)_n}{n!}(2ikr)^{-n} \pm \sum_{n=0}\frac{(\gamma+1+i\eta)_n(-\gamma+1+i\eta)_n}{n!}(-2ikr)^{-n} \tag{28}$$

where $(a)_n = a(a+1)(a+2)\dots(a+n-1)$, $(a)_0 = 1$. Similarly to the non-relativistic case $\bar{g}_{\text{large/small}} \to 0$ and $\bar{f}_{\text{large/small}} \to 1$ when $r \to \infty$. Thus the upper regular, and the lower irregular, approach $\cos\Delta$, whereas the lower regular and the upper irregular tend to $\sin\Delta$. The asymptotic expressions are thus

$$\tilde{F}_\gamma(\eta, kr) \to \sqrt{\frac{E+mc^2}{2E}}\begin{pmatrix} \cos\Delta \\ -\zeta\sin\Delta \end{pmatrix}, \tag{29}$$

$$\tilde{G}_\gamma(\eta, kr) \to \sqrt{\frac{E+mc^2}{2E}} \begin{pmatrix} -\sin\Delta \\ -\zeta\cos\Delta \end{pmatrix}, \tag{30}$$

when $kr \to \infty$, and the combination

$$\tilde{F}_\gamma(\eta, kr) - i\tilde{G}_\gamma(\eta, kr) \to \sqrt{\frac{E+mc^2}{2E}} \begin{pmatrix} 1 \\ i\zeta \end{pmatrix} e^{i\Delta} \tag{31}$$

will, in close analogy with the non-relativistic expression in Equation (19), asymptotically approach an outgoing wave, with modulus square unity. The energy normalized functions are again obtained by multiplication with $\sqrt{2m/\pi k\hbar^2}$. We note finally that Equation (20) holds also in a relativistic framework. It provides the irregular solution, $\tilde{G}_\gamma$ from $\tilde{F}_\gamma$, if $\bar{f}$ and $\bar{g}$ are just replaced with $\bar{f}_{\text{large}}$ and $\bar{g}_{\text{large}}$ or $\bar{f}_{\text{small}}$ and $\bar{g}_{\text{small}}$ for the upper and lower components respectively.

3. Delay in Photoionization

We will here briefly discuss the calculation of delays in laser-assisted photoionization, emphasizing the differences compared to the non-relativistic description. A detailed account of the latter can be found in refs. [42,44].

3.1. The Wigner Delay

The concept of delay was introduced by Wigner [35], Smith [36] and Eisenbud [37] as the derivative of the scattering phase with respect to energy. With a finite difference approximation of the derivative $\Delta\omega = 2\omega$, the Wigner contribution to the atomic delay measured in a RABITT experiemnt is

$$\tau_W = \frac{\phi_> - \phi_<}{2\omega}, \tag{32}$$

where $\phi_{>/<}$ refer to the phases acquired in the XUV absorption step in the two paths where either the higher or the lower harmonic is absorbed. Non-relativistically, and for detection of the photoelectron in the $\hat{\mathbf{z}}$ direction, these phases are

$$\begin{aligned} \phi_>^{\text{nrel}} &= \arg\left(\sum_\ell M_>^{\text{nrel}}(\ell) e^{i\left(-\ell\frac{\pi}{2} - \sigma_{Z,k_>,\ell} + \delta_{k_>,\ell}\right)} Y_{\ell,0}(\hat{\mathbf{z}})\right) \\ \phi_<^{\text{nrel}} &= \arg\left(\sum_\ell M_<^{\text{nrel}}(\ell) e^{i\left(-\ell\frac{\pi}{2} - \sigma_{Z,k_<,\ell} + \delta_{k_<,\ell}\right)} Y_{\ell,0}(\hat{\mathbf{z}})\right), \end{aligned} \tag{33}$$

where the short-hand notation for the one-photon matrix elements, $M_{>/<}(\ell) \equiv M^{(1)}(q_{>/<}, \Omega_{>/<}, b)$, with final photoelectron wave number $k_{>/<}$ and angular momentum ℓ, after absorption of a photon with angular frequency $\Omega_{>/<}$, is used. Relativistically the corresponding amplitudes have two components and it is more appropriate to define the Wigner delay as

$$\begin{aligned} \tau_W = \frac{1}{2\omega} \arg\Bigg[\sum_{m=\pm\frac{1}{2}} & \left(\sum_\kappa M_< \begin{pmatrix} \chi_{\kappa m}(\hat{\mathbf{z}}) \\ i\zeta\chi_{-\kappa m}(\hat{\mathbf{z}}) \end{pmatrix} e^{i\left(-\gamma\frac{\pi}{2} - \tilde{\sigma}_{Z,k,\gamma} + \nu + \tilde{\delta}_{Z,k,\gamma}\right)}\right)^\dagger \\ & \left(\sum_{\kappa'} M_> \begin{pmatrix} \chi_{\kappa' m}(\hat{\mathbf{z}}) \\ i\zeta\chi_{-\kappa' cm}(\hat{\mathbf{z}}) \end{pmatrix} e^{i\left(-\gamma'\frac{\pi}{2} - \tilde{\sigma}_{Z,k,\gamma'} + \nu' + \tilde{\delta}_{Z,k,\gamma'}\right)}\right)\Bigg], \end{aligned} \tag{34}$$

where the calculation of the delay of electrons emitted along the z-axis requires an incoherent sum over $m = \pm 1/2$. The two incoherent contributions to the Wigner delay are due to unresolved photoelectron spin in the final state.

3.2. The Atomic Delay

We now consider measurements that employ the RABBIT technique [2], where an XUV comb of odd-order harmonics of a fundamental laser field with angular frequency ω, is combined with a synchronized, weak laser field with the same angular frequency. In RABBIT, the one-photon ionization process is assisted by an IR photon that is either absorbed or emitted. The same final state is reached when both an XUV harmonic with energy $\hbar\Omega_< = (2n-1)\hbar\omega$ and an IR photon is absorbed, as when the next XUV harmonic, $\hbar\Omega_> = (2n+1)\hbar\omega$, is absorbed while an IR photon is emitted. This gives rise to modulated sidebands in the photoelectron spectrum at energies corresponding to the absorption of an even number of IR photons. Schematically the intensity of such a sideband can be written as [25]

$$S = \mid A_{\rm a} + A_{\rm e} \mid^2 = \mid A_{\rm a} \mid^2 + \mid A_{\rm e} \mid^2 + A_{\rm e}^* A_a + A_{\rm e} A_{\rm a}^*$$
$$\mid A_{\rm a} \mid^2 + \mid A_{\rm e} \mid^2 + 2 \mid A_{\rm e} \mid\mid A_{\rm a} \mid \cos[\arg(A_{\rm e}) - \arg(A_{\rm a})], \quad (35)$$

where $A_{\rm a/e}$ are the complex quantum amplitudes for the two-photon processes involving absorption (a) or emission (e) of an IR photon, and leading to the same final energy. The modulation arises from the last term in Equation (35) and can be shown to be governed by the delay between the IR and XUV pulses, τ, the group delay of the attosecond pulses in the train, τ_{XUV}, and by a contribution from the atomic system which is due the phase difference between the emission and the absorption paths in the atom:

$$\cos[\arg(A_{\rm e}) - \arg(A_{\rm a})] = \cos[2\omega(\tau - \tau_{XUV}) + \phi_{\rm e} - \phi_{\rm a}]. \quad (36)$$

The atomic contribution can be interpreted as an atomic delay: $\tau_A = (\phi_{\rm e} - \phi_{\rm a})/2\omega$. Because the delay between the two light fields is controlled in the experiments and the pulse train group delay can be canceled through relative measurements, the atomic contribution can be extracted. A recent review of the experimental method can be found in [84]. In the following, we discuss the determination of $\phi_{\rm a}$ and $\phi_{\rm e}$.

The outgoing radial wave function for the large component, after interaction with two photons, will, in accordance with the one-photon situation in Equation (8), have the asymptotic form

$$u^{(f,2)}_{q,\omega,\Omega,b}(r) \approx -\pi M^{(2)}(q,\omega,\Omega,b)\sqrt{\frac{2m}{\pi k\hbar^2}\left(1+\frac{\varepsilon}{2mc^2}\right)}\, e^{i\left(kr+\eta \ln 2kr - \gamma\frac{\pi}{2} - \bar{\sigma}_{Z,k,\gamma} + \nu + \bar{\delta}_{Z,k,\gamma}\right)}, \quad (37)$$

where the important difference compared to the one-photon case lies in the presence of the two-photon transition element $M^{(2)}$, which connects the initial state b to the continuum state q through all dipole-allowed intermediate states. The small component follows as in Equation (8). The phases acquired in the absorption and emission paths (cf. Equation (36)), are given by the corresponding two-photon matrix element and the phase of the photoelectron. In the non-relativistic case, and for photoelectrons with momentum along the common polarization axis of the fields, $\hat{\mathbf{z}}$, they are given as

$$\phi_{\rm a}^{\rm nrel} = \arg\left(\sum_\ell M_{\rm a,nrel}(\ell) e^{i\left(-\ell\frac{\pi}{2} - \sigma_{Z,k,\ell} + \delta_{k,\ell}\right)} Y_{\ell,0}(\hat{\mathbf{z}})\right)$$
$$\phi_{\rm e}^{\rm nrel} = \arg\left(\sum_\ell M_{\rm e,nrel}(\ell) e^{i\left(-\ell\frac{\pi}{2} - \sigma_{Z,k,\ell} + \delta_{k,\ell}\right)} Y_{\ell,0}(\hat{\mathbf{z}})\right), \quad (38)$$

where the short-hand notation

$$M_{\text{a,nrel}}(\ell) = M^{(2)}_{\text{nrel}}(q, \omega, \Omega_<, b),$$
$$M_{\text{e,nrel}}(\ell) = M^{(2)}_{\text{nrel}}(q, -\omega, \Omega_>, b) \quad (39)$$

has been used and the subscripts a and e stand for IR absorption and emission, respectively. For photoelectron emission along the $\hat{\mathbf{z}}$-direction, i.e., $\theta = 0$, the spherical harmonic is non-zero only for azimutal quantum number $m_\ell = 0$. The atomic delay, defined as the phase difference divided by 2ω, can subsequently be calculated as

$$\tau_A = \frac{\phi_{\text{e}} - \phi_{\text{a}}}{2\omega}. \quad (40)$$

In the Dirac case, there are two distinct differences. First, a sum over $m = \pm 1/2$ is required, because both spin-directions contribute to the emission along the $\hat{\mathbf{z}}$-direction. Second, due to the multi-component wave function the Dirac case the expression gets more involved:

$$\tau_A = \frac{1}{2\omega} \arg \Bigg[\sum_{m=\pm\frac{1}{2}} \left(\sum_\kappa M_{\text{a}} \begin{pmatrix} \chi_{\kappa m}(\hat{\mathbf{z}}) \\ i\zeta\chi_{-\kappa m}(\hat{\mathbf{z}}) \end{pmatrix} e^{i\left(-\gamma\frac{\pi}{2} - \bar{\sigma}_{Z,k,\gamma} + \nu + \bar{\delta}_{Z,k,\gamma}\right)} \right)^\dagger \times \left(\sum_{\kappa'} M_{\text{e}} \begin{pmatrix} \chi_{\kappa' m}(\hat{\mathbf{z}}) \\ i\zeta\chi_{-\kappa' m}(\hat{\mathbf{z}}) \end{pmatrix} e^{i\left(-\gamma'\frac{\pi}{2} - \bar{\sigma}_{Z,k,\gamma'} + \nu + \bar{\delta}_{Z,k,\gamma'}\right)} \right) \Bigg], \quad (41)$$

where the sum over $m = \pm 1/2$ is done incoherently (see e.g., the discussion in ref. [85]) , whereas the sum over κ is done coherently. The two incoherent contributions to the atomic delay are due to unresolved photoelectron spin in the final state. The expression for the Wigner and atomic delay for electrons detected along an arbitrary direction have been discussed in ref. [86]

4. Method

In the following, we label the full four-component "perturbed wave function", associated with absorption of one photon with angular frequency Ω and a hole in orbital b, by $|\rho_{\Omega,b}\rangle$, including both radial and spin-angular parts implicitly. As in [41–43], we use here the RPAE-approximation for the many-body response to the absorption of an XUV-photon, albeit within a relativistic framework.

4.1. The Form of the Light–Matter Interaction

The standard expression for light–matter interaction,

$$h_I = ec\boldsymbol{\alpha} \cdot \mathbf{A}(\mathbf{r}, t), \quad (42)$$

comes from applying minimal coupling: $\mathbf{p} \to \mathbf{p} + e\mathbf{A}$ to the Dirac Hamilonian in Equation (1). Within the dipole approximation, the vector potential is assumed to be space-independent: $\mathbf{A}(\mathbf{r}, \mathbf{t}) \to \mathbf{A}(\mathbf{t})$. This is often referred to as the "velocity gauge" expression for light—matter interaction:

$$h_I^{\text{velocity}} = ec\boldsymbol{\alpha} \cdot \mathbf{A}(t). \quad (43)$$

A unitary transformation can be made to recast the interaction in the alternative "length gauge" form

$$h_I^{\text{length}} = e\mathbf{r} \cdot \mathbf{E}. \tag{44}$$

For details see e.g., ref. [87]. Because our interest here is low-energy photoelectrons, we will stick to the dipole approximation. It is well known that the two gauge forms give identical results when evaluated by an exact wave function, but also for approximations that employ a local potential to describe electron–electron interaction. The non-local exchange potential in the Hartree–Fock approximation can lead to different results in the two gauges when static orbitals are assumed [65,66]. As was shown in the 1970s, the gauge invariance for one-photon processes is restored by the RPAE class of many-body effects [88]. Recently, this was discussed in connection with the calculation of two-photon processes, as needed for the calculation of atomic delays [44], and it was shown that gauge invariance required a full two-photon RPAE treatment. Because ref. [44] also showed that the length gauge results are completely dominated by the time-order where the XUV photon is absorbed first and much less sensitive to final state interactions (after absorption of two photons) than velocity gauge, only the length form will be used here.

With linearly polarized light, we may now write the lowest order approximation of the transition matrix elements from Equation (8) as

$$M^{(1)}(q, \Omega, b) = \langle q \mid ez \mid b \rangle E_{\Omega}, \tag{45}$$

and similarly the two-photon matrix element in Equation (37) as

$$M^{(2)}(q, \omega, \Omega, b) = \lim_{\xi \to 0^+} \sum\!\!\!\!\!\!\int_p \frac{\langle q \mid ez \mid p \rangle \langle p \mid ez \mid b \rangle}{\epsilon_b + \hbar\Omega - \epsilon_p + i\xi} E_{\omega} E_{\Omega}, \tag{46}$$

where intermediate states, p, are to be summed and integrated over for the bound and continuum part of the spectrum respectively. An important difference compared to the one-photon matrix element is that the two-photon matrix element is intrinsically complex for the above threshold ionization, i.e., when $\hbar\Omega$ exceeds the binding energy, even if correlation effects are neglected.

4.2. Diagrammatic Perturbation Theory

The approximation is illustrated by the diagrams in Figure 1, and a detailed derivation can be found in ref. [44]. The solution of the RPAE equations is done iteratively as indicated in Figure 1 and includes the linear response to the interaction with the XUV photon,

$$\left(\epsilon_b \pm \hbar\Omega - h_{\kappa}^{DF}\right) \mid \rho_{\Omega,b}^{\pm} \rangle = \sum_p^{exc} \mid p \rangle \langle p \mid \left(d_{\Omega_j} + \delta u_{\Omega}^{\pm}\right) \mid b \rangle, \tag{47}$$

where $\delta u_{\Omega}^{\pm}$ is the (linearized) corrections to the Dirac–Fock potential induced by the electromagnetic field (cf. Figure 1c–f,i–l). The Dirac–Fock potential, cf. Equation (4) is defined from its matrix element between orbitals m, n (occupied or unoccupied):

$$\langle m \mid u_{DF} \mid n \rangle = \sum_c^{core} \langle \{mc\} \mid V_{12} \mid \{nc\} \rangle, \tag{48}$$

where curly brackets denote anti-symmetrization. V_{12} denotes here the Coulomb interaction. It is also possible to define a Hartree–Fock type potential for the Breit interaction [89,90], but this aspect of the electron–electron interaction is neglected here. In addition to the Dirac–Fock potential, we usually add a so-called projected potential, u_{proj}, to the Hamiltonian in Equation (4). Aiming for a final state with a hole in one of the originally occupied orbitals, the projected potential cancels the removed electron's monopole interaction with all unoccupied orbitals, without affecting the interaction between the electrons in the ground state. More details can be found in [44]. Through this extra potential, some of the contributions from Figure 1c, precisely those which ensure that the photoelectron feels the correct long-range potential, are accounted for already in lowest order. When converged, the iterative procedure gives the same results if the projected potential is used or not, but the convergence is often much improved in the latter case, especially close to ionization thresholds.

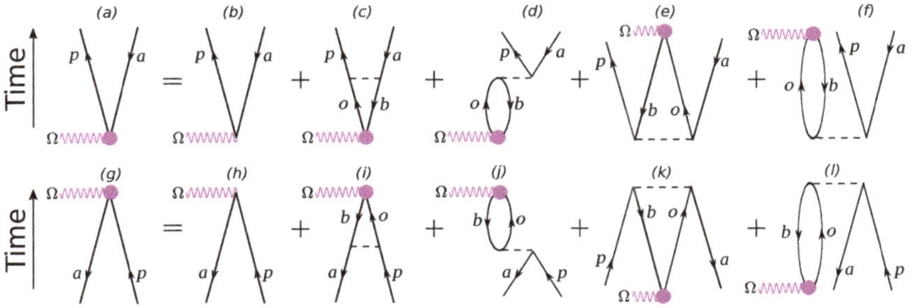

Figure 1. Goldstone diagrams illustrating RPAE for the many-body screening of the photon interaction. The set labelled with **(a,g)** are forward and backward propagation, respectively. Diagrams **(b,h)** are the lowest order contributions, while **(c–f)** and **(i–l)** give the many-body response. The sphere indicates the correlated interaction to infinite order. The wavy line indicates the photon interaction and the dashed line the Coulomb interaction. Downward lines (labelled with a, b) stand for holes created when electrons are removed from initially occupied orbitals, and upward lines (labelled with o, p) for initially unoccupied orbitals.

The calculations are performed with a basis set obtained through diagonalization of the radial one-particle Dirac–Fock Hamiltonians in a primitive basis of B-splines [91], defined on a knot sequence in a spherical box. B-splines are piecewise polynomials of a given order k. The radial components f and g of the relativistic wave function, (cf. Equation (5)) are expanded in B-splines of different orders: typically $k = 7$ and $k = 8$ respectively. It has been shown by Froese, Fischer, and Zatsarinny [92] that the use of different B-spline orders is a way to get rid of the so-called spurious states, which are known to appear in the numerical spectrum after discretization of the Dirac Hamiltonian. Details of the use of B-splines to solve the Dirac equation can be found in ref. [93].

We use further exterior complex scaling (ECS)

$$r \to \begin{cases} r, & 0 < r < R_C \\ R_C + (r - R_C)e^{i\varphi}, & r > R_C, \end{cases} \tag{49}$$

and thus the eigenenergies of the virtual orbitals are complex in general. As a consequence, the energy integration path avoids the pole in Equation (46) and thus the sum and integration over unoccupied states p can be represented by a finite sum [41,42].

With converged RPAE results the two-photon matrix elements in Equation (37) can be calculated for the absorption as well as the emission path to sideband n:

$$M^{(2)}_{\mathrm{a/e}} = \langle q \mid ez \mid \rho^{+}_{(2n\mp1)\omega,b} \rangle, \text{ where } \epsilon_q = \epsilon_b + 2n\omega, \tag{50}$$

where length gauge has been assumed. The integration in Equation (50) involves two continuum functions and will not converge on any finite interval. The integrand is therefore divided into two parts. The first is an inner region $0 \leq r < R < R_C$, where the perturbed wave function and final state can be determined numerically on the B-spline basis. The second is an outer region $R \leq r < \infty$ where the functions can be assumed to be solutions to the pure Coulomb problem, albeit with a possible phase shift. By using different breakpoints

R, we can check that the result is independent on where the change from numerical to the analytical integration is done. This procedure was described in ref. [42] but has to be slightly changed for the relativistic case, as will be discussed in the next subsection.

4.3. The Continuum–Continuum Transition

To evaluate Equation (50), we need the final continuum state, q, for a photoelectron of energy ϵ_q, obtained in a relativistic framework and with the phase shift it gets from the many-body environment. A good approximation is found as the solution of

$$h\psi_q = \epsilon_q \psi_q \tag{51}$$

where $h = h_\kappa^{DF} + u_{proj}$. By expanding the radial functions f and g (cf. Equation (5)), in B-splines, $f_q(r) = \sum c_i B_i(r)$, and vice versa for g, we can reformulate Equation (51) to a system of linear equations for the coefficients c_i. Exclusion of the first B-spline yields a regular solution, that is zero at the origin. This determines ψ_q up to a normalization constant. After normalization, which will be discussed below, ψ_q is used for the first part of the integration in Equation (50), i.e., from zero to R. We note in passing that in practice it is enough to obtain the large component of the relativistic wave function for a specific energy, because in the region dominated by the Coulomb potential, the Dirac equation gives the small component directly from the large one:

$$u_g(r) = \frac{c\hbar\left(\frac{d}{dr} + \frac{\kappa}{r}\right)u_f(r)}{\epsilon + 2mc^2 + \frac{e^2 Z_{eff}}{4\pi\epsilon_0}\frac{1}{r}}, \tag{52}$$

where Z_{eff} is the effective Coulomb potential felt by the escaping electron.

For the second part of the integration, from R to infinity, we need to extract information from the numerical representations of q and ρ to perform the rest of the integral in Equation (50) analytically as was described in ref. [42]. The final state q, is a phase shifted regular solution to the Coulomb problem, which should be correctly normalized, and the perturbed wave function ρ, is a phase-shifted outgoing solution with an amplitude determined by the photoionization process.

The outgoing solution ρ well outside the ionic core can easily be compared with the pure Coulomb solutions, Equations (23) and (24), combined as in Equation (31), to determine the phase shift, $\tilde{\delta}$ in Equation (25). It can easily be checked that the obtained phase shift is independent of r, and then Equations (23) and (24) can be used again to construct the solution at any large r.

The final state phase-shifted regular solution from Equation (51) can be complimented by its irregular counterpart through Equation (20), evaluated with the relativistic forms of $\bar{f}$ and $\bar{g}$, and then again the phase shift can be determined from comparision with Equations (23) and (24), combined as in Equation (31), and finally Equation (23) can be used to construct the final state at any r.

An additional advantage with the possibility to complement a regular solution with its corresponding irregular solution, and be able to construct the outgoing function, is that it is easy to normalize. The probability flux through the surface of a sphere of radius R is

$$\mathcal{J}(R) = ic\left(u^f(r)^* u^g(r) - u^g(r)^* u^f(r)\right)_{r=R} \tag{53}$$

and is constant for any large value of R, far outside the core. Because the asymptotic expressions for the large and small components are simple oscillating waves and their relation is ζ (cf. Equation (8)), the rate should be $2c\zeta|A|^2$ and from that we can determine the amplitude A. From the expression for ζ in Equation (9), we note the close resemblance with the non-relativistic rate $\hbar k|A|^2/m$, just slightly adjusted for relativistic effects.

The last part of the integral, from R to $r \to \infty$, in Equation (50) can now be calculated as was described in ref. [42], but now with continuum solutions obtained from Equations (23) and (24).

5. Results

The two-photon matrix elements for the absorption, M_a, and emission, M_e, paths are calculated as indicated in Equation (50) and then the atomic delay for electrons emitted in the direction of the laser field polarization is obtained from Equation (41). The Wigner delays are calculated as in Equation (34).

5.1. A Light Element: Argon

Results for ionization of argon atoms to the outermost p doublet $3p_{1/2}^{-1}$ and $3p_{3/2}^{-1}$, are shown in Figure 2. The two curves for the atomic delays are, more or less, indistinguishable. The negative atomic delay peak at 50 eV is due to the $3p$-Cooper minimum in the cross section of argon. A slight shift of the negative atomic delays peaks of a few meV is observed. The similarity of the two fine-structure split channels is expected for such a light system with $\Delta E_{\mathrm{FS}}^{Ar:3p_j} = 0.18$ meV. The Wigner delays from the two fine-structure channels are also mostly indistinguishable. Just below the threshold for release from the 3s-orbital, ~30 eV, there are narrow resonances that are not fully resolved in the present calculation.

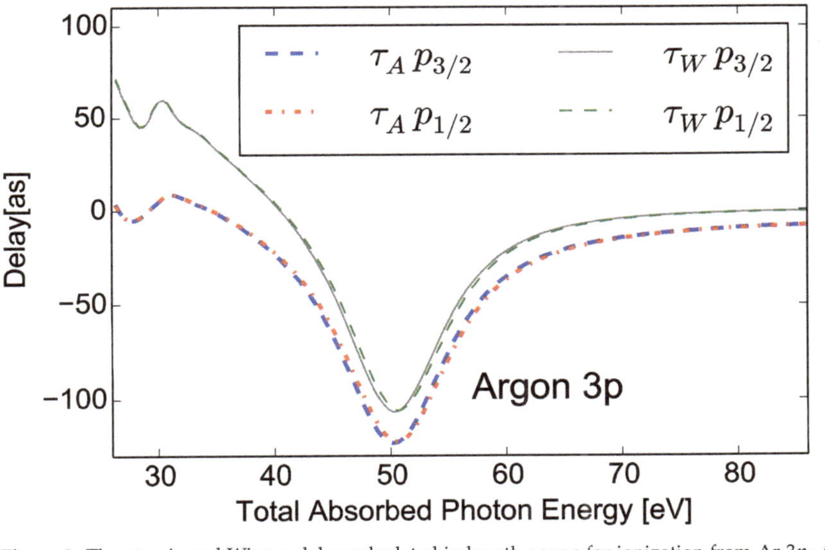

Figure 2. The atomic and Wigner delay calculated in length gauge for ionization from Ar $3p_j$, for electrons emitted along the polarization axis. The figure shows the region in the vicinity of the Cooper minimum. The thick dashed blue line shows the atomic delay for electrons ionized from $3p_{3/2}$. It is hardly distinguishable from the dashed–dotted red line that shows the atomic delay for electrons ionized from $3p_{1/2}$. The thin dashed green and solid grey lines show the Wigner delay for electrons ionized from $3p_{1/2}$ and $3p_{3/2}$ respectively. Dirac–Fock orbital energies have been replaced with experimental ionization energies: For $3p_{3/2}$ the binding energy is 15.76 eV, and for $3p_{1/2}$ it is 15.94 eV.

5.2. Heavy Elements: Krypton and Xenon

Atomic and Wigner delays for ionization to the outermost p-doublet in krypton and xenon are shown in Figures 3 and 4, respectively. Here the delay differences between the electrons ionized to the $4p_{1/2}^{-1}$ and $4p_{3/2}^{-1}$ ($5p_{1/2}^{-1}$ and $5p_{3/2}^{-1}$) in krypton (xenon) show that relativistic effects are important. Differences between the delays are clearly visible on the order of a few eV at the Cooper minima. Such shifts can be expected because the fine-structure shifts are $\Delta E_{\mathrm{FS}}^{4p_j} = 0.67$ eV for krypton ($\Delta E_{\mathrm{FS}}^{5p_j} = 1.3$ eV for xenon). Furthermore, a difference between the doublet channels is observed at low energies, where xenon shows the largest delay difference that exceeds 10 as.

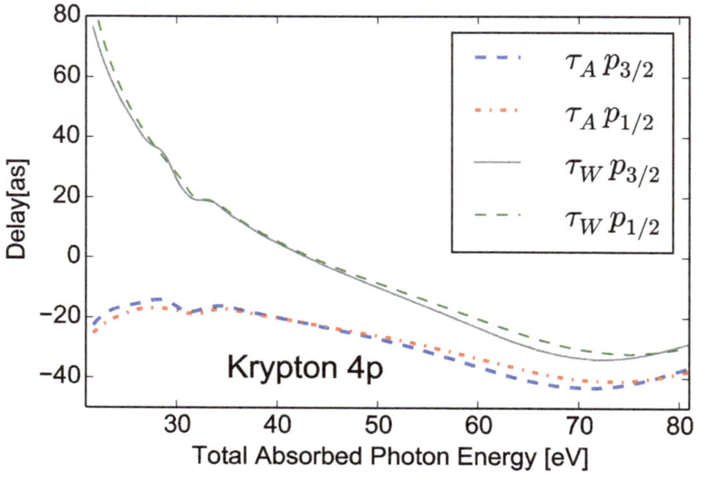

Figure 3. The atomic and Wigner delay calculated in length gauge for ionization from Kr $4p_j$, for electrons emitted along the polarization axis. The thick dashed blue line shows the atomic delay for electrons ionized from $4p_{3/2}$, and the dotted–dashed red line shows the atomic delay for electrons ionized from $4p_{1/2}$. The thin dashed green and solid grey lines show the Wigner delay for electrons ionized from $4p_{1/2}$ and $4p_{3/2}$ respectively. Dirac–Fock orbital energies have been replaced with experimental ionization energies: For $4p_{3/2}$ the binding energy is 14.00 eV, and for $4p_{1/2}$ it is 14.67 eV [94]. Dirac–Fock orbital energies are used for the deeper lying orbitals.

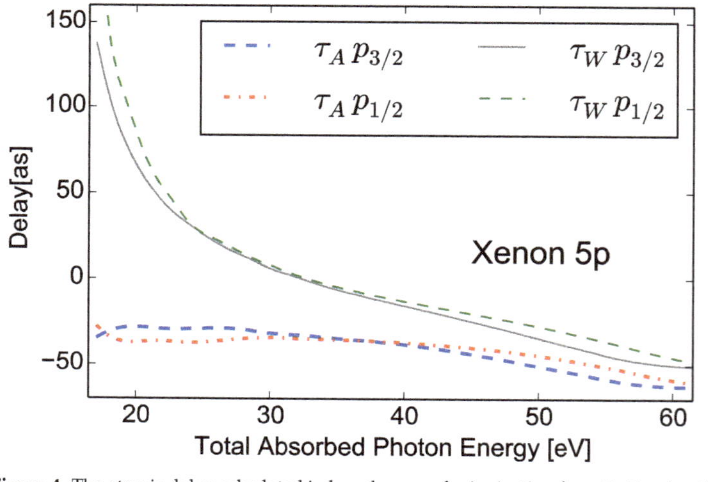

Figure 4. The atomic delay calculated in length gauge for ionization from Xe $5p_j$, for electrons emitted along the polarization axis. The thick dashed blue line shows the atomic delay for electrons ionized from $5p_{3/2}$, and the dotted–dashed red line shows the atomic delay for electrons ionized from $5p_{1/2}$. The thin dashed green and solid grey lines show the Wigner delay for electrons ionized from $5p_{1/2}$ and $5p_{3/2}$ respectively. Dirac–Fock orbital energies have been replaced with experimental ionization energies. For $5p_{3/2}$, the binding energy is 12.13 eV, and for $5p_{1/2}$ it is 13.44 eV [94]. Dirac–Fock orbital energies are used for the deeper lying orbitals.

5.3. Study of Continuum–Continuum Delay

The difference between the atomic and the Wigner delay is plotted for argon, krypton, xenon, and radon in Figure 5. For all the elements, and all fine-structure components, the results are very similar. This is in accordance with earlier findings, using non-relativistic calculations [41,42,44], and the corresponding numerical continuum–continuum delay: $\tau_{cc}^{\mathrm{MBPT}}$, is shown as a dotted line in Figure 5 for comparison with the relativistic results. Thus, the contribution from the second photon depends on the kinetic energy and the long-range potential, but only weakly, or not at all, on the structure of the remaining ion, or its angular momentum, for photoelectrons emitted along the polarization axis.

Only in the vicinity of Cooper minima, or close to resonances, is there are a deviation from this general trend. We stress that non-relativistic deviations, of a few attoseconds, have also been found for Ar3p at the Cooper minimum by using the effective one-body potential for the final state [44]. In that case, however, the complete 2P2C-RPAE method was used to show that these deviations could be reduced, as shown Figure 9b of ref. [44]. Thus, we may speculate that the present relativistic deviations at the Ar3p_j Cooper minima could be reduced by turning to 2P2C-RRRA theory. On the other hand, the correlation-induced 3s-minimum was shown to be non-separable by using the 2P2C-RPAE method, as shown in Figure 9a of [44]. Obviously, fast photoelectrons are also well described by the analytical cc-delay in ref. [33], but more importantly, Figure 5 demonstrates that a universal behaviour extends to much lower energies than expected from the asymptotic theory (>20 eV) [33], in good agreement with non-relativistic 2P2C-RPAE matrix elements [44].

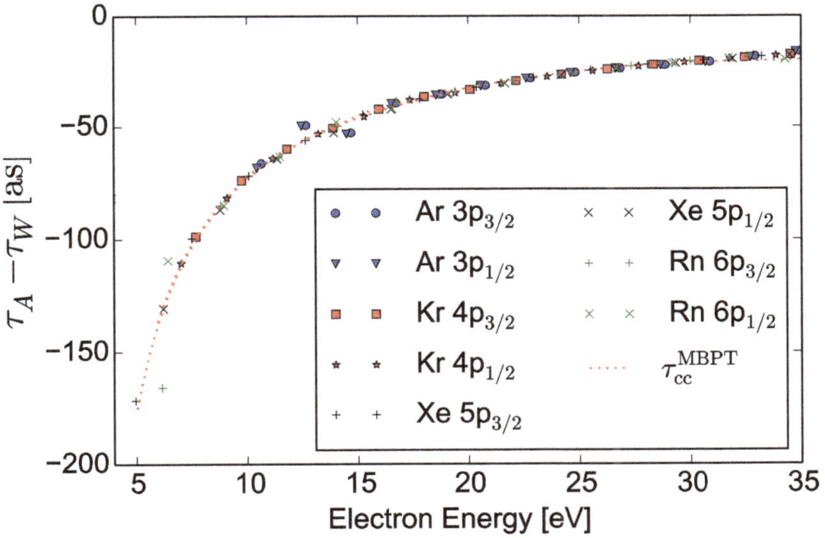

Figure 5. The difference between the atomic delay and the Wigner delay for the two outermost orbitals in Ar, Kr, Xe, and Rn calculated in length gauge and for for electrons emitted along the polarization axis. The red dotted line shows the non-relativistic result calculated for Ne 2p, i.e., the numerically obtained continuum–continuum delay discussed in the introduction.

5.4. Comparison with Experiments

The delay difference between photoelectrons originating from the outermost $p_{3/2}$ and $p_{1/2}$ orbitals in krypton and xenon have been studied in refs. [71,72] by using the interferometric RABBITT technique. In Figure 6, this difference, as calculated here, is shown for xenon. The experiment from ref. [71] includes one data point at 18.6 eV and one at 24.8 eV which are in qualitative agreement both with the calculation presented here, and with accompanying calculations in ref. [71], based on the Wigner delay from RRPA augmented with the the cc-delay from ref. [33]. Three other data points, on the other hand, differ markedly from both theoretical results. Especially striking are the large measured delays for higher energies (around 30 as at 30 eV), where the calculated result is very small. This might be due to resonances, not fully accounted for in the calculations, as discussed in ref. [71].

Moreover, a higher energy region has been explored. Ref. [72] measured the the delay difference for the xenon $5p$ fine-structure components for the sideband at 90 eV (with IR photon energies of 1.55 eV) to $\tau_A(5p_{3/2}) - \tau_A(5p_{1/2}) = 14.5 \pm 9.3$ as. Moreover, the calculated delay is much smaller, around 2 as (not shown in the figures). We note that the cross section to produce photoelectrons in the $5\ell_j$ channels at around 90 eV photon energy is comparable to those for $4d$ and shake-up satellites [95]. Because shake-up channels can have significantly larger delays [12], this region might need a more careful investigation of all competing channels.

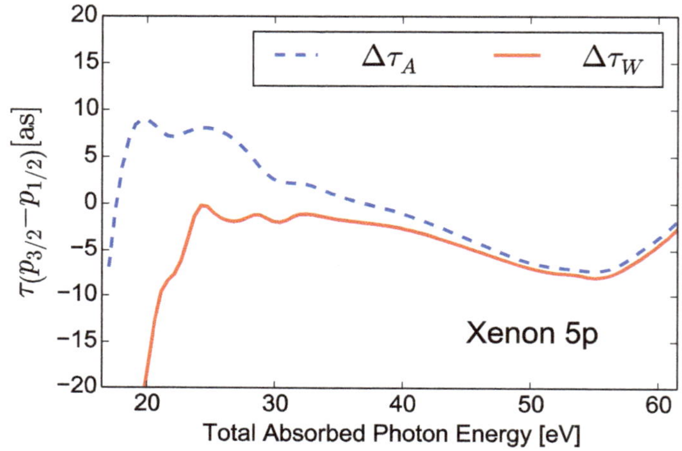

Figure 6. The delay difference between photoelectrons originating from the $5p_{3/2}$ and $5p_{1/2}$ orbitals in xenon. The dashed blue line shows the atomic delay, and the solid red the Wigner delay.

Figure 7 shows finally the atomic and Wigner delay for photoelectrons released from the xenon $4d$ orbitals. The result agrees within error bars with the measurement, from threshold up to ∼100 eV, in ref. [74]. It is interesting to note the large difference between the two channels, defined by the two fine-structure components, in the region just above the $4d$ thresholds at 67.5 and 69.5 eV, and the rapid variation of the delay with photon energy. The experiments in refs. [96,97] have shown that also the cross-section branching ratio (for leaving the ion with $4d_{3/2}^{-1}$ or $4d_{5/2}^{-1}$) shows a rapid variation in this region. In both cases, this behaviour can be traced back to the presence of two resonances close to threshold. They are of ^{3}D and ^{3}P character and cannot be populated by one-photon absorption in a non-relativistic description. The spin-orbit-induced singlet-triplet mixing opens, however, the route to ionization through these resonances, and thus for a population transfer from one final channel to the other. This has been further discussed in refs. [74,98]. We note that although the resonances in argon, mentioned above, are just unresolved in the calculation, the reason that these xenon resonances are not seen directly is not a question of resolution. The cross section in this region is completely dominated by the so-called giant resonance of ^{1}P character and the spin-orbit-induced resonances can simply not be seen against this background. Still, their mark in the more sensitive observables, such as atomic delays and branching ratios, is clearly seen.

Also for xenon $4d$ ref. [72] gives a value at 90 eV: $\tau_A(4d_{5/2}) - \tau_A(4d_{3/2}) = -4.0 \pm 4.1$ as, which agrees with our value of -1.2 as.

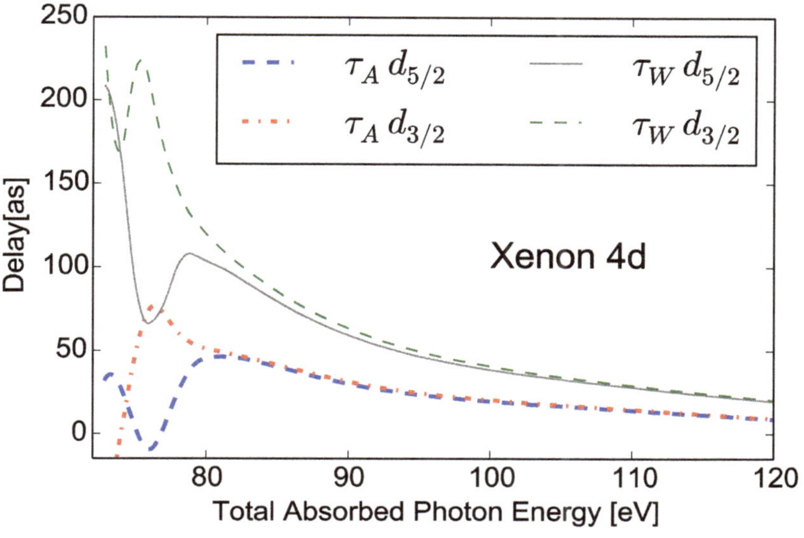

Figure 7. The thick dashed blue line shows the atomic delay for electrons ionized from $4d_{3/2}$, and the dotted–dashed red line shows the atomic delay for electrons ionized from $4d_{5/2}$. The thin dashed green and solid grey lines show the Wigner delay for electrons ionized from $4d_{3/2}$ and $4d_{5/2}$ respectively. Dirac–Fock orbital energies have been replaced with experimental ionization energies. For $4d_{5/2}$ the binding energy is 67.5 eV, and for $4d_{3/2}$ it is 69.5 eV [99].

6. Conclusions

We have shown how two-photon above-threshold ionization can be treated in a relativistic framework from first principles. Correlation is included in the photoionization process at the level of the relativistic random phase approximation. As in the non-relativistic case, the calculation of the subsequent continuum–continuum transition relies on knowledge of the form of the intermediate wave packet when it is well outside the atomic core. For this purpose, we present a convenient recursive formula for both the large and small component of the regular and irregular solution to the relativistic Coulomb problem. The procedure have been applied to a few heavy elements, and it is shown that the separation of the atomic delay into a Wigner–Smith–Eisenbud delay and a universal continuum–continuum works reasonably well also for these systems.

We have further demonstrated qualitative agreement with existing experimental photoionization-delay data for ionization from the $4d$-orbitals in xenon, and with lower energy results from the outermost orbitals in xenon and krypton. For higher photon energies, experiments report considerably larger delay differences between the fine-structure split channels than supported by the calculations. This might be connected to resonances or interfering shake-up channels, which can hopefully be resolved in the future.

Author Contributions: Conceptualization, E.L. and J.M.D.; methodology, E.L., J.M.D. and J.V.; software, J.V. and E.L.; validation, E.L., J.M.D. and J.V.; formal analysis, E.L., J.M.D. and J.V.; investigation, J.V.; resources, E.L. and J.M.D.; data curation, J.V.; writing—original draft preparation, E.L.; writing—review and editing, E.L. and J.M.D.; visualization, J.V. and E.L.; supervision, E.L. and J.M.D.; project administration, E.L. and J.M.D.; funding acquisition, E.L. and J.M.D. All authors have read and agreed to the published version of the manuscript.

Funding: This research was funded by the Knut and Alice Wallenberg Foundation: Grant No. 2017.0104 and 2019.0154, from the Swedish Research Council: GrantNo. 2018-03845 and 2020-03315, and from the Olle Engkvist Foundation: Grant No. 194-0734.

Institutional Review Board Statement: Not applicable.

Informed Consent Statement: Not applicable.

Data Availability Statement: Not applicable.

Conflicts of Interest: The authors declare no conflict of interest.

References

1. Lewenstein, M.; Balcou, P.; Ivanov, M.Y.; L'Huillier, A.; Corkum, P.B. Theory of high-harmonic generation by low-frequency laser fields. *Phys. Rev. A* **1994**, *49*, 2117–2132. [CrossRef]
2. Paul, P.M.; Toma, E.S.; Breger, P.; Mullot, G.; Augé, F.; Balcou, P.; Muller, H.G.; Agostini, P. Observation of a Train of Attosecond Pulses from High Harmonic Generation. *Science* **2001**, *292*, 1689–1692. [CrossRef] [PubMed]
3. Itatani, J.; Quéré, F.; Yudin, G.L.; Ivanov, M.Y.; Krausz, F.; Corkum, P.B. Attosecond Streak Camera. *Phys. Rev. Lett.* **2002**, *88*, 173903. [CrossRef] [PubMed]
4. Schultze, M.; Fiess, M.; Karpowicz, N.; Gagnon, J.; Korbman, M.; Hofstetter, M.; Neppl, S.; Cavalieri, A.L.; Komninos, Y.; Mercouris, T.; et al. Delay in Photoemission. *Science* **2010**, *328*, 1658–1662. [CrossRef] [PubMed]
5. Klünder, K.; Dahlström, J.M.; Gisselbrecht, M.; Fordell, T.; Swoboda, M.; Guénot, D.; Johnsson, P.; Caillat, J.; Mauritsson, J.; Maquet, A.; et al. Probing Single-Photon Ionization on the Attosecond Time Scale. *Phys. Rev. Lett.* **2011**, *106*, 143002. [CrossRef] [PubMed]
6. Guénot, D.; Klünder, K.; Arnold, C.L.; Kroon, D.; Dahlström, J.M.; Miranda, M.; Fordell, T.; Gisselbrecht, M.; Johnsson, P.; Mauritsson, J.; et al. Photoemission-time-delay measurements and calculations close to the 3*s*-ionization-cross-section minimum in Ar. *Phys. Rev. A* **2012**, *85*, 053424. [CrossRef]
7. Palatchi, C.; Dahlström, J.M.; Kheifets, A.S.; Ivanov, I.A.; Canaday, D.M.; Agostini, P.; DiMauro, L.F. Atomic delay in helium, neon, argon and krypton. *J. Phys. B At. Mol. Opt. Phys.* **2014**, *47*, 245003. [CrossRef]
8. Guénot, D.; Kroon, D.; Balogh, E.; Larsen, E.W.; Kotur, M.; Miranda, M.; Fordell, T.; Johnsson, P.; Mauritsson, J.; Gisselbrecht, M.; et al. Measurements of relative photoemission time delays in noble gas atoms. *J. Phys. B At. Mol. Opt. Phys.* **2014**, *47*, 245602. [CrossRef]
9. Sabbar, M.; Heuser, S.; Boge, R.; Lucchini, M.; Carette, T.; Lindroth, E.; Gallmann, L.; Cirelli, C.; Keller, U. Resonance Effects in Photoemission Time Delays. *Phys. Rev. Lett.* **2015**, *115*, 133001. [CrossRef]
10. Kotur, M.; Guénot, D.; Jimenez-Galan, A.; Kroon, D.; Larsen, E.W.; Louisy, M.; Bengtsson, S.; Miranda, M.; Mauritsson, J.; Arnold, C.L.; et al. Spectral phase measurement of a Fano resonance using tunable attosecond pulses. *Nat. Commun.* **2016**, *7*, 10566. [CrossRef]
11. Gruson, V.; Barreau, L.; Jiménez-Galan, Á.; Risoud, F.; Caillat, J.; Maquet, A.; Carré, B.; Lepetit, F.; Hergott, J.F.; Ruchon, T.; et al. Attosecond dynamics through a Fano resonance: Monitoring the birth of a photoelectron. *Science* **2016**, *354*, 734–738. [CrossRef] [PubMed]
12. Isinger, M.; Squibb, R.; Busto, D.; Zhong, S.; Harth, A.; Kroon, D.; Nandi, S.; Arnold, C.L.; Miranda, M.; Dahlström, J.; et al. Photoionization in the time and frequency domain. *Science* **2017**, *358*, 893–896. [CrossRef] [PubMed]
13. Cirelli, C.; Marante, C.; Heuser, S.; Petersson, C.; Galán, Á.J.; Argenti, L.; Zhong, S.; Busto, D.; Isinger, M.; Nandi, S.; et al. Anisotropic photoemission time delays close to a Fano resonance. *Nat. Commun.* **2018**, *9*, 955. [CrossRef] [PubMed]
14. Barreau, L.; Petersson, C.L.M.; Klinker, M.; Camper, A.; Marante, C.; Gorman, T.; Kiesewetter, D.; Argenti, L.; Agostini, P.; González-Vázquez, J.; et al. Disentangling Spectral Phases of Interfering Autoionizing States from Attosecond Interferometric Measurements. *Phys. Rev. Lett.* **2019**, *122*, 253203. [CrossRef]
15. Alexandridi, C.; Platzer, D.; Barreau, L.; Busto, D.; Zhong, S.; Turconi, M.; Neoričić, L.; Laurell, H.; Arnold, C.L.; Borot, A.; et al. Attosecond photoionization dynamics in the vicinity of the Cooper minima in argon. *Phys. Rev. Res.* **2021**, *3*, L012012. [CrossRef]
16. Drescher, L.; Witting, T.; Kornilov, O.; Vrakking, M.J.J. Phase dependence of resonant and antiresonant two-photon excitations. *Phys. Rev. A* **2022**, *105*, L011101. [CrossRef]
17. Caillat, J.; Maquet, A.; Haessler, S.; Fabre, B.; Ruchon, T.; Salières, P.; Mairesse, Y.; Taïeb, R. Attosecond Resolved Electron Release in Two-Color Near-Threshold Photoionization of N2. *Phys. Rev. Lett.* **2011**, *106*, 093002. [CrossRef] [PubMed]
18. Huppert, M.; Jordan, I.; Baykusheva, D.; von Conta, A.; Wörner, H.J. Attosecond Delays in Molecular Photoionization. *Phys. Rev. Lett.* **2016**, *117*, 093001. [CrossRef]
19. Loriot, V.; Marciniak, A.; Nandi, S.; Karras, G.; Hervé, M.; Constant, E.; Plésiat, E.; Palacios, A.; Martín, F.; Lépine, F. High harmonic generation-2ω attosecond stereo-photoionization interferometry in N 2. *J. Phys. Photonics* **2020**, *2*, 024003. [CrossRef]
20. Nandi, S.; Plésiat, E.; Zhong, S.; Palacios, A.; Busto, D.; Isinger, M.; Neoričić, L.; Arnold, C.L.; Squibb, R.J.; Feifel, R.; et al. Attosecond timing of electron emission from a molecular shape resonance. *Sci. Adv.* **2020**, *6*, eaba7762. [CrossRef]
21. Kamalov, A.; Wang, A.L.; Bucksbaum, P.H.; Haxton, D.J.; Cryan, J.P. Electron correlation effects in attosecond photoionization of CO_2. *Phys. Rev. A* **2020**, *102*, 023118. [CrossRef]
22. Cavalieri, A.L.; Muller, N.; Uphues, T.; Yakovlev, V.S.; Baltuska, A.; Horvath, B.; Schmidt, B.; Blumel, L.; Holzwarth, R.; Hendel, S.; et al. Attosecond spectroscopy in condensed matter. *Nature* **2007**, *449*, 1029–1032. [CrossRef]
23. Siek, F.; Neb, S.; Bartz, P.; Hensen, M.; Strüber, C.; Fiechter, S.; Torrent-Sucarrat, M.; Silkin, V.M.; Krasovskii, E.E.; Kabachnik, N.M.; et al. Angular momentum–induced delays in solid-state photoemission enhanced by intra-atomic interactions. *Science* **2017**, *357*, 1274–1277. [CrossRef] [PubMed]

24. Heinrich, S.; Saule, T.; Högner, M.; Cui, Y.; Yakovlev, V.S.; Pupeza, I.; Kleineberg, U. Attosecond intra-valence band dynamics and resonant-photoemission delays in W(110). *Nat. Commun.* **2021**, *12*, 3404. [CrossRef] [PubMed]
25. Véniard, V.; Taïeb, R.; Maquet, A. Phase dependence of (N + 1)-color (N >1) ir-uv photoionization of atoms with higher harmonics. *Phys. Rev. A* **1996**, *54*, 721–728. [CrossRef]
26. Toma, E.S.; Muller, H.G. Calculation of matrix elements for mixed extreme-ultraviolet–infrared two-photon above-threshold ionization of argon. *J. Phys. B At. Mol. Opt. Phys.* **2002**, *35*, 3435. [CrossRef]
27. Mauritsson, J.; Gaarde, M.B.; Schafer, K.J. Accessing properties of electron wave packets generated by attosecond pulse trains through time-dependent calculations. *Phys. Rev. A* **2005**, *72*, 013401. [CrossRef]
28. Nagele, S.; Pazourek, R.; Feist, J.; Doblhoff-Dier, K.; Lemell, C.; Tőkési, K.; Burgdörfer, J. Time-resolved photoemission by attosecond streaking: extraction of time information. *J. Phys. B At. Mol. Opt. Phys.* **2011**, *44*, 081001. [CrossRef]
29. Dahlström, J.M.; L'Huillier, A.; Maquet, A. Introduction to attosecond delays in photoionization. *J. Phys. B At. Mol. Opt. Phys.* **2012**, *45*, 183001. [CrossRef]
30. Pazourek, R.; Nagele, S.; Burgdörfer, J. Attosecond chronoscopy of photoemission. *Rev. Mod. Phys.* **2015**, *87*, 765–802. [CrossRef]
31. Klarsfeld, S.; Maquet, A. Pade approximants and multiphonon ionisation of atomic hydrogen. *J. Phys. B At. Mol. Phys.* **1979**, *12*, L553. [CrossRef]
32. Aymar, M.; Crance, M. Two-photon ionisation of atomic hydrogen in the presence of one-photon ionisation. *J. Phys. B At. Mol. Phys.* **1980**, *13*, L287. [CrossRef]
33. Dahlström, J.; Guénot, D.; Klünder, K.; Gisselbrecht, M.; Mauritsson, J.; L'Huillier, A.; Maquet, A.; Taïeb, R. Theory of attosecond delays in laser-assisted photoionization. *Chem. Phys.* **2013**, *414*, 53 – 64. [CrossRef]
34. Ivanov, M.; Smirnova, O. How Accurate Is the Attosecond Streak Camera? *Phys. Rev. Lett.* **2011**, *107*, 213605. [CrossRef] [PubMed]
35. Wigner, E.P. Lower Limit for the Energy Derivative of the Scattering Phase Shift. *Phys. Rev.* **1955**, *98*, 145–147. [CrossRef]
36. Smith, F.T. Lifetime Matrix in Collision Theory. *Phys. Rev.* **1960**, *118*, 349–356. [CrossRef]
37. Eisenbud, L. The Formal Properties of Nuclear Collisions. Ph.D. Thesis, Princeton University, Princeton, NJ, USA, 1948.
38. Cooper, J.W. Photoionization from Outer Atomic Subshells. A Model Study. *Phys. Rev.* **1962**, *128*, 681–693. [CrossRef]
39. Baykusheva, D.; Wörner, H.J. Theory of attosecond delays in molecular photoionization. *J. Chem. Phys.* **2017**, *146*, 124306. [CrossRef]
40. Benda, J.; Mašín, Z.; Gorfinkiel, J.D. Analysis of RABITT time delays using the stationary multiphoton molecular *R*-matrix approach. *Phys. Rev. A* **2022**, *105*, 053101. [CrossRef]
41. Dahlström, J.M.; Carette, T.; Lindroth, E. Diagrammatic approach to attosecond delays in photoionization. *Phys. Rev. A* **2012**, *86*, 061402. [CrossRef]
42. Dahlström, J.M.; Lindroth, E. Study of attosecond delays using perturbation diagrams and exterior complex scaling. *J. Phys. B At. Mol. Opt. Phys.* **2014**, *47*, 124012. [CrossRef]
43. Lindroth, E.; Dahlström, J.M. Attosecond delays in laser-assisted photodetachment from closed-shell negative ions. *Phys. Rev. A* **2017**, *96*, 013420. [CrossRef]
44. Vinbladh, J.; Dahlström, J.M.; Lindroth, E. Many-body calculations of two-photon, two-color matrix elements for attosecond delays. *Phys. Rev. A* **2019**, *100*, 043424. [CrossRef]
45. Kheifets, A.S.; Ivanov, I.A. Delay in Atomic Photoionization. *Phys. Rev. Lett.* **2010**, *105*, 233002. [CrossRef] [PubMed]
46. Kheifets, A.S. Time delay in valence-shell photoionization of noble-gas atoms. *Phys. Rev. A* **2013**, *87*, 063404. [CrossRef]
47. Amusia, M.Y. *Atomic Photoeffect*; Plenum Press: New York, NY, USA, 1990.
48. Heuser, S.; Jiménez Galán, A.; Cirelli, C.; Marante, C.; Sabbar, M.; Boge, R.; Lucchini, M.; Gallmann, L.; Ivanov, I.; Kheifets, A.S.; et al. Angular dependence of photoemission time delay in helium. *Phys. Rev. A* **2016**, *94*, 063409. [CrossRef]
49. Busto, D.; Vinbladh, J.; Zhong, S.; Isinger, M.; Nandi, S.; Maclot, S.; Johnsson, P.; Gisselbrecht, M.; L'Huillier, A.; Lindroth, E.; et al. Fano's Propensity Rule in Angle-Resolved Attosecond Pump-Probe Photoionization. *Phys. Rev. Lett.* **2019**, *123*, 133201. [CrossRef]
50. Fuchs, J.; Douguet, N.; Donsa, S.; Martin, F.; Burgdörfer, J.; Argenti, L.; Cattaneo, L.; Keller, U. Time delays from one-photon transitions in the continuum. *Optica* **2020**, *7*, 154–161. [CrossRef]
51. Boll, D.I.R.; Martini, L.; Fojón, O.A. Analytical model for attosecond time delays and Fano's propensity rules in the continuum. *arXiv* **2022**. [CrossRef]
52. Argenti, L.; Jiménez-Galán, A.; Caillat, J.; Taïeb, R.; Maquet, A.; Martín, F. Control of photoemission delay in resonant two-photon transitions. *Phys. Rev. A* **2017**, *95*, 043426. [CrossRef]
53. Swoboda, M.; Fordell, T.; Klünder, K.; Dahlström, J.M.; Miranda, M.; Buth, C.; Schafer, K.J.; Mauritsson, J.; L'Huillier, A.; Gisselbrecht, M. Phase Measurement of Resonant Two-Photon Ionization in Helium. *Phys. Rev. Lett.* **2010**, *104*, 103003. [CrossRef]
54. Kheifets, A.S.; Bray, A.W. RABBITT phase transition across the ionization threshold. *Phys. Rev. A* **2021**, *103*, L011101. [CrossRef]
55. Zhao, Z.X.; Lin, C.D. Theory of laser-assisted autoionization by attosecond light pulses. *Phys. Rev. A* **2005**, *71*, 060702. [CrossRef]
56. Jimenez-Galan, A.; Argenti, L.; Martin, F. Modulation of Attosecond Beating in Resonant Two-Photon Ionization. *Phys. Rev. Lett.* **2014**, *113*, 263001. [CrossRef]
57. Jiménez-Galán, A.; Martín, F.; Argenti, L. Two-photon finite-pulse model for resonant transitions in attosecond experiments. *Phys. Rev. A* **2016**, *93*, 023429. [CrossRef]

58. Carette, T.; Dahlström, J.M.; Argenti, L.; Lindroth, E. Multiconfigurational Hartree-Fock close-coupling ansatz: Application to the argon photoionization cross section and delays. *Phys. Rev. A* **2013**, *87*, 023420. [CrossRef]
59. Magrakvelidze, M.; Madjet, M.E.A.; Dixit, G.; Ivanov, M.; Chakraborty, H.S. Attosecond time delay in valence photoionization and photorecombination of argon: A time-dependent local-density-approximation study. *Phys. Rev. A* **2015**, *91*, 063415. [CrossRef]
60. Sato, S.A.; Hübener, H.; Rubio, A.; De Giovannini, U. First-principles simulations for attosecond photoelectron spectroscopy based on time-dependent density functional theory. *Eur. Phys. J. B* **2018**, *91*, 126. [CrossRef]
61. Amusia, M.; Ivanov, V.; Cherepkov, N.; Chernysheva, L. Interference effects in photoionization of noble gas atoms outer s-subshells. *Phys. Lett. A* **1972**, *40*, 361–362. [CrossRef]
62. Kheifets, A.S.; Toffoli, D.; Decleva, P. Angular dependent time delay near correlation induced Cooper minima. *J. Phys. B At. Mol. Opt. Phys.* **2020**, *53*, 115201. [CrossRef]
63. Wijesundera, W.; Kelly, H.P. Correlation satellites in the photoelectron spectrum of argon. *Phys. Rev. A* **1989**, *39*, 634–643. [CrossRef] [PubMed]
64. Amusia, M.; Kheifets, A. The influence of "two-electron-two-hole" excitations on the 3s–14p autoionization profile in Ar atoms. *Phys. Lett. A* **1981**, *82*, 407–411. [CrossRef]
65. Kobe, D.H.; Smirl, A.L. Gauge Invariant Formulation of the Interaction of Electromagnetic Radiation and Matter. *Am. J. Phys.* **1978**, *46*, 624–633. [CrossRef]
66. Kobe, D.H. Gauge-Invariant Resolution of the Controversy over Length versus Velocity Forms of the Interaction with Electric Dipole Radiation. *Phys. Rev. A* **1979**, *19*, 205–214. [CrossRef]
67. Saha, S.; Mandal, A.; Jose, J.; Varma, H.R.; Deshmukh, P.C.; Kheifets, A.S.; Dolmatov, V.K.; Manson, S.T. Relativistic effects in photoionization time delay near the Cooper minimum of noble-gas atoms. *Phys. Rev. A* **2014**, *90*, 053406. [CrossRef]
68. Kheifets, A.S.; Saha, S.; Deshmukh, P.C.; Keating, D.A.; Manson, S.T. Dipole phase and photoelectron group delay in inner-shell photoionization. *Phys. Rev. A* **2015**, *92*, 063422. [CrossRef]
69. Johnson, W.R.; Cheng, K.T. Photoionization of the outer shells of neon, argon, krypton, and xenon using the relativistic random-phase approximation. *Phys. Rev. A* **1979**, *20*, 978–988. [CrossRef]
70. Johnson, W.R.; Lin, C.D. Multichannel relativistic random-phase approximation for the photoionization of atoms. *Phys. Rev. A* **1979**, *20*, 964–977. [CrossRef]
71. Jordan, I.; Huppert, M.; Pabst, S.; Kheifets, A.S.; Baykusheva, D.; Wörner, H.J. Spin-orbit delays in photoemission. *Phys. Rev. A* **2017**, *95*, 013404. [CrossRef]
72. Jain, A.; Gaumnitz, T.; Kheifets, A.; Wörner, H.J. Using a passively stable attosecond beamline for relative photoemission time delays at high XUV photon energies. *Opt. Express* **2018**, *26*, 28604–28620. [CrossRef]
73. Jain, A.; Gaumnitz, T.; Bray, A.; Kheifets, A.; Wörner, H.J. Photoionization delays in xenon using single-shot referencing in the collinear back-focusing geometry. *Opt. Lett.* **2018**, *43*, 4510–4513. [CrossRef] [PubMed]
74. Zhong, S.; Vinbladh, J.; Busto, D.; Squibb, R.J.; Isinger, M.; Neoričić, L.; Laurell, H.; Weissenbilder, R.; Arnold, C.L.; Feifel, R.; et al. Attosecond electron–spin dynamics in Xe 4d photoionization. *Nat. Commun.* **2020**, *11*, 5042. [CrossRef] [PubMed]
75. Amusia, M.Y.; Cherepkov, N.A.; Chernysheva, L.V. Cross Section for the Photoionization of Noble-gas Atoms with Allowance for Multielectron Correlations *Sov. Phys. JETP* **1971**, *33*, 90.
76. Wendin, G. Collective resonance in the 4d10 shell in atomic Xe. *Phys. Lett. A* **1971**, *37*, 445–446. [CrossRef]
77. Zapata, F.; Vinbladh, J.; Ljungdahl, A.; Lindroth, E.; Dahlström, J.M. Relativistic time-dependent configuration-interaction singles method. *Phys. Rev. A* **2022**, *105*, 012802. [CrossRef]
78. Lindgren, I.; Rosén, A. Relativistic self-consistent-field calculations with application to atomic hyperfine interaction, Part I, relativistic self-consistent fields. *Case Stud. At. Phys.* **1974**, *4*, 93–149.
79. Jamieson, M.J. Time-Dependent Hartree-Fock Theory for Atoms. *Int. J. Quantum Chem.* **1970**, *5*, 103–115. [CrossRef]
80. Johnson, W.R.; Cheng, K.T. Quantum defects for highly stripped ions. *J. Phys. B At. Mol. Phys.* **1979**, *12*, 863–879. [CrossRef]
81. Friedrich, H. *Theoretical Atomic Physics*; Springer: Berlin/Heidelberg, Germany, 2006.
82. Abramowitz, M. *Handbook of Mathematical Functions*, 4th ed.; 14. Coulomb Wave Functions; Abramowitz, M., Stegun, I.A., Eds.; National Bureau of Standards: Washington, DC, USA, 1964; p. 537.
83. Slater, L.J. *Handbook of Mathematical Functions*, 4th ed.; 13. Confluent Hypergeometric Functions; Abramowitz, M., Stegun, I.A., Eds.; National Bureau of Standards: Washington, DC, USA, 1964; p. 503.
84. Isinger, M.; Busto, D.; Mikaelsson, S.; Zhong, S.; Guo, C.; Salières, P.; Arnold, C.L.; L'Huillier, A.; Gisselbrecht, M. Accuracy and precision of the RABBIT technique. *Philos. Trans. R. Soc. A Math. Phys. Eng. Sci.* **2019**, *377*, 20170475. [CrossRef] [PubMed]
85. Sörngård, J.; Dahlström, J.M.; Lindroth, E. Study of the possibilities with combinations of circularly and linearly polarized light for attosecond delay investigations. *J. Phys. B At. Mol. Opt. Phys.* **2020**, *53*, 134003. [CrossRef]
86. Saha, S.; Vinbladh, J.; Sörngård, J.; Ljungdahl, A.; Lindroth, E. Angular anisotropy parameters for photoionization delays. *Phys. Rev. A* **2021**, *104*, 033108. [CrossRef]
87. Selstø, S.; Lindroth, E.; Bengtsson, J. Solution of the Dirac equation for hydrogenlike systems exposed to intense electromagnetic pulses. *Phys. Rev. A* **2009**, *79*, 043418. [CrossRef]
88. Lin, D.L. Gauge properties of the Hartree-Fock and random-phase approximations. *Phys. Rev. A* **1977**, *16*, 600–605. [CrossRef]
89. Breit, G. Dirac's Equation and the Spin-Spin Interactions of Two Electrons. *Phys. Rev.* **1932**, *39*, 616–624. [CrossRef]

90. Lindroth, E.; Mårtensson-Pendrill, A.M.; Ynnerman, A.; Öster, P. Self-consistent treatment of the Breit interaction, with application to the electric dipole moment in thallium. *J. Phys. B* **1989**, *22*, 2447–2464. [CrossRef]
91. deBoor, C. *A Practical Guide to Splines*; Springer: New York, NY, USA, 1978.
92. Froese Fischer, C.; Zatsarinny, O. A B-spline Galerkin method for the Dirac equation. *Comput. Phys. Commun.* **2009**, *180*, 879. [CrossRef]
93. Kjellsson, T.; Selstø, S.; Lindroth, E. Relativitic ionization dynamics for a hydrogen atom exposed to super-intense XUV laser pulses. *Phys. Rev. A* **2017**, *95*, 043403. [CrossRef]
94. NIST. Atomic Spectra Database. Available online: http://physics.nist.gov/cgi-bin/AtData/main_asd (accessed on 20 July 2022).
95. Becker, U.; Szostak, D.; Kerkhoff, H.G.; Kupsch, M.; Langer, B.; Wehlitz, R.; Yagishita, A.; Hayaishi, T. Subshell photoionization of Xe between 40 and 1000 eV. *Phys. Rev. A* **1989**, *39*, 3902–3911. [CrossRef]
96. Shannon, S.P.; Codling, K.; West, J.B. The absolute photoionization cross sections of the spin-orbit components of the xenon 4d electron from 70–130 eV. *J. Phys. B At. Mol. Phys.* **1977**, *10*, 825. [CrossRef]
97. Adam, M.; Wuilleumier, F.; Krummacher, S.; Sandner, N.; Schmidt, V.; Mehlhorn, W. Recent progress in the study of photoionization processes of atomic species by electron spectroscopy using synchrotron radiation. *J. Electron Spectrosc. Relat. Phenom.* **1979**, *15*, 211–224. doi: 10.1016/0368-2048(79)87035-8. [CrossRef]
98. Cheng, K.T.; Johnson, W.R. Orbital collapse and the photoionization of the inner 4*d* shells for Xe-like ions. *Phys. Rev. A* **1983**, *28*, 2820–2828. [CrossRef]
99. King, G.C.; Tronc, M.; Read, F.H.; Bradford, R.C. An investigation of the structure near the L2,3edges of argon, the M4,5edges of krypton and the N4,5edges of xenon, using electron impact with high resolution. *J. Phys. B At. Mol. Phys.* **1977**, *10*, 2479–2495. [CrossRef]

atoms

MDPI

Article

Photoionization of Atomic Systems Using the Random-Phase Approximation Including Relativistic Interactions

Pranawa C. Deshmukh [1,2] **and Steven T. Manson** [3,*]

[1] Department of Physics and CAMOST, Indian Institute of Technology Tirupati, Tirupati 517506, India; pcd@iittp.ac.in
[2] Depatrment of Physics, Dayananda Sagar University, Bangalore 560114, India
[3] Department of Physics and Astronomy, Georgia State University, Atlanta, GA 30030, USA
* Correspondence: smanson@gsu.edu

Abstract: Approximation methods are unavoidable in solving a many-electron problem. One of the most successful approximations is the random-phase approximation (RPA). Miron Amusia showed that it can be used successfully to describe atomic photoionization processes of many-electron atomic systems. In this article, the historical reasons behind the term "random-phase approximation" are revisited. A brief introduction to the relativistic RPA (RRPA) developed by Walter Johnson and colleagues is provided and some of its illustrative applications are presented.

Keywords: photoionization; random-phase approximation; relativistic-random-phase approximation

Citation: Deshmukh, P.C.; Manson, S.T. Photoionization of Atomic Systems Using the Random-Phase Approximation Including Relativistic Interactions. *Atoms* **2022**, *10*, 71. https://doi.org/10.3390/atoms10030071

Academic Editor: Eugene T. Kennedy

Received: 31 May 2022
Accepted: 1 July 2022
Published: 11 July 2022

1. Introduction

About a half-century ago, Miron Amusia showed that the photoionization of the noble gas atoms could be described quite accurately using the random-phase approximation with exchange (RPAE); see [1] and references therein. RPAE includes two-particle two-hole correlations in the initial state of the photoionization process and coupling among the ionization channels in the final state, essentially configuration interaction in the continuum. The methodology did not include relativistic interactions. Building upon Amusia's work, Walter Johnson and his co-workers developed a relativistic version, the relativistic-random-phase approximation (RRPA) which is based on the Dirac equation rather than the Schrödinger equation [2–5]. This advance allowed us to study heavier systems, where relativistic effects are large, and processes that are only possible owing to relativistic interactions. In this paper, an exposition of the random-phase approximation is presented along with some of the important advances using the relativistic version of Amusia's methodology.

2. The Random-Phase Approximation

The random-phase approximation can be formulated in a few alternative ways. An extensive review of these methods and their equivalence is not attempted in this article. Instead, we limit ourselves to comment on the expression "random-phase approximation" and the "linearization" process that it entails. These expressions are widely used in the literature, but the historical reasons behind this terminology are seldom discussed. The scope of this article is further limited to only illustrating some of the applications of the relativistic RPA, RRPA, for which the earlier nonrelativistic work of [1] set the stage.

2.1. Beyond the Hartree–Fock Method

The simplest N-electron atomic problem, even if we leave out relativistic effects such as the screen-orbit interaction, is described by the Hamiltonian

$$H^{(N)}(q_1, q_2, \ldots, q_N) = \left\{ \sum_{i=1}^{N} \left(-\frac{\hbar^2}{2m}\nabla_i^2 - \frac{Ze^2}{r_i} \right) \right\}$$
$$+ \left\{ \sum_{i<j=1}^{N} \frac{e^2}{r_{ij}} \right\} = \left\{ \sum_{i=1}^{N} f(\vec{r}_i) \right\} + \left\{ \frac{1}{2} \sum_{i=1;i\neq j}^{N} \sum_{j=1}^{N} \frac{e^2}{r_{ij}} \right\}, \quad (1)$$

consisting of one- and two- electron operators. The electron–electron Coulomb interaction must be described formally in terms of charge densities which require the electron wavefunctions, which is to say that one needs the solution of the N-electron Schrödinger equation even before the differential equation is formulated. The single electron atomic problem has analytical solutions at both non-relativistic and relativistic levels, but approximation methods are necessary to solve the N-electron problem for $N \geq 2$. The many-electron problem is a vexing one on account of two types of correlations between the electrons: (a) statistical (also called Fermi–Dirac or exchange) correlations and (b) Coulomb correlations. The Hartree–Fock self-consistent field (HF-SCF) provides an effective strategy to obtain solutions that are excellent approximations [6–8] to the required solutions using an *antisymmetrized* product of N single-electron wave wavefunctions.

The two-electron terms render the electron–electron potential non-local. The Hartree–Fock method obtains atomic wavefunctions employing numerical procedures, using variational calculus to obtain a self-consistent field in the *frozen orbital approximation*.

Since the wavefunction employed in the HF method is antisymmetrized, statistical correlations are accounted for, but not the *Coulomb correlations*. In fact, the Coulomb correlations are *defined* to be just those that are left out of the HF method. There is no method available to account for the Coulomb correlations exactly.

One must employ approximation methods. Various approximation methods have been developed to address the Coulomb correlations, such as the Multi-Configuration Hartree–Fock (MCHF), also called the Configuration Interaction (CI) method [9], diagrammatic perturbation theory [10], Greens function method [11], etc. A very successful approximation method to treat the electron correlations is the random-phase approximation (RPA). There are various routes to the RPA such as the method of canonical transformations [12], the equation of motion method [13], and the linearized Time-Dependent Hartree–Fock method—*commonly known as the random-phase approximation with exchange*, or RPAE, for short [1,14,15]. All of these routes to RPA are equivalent; they depend on employing a linear approximation to the electron correlations.

We shall first briefly visit salient features of the method of canonical transformation of the Hamiltonian employed by Bohm and Pines [12,16–19] since: (i) their method lucidly illustrates the *linear* approximation to electron correlations and (ii) explains the RPA which involves cancellation of terms associated with the term *random-phase*, arrived at using a *linearization process. This linearization process is the heart of the RPA*. We shall then review the linearization of the Time-Dependent Hartree–Fock system (TDHF) of equations developed by Dalgarno and Victor (1968). This approach is equivalent to that of Bohm–Pines on account of the linearization process that drives it. It is especially insightful toward appreciating the treatment of the many-electron system *beyond* the Hartree–Fock model. The *linearized* TDHF thus provides the essential platform toward methods in the analysis of atomic collisions and photoionization processes [1] employing the technique known as the random-phase-approximation with exchange (RPAE). Finally, we shall summarize the relativistic improvisation of the RPAE, called the relativistic-random-phase approximation (RRPA) developed by Johnson, Lin, and Dalgarno [2–5,14] which is arrived at by linearizing the Time-Dependent Dirac–Hartree–Fock (TDDHF, often abbreviated as TDDF) and illustrate some of its applications.

2.2. Linear approximation to Coulomb correlations

Strong electron–electron correlations in a free electron gas were dealt with by Bohm and Pines by subjecting the many-electron Hamiltonian to a series of canonical transformations. These transformations result in weakly interacting *elementary excitations* (plasmons) which represent collective elementary excitations of the electron gas.

Using the second quantized notation [20] for the electron creation ($c^{\dagger}$) and annihilation operator (c), the Schrödinger equation for an N-electron free electron gas is

$$\frac{\partial}{\partial t}|\Psi(t)\rangle = \Big[\sum_{i}\sum_{j} c_i^{\dagger}\langle i|f|j\rangle c_j + \frac{1}{2}\sum_{i}\sum_{j}\sum_{k}\sum_{l} c_i^{\dagger}c_j^{\dagger}\langle ij|v|kl\rangle c_l c_k\Big]|\Psi(t)\rangle, \tag{2a}$$

where

$$\langle ij|v|kl\rangle = \int dq_1 \int dq_2 \psi_i^*(q_1)\psi_j^*(q_2)v(q_1,q_2)\psi_k(q_1)\psi_l(q_2), \tag{2b}$$

with the subscripts i, j, k, l denoting the set of four one-electron quantum numbers and the arguments (q_r) denoting the four coordinates of the rth electron, three of which being the space coordinates and the fourth being the spin coordinate. The operator f is a *single-electron* operator, similar to that in Equation (1), but consisting of only the kinetic energy terms, the electrons being *free*. In Equation (2b), a typical spin-orbital is represented by

$$\psi_i(q) = \psi_i(\vec{r})\chi_i(\zeta), \tag{2c}$$

wherein the spin part is either

$$\chi_i(\zeta) = \alpha = \begin{bmatrix}1\\0\end{bmatrix}, \tag{2d}$$

for $m_{s_i} = +\frac{1}{2}$ i.e., $\uparrow$, or

$$\chi_i(\zeta) = \beta = \begin{bmatrix}0\\1\end{bmatrix}, \tag{2e}$$

for $m_{s_i} = -\frac{1}{2}$ i.e., $\downarrow$.

The second quantized Hamiltonian in Equation (2a) is equivalently written as

$$H = \Big[\sum_{i,\alpha}\sum_{j,\beta} c_{i\alpha}^{\dagger}\Big(\int \psi_{i\alpha}^*(q)f(q)\psi_{j\beta}^*(q)dq\Big)c_{j\beta} + \frac{1}{2}\sum_{i\alpha}\sum_{j\beta}\sum_{k\gamma}\sum_{l\delta} c_{i\alpha}^{\dagger}c_{j\beta}^{\dagger} \times\Big(\int dqdq'\psi_{i\alpha}^*(q)\psi_{j\beta}^*(q)v(q,q')\psi_{l\delta}(q)\psi_{k\gamma}(q)\Big)c_{k\gamma}c_{l\delta}\Big], \tag{3a}$$

or more compactly using the Dirac notation for the integrals as

$$H = \sum_{i,\alpha}\sum_{j,\beta} c_{i\alpha}^{\dagger}\langle i\alpha|f|j\beta\rangle c_{j\beta} + \frac{1}{2}\sum_{i\alpha}\sum_{j\beta}\sum_{k\gamma}\sum_{l\delta} c_{i\alpha}^{\dagger}c_{j\beta}^{\dagger} \times\langle i\alpha, j\beta|v|l\delta, k\gamma\rangle c_{k\gamma}c_{l\delta}. \tag{3b}$$

Without compromising the above Hamiltonian in any way, we can place the *most part* of the two-electron interactions

$$\left\{\frac{1}{2}\sum_{i=1;i\neq j}^{N}\sum_{j=1}^{N}v(\vec{r}_i,\vec{r}_j)\right\}$$

in Equation (1) in a one-electron operator

$$\left\{\sum_{i=1}^{N}F(\vec{r}_i)\right\}$$

by writing

$$\begin{aligned}H^{(N)}(q_1,q_2,\ldots,q_N) &= \sum_{i=1}^{N}f(\vec{r}_i)+\sum_{i=1}^{N}F(\vec{r}_i)\\ &+\frac{1}{2}\sum_{i=1;i\neq j}^{N}\sum_{j=1}^{N}v(\vec{r}_i,\vec{r}_j)-\sum_{i=1}^{N}F(\vec{r}_i),\end{aligned} \tag{4a}$$

or briefly as

$$\begin{aligned}&H^{(N)}(q_1,q_2,\ldots,q_N) =\\ &(f+F)+(H_2-F)=O_1+O',\end{aligned} \tag{4b}$$

where O_1 is a one-electron operator and O' is only a small part of the full Hamiltonian. The notations employed in Equations (4a) and (4b) are self-explanatory. Essentially, the N-electron Hamiltonian is re-written such that it can be approximated by $(f+F)$, since

$$\left\{\frac{1}{2}\sum_{i=1;i\neq j}^{N}\sum_{j=1}^{N}v(\vec{r}_i,\vec{r}_j)-\sum_{i=1}^{N}F(\vec{r}_i)\right\}$$

is small; we shall treat it perturbatively. The choice of the operator F is so made that the total energy functional

$$E^{(N)}=\langle\Psi^{(N)}|H^{(N)}|\Psi^{(N)}\rangle, \tag{5}$$

is minimized.

When O' is neglected, the unperturbed ground state wavefunction of the N-electron system is expressible as a determinant:

$$\begin{aligned}\Phi^{(N)} = \frac{1}{\sqrt{N!}}&\begin{bmatrix}\psi_{1\uparrow}(1) & \ldots & \ldots & \psi_{1\uparrow}(N)\\ \psi_{1\downarrow}(1) & \ldots & \ldots & \psi_{1\downarrow}(N)\\ \ldots & \ldots & \ldots & \ldots\\ \psi_{\frac{N}{2}\uparrow}(1) & \ldots & \ldots & \psi_{\frac{N}{2}\uparrow}(N)\\ \psi_{\frac{N}{2}\downarrow}(1) & \ldots & \ldots & \psi_{\frac{N}{2}\downarrow}(N)\end{bmatrix}_{N\times N} =\\ \frac{1}{\sqrt{N!}}&\begin{bmatrix}\psi_1(1) & \ldots & \ldots & \psi_1(N)\\ \psi_1(1) & \ldots & \ldots & \psi_1(N)\\ \ldots & \ldots & \ldots & \ldots\\ \psi_{N-1}(1) & \ldots & \ldots & \psi_{N-1}(N)\\ \psi_N(1) & \ldots & \ldots & \psi_N(N)\end{bmatrix}_{N\times N},\end{aligned} \tag{6}$$

wherein

$$[f(\vec{r})+F(\vec{r})]\psi_{i\sigma}(\vec{r})=\varepsilon_i\psi_{i\sigma}(\vec{r}), \tag{7}$$

with $\psi_{i\sigma}(\vec{r})=\psi_{i\downarrow}(\vec{r})$ or $\psi_{i\uparrow}(\vec{r})$. Observe that the one-particle eigenvalue ε_i is doubly degenerate with the spin-orbitals for spin up and down being linearly independent. In the second determinant in Equation (6), we have only re-designated the single-electron spin-orbitals.

$$\Phi^{(N)} = \frac{1}{\sqrt{N!}} \begin{bmatrix} \psi_1(1) & \dots & \dots & \dots & \psi_1(N) \\ \psi_1(1) & \dots & \dots & \dots & \psi_1(N) \\ \dots & \dots & \dots & \langle j|i\rangle = \psi_i(q_j) & \dots \\ \psi_{N-1}(1) & \dots & \dots & \dots & \psi_{N-1}(N) \\ \psi_N(1) & \dots & \dots & \dots & \psi_N(N) \end{bmatrix}_{N\times N} . \tag{8}$$

Re-designation of the single-electron spin-orbitals in Equation (6) is appropriate for closed shell atoms for which the random-phase approximation is applicable. Wave functions of the *excited* unperturbed states are also Nth order determinants made up of eigenfunctions of Equation (7), but with one or more $\varepsilon_i > \varepsilon_{N/2}$. In an ordered set of spin-orbitals, let us denote $p \leq N$ and $q > N$. Thus, we denote a typical spin-orbital in the ground state Slater determinant by p and an excited state spin-orbital corresponding to a single excitation by q. The operator $(f+F)$ is diagonal with respect to one-electron functions, and $q \neq p$. The choice of the operator $F(\vec{r})$ which makes the energy functional (Equation (5)) a minimal is the one for which the matrix element

$$\langle q|F|p\rangle = \sum_{i=1}^{N} \Big[\langle iq|v|ip\rangle - \langle qi|v|ip\rangle \Big], \tag{9}$$

as can be shown using a variational method under the *frozen orbital approximation*; i.e., for the excited state we use a Slater determinant only the pth spin-orbital replaced by the excited q^{th}—*all other spin-orbitals in the Slater determinant (8) retain their occupancies.*

The one-electron Hartree–Fock equation satisfied by the SCF ground-state spin-orbitals is

$$\begin{aligned} &\Bigg[\left(-\frac{\hbar^2}{2m}\nabla^2 - \frac{Ze^2}{r} \right) \psi_p(\xi) \\ &+ \sum_{i=1}^{N} \left[\int d^4V' \frac{\psi_i^*(\xi')\psi_i(\xi')\psi_p(\xi)e^2}{\vec{r}-\vec{r'}} \right] \\ &- \sum_{i=1}^{N} \delta(m_{s_p}, m_{s_i}) \left[\int d^4V' \frac{\psi_i^*(\xi')(\psi_p(\xi')\psi_i(\xi))e^2}{\vec{r}-\vec{r'}} \right] \Bigg] = \\ &\varepsilon_p \psi_p(\xi). \end{aligned} \tag{10a}$$

The four-dimensional integration $\int d^4V'$ in Equation (10a) includes integration over the three (continuous) space coordinates and the discrete summation over the spin-coordinate. The second and the third terms on the left-hand side of Equation (10a) involve two-electron terms; the second is the Coulomb term and the third is the exchange term. We consider non-ferromagnetic systems so that the number of spin-up and -down terms is equal. The energies ε_p are doubly degenerate with two linearly independent functions corresponding to up- and down-spins. Carrying out the summation over the spin variable, Equation (10a) simplifies to

$$\begin{aligned} &\Bigg[\left(-\frac{\hbar^2}{2m}\nabla^2 - \frac{Ze^2}{r} \right) \psi_p(\vec{r}) \\ &+2 \sum_{i=1}^{N/2} \left[\int dV' |\psi_i(\vec{r'})|^2 v(\vec{r},\vec{r'}) \right] \psi_p(\vec{r}) \\ &- \sum_{i=1}^{N/2} \psi_i(\vec{r}) \left[\int dV' \psi_i^*(\vec{r'})\psi_p(\vec{r'}) v(\vec{r},\vec{r'}) \right] \Bigg] = \\ &\varepsilon_p \psi_p(\vec{r}), \end{aligned} \tag{10b}$$

in which we have written the inter-electron Coulomb interaction as

$$\frac{e^2}{\vec{r}-\vec{r'}} = v(\vec{r},\vec{r'}). \tag{11}$$

Now,

$$F\psi_p(\vec{r_2}) = 2\sum_{i=1}^{N/2}\int d^3\vec{r_1}|\psi_i(\vec{r_1})|^2 v(\vec{r_1},\vec{r_2})\psi_p(\vec{r_2}) - \sum_{i=1}^{N/2}\int d^3\vec{r_1}\psi_i^*(\vec{r_1})v(\vec{r_1},\vec{r_2})\psi_p(\vec{r_1})\psi_i(\vec{r_2}),$$

hence the HF SCF equation becomes

$$\Bigg[\Bigg(-\frac{\hbar^2}{2m}\nabla^2 - \frac{Ze^2}{r}\Bigg)\psi_p(\vec{r}) + F\psi_p(\vec{r}) = (f+F)\psi_p(\vec{r}) = \varepsilon_p\psi_p(\vec{r}). \tag{12}$$

We write the momentum-dependent energies as $\varepsilon(\vec{k})$ or equivalently as $\varepsilon(\vec{k})$, since $\vec{p} = \hbar\vec{k}$. The free electron gas is the only many-electron system for which the HF SCF equation can be obtained analytically. The linearization of the Coulomb approximation was developed by Bohm and Pines [17] in which the positive charges of the nuclei were considered to be spread out uniformly over a volume V as jellium (Figure 1). The electron gas is also spread out over the volume V in which the electron wavefunction is box-normalized.

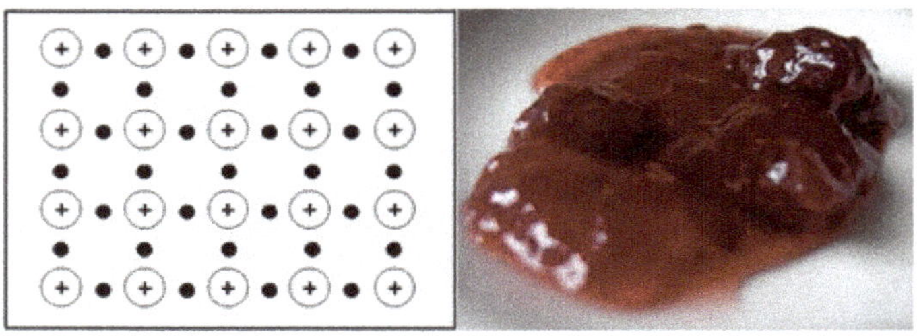

Figure 1. The uniform spread of the potential generated by the positive nuclei as a jellium.

Adding the jellium potential $V(\vec{r})$ in the HF SCF equation we get

$$\Bigg[-\frac{\hbar^2}{2m}\nabla^2\psi_p(\vec{r}) + \cancel{V(\vec{r})\psi_p(\vec{r})} \cancel{+2\sum_{i=1}^{N/2}\Big[\int dV'|\psi_i(\vec{r'})|^2 v(\vec{r},\vec{r'})\Big]\psi_p(\vec{r})} - \sum_{i=1}^{N/2}\psi_i(\vec{r})\Big[\int dV'\psi_i^*(\vec{r'})\psi_p(\vec{r'})v(\vec{r},\vec{r'})\Big]\Bigg] = \varepsilon_p\psi_p(\vec{r}). \tag{13}$$

The attractive jellium potential (second term on the left-hand side of Equation (13)) exactly cancels the electron–electron Coulomb repulsion term (third term). We are then left with

$$\Bigg[-\frac{\hbar^2}{2m}\nabla^2 - V^{exchange}(q)\Bigg]\psi_p(q) = \varepsilon_p\psi_p(\vec{r}), \tag{14}$$

where

$$V^{exchange}(q)\psi_p(q) = \sum_{i=1}^{N} V_i^{exchange}(q)\psi_p(q) = \sum_{i=1}^{N} \psi_i(q)\left[\int dq' \frac{\psi_i^*(q')\psi_p(q')}{|\vec{r}-\vec{r}'|}\right], \tag{15a}$$

i.e.,

$$V^{exchange}(q)\psi_p(q) = \sum_{i=1}^{N/2} \psi_i(\vec{r})\left[\psi_i^*(\vec{r}')\psi_p(\vec{r}')v(\vec{r},\vec{r}')\right]. \tag{15b}$$

Using (i) box-normalized wavefunctions

$$\psi_{\vec{k}\sigma}(\vec{r}) = L^{-3/2}e^{i\vec{k}\cdot\vec{r}}\chi_\sigma(\zeta), \tag{16}$$

and (ii)

$$\phi(\vec{r_1}) = \int d^3\vec{r_2}\frac{e^{i(\vec{k}-\vec{k}')\cdot\vec{r_2}}}{r_{12}} = \frac{4\pi e^{i(\vec{k}-\vec{k}')\cdot\vec{r_1}}}{|\vec{k}-\vec{k}'|^2}, \tag{17}$$

the Hartree–Fock equation for the free-electron gas (with exchange) in the positive jellium potential becomes

$$-\frac{\hbar^2}{2m}\nabla^2\psi_{\vec{k}}(\vec{r_1}) + \varepsilon_{\vec{k}}\psi_{\vec{k}}(\vec{r_1}) = \varepsilon(\vec{k})\psi_{\vec{k}}(\vec{r_1}), \tag{18}$$

with the exchange term given by

$$\varepsilon_{\vec{k}} = \frac{4\pi e^2}{L^3}\sum_{\vec{k}'}\frac{1}{|\vec{k}-\vec{k}'|^2}. \tag{19}$$

It follows that

$$\varepsilon(\vec{p}) = \frac{p^2}{2m}\varepsilon_{exchange}(\vec{p}), \tag{20}$$

where

$$\varepsilon_{exchange}(\vec{p}) = \frac{-e^2p_f}{\hbar\pi}\left[1 + \frac{p_f^2-p^2}{2p_fp}\ln\left|\frac{p+p_f}{p-p_f}\right|\right], \tag{21}$$

p_f being the electron momentum at the highest occupied energy level; viz., Fermi level. Whereas the HF SCF energy of an atom described by the Hamiltonian (Equation (1)) is

$$E_{HF}^{atom} = \langle\psi^{(N)}|H|\psi^{(N)}\rangle = \sum_{i=1}^{N}\langle i|f|i\rangle + \frac{1}{2}\sum_{j=1}^{N}\sum_{i=1}^{N}\left[\langle ij|g|ij\rangle - \langle ij|g|ji\rangle\right], \tag{22}$$

that of an electron gas in the jellium potential of the positive charges is

$$E_{HF}^{electron\ gas\ in\ jellium\ potential} = E_{KE} + E_{exchange\ correlation}, \tag{23}$$

where

$$E_{KE} = 2\frac{L^3}{(2\pi\hbar^2)}\int_{p=0}^{p=p_f} p^2 dp \int_{\theta=0}^{\theta=\pi} sin\theta d\theta \times \int_{\phi=0}^{\phi=2\pi} d\phi\left[\frac{\vec{p}.\vec{p}}{2m}\right] = \frac{\hbar^2 L^3}{10\pi^2 m}k_f^5, \quad (24)$$

with

$$k_f = \left(\frac{3\pi^2 N}{V}\right)^{\frac{1}{3}}, \quad (25)$$

and

$$E_{exchange\,correlation} = 2\frac{L^3}{(2\pi\hbar^2)} \times \int_{p=0}^{p=p_f} p^2 dp \int_{\theta=0}^{\theta=\pi} sin\theta d\theta \int_{\phi=0}^{\phi=2\pi} d\phi\left[\frac{1}{2}\varepsilon_{exchange}(\vec{p})\right] = \frac{\hbar^2 L^3}{10\pi^2 m}k_f^5, \quad (26a)$$

$$E_{exchange\,correlation} = -\frac{Ve^2}{4\pi^3}\int_{k=0}^{k=k_f} dk\left[2k_f k^2 + k(k_f^2 - k^2)ln\left(\frac{k_f + k}{k_f - k}\right)\right]. \quad (26b)$$

If we now consider the entire physical volume under consideration to consist of spheres of radius r_s (Seitz parameter in Bohr units), each having *one* unit of electron charge, then

$$N \times \left(\frac{4}{3}\pi r_s^3\right) = V = \frac{3\pi^2 N}{k_f^3}, \quad (27)$$

then the K.E. contribution to the average HF ground state energy *per electron* in a free-electron gas is

$$\frac{E_{KE}}{N} = \frac{3\hbar^2}{10m}\left(\frac{9\pi}{4}\right)^{\frac{2}{3}}\frac{1}{r_s^2} = \frac{2.21}{r_s^2}Ryd, \quad (28a)$$

and the exchange correlation energy per electron is

$$\frac{E_{exchange\,correlation}}{N} = -\frac{0.916}{r_s}Ryd. \quad (28b)$$

Thus, for electron gas in the SCF jellium potential, the average Hartree–Fock energy per electron is

$$\frac{E_{HF}}{N} = \left(\frac{2.21}{r_s^2} - \frac{0.916}{r_s}\right)Ryd. \quad (29)$$

A first order perturbative treatment gives essentially the same result as above. electron–electron exchange interactions reduce the energy *below* that of the Sommerfeld gas in a positive jellium potential; exchange energy is negative.

In the mid-1950s, Bohm and Pines improvised on the above model by considering a random mutual displacement of the centers of the positive and negative charge densities (Figure 2). In the jellium potential, these are coincident; their mutual displacement can be considered to have been triggered by a random event, but once displaced, the positive and negative charges are set in oscillations of the plasma as the system seeks its original configuration. Bohm and Pines modeled these oscillations using a harmonic oscillator potential, inclusive of a dispersive wavelength-dependent frequency of the plasma oscillations.

The Hamiltonian for N electrons in a volume V together with a uniform positive charge background jellium distribution is

$$H_0 = H_{el} + H_b + H_{el-b}, \tag{30}$$

where

$$H_{EL} = \sum_{i=1}^{N} \frac{p_i^2}{2m} + \frac{1}{2}e^2 \sum_{j=1;j\neq i}^{N} \sum_{i=1}^{N} \frac{exp(-\mu|\vec{r}_i - \vec{r}_j|)}{|\vec{r}_i - \vec{r}_j|}, \tag{31}$$

represents the many-electron part,

$$H_b = \frac{1}{2}e^2 \int\int\int d^3\vec{x} \int\int\int d^3\vec{x}' \frac{\rho^+_{\vec{x}} \rho^+_{\vec{x}'} exp(-\mu|\vec{x} - \vec{x}'|)}{|\vec{x} - \vec{x}'|}, \tag{32}$$

represents the jellium background and

$$H_{el-b} = -e^2 \sum_{i=1}^{N} \int\int\int d^3\vec{x} \frac{\rho^+_{\vec{x}} exp(-\mu|\vec{x} - \vec{r}_i|)}{|\vec{x} - \vec{r}_i|}, \tag{33}$$

represents the interaction between the electrons and the jellium background.The N-electron system in the jellium background potential constitutes an electrically neutral system, but the relative displacements of the positive and negative charges allow for plasma oscillations of the electron gas. A mathematical device using the coefficient μ in the exponential terms in Equations (31)–(33) is introduced to avoid some divergences; solutions are finally sought in the limit $\mu \to 0$. As a result of carrying out the integrals in Equations (32) and (33), the Hamiltonian (Equation (30)) turns out to be

$$H = H_{el} - \frac{1}{2}e^2 \frac{N^2}{V} \frac{4\pi}{\mu^2}, \tag{34}$$

which manifestly diverges in the limit $\mu \to 0$. This is commonly referred to as the μ^2-divergence.

It is most convenient to: (i) use the second quantized representations of the Hamiltonian for the bulk electron gas in a uniform positive background jellium potential (Equation (30)) using electron creation operator $c^+_{\vec{k}_i\sigma_i}$ and the annihilation operator $c_{\vec{k}_i\sigma_i}$, $\vec{k}_i\sigma_i$ being the momentum (in units of $\hbar$) and spin quantum numbers; (ii) employ the Fourier representation of the screened Coulomb interaction that appears in Equations (31)–(33); and finally (iii) seek the limits $(L^3 = V) \to \infty$ (specifically in this order, with $L^{-1} << \mu$). Using the three steps described, after some tedious algebra, one finds that terms corresponding to momentum transfer $\vec{q}$ in the two-electron interactions term for which $\vec{q} = \vec{0}$ in the H_{el} part of the Hamiltonian cancels the abovementioned μ^2-divergence, and along with the limits sought as per (iii), the Hamiltonian in the second quantized notation is

$$\begin{aligned} H_0 = &\sum_{\vec{k}} \sum_{\sigma} \frac{\hbar^2 \vec{k}^2}{2m} c^+_{\vec{k}\sigma} c_{\vec{k}\sigma} \\ &+ \frac{1}{2}\frac{e^2}{V} \sum_{\vec{k},\sigma_1} \sum_{\vec{p},\sigma_2} \sum_{\vec{q}\neq\vec{0}} \left(\frac{4\pi}{q^2} c^+_{\vec{k}+\vec{q}\sigma_1} c^+_{\vec{p}-\vec{q}\sigma_2} c_{\vec{p}\sigma_2} c_{\vec{k}\sigma_1} \right). \end{aligned} \tag{35}$$

Scaling

$$\tilde{\vec{k}} = r_s \vec{k}, \tag{36a}$$

$$\tilde{\vec{p}} = r_s \vec{p}, \tag{36b}$$

$$\tilde{V} = \frac{V}{r_s^3}, \tag{36c}$$

and

$$\tilde{q} = r_s q. \tag{36d}$$

allows us to write the Hamiltonian as

$$H_0 = \left(\frac{e^2}{a_0 r_0^2}\right)\left[\sum_{\tilde{\vec{k}}}\sum_{\sigma}\frac{\tilde{\vec{k}}^2}{2}c^{\dagger}_{\tilde{\vec{k}}\sigma}c_{\tilde{\vec{k}}\sigma} + \frac{1}{2}\frac{r_0}{\tilde{V}}\sum_{\tilde{\vec{k}}}\sum_{\tilde{\vec{p}}}\sum_{\tilde{\vec{q}}\neq\vec{0}}\sum_{\sigma_1}\sum_{\sigma_2}\left(\frac{4\pi}{\tilde{q}^2}c^{\dagger}_{\tilde{\vec{k}}+\tilde{\vec{q}}\sigma_1}c^{\dagger}_{\tilde{\vec{p}}-\tilde{\vec{q}}\sigma_2}c_{\tilde{\vec{p}}\sigma_2}c_{\tilde{\vec{k}}\sigma_1}\right)\right], \tag{37}$$

where we have introduced a dimensionless variable $r_0 = \frac{r_s}{a_0}$, a_0 being the Bohr radius. In the high density limit $r_0 \to \infty$, the second term in Equation (37) can be treated using first order perturbation theory even if the electron–electron interactions in the second term are quite strong. The result in the first order turns out to be essentially the same as in Equation (29), but higher order perturbation theory does not converge. Therefore, Bohm and Pines developed a non-perturbative approximation by carrying out canonical transformation of the Hamiltonian to represent pseudoparticles (elementary excitations of the many-electron gas) called *plasmons* which represent collective oscillations of the electron gas. The approximation involves linearization of the Hamiltonian concomitant with the neglect of certain terms whose phases are random and hence cancelable. Prior to discussing the canonical transformation of the Hamiltonian, we briefly visit their earlier semi-classical model which helps build insight in the *linearization* process and also in the *approximation* involved in the concomitant cancellation of terms having *random phases*.

In the semi-classical model, both the electron gas and the positive charge in the bulk medium are considered to be uniformly spread over the entire volume with their collective centers coincident. An incidental movement of the electron density ρ (Figure 2) sets in oscillations of the electron gas described by the classical equation of motion:

$$m\frac{d^2\xi}{dt^2} = \left(\frac{1}{\varepsilon_0}e\bar{\rho}\xi\right)(-e), \tag{38}$$

wherein $e\bar{\rho}\xi$ denotes the surface charge density σ, the static average volume charge density being written as

$$\bar{\rho} = \frac{N}{N\frac{4}{3}\pi r_s^3} = \frac{3}{4\pi r_s^3}. \tag{39}$$

The zero-point energy of the plasma oscillations is $\frac{1}{2}\hbar\omega_p$ wherein the natural frequency of the plasma oscillations is

$$\omega_p = \sqrt{\frac{\bar{\rho}e^2}{m\varepsilon_0}} = \sqrt{\frac{3e^2}{mr_s^3}}. \tag{40}$$

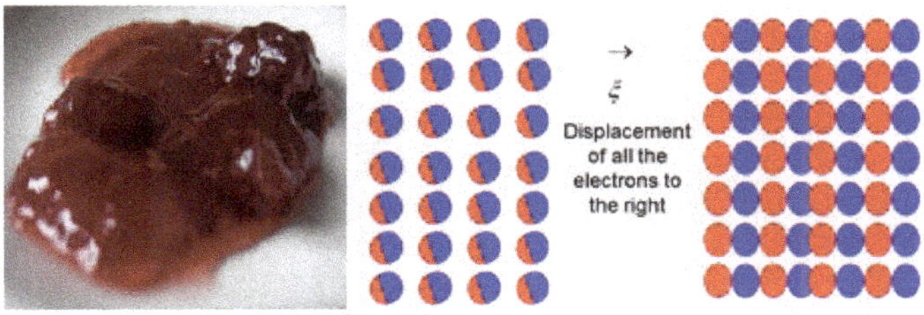

Figure 2. The 'jellium model' considers the collection of positive nuclear charges and negative electron charges smeared out uniformly across the region of a metal; the centers of positive and negative charges being coincident. A slight relative displacement of the centers of positive and negative parts of the total charge density sets in plasma oscillations.

Using the Fourier decomposition of the inter-electron interaction,

$$\frac{e^2}{r_{ij}} = \frac{1}{V}\sum_{\vec{k}} c_{\vec{k}} e^{i\vec{k}\cdot(\vec{r}_i-\vec{r}_i)},$$

the potential energy of the ith electron due to one electron charge *uniformly smeared* throughout the box is $e^2 \iiint d^3\vec{r}_j$, i.e., $e^2\sum_{\vec{k}} c_{\vec{k}}\{\frac{1}{V}\iiint d^3\vec{r}_j e^{i\vec{k}\cdot(\vec{r}_i-\vec{r}_i)}\}$.

Hence, the potential energy due to *all* the electrons is:

$$P(\vec{r}_i) = \sum_{\substack{j=1\\j\neq i}}^{N}\frac{e^2}{r_{ij}} = \frac{1}{V}\sum_{\substack{j=1\\j\neq i}}^{N}\sum_{\vec{k}} c_{\vec{k}} e^{i\vec{k}\cdot(\vec{r}_i-\vec{r}_i)}, \tag{41a}$$

where $c_{\vec{k}} = \frac{4\pi e^2}{|\vec{k}^2|}$, except for $\vec{k}=\vec{0}$. The term corresponding to $\vec{k}=\vec{0}$ cancels the positive jellium; hence the potential energy of the ith electron due to all the electrons *and* the positive background is

$$U(\vec{r}_i) = \frac{1}{V}\sum_{\substack{j=1\\j\neq i}}^{N}\sum_{\substack{\vec{k}\\\vec{k}\neq\vec{0}}}\frac{4\pi e^2}{k^2} e^{i\vec{k}\cdot(\vec{r}_i-\vec{r}_i)}. \tag{41b}$$

Now, in terms of the electron field operators, the total number of electrons is

$$N = \sum_{\zeta}\iiint d^3\vec{r}\psi^{\dagger}(q)\psi(q) = \sum_{\zeta}\iiint d^3\vec{r}\rho(q) = \iiint d^3\vec{r}\rho(\vec{r}) = \sum_{i=1}^{N}\iiint d^3\vec{r}\delta(\vec{r}-\vec{r}_i), \tag{42}$$

and the electron density is

$$\rho(\vec{r}) = \sum_{i=1}^{N}\delta(\vec{r}-\vec{r}_i) = \frac{1}{V}\sum_{k=1}^{N}\rho_{\vec{k}} e^{i\vec{k}\cdot\vec{r}}, \tag{43}$$

wherein we have used the Fourier expansion of the charge density with the Fourier components being given by

$$\rho_{\vec{k}} = \sum_{i=1}^{N} e^{-i\vec{k}\cdot\vec{r}_i}. \tag{44}$$

Identifying the force on the electron force as the negative gradient of the potential in Equation (41b), we arrive at the semi-classical equation of motion for the harmonic oscillator

$$m\ddot{\vec{r}}_i = m\dot{\vec{v}}_i = -\vec{\nabla}_i U(\vec{r}_i),$$

which translates to the equation of motion for density fluctuations of the Fourier components in Equation (43):

$$\ddot{\rho}_{\vec{k}} = \Big(-\sum_{i=1}^{N} \big(\vec{k}\cdot\dot{\vec{r}}_i\big)^2 e^{-i\vec{k}\cdot\vec{r}_i} - \frac{1}{V}\frac{4\pi N e^2}{m}\rho_{\vec{k}} - \frac{1}{V}\frac{4\pi e^2}{m}\sum_{\substack{\vec{k}'\neq\vec{k}\\ \vec{k}'\neq\vec{0}}} \rho_{\vec{k}'}\big(\rho_{\vec{k}} - \rho_{\vec{k}'}\big). \tag{45}$$

The first term in Equation (45) is quadratic in k. It can be ignored if $\big\langle \big(\vec{k}\cdot\dot{\vec{r}}_i\big)^2 \big\rangle_{average} \ll \omega_p^2$; i.e., if one limits k to be small. The 'upper bound' on the wave number, denoted by k_c, of the plasma oscillations is

$$k_{max} = k_c \approx \frac{\omega_p}{v_f}. \tag{46}$$

Now, the integral of the charge density over the entire space adds up to the total number of electrons N, i.e.,

$$\iiint d^3\vec{r}\rho(\vec{r}) = \sum_{i=1}^{N} \iiint d^3\vec{r}\delta(\vec{r}-\vec{r}_i) = N, \tag{47a}$$

corresponding to

$$\rho(\vec{r}) = \sum_{i=1}^{N} \delta(\vec{r}-\vec{r}_i). \tag{47b}$$

Using the fact that the Fourier expansion of the charge density

$$\rho(\vec{r}) = \frac{1}{V}\sum_{\vec{k}=1}^{N} \rho_{\vec{k}} e^{i\vec{k}\cdot\vec{r}}, \tag{48}$$

with $\rho_{\vec{k}} = \sum_{i=1}^{N} e^{-i\vec{k}\cdot\vec{r}_i}$ and $\rho_{\vec{k}}^* = \sum_{j=1}^{N} e^{+i\vec{k}\cdot\vec{r}_j}$.

In the third term on the right-hand side of Equation (45), we have $\rho_{\vec{k}} = \sum_{i=1}^{N} e^{-i\vec{k}\cdot\vec{r}_i}$ and $\rho_{\vec{k}-\vec{k}'} = \sum_{i=1}^{N} e^{i\vec{k}'-\vec{k}\cdot\vec{r}_i}$, which involve oscillatory terms consisting of phase factors of modulus unity. It is like carrying out a sum of vectors in a complex plane whose directions are *random*, and one expects this to be a zero-sum addition. Thus: (i) neglecting the first term (enabled by placing an upper limit on k) and (ii) *linearizing* Equation (45) (i.e., *neglecting the quadratic terms*, taking advantage of the *random-phases*), we obtain

$$\ddot{\rho}_{\vec{k}} = -\frac{4\pi\bar{\rho}e^2}{m}\rho_{\vec{k}} = \omega_p^2\rho_k, \tag{49}$$

which essentially is an equation of motion for a simple harmonic oscillator. Quantized collective excitations of this system are elementary excitations. They are pseudo-particles called plasmons. The frequency of plasma oscillations is

$$\omega_p = \sqrt{\frac{4\pi\bar{\rho}e^2}{m}} = \sqrt{\frac{4\pi\left(\frac{3}{4\pi r_s^3}\right)e^2}{m}} = \sqrt{\frac{3e^2}{mr_s^3}}. \tag{50a}$$

The zero-point energy of the plasma oscillations is $\frac{1}{2}\hbar\omega_p$, where

$$\hbar\omega_p = \frac{2\sqrt{3}}{r_s^{\frac{3}{2}}}\text{Ryd}. \tag{50b}$$

In the Bohm–Pines method of canonical transformation of the Hamiltonian discussed below, the significance of the approximation involving *linearization* of the Hamiltonian concomitant with the neglect of terms having *random phases* gets *further* accentuated.

We have seen that the Hamiltonian for a bulk electron gas in a uniform positive background jellium potential is

$$H_0 = \sum_{i=1}^{N} \frac{p_i^2}{2m} + \frac{1}{2}\frac{e^2}{V} \sum_{\substack{j=1 \\ j\neq i}}^{N} \sum_{i=1}^{N} \sum_{\substack{\vec{k} \\ \vec{k}\neq\vec{0}}} \frac{4\pi e^2}{k^2} e^{i\vec{k}\cdot(\vec{r}_i-\vec{r}_i)}, \tag{51a}$$

and noting that the $j = i$ term adds up to N, the total number of electrons, we arrive at

$$H_0 = \sum_{i=1}^{N} \frac{p_i^2}{2m} + \frac{2\pi e^2}{V} \sum_{\substack{\vec{k} \\ \vec{k}\neq\vec{0}}} \frac{1}{k^2} \sum_{i=1}^{N} e^{i\vec{k}\cdot\vec{r}_i} \sum_{\substack{j=1 \\ j\neq i}}^{N} e^{-i\vec{k}\cdot\vec{r}_j} = \sum_{i=1}^{N} \frac{p_i^2}{2m} + \frac{2\pi e^2}{V} \sum_{\substack{\vec{k} \\ \vec{k}\neq\vec{0}}} \frac{1}{k^2} (\rho_{\vec{k}}^* \rho_{\vec{k}} - N), \tag{51b}$$

where the last form is obtained by adding and subtracting the term corresponding to the $j = i$.

The quantum problem to be solved for the above Hamiltonian is

$$H_0\psi = E\psi. \tag{52}$$

Bohm and Pines recognized that the classical model which yielded plasma oscillations described by Equation (49) would be an approximation to a quantum model. One ought to seek a transformation of the Hamiltonian (Equations (51a) and (51b)) such that plasma oscillations appear explicitly as a set of *Hamiltonians for simple harmonic oscillators* for various $\vec{k}$ values limited by Equation (46). They therefore proposed canonical transformations $(q, p) \to (Q, P)$ of the Hamiltonian in Equations (51a) and (51b) to a *new* set of generalized coordinates Q and momenta P such that the new quantum Hamiltonian would have the *form*

$$H_k = \frac{P_k^\dagger P_k}{2} + \frac{1}{2}\omega^2 Q_k^\dagger Q_k, \tag{53a}$$

which is characteristic of the Hamiltonian for a simple harmonic oscillator represented by the Hamiltonian (in units of $m = 1$)

$$h_{SHO} = \frac{p^2}{2} + \frac{1}{2}\omega^2 q^2. \tag{53b}$$

The transformation we seek is not inspired by actual measurements of the new coordinates and momenta; it is motivated only by seeking the form in Equation (53a). Hence, the operators Q & P need not necessarily be Hermitian. The Bohm–Pines strategy consists of starting with an *auxiliary* Hamiltonian

$$H_1 = \sum_{\substack{\vec{k} \\ \vec{k}<\vec{k}_c}} \frac{1}{2} P_{\vec{k}}^\dagger P_{\vec{k}} - M_k P_{\vec{k}}^\dagger \rho_{\vec{k}}, \tag{54a}$$

with

$$M_k = \sqrt{\frac{4\pi e^2}{Vk^2}}. \tag{54b}$$

We do not demand the operators Q & P to be Hermitian. Instead, by choosing

$$P_{\vec{k}}^\dagger = P_{-\vec{k}}, \tag{55a}$$

and

$$Q^{\dagger}_{\vec{k}} = Q_{-\vec{k}}, \tag{55b}$$

we see on recognizing that the summation over $\vec{k}$ and that over $-\vec{k}$ is equivalent considering the symmetry in the momentum space that, H_1 is Hermitian, even if Q and P are not. The wavefunction depends only on the original set of electron coordinates; it cannot depend on any additional degrees of freedom. It is therefore judicious to employ subsidiary conditions

$$\frac{\partial \psi}{\partial Q_{\vec{k}}} = 0, \tag{56a}$$

however, limited by $k < k_c$. The derivative operator is the momentum,

$$P_k = -i\hbar \frac{\partial}{\partial Q_{\vec{k}}}, \tag{56b}$$

hence

$$P_k \psi = 0 \quad \text{for} \quad k < k_c, \tag{56c}$$

and

$$[Q_k, P_{k'}] = i\hbar \delta_{k,k'}, \tag{57}$$

which is just the uncertainty relation for canonically conjugate coordinates and momenta. Use of Equation (56c) in Equation (54a) ensures that

$$(H_0 + H_1)\psi = E\psi, \tag{58}$$

and the Hamiltonian $(H_0 + H_1)$ describes the same quantum system. We seek a transformation affected by an operator

$$U = e^{\frac{i}{\hbar}S}, \tag{59a}$$

with

$$S = \sum_{\vec{k};\vec{k}<\vec{k}_c} M_k Q_{\vec{k}} \rho_{\vec{k}}, \tag{59b}$$

and

$$S^{\dagger} = \sum_{\vec{k};\vec{k}<\vec{k}_c} M_k Q^{\dagger}_{\vec{k}} \rho^{*}_{\vec{k}} = \sum_{\vec{k};\vec{k}<\vec{k}_c} M_k Q_{-\vec{k}} \rho_{-\vec{k}} = S, \tag{59c}$$

which gives

$$U^{\dagger} = e^{-\frac{i}{\hbar}S} = U^{-1}, \tag{59d}$$

and we see that the transformation is unitary. It follows that

$$(P_k)_{new} = U^{-1}\Big(-i\hbar \frac{\partial U}{\partial Q_{\vec{k}}} + UP_k\Big) = P_k - i\hbar U^{-1}\frac{\partial U}{\partial Q_{\vec{k}}} = P_k + U^{-1}[P_k, U]_{-} = P_k + M_k \rho_{\vec{k}}, \tag{60a}$$

and the ith component of the operator is

$$(\vec{p}_i)_{new} = \vec{p}_i - i \sum_{\vec{k};\vec{k}<\vec{k}_c} M_{\vec{k}} Q_{\vec{k}} \vec{k} e^{i\vec{k}\cdot\vec{r}_i}. \tag{60b}$$

Essentially, under the transformation under consideration, the operators $\vec{r}_i, Q_{\vec{k}}, \rho_{\vec{k}}$ remain invariant, but $(\vec{p})_{new}$ and $(P_k)_{new}$ are different. We now ask what the transformed Hamiltonian,

$$H_{new} = U^{-1}(H_0 + H_1)U, \tag{61}$$

is. After some tedious algebra, it turns out to be

$$\mathfrak{H} = H_{new} = \sum_{i=1}^{N} \frac{p_i^2}{2m} + \sum_{\substack{\vec{k} \\ \vec{k}<\vec{k}_c}} \frac{1}{2}(P_{\vec{k}}^\dagger P_{\vec{k}} + \omega_p^2 Q_{\vec{k}}^\dagger Q_{\vec{k}}) - \sum_{\vec{k};\vec{k}\neq\vec{0}} \frac{2\pi e^2}{Vk^2} N + H_{s.r.} + H_{int} + K, \tag{62}$$

where

$$H_{s.r.} = \frac{1}{2} \sum_{\vec{k};\vec{k}\neq\vec{0},\vec{k}>\vec{k}_c} (\rho_{\vec{k}}^* \rho_{\vec{k}} - N), \tag{63}$$

$$H_{int} = -\frac{i}{2m} \sum_j \sum_{\substack{\vec{k} \\ \vec{k}\neq\vec{0}}} M_{\vec{k}} Q_{\vec{k}} \vec{k} \cdot (2\vec{p}_j + \hbar\vec{k}) e^{-i\vec{k}\cdot\vec{r}_j}, \tag{64}$$

and

$$K = \frac{1}{2m} \sum_{\substack{-\vec{k} \\ \vec{k}<\vec{k}_c}}^{k\neq l} \sum_{\substack{\vec{l} \\ \vec{l}<\vec{k}_c}} M_{-\vec{k}} M_{\vec{k}} (\vec{k}\cdot\vec{l}) \Big\{ \sum_j (Q_{-\vec{k}} e^{+i\vec{k}\cdot\vec{r}_j} \times Q_{\vec{k}} e^{-i\vec{k}\cdot\vec{r}_j}) \Big\}, \tag{65}$$

The new Hamiltonian has a manifestly complicated form. The term K (Equation (65)) is quadratic in the new coordinates and has *random phases* which would cancel out in a *linearization* process, as explained earlier in the context of the classical model and arrived at Equation (49). Linearization of the $\mathfrak{H} = H_{new}$ makes it possible to drop the operator K and justifies the term *random-phase approximation*. The rest of the Hamiltonian is

$$\mathfrak{H} = H_{new} = \sum_{i=1}^{N} \frac{p_i^2}{2m} + \sum_{\substack{\vec{k} \\ \vec{k}<\vec{k}_c}} \frac{1}{2}(P_{\vec{k}}^\dagger P_{\vec{k}} + \omega_p^2 Q_{\vec{k}}^\dagger Q_{\vec{k}}) - \frac{N}{V} \sum_{\vec{k};\vec{k}\neq\vec{0}} \frac{2\pi e^2}{k^2} + H_{s.r.} + H_{int}, \tag{66}$$

in which $H_{s.r.}$ (Equation (63)) represents a set of *quasi-particles* interacting via *short-range* screened-Coulomb potential and given by

$$H_{s.r.} = \frac{1}{V} \sum_{\vec{k};\vec{k}\neq\vec{0},\vec{k}>\vec{k}_c} \frac{2\pi e^2}{k^2} (\rho_{\vec{k}}^* \rho_{\vec{k}} - N), \tag{67}$$

and

$$-\frac{N}{V} \sum_{\vec{k};\vec{k}\neq\vec{0}} \frac{2\pi e^2}{k^2} \tag{68}$$

is the self-energy of the electron gas.

H_{int} is accounted for by a further canonical transformation of the Hamiltonian (in which K is ignored) written in terms of transformed coordinates and momenta. Using the *random-phase approximation* concomitant with *linearization of the transformed Hamiltonian* (i.e., neglect of quadratic terms in the newer set of coordinates), the H_{int} term gets dropped, but in the process, the first two terms get somewhat modified, and the new approximate Hamiltonian becomes

$$\mathfrak{H} = H_{new} = \{ \sum_{\vec{k};\vec{k}<\vec{k}_c} \frac{1}{2}(P_{\vec{k}}^\dagger P_{\vec{k}} + \omega_p^2 Q_{\vec{k}}^\dagger Q_{\vec{k}}) \} + \{ \sum_{i=1}^{N} \frac{p_i^2}{2m}(1 - \frac{\beta^2}{6}) + H_{s.r.} \} - \{ \sum_{\vec{k};\vec{k}\neq\vec{0}} \frac{2\pi e^2}{Vk^2} N \}, \tag{69}$$

where

$$\beta = \frac{k_c}{k_F}, \tag{70}$$

and

$$\omega_k^2 = \omega_p^2 + \frac{2}{m} E_F k^2 \tag{71}$$

expresses a weak k-dependent dispersion of the plasma frequency.

We have another subsidiary condition, similar to Equation (56c):

$$(P_k + M_K \rho_{\vec{k}}) \psi_{new} = 0 \quad \textbf{for} \quad k < k_c. \tag{72}$$

The kinetic energy part in the newer Hamiltonian is diminished by the factor $(1 - \frac{\beta^2}{6})$; actual calculations show that $\beta \simeq 0.7$ and hence the kinetic energy part is reduced by about 8%. The long-range part of the interaction is what leads to the plasma oscillations corresponding to the first curly bracket in Equation (69). $H_{s.r.}$ denotes the short-range screened-Coulomb interaction between the new pseudo-particles, which are elementary excitations called plasmons. The Hartree–Fock approximation accounted for only the static part of the density fluctuations of the collective behavior of an electron gas. The frozen-orbital approximation that leads us to the Koopmans theorem highlights this approximation which limits it to the neglect of the Coulomb correlations. The method of canonical transformation of the Hamiltonian enables us address the Coulomb correlations albeit in an approximate manner by systematic and straightforward interpretation of Equation (69) in which the first curly bracket represents the collective oscillations of the electron gas resulting from the long-range part of the Coulomb interaction. A quantum of these oscillations is the plasmon. The second curly bracket represents the Hamiltonian for the screened Coulomb interaction, and the third represents the self-energy of the electron gas.

$$E_{BP} = \frac{2.21}{r_s^2} - \frac{0.916}{r_s} + \frac{\sqrt{3}}{2r_s^{3/2}}\beta^2 - \frac{0.916}{r_s}\left(\frac{\beta^2}{2} - \frac{\beta^4}{48}\right).$$

The Bohm–Pines method elucidates the physical content of the *random-phase approximation (RPA) and the linearization process* it involves. There are other methods of arriving at the RPA, such as the Equation of Motion method [13] and the Greens function method [11]. The approximation is equivalent to summing over all the ring diagrams (along with the diagrams for the exchange interaction corresponding to each Coulomb term) in Feynman diagrammatic perturbation theory. Another equivalent approach to the RPA(E) results from the linearization of the Time-Dependent Hartree–Fock (TDHF) method developed by Dalgarno and Victor [14] and Amusia [15], and its relativistic version, namely the linearized Time Dependent Dirac–Hartree–Fock (TDDHF, often briefly denoted as TDDF) developed by Johnson and Lin [4]. In the next section, we summarize the linearized TDHF/TDDF approximations.

2.3. Linearization of TDHF and that of TDDF Formalism

The Hartree–Fock self-consistent field (HF-SCF) method accounts for correlations in many-electron dynamics that result by demanding that a many-fermion wavefunction must be anti-symmetric with respect to every exchange of pairs of the elementary particles. These correlations are therefore equivalently referred to as exchange correlations or as statistical (Fermi–Dirac) correlations. The Pauli Exclusion Principle automatically follows from it; hence, they are also sometimes called the Pauli correlations. The HF-SCF, however, only accounts for a static average of the density fluctuations of the many-electron system and thus leaves out what are known as *Coulomb correlations*. In the previous section, we discussed the RPA which employs a *linearization* technique and provides for a very successful methodology to account for the Coulomb correlations. We now proceed to discuss the RRPA [4], which is essentially based on *linearization* of the Time-Dependent Dirac–Hartree–Fock (TD-DHF, or just TDDF) family of coupled integro-differential equations. The linearized TDHF (RPAE) approximation [Amusia] was the precursor to the RRPA; it employs the *linearization* of the TDHF family of equations. The RPAE employs two-component spin-orbitals obtained as SCF solutions to the non-relativistic Schrödinger equation for the many-electron system. These spin-orbitals go into the construct of the Slater determinantal single-configuration wavefunctions. The RRPA is the relativistic extension of the RPAE. It employs four-component bi-spinor SCF solutions to the Dirac equation for the many-electron system which appear in the Slater determinant. The bi-spinors (spin-orbitals) are, however, admittedly time-dependent allowing for density fluctuations of the many-electron system. The resulting TD-DHF (TDHF) equations are non-linear; making a *linear approximation* to the TD-DHF (TDHF) equations result in the RRPA (RPAE).

The time-independent DHF equations for an N-electron closed-shell atomic system are

$$(h_0 + V_{DHF})u_i = \varepsilon_i u_i \qquad \ldots i = 1,2,\ldots,N, \tag{73}$$

where u_i represent the four-component bispinor, h_0 represents the Dirac Hamiltonian,

$$h_0 = \vec{\alpha}\cdot\vec{p} + \beta m - \frac{Ze^2}{r} \quad (\hbar = 1, c = 1), \tag{74}$$

ε_i represent the DHF eigenvalue and $V_{DHF}(\vec{r})$ represents the DHF potential, given by

$$V_{DHF}u(\vec{r}) = \sum_{j=1}^{N} e^2 \int d^3r' \frac{\left\{(u_j^\dagger u_j)' u - (u_j^\dagger u)' u_j\right\}}{|\vec{r} - \vec{r'}|}, \tag{75}$$

and the prime denotes the argument over which integration is carried out. Solutions to the DHF equations are best represented by a Slater determinant

$$\psi^{(N)} = \frac{1}{\sqrt{N!}} \begin{vmatrix} u_1(1) & \ldots & \ldots & \ldots & u_1(N) \\ u_1(1) & \ldots & \ldots & \ldots & u_1(N) \\ \ldots & \ldots & \ldots & \ldots & \langle N|i\rangle \\ \ldots & \ldots & \ldots & \ldots & \ldots \\ \ldots & \ldots & \ldots & \ldots & \ldots \\ u_N(1) & \ldots & \ldots & \ldots & u_N(N) \end{vmatrix}, \tag{76a}$$

where

$$u_{n\kappa m} = \frac{1}{r}\begin{pmatrix} iG_{n\kappa}(r)\Omega_{\kappa m}(\hat{r}) \\ F_{n\kappa}(r)\Omega_{-\kappa m}(\hat{r}) \end{pmatrix} = \begin{pmatrix} u_+ \\ u_- \end{pmatrix}, \tag{76b}$$

with

$$\text{for} \quad j = \ell + \frac{1}{2}, \quad \Omega_{\kappa m} = \begin{pmatrix} \sqrt{(\frac{j+m}{2j})} Y_{\left(\ell=j-\frac{1}{2}\right)\left(m_{\ell'=m-\frac{1}{2}}\right)(\hat{r})} \\ \sqrt{(\frac{j-m}{2j})} Y_{\left(\ell=j-\frac{1}{2}\right)\left(m_{\ell'=m+\frac{1}{2}}\right)(\hat{r})} \end{pmatrix}$$

$$\text{for} \quad j = \ell - \frac{1}{2}, \quad \Omega_{\kappa m} = \begin{pmatrix} -\sqrt{(\frac{j-m+1}{2j+2})} Y_{\left(\ell=j+\frac{1}{2}\right)\left(m_{\ell'=m-\frac{1}{2}}\right)(\hat{r})} \\ \sqrt{(\frac{j+m+1}{2j+2})} Y_{\left(\ell=j+\frac{1}{2}\right)\left(m_{\ell'=m+\frac{1}{2}}\right)(\hat{r})} \end{pmatrix}. \tag{76c}$$

The electron densities of the DHF one-electron bispinors (spin-orbitals) represent only a time-average since the DHF model ignores electron correlations. Due to the electron correlations in the initial and the final state of a transition affected by what may be represented by an interaction operator

$$\Omega = v_+ e^{-i\omega t} + v_- e^{+i\omega t}, \tag{77}$$

with the positive and negative frequency driving terms respectively denoted by

$$v_+ = \vec{\alpha} \cdot \vec{A}; \quad v_- = v_+^\dagger. \tag{78}$$

$\vec{A}$ being the vector potential of the electromagnetic field, the DHF orbitals must be represented by time-dependent functions described by

$$u_i(\vec{r}) \to u_i(\vec{r}) + w_{i-}(\vec{r})e^{i\omega t} + w_{i+}(\vec{r})e^{-i\omega t} + \cdots \tag{79}$$

The dots ... at the end in Equation (79) represent higher harmonics. If we *rebuild* the (Time-Dependent) Dirac–Hartree–Fock scheme with all the higher harmonics, we obtain Non-Linear Time-Dependent Dirac–Hartree–Fock equations. Dalgarno and Victor [14] proposed the RPA linearization of the (non-relativistic TD-HF equations by dropping the higher harmonics. Following a similar logic, Johnson, Lin and Dalgarno [2–5] introduced the very same linearization in the TD-DHF system of coupled integro-differential equations for the orbitals $w_{i\pm}$.The orbitals w_{i+} represent perturbation of the DHF orbitals due to the positive frequency part of the perturbation, and the orbitals w_{i-} represent perturbation of the DHF orbitals due to the negative frequency part. The linearized Time-Dependent Dirac–Hartree–Fock (L-TslatdetD-DHF) equations are

$$(h_0 + V_{DHF} - \varepsilon_i \mp \omega)w_{i\pm} = \left(v_\pm - V_\pm^{(1)}\right)u_i + \sum_j \lambda_{ij\pm}u_j; \quad i = 1, 2, \ldots, N, \tag{80}$$

with

$$V_\pm^{(1)}u_i(\vec{r}) = \sum_{j=1}^{N} \int d^3r' \frac{\left[(u_j^\dagger w_{j\pm})'u_i + (w_{j\mp}^\dagger u_j)'u_i - (w_{j\mp}^\dagger u_i)'u_j - (u_j^\dagger u_i)'u_j w_{j\pm}\right]}{|\vec{r} - \vec{r}'|}, \tag{81}$$

which includes the Coulomb correlations that are omitted in the DHF method. The factors $\lambda_{ij\pm}$ in Equation (80) are the Lagrange's variational multipliers, introduced in the algebraic equations to ensure orthogonality of the perturbed orbitals $w_{j\mp}$ with respect to the unperturbed ones u_j. Omission of the driving terms $v_\pm$ gives us the fundamental RRPA equations:

$$\pm (h_0 + V - \varepsilon_i)w_{i\pm} + V_\pm^{(1)}u_i \mp \sum_j \lambda_{ij\pm}u_j = \omega w_{i\pm}; \quad i = 1, 2, \ldots, N, \tag{82}$$

The eigenvalues of the RRPA equations provide the linear approximation to the excitation spectra in both the discrete and the continuum. The positive and negative components $w_{i\pm}$ of the eigenfunctions describe the correlations which are omitted in the DHF formalism, respectively, in the excited state (both discrete and continuum) and in the initial states. The amplitude of transition from an initial state to the excited state described by the RRPA function $w_{i\pm}(\vec{r})$ corresponding to the frequency ω brought about by the interaction (Equation (76)) is

$$T = \sum_{j=1}^{N} \int d^3r\left(w_{i+}^\dagger v_+ u_i + w_{i-}^\dagger v_- u_i\right), \tag{83a}$$

i.e.,

$$T = \sum_{j=1}^{N} \int d^3r\left(w_{i+}^\dagger \vec{\alpha} \cdot \vec{A}u_i + w_{i-}^\dagger \vec{\alpha} \cdot \vec{A}u_i\right). \tag{83b}$$

It is a very important property of the (R)RPA equations that the transition matrix element is invariant under gauge transformations of the electromagnetic potentials. In actual calculations, one often employs truncated RRPA in which only the most important interchannel coupling is used. This leads to a slight disagreement in the estimation of the transition matrix elements in the length gauge and in the velocity gauge.

It is very convenient to have a pictorial representation of the correlations that are addressed in a many-electron theory. The diagrammatic representation developed by Feynman, first presented in the Spring of 1948 at the Pocono Conference, is suitable for our purpose. In AMO sciences, we (normally) do not work with positrons, but there are 'hole' states which are vacant states normally occupied by electrons. Thus we represent evolution of atomic states by vertical solid lines with reference to a time-axis going from the bottom to the top (left to right is an alternative convention). The atomic state lines are sometimes referred to as the 'trunk' of the diagram. A vertex in the diagram represents an intersection of a photon wavy line and the trunk. Particle lines point upwards and hole lines point downwards. Summing over only the ring graphs, as shown by Gell-Mann and Brueckner [21] has precisely the same effect as the linearization approximation that results in the random-phase approximation introduced by Bohm and Pines. Electron correlations are interpreted by recognizing that electrons exchange virtual photons which mediate the interaction between the electrons. The electromagnetic interaction is treated at the level of quantum theory. A positron is anelectron propagating backward in time. Figure 3 shows some of the lowest order diagrams [22] which contribute to the RRPA matrix elements.

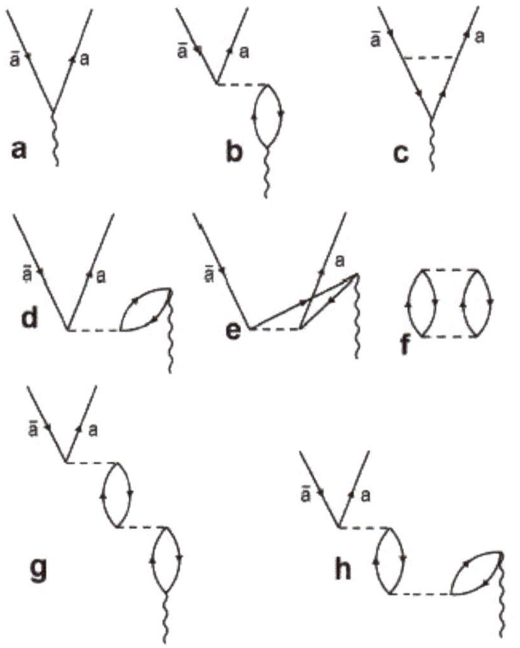

Figure 3. Lowest order Feynman diagrams which contribute to the RRPA transition amplitude for the transition $a \rightarrow \bar{a}$. Time axis is from the bottom to the top of the page. The dashed lines represent electron–electron correlations, and the wiggly lines correspond to the photon operator. Arrows pointing upward (downward) are the electron (hole): (**a**) represents uncorrelated transition matrix element; (**b**,**c**) represent, respectively, the first order time-forward (i.e., final state) Coulomb and exchange terms; and (**d**,**e**) represent, respectively, the first order time-backward (i.e., initial state) Coulomb and exchange terms. (**f**–**h**) represent higher order ring diagrams.

Time-forward diagrams represent correlations in the final state in which configuration-interaction in the continuum is taken care of. In the RRPA, this corresponds to interchannel coupling. The time-backward diagrams represent correlations in the initial state. Variants of the RRPA that offer some advantages include the R-MCTD (Relativistic Multi-Configuration Tamm–Dancoff) method [23] and the RRPA-with-relaxation (RRPA-R) method [24]. The particle and the hole creation and annihilation operators that are referenced in the ring diagrams (Figure 3) are defined with respect to a vacuum consisting of a closed-shell Hartree–Fock (or Dirac–Hartree–Fock) fermion system, such as that described by Equation (8). *Two-particle two-hole* correlations are included, accommodating both creation and annihilation of a particle-hole pair. The focus of this article is to discuss the linearization approximation involved in the RPA and illustrate a few applications of the RRPA; hence, we omit an elaboration of the RRPA-R and R-MCTD techniques.

3. Illustrative Examples

As a first example, consider the dipole photoelectron angular distribution resulting from the photoionization of the 5s subshell of the closed-shell Xe atom. The general form of the dipole angular distribution for linearly polarized incident radiation for subshell i is given as the differential photoionization cross section by [25]

$$\frac{d\sigma_i}{d\Omega} = \frac{\sigma_i}{4\pi}[1 + \beta_i P_2(\cos\theta)], \tag{84}$$

where σ_i is the total subshell cross section, $P_2(x) = (3x^2 - 1)/2$, θ is the angle between photon polarization and photoelectron momentum directions, and β_i is the dipole angular distribution asymmetry parameter. Nonrelativistically, for a closed shell atom, $\beta_i = 2$ and is independent of energy for an initial ns subshell [24]. From a physical point of view, this occurs because, nonrelativistically, there is only a single final state partial wave, characterized by ns $\rightarrow$ εp, so that there is nothing to interfere with, and the angular distribution is just determined by the symmetry of the εp-wave. When relativistic interactions are included, the situation is changed in that there are then two possible photoionizing transitions ns $\rightarrow \varepsilon p_{1/2}$ and ns $\rightarrow \varepsilon p_{3/2}$, and these can interfere with one another, giving rise to both a deviation of β_i from the value two, and an energy dependence. This behavior will be most evident near Cooper minima [26] owing to the fact that the two relativistic channels exhibit the minima at slightly different energies [27], so that in this region the matrix elements for the two relativistic channels can be vastly different, both in magnitude and phase.

The Xe 5s photoionization cross section and β parameter calculated using RRPA, including correlation in the form of interchannel coupling with 5p and 4d relativistic photoionization channels, are depicted in Figure 4. The cross section exhibits a deep minimum, indicative of the Cooper minima in the 5s $\rightarrow \varepsilon p_{1/2}$ and 5s$\rightarrow \varepsilon p_{3/2}$ channels. Correspondingly, the β parameter shows a deep minimum around that energy, and over a significant energy range, β is energy-dependent. It is to be emphasized, that, while the nonrelativistic RPAE does pretty well on the cross section [1], the β parameter predicted by the nonrelativistic calculation is constant and equal to two at all energies. Thus, it is evident that the addition of relativistic effects to the original nonrelativistic RPAE brings out additional physical effects. It should also be noted that, although the RRPA result for the cross section is rather good in the Cooper minimum region, the predictions for β are not, owing to the mission of satellite photoionization channels in the RRPA [28].

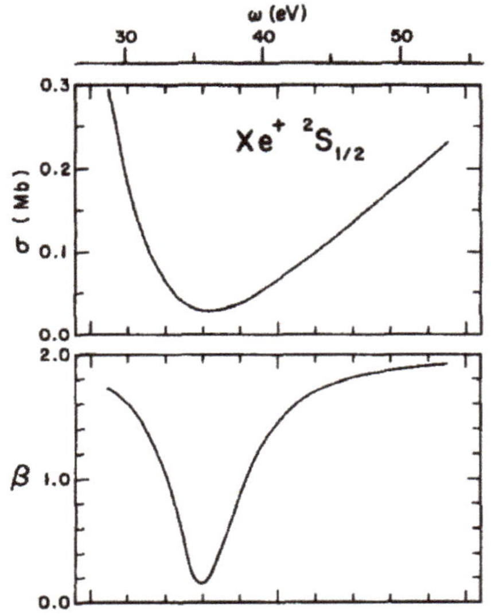

Figure 4. Xe 5s photoionization cross section (upper curve) and β parameter vs. photon energy, ω, calculated using RRPA [29].

The next example is the phenomenon of spin-orbit-activated interaction interchannel coupling (SOIAIC). Basically, this results from the interchannel coupling among the photoionization channels emanating from the two members of a spin-orbit doublet, nl, with $j = l \pm \frac{1}{2}$. The phenomenon was discovered experimentally in the photoionization of the 3d subshell of Xe [30] and subsequently explained theoretically [31]. The explanation given in [31] was based upon the nonrelativistic RPAE that was artificially made to include the spin-orbit splitting of the Xe 3d spin-orbit split levels. Subsequently, RRPA was applied and verified the explanation [32]. The results, both experimental and theoretical, are shown in Figure 5.

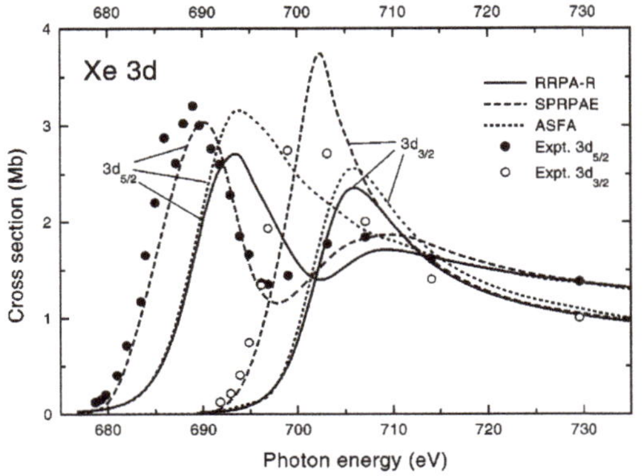

Figure 5. Photoionization cross sections for the 3d subshell of xenon. Solid lines represent calculated 13-channel RRPA (with relaxation) cross sections for the respective $3d_{5/2}$ and $3d_{3/2}$ subshells [32]. The calculated cross sections in length and velocity gauge differ by at most 3%, so that the corresponding curves are almost indistinguishable in the scale of the figure. Dashed lines represent the results of SPRPAE results of [31]. Dotted-dashed lines represent results of the ASFA calculations [30]. Closed and open circles denote the respective experimental results for $3d_{5/2}$ and $3d_{3/2}$ subshells [30].

The key point here is that both the $3d_{5/2}$ and $3d_{3/2}$ cross sections exhibit shape resonances [33] just above their respective thresholds. As a result, the $3d_{3/2}$ cross section is significantly larger than the $3d_{5/2}$ cross section just above the $3d_{3/2}$ threshold. As a general rule, when a large photoionization cross section is degenerate with a small one, the small cross section is altered owing to interchannel coupling (configuration interaction in the continuum) [34]. This is exactly what is seen here in the form of the structure in the $3d_{5/2}$ cross section just above the $3d_{3/2}$ threshold. It is evident that the relativistic spin-orbit splitting of the 3d thresholds is crucial to the existence of this SOIAIC phenomenon, and it is also clear that SOIAIC is likely to be in evidence for the photoionization of inner subshells that have a significant spin-orbit splitting and exhibit a near-threshold shape resonance. Furthermore, as a result of the fact that the photoionization cross sections of inner shells of confined atoms often have significant near-threshold maxima [35], owing to the phenomenon of confinement resonances [36], the SOIAIC should be much more generally exhibited [37].

The asymptotic branching ratios of spin-orbit doublets have been a topic of interest since the 1960s. Earlier, it was expected that far above thresholds, the ratio of cross sections for the members of a spin-orbit doublet, nl, with $j = l \pm \frac{1}{2}$, respectively, should approach $(l+1)/l$, known as the statistical value, at asymptotically high energies [38]. However, it was later shown that relativistic effects on the initial state wave functions would cause the ratio to drop below the statistical [39,40]. Recently, experimental technology has improved to the point that this prediction has been verified experimentally [41]. In addition, it had been found that, in the vicinity of inner shell thresholds, there can be very large swings of the branching ratio over relatively small energy ranges [41,42]. However, all of this phenomenology cannot occur within a nonrelativistic framework; relativistic interactions are required.

As a particular example, the calculated branching ratio for Xe 5p using the RRPA with coupling among relativistic channels from all subshells except 1s [42] is depicted in Figure 6 over a very large energy range, from threshold to 500 a.u. (approximately 13.6 keV), where it is seen that, at the highest energies, that branching ratio is about 1.6, well below the statistical value for an initial np doublet, in keeping with the earlier predictions. In addition, there are seen to be significant excursions from monotone decreasing results in the vicinity of the n = 4, n = 3 and n = 2 thresholds. These excursions are the result of interchannel coupling between the relativistic photoionization channels resulting from the $5p_j$ initial states and the channels from the inner shells. The fact that there are significant excursions of the branching ratios in these energy ranges means that the actual interchannel coupling matrix elements are themselves dependent upon relativistic interactions.

Figure 6 also shows the results of RRPA calculation including only coupling among the five 5p relativistic photoionization channels. It is evident that the truncated RRPA results agree quite well with the fully coupled branching ratios away from the inner shell thresholds but do not reproduce the significant excursions from the smooth curve in the vicinity of the inner shell thresholds, thereby showing conclusively that these excursions are the result of interchannel coupling with the inner shell photoionization channels.

The Cooper minimum [26] mentioned above in connection with Xe 5s photoionization, is a ubiquitous phenomenon that pervades the photoionization of outer and/or near-outer subshells of all of the elements of the periodic table [43,44]. Among the interesting facets of the influence of relativistic interactions which cause a single nonrelativistic Cooper minimum to be split into several relativistic Cooper minima dependent upon the total angular momentum, j, is of the initial and final states of the relativistic photoionizing transition [45,46]. In addition, the locations of Cooper minima depend very sensitively on many-body correlations, in addition to relativistic interactions; as a matter of fact, the Xe 5s Cooper minimum is below threshold in the discrete region at the level of single particle calculations [47] but appear in the continuum in calculations that include significant many-body effects [5]. Thus, RRPA is an ideal formalism to study these relativistic effects in Cooper minima. For completeness, it should be pointed out here that in the neighborhood of the Xe 5s Cooper minimum, quadrupole effects become important in the angular distribution, although the total subshell cross section is virtually unaffected. The interference between dipole and quadrupole photoionizing transitions leads to extra terms in the expression for the differential cross section, but β remains unaffected to the first order [29]. In addition, calculations of the effects of quadrupole transitions on the differential cross section (photoelectron angular distribution) using the RRPA methodology has been shown to be in good agreement with the experiment [48].

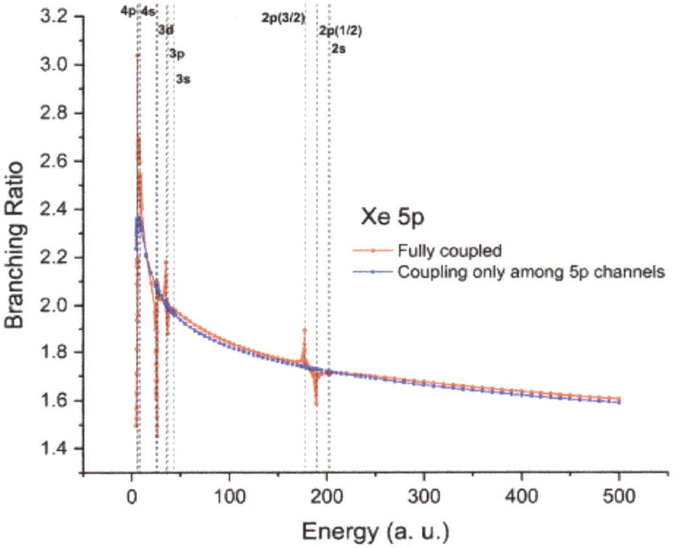

Figure 6. Photoionization cross section branching ratios for Xe $5p_{3/2}/5p_{1/2}$ calculated using RRPA with full coupling (red dots) and with only coupling among the 5p photoionization channels (blue squares) [41,42]. The vertical dashed lines indicate the thresholds.

Recent work on the splitting of Cooper minima (CM) for heavy and superheavy atoms has illustrated the importance of including correlation in the calculations [49]. As an example, given in Table 1 are the positions of the various 6s CM for six elements obtained at three levels of calculation: independent particle Dirac–Fock (DF), two-channel RRPA coupling on the two relativistic channels arising from 6s photoionization, and RRPA with full coupling of all the channels that might affect the result. To begin with, there are huge differences between the positions of the 6s $\rightarrow \varepsilon p_{3/2}$ CM and the 6s $\rightarrow \varepsilon p_{1/2}$ CM, and the differences increase with Z to an astounding degree in Og—a splitting of more than 4.5 keV. This comes about owing to the spin-orbit force which is attractive for the 6s $\rightarrow \varepsilon p_{1/2}$ final state but repulsive for the 6s $\rightarrow \varepsilon p_{3/2}$ final state. In addition, it is clear from Table 1 that correlation in the form of interchannel coupling induces rather significant changes in the location of 6s $\rightarrow \varepsilon p_{3/2}$ CM compared to the two-channel and DF results, changes that generally increase with Z, indicating the crucial importance of the interchannel coupling in the determination of the position of the CM in these heavy and superheavy elements.

Table 1. Positions of the Cooper minima (CM) in Dirac–Fock (DF), two-channel RRPA and RRPA with full coupling in photoelectron energy in (a.u.) [5].

Atoms	Photoelectron Energy (a.u.) of the 6s $\rightarrow \varepsilon p_{3/2}$ CM			Photoelectron Energy (a.u.) of the 6s $\rightarrow \varepsilon p_{1/2}$ CM		
	DF	RRPA (2 Channel)	RRPA (Full Coupling)	DF	RRPA (2 Channel)	RRPA (Full Coupling)
Hg (Z = 80)	2.73	2.17	4.17	0.55	0.67	3.67
Rn (Z = 86)	3.63	1.43	5.93			4.43
Ra (Z = 88)	4.01	1.38	6.38			3.88
No (Z = 102)	10.91	5.70	11.70			6.70
Cn (Z = 112)	32.53	29.32	24.82			4.82
Og (Z = 118)	62.12	63.02	171.02			3.52

4. Concluding Remarks

It is clear from the above exposition and examples that RRPA, which includes significant many-body correlation including initial state two-particle two-hole terms and final state interchannel coupling, can be suitably applied to a number of aspects of atomic photoionization and give physical insight into what makes the results for these processes what they are. Furthermore, it must be noted that only a few of the many examples that have been studied over the years are presented above. Thus, it is evident that the RRPA methodology has contributed greatly to our understanding of atomic photoionization as well as many other atomic processes [50]. However, it must be recalled that all of this was made possible by the pioneering work on the RPAE method by Miron Amusia and his collaborators.

Author Contributions: Both authors contributed equally to the writing. All authors have read and agreed to the published version of the manuscript.

Funding: This work of STM was supported by the US Department of Energy, Office of Science, Basic Energy Sciences under Award Number DE-FG02-03ER15428.

Institutional Review Board Statement: Not applicable.

Informed Consent Statement: Not applicable.

Data Availability Statement: All data is available from the authors upon request.

Acknowledgments: This work of STM was supported by the US Department of Energy, Office of Science, Basic Energy Sciences under Award Number DE-FG02-03ER15428. We are thankful to Sourav Banerjee and Soumyajit Saha for their help in the preparation of this article.

Conflicts of Interest: The authors declare no conflict of interest.

References

1. Amusia, M.Y. *Atomic Photoeffect*, 1st ed.; Springer: New York, NY, USA, 1990.
2. Johnson, W.R.; Lin, C.D. Relativistic random-phase approximation applied to atoms of the He isoelectronic sequence. *Phys. Rev. A* **1976**, *14*, 565. [CrossRef]
3. Johnson, W.R.; Lin, C.D.; Dalgarno, A. Allowed and Forbidden Transitions of Helium-like Ions. *J. Phys. B* **1976**, *9*, L303. [CrossRef]
4. Johnson, W.R.; Lin, C.D. Multichannel relativistic random-phase approximation for the photoionization of atoms. *Phys. Rev. A* **1979**, *20*, 964. [CrossRef]
5. Johnson, W.R.; Lin, C.D.; Cheng, K.T.; Lee, C.M. Relativistic Random-Phase Approximation. *Phys. Scr.* **1980**, *21*, 409. [CrossRef]
6. Hartree, D.R. The Wave Mechanics of an Atom with a Non-Coulomb Central Field. Part I. Theory and Methods. *Proc. Camb. Philos. Soc.* **1928**, *24*, 89. [CrossRef]
7. Fock, V. Näherungsmethode zur Lösung des quantenmechanischen Mehrkörperproblems. *Z. Phys.* **1930**, *61*, 126. [CrossRef]
8. Slater, J.C. A simplification of the Hartree–Fock method. *Phys. Rev.* **1951**, *81*, 385. [CrossRef]
9. Fischer, C.; Brage, T.; Jönsson, P. *Computational Atomic Structure: An MCHF Approach*; Taylor & Francis, Inc.: London, UK, 1997.

10. Wilson, S. Diagrammatic many-body perturbation theory of atomic and molecular electronic structure. *Comput. Phys. Rep.* **1985**, *2*, 391. [CrossRef]
11. Thouless, D.J. Green functions in low-energy nuclear physics. *Rep. Prog. Phys.* **1964**, *27*, 53. [CrossRef]
12. Bohm, D.; Pines, D. Application of Collective Treatment of Electron and Ion Vibrations to Theories of Conductivity and Superconductivity. *Phys. Rev.* **1951**, *82*, 625. [CrossRef]
13. Rowe, D.J. Equations-of-Motion Method and the Extended Shell Model. *Rev. Mod. Phys.* **1968**, *40*, 153. [CrossRef]
14. Dalgarno, A.; Victor, G.A. The time-dependent coupled Hartree–Fock approximation. *Proc. Roy. Soc. A* **1966**, *A291*, 291.
15. Amusia, M.Y.; Cherepkov, N.A. Many-electron correlations in scattering processes. *Case Stud. Atom. Phys.* **1975**, *5*, 47.
16. Pines, D.; Bohm, D. A Collective Description of Electron Interactions: II. Collective vs Individual Particle Aspects of the Interactions. *Phys. Rev.* **1952**, *85*, 338. [CrossRef]
17. Bohm, D.; Pines, D. A Collective Description of Electron Interactions: III. Coulomb Interactions in a De-generate Electron Gas. *Phys. Rev.* **1953**, *92*, 609. [CrossRef]
18. Pines, D. Collective Energy Losses in Solids. *Rev. Mod. Phys.* **1956**, *28*, 184. [CrossRef]
19. Raimes, S. The theory of plasma oscillations in metals. *Rep. Prog. Phys.* **1957**, *20*, 1. [CrossRef]
20. Raimes, S. *The Wave Mechanics of Electrons in Metals*; North-Holland: Amsterdam, The Netherlands, 1961; p. 170.
21. Gell-Mann, M.; Brueckner, K.A. Correlation Energy of an Electron Gas at High Density. *Phys. Rev.* **1957**, *102*, 364. [CrossRef]
22. Grant, I.P. *Relativistic Quantum Theory of Atoms and Molecules*; Springer: New York, NY, USA, 2007; Chapter 9.
23. Radojević, V.; Johnson, W.R. Multiconfiguration Tamm-Dancoff approximation applied to photoioniza-tion of the outer shells of Be and Mg. *Phys. Rev. A* **1985**, *31*, 2991. [CrossRef]
24. Radojević, V.; Kutzner, M.; Kelly, H.P. Photoionization of the barium 4d subshell including relativistic and relaxation effects. *Phys. Rev. A* **1989**, *40*, 727. [CrossRef]
25. Manson, S.T.; Starace, A.F. Photoelectron angular distributions: Energy dependence for s subshells. *Rev. Mod. Phys.* **1982**, *54*, 389. [CrossRef]
26. Cooper, J.W. Photoionization from Outer Atomic Subshells. A Model Study. *Phys. Rev.* **1962**, *128*, 681. [CrossRef]
27. Seaton, M.J. A comparison of theory and experiment for photo-ionization cross-sections II. Sodium and the alkali metals. *Proc. RŚoc. A* **1951**, *208*, 418.
28. Aarthi, G.; Jose, J.; Deshmukh, S.; Radojevic, V.; Deshmukh, P.C.; Manson, S.T. Photoionization study of Xe 5s: Ionization cross sections and photoelectron angular distributions. *J. Phys. B* **2014**, *47*, 025004. [CrossRef]
29. Huang, K.N.; Johnson, W.R.; Cheng, K.T. Theoretical photoionization parameters for the noble gases argon, krypton, and xenon. *At. Data Nucl. Data Tables* **1985**, *26*, 33–45. [CrossRef]
30. Kivimäki, A.; Hergenhahn, U.; Kempgens, B.; Hentges, R.; Piancastelli, M.N.; Maier, K.; Rüdel, J.; Telkki, J.J.; Bradshaw, A.M. Near-threshold study of Xe 3d photoionization. *Phys. Rev. A* **2000**, *63*, 012716. [CrossRef]
31. Amusia, M.Y.; Chernysheva, L.V.; Manson, S.T.; Msezane, A.M.; Radojević, V. Strong electron correlation in photoionization of spin-orbit doublets. *Phys. Rev. Lett.* **2002**, *88*, 093002. [CrossRef]
32. Radojević, V.; Davidović, D.M.; Amusia, M.Y. Near-threshold photoionization of the Xe 3 d spin-orbit doublet: Relativistic, relaxation, and intershell interaction effects. *Phys. Rev. A* **2003**, *67*, 022719. [CrossRef]
33. Manson, S.T.; Cooper, J.W. Photo-Ionization in the Soft x-Ray Range: Z Dependence in a Central-Potential Model. *Phys. Rev.* **1968**, *165*, 126. [CrossRef]
34. Hansen, D.L.; Hemmers, O.; Wang, H.; Lindle, D.W.; Focke, P.; Sellin, I.A.; Chakraborty, H.S.; Deshmukh, P.C.; Manson, S.T. Validity of the independent-particle approximation in x-ray photoemission: The exception, not the rule. *Phys. Rev. A* **1999**, *60*, R2641. [CrossRef]
35. Dolmatov, V.K.; Baltenkov, A.S.; Connerade, J.P.; Manson, S.T. Structure and photoionization of confined atoms. *Radiat. Phys. Chem.* **2004**, *70*, 417. [CrossRef]
36. Connerade, J.P.; Dolmatov, V.K.; Manson, S.T. On the nature and origin of confinement resonances. *J. Phys. B* **2000**, *33*, 2279. [CrossRef]
37. Keating, D.A.; Deshmukh, P.C.; Manson, S.T. Wigner time delay and spin–orbit activated confinement res-onances. *J. Phys. B* **2017**, *50*, 175001. [CrossRef]
38. Samson, J.A.; Gardner, J.L.; Starace, A.F. $^2P_{3/2}$: $^2P_{1/2}$ partial photoionization cross-section ratios in the rare gases. *Phys. Rev. A* **1975**, *12*, 1459. [CrossRef]
39. Ron, A.; Kim, Y.S.; Pratt, R.H. Subshell branching ratios of partial photoionization cross sections. *Phys. Rev. A* **1981**, *24*, 1260. [CrossRef]
40. Kim, Y.S.; Pratt, R.H.; Ron, A. Nonstatistical behavior of photoeffect subshell branching ratios at high energies. *Phys. Rev. A* **1981**, *24*, 1889. [CrossRef]
41. Püttner, R.; Martins, J.B.; Marchenko, T.; Travnikova, O.; Guillemin, R.; Journel, L.; Ismail, I.; Goldsztejn, G.; Koulentianos, D.; Céolin, D.; et al. Nonstatistical behavior of the photoionization of spin–orbit doublets. *J. Phys. B* **2021**, *54*, 085001. [CrossRef]
42. Munasinghe, C.R.; Deshmukh, P.C.; Manson S.T. Photoionization Branching Ratios of Spin-Orbit Doublets Far above Thresholds: Interchannel and Relativistic Effects in the Noble Gases. *Phys. Rev. A* **2022**, *106*, 013102. [CrossRef]
43. Manson, S.T. Systematics of zeros in dipole matrix elements for photoionizing transitions: Nonrelativistic calculations. *Phys. Rev. A* **1985**, *31*, 3698. [CrossRef]

44. Yin, R.Y.; Pratt, R.H. Survey of relativistic Cooper minima. *Phys. Rev. A* **1987**, *35*, 1149. [CrossRef]
45. Kim, Y.S.; Ron, A.; Pratt, R.H.; Tambe, B.R.; Manson, S.T. Relativistic effects in the photoionization of high-Z elements: Splittings and shifts of minima. *Phys. Rev. A* **1981**, *46*, 1326.
46. Deshmukh, P.C.; Tambe, B.R.; Manson, S.T. Relativistic Effects in the Photoionisation of Heavy Atoms: Cooper Minima. *Austral. J. Phys.* **1986**, *39*, 679. [CrossRef]
47. Kennedy, D.J.; Manson, S.T. Photoionization of the noble gases: Cross sections and angular distributions. *Phys. Rev. A* **1972**, *5*, 227. [CrossRef]
48. Hemmers, O.; Guillemin, R.; Kanter, E.P.; Krässig, B.; Lindle, D.W.; Southworth, S.H.; Wehlitz, R.; Baker, J.; Hudson, A.; Lotrakul, M.; et al. Dramatic Nondipole Effects in Low Energy Photoionization: Experimental and Theoretical Study of Xe 5s. *Phys. Rev. Lett.* **2003**, *91*, 053002. [CrossRef] [PubMed]
49. Baral, S.; Saha, S.; Dubey, K.A.; Jose, J.; Deshmukh, P.C.; Razavi, A.K.; Manson, S.T. Unusual behavior of Cooper minima of ns subshells in high-Z atoms. *Phys. Rev. A* **2022**, *105*, 062819. [CrossRef]
50. Manson, S.T. Relativistic-random-phase approximation calculations of atomic photoionization: What we have learned. *Can. J. Phys.* **2009**, *87*, 5. [CrossRef]

atoms

MDPI

Article

Spin Polarization of Electrons in Two-Color XUV + Optical Photoionization of Atoms

Nikolay M. Kabachnik [1,2,*] and Irina P. Sazhina [2]

[1] Donostia International Physics Center, E-20018 Donostia-San Sebastian, Spain
[2] Skobeltsyn Institute of Nuclear Physics, Lomonosov Moscow State University, 119991 Moscow, Russia; isazhina@mail.ru
* Correspondence: nkabach@gmail.com

Abstract: The spin polarization of photoelectrons in two-color XUV + optical multiphoton ionization is theoretically considered using strong field approximation. We assume that both the XUV and the optical radiation are circularly polarized. It is shown that the spin polarization is basically determined by the XUV photoabsorption and that the sidebands are spin polarized as well. Their polarization may be larger or smaller than that of the central photoelectron line depending on the helicity of the dressing field.

Keywords: multiphoton ionization; two-color; photoelectrons; spin polarization; femtosecond pulses; xenon atom

Citation: Kabachnik, N.M.; Sazhina, I.P. Spin Polarization of Electrons in Two-Color XUV + Optical Photoionization of Atoms. *Atoms* **2022**, *10*, 66. https://doi.org/10.3390/atoms10020066

Academic Editors: Anatoli Kheifets, Gleb Gribakin and Vadim Ivanov

Received: 6 May 2022
Accepted: 11 June 2022
Published: 20 June 2022

1. Introduction

The spin polarization of photoelectrons is one of the fundamental characteristics of photoionization processes. Starting from the pioneering paper by U. Fano [1] and the seminal papers by M. Amusia's pupil N. Cherepkov [2,3], it was realized that in spite of the weakness of the spin-orbit interaction, the spin polarization of photoelectrons may be large, of the order of unity. During the last 50 years, a large number of experimental and theoretical works have been devoted to studying the spin polarization of photoelectrons in the photoemission from atoms, molecules and solids (see [3–5] and references therein). There are two main reasons why these investigations are considered important. One is that a high degree of polarization of photoelectrons is an important prerequisite for creating sources of polarized electrons, which in turn may serve as a tool for investigating various aspects of magnetism in solids [6,7]. Another reason is that measurements of spin polarization provide additional information about the mechanism of the photoemission; in particular, they are necessary for the realization of the so-called complete experiment, i.e., the experimental determination of the complex amplitudes of photoionization [8].

Until quite recently, the majority of the experimental investigations of spin polarization of photoelectrons have been performed at synchrotron radiation sources. Since the intensity of the sources is rather low, the interpretation of these experiments was based on the linear single-photon approximation describing the interaction of the photons with a quantum system. Spin polarization in multiphoton processes has been considered theoretically first for relatively weak laser fields where perturbation theory is applicable [9–11]. Here, it was demonstrated that the degree of polarization may be high also in the multiphoton processes. With the advent of free-electron lasers with extremely high intensities of the photon beam, the possibility arises to study experimentally the spin polarization in multiphoton ionization in a wide range of photon wavelengths. A theoretical prediction of high degree of spin polarization in multiphoton strong-field ionization was presented in Ref. [12]. Recently, first experiments of this kind were reported [13–15]. Photoelectron spin polarization, in an interesting particular case of the three-photon bichromatic ($\omega + 2\omega$) ionization, was theoretically considered in papers [16,17].

A special case of multiphoton processes is the photoionization of atoms and molecules by extreme ultraviolet (XUV) or soft X-ray pulses in the presence of infrared (IR) or optical radiation [18]. In the following, in order to shorten the explanations, we shall discuss the XUV+IR two-color case, although all discussed properties and conclusions are also valid for ionizing by soft X-rays and for dressing by an optical pulse (OP). If the energy of the XUV photons in these two-color experiments is sufficient to ionize the atom, a series of sidebands appear in the photoelectron spectrum at both sides of the photoline, due to the simultaneous emission or absorption of the IR photons [19]. The energy separation between the sidebands is equal to the photon energy of the IR field. This process is convenient for studying the photoinduced transitions in the continuum. The appearance of the sidebands was used for measuring the duration and the arrival time of the XUV pulses from the free-electron lasers (FELs) [20,21]. Sidebands were also used for the determination of the circular polarization of the FEL beams [22,23] by measuring the circular dichroism of the sidebands.

As far as we know, there are no investigations of the spin polarization of photoelectrons generated in two-color XUV+IR processes, although it is clear that the photoelectrons both in the central line and in the sidebands should be spin polarized, provided that the fine structure of the lines is resolved in the experiment [3]. In this paper, we report a theoretical investigation of the spin polarization in such processes. We suggest a simple theoretical model based on the description of the XUV+IR processes in the strong field approximation (SFA) [24], which was widely used in the description of multiphoton processes [25]. We analyze the spin polarization of photoelectrons induced by circularly polarized XUV photons in the presence of collinear circularly polarized IR beams. In particular, we consider the component of the spin polarization parallel to the light helicity, the so-called "polarization transfer". This component is non-zero not only in angle-resolved measurements, but also in angle-integrated experiments [4]. We show that the spin polarization of sidebands is different from the spin polarization of the central photoline and changes with the sideband order. It strongly depends on the helicity of the IR beam.

As a particular example, we chose a short-pulse two-color photoionization of Xe atoms. The ground state of Xe contains a closed 5p electronic subshell. Upon photoionization, the lowest state of the Xe^+ ion is a spin-orbit doublet $^2P_{3/2}$ and $^2P_{1/2}$. The spin-orbit splitting is sufficiently large at 1.3 eV, which simplifies the spin-polarization measurements in which it is necessary to resolve the fine-structure components [4]. The spin polarization of photoelectrons from the single-photon ionization of Xe is well investigated both theoretically and experimentally (see, for example, [4] and references therein). Additionally, the spin polarization of emitted electrons in single-color multiphoton ionization of Xe was investigated [13–15] .

In this paper, we consider the two-color multiphoton ionization of Xe by ultrashort (femtosecond) XUV and optical pulses. In the following Section 2.1, a short description of the theoretical approach is given, which is based on the strong field approximation (SFA) [24,25]. Section 2.2 contains the parameters and details of the calculation. Section 3 presents the results of the calculations as well as a discussion including simple approximate formulas for the spin polarization of the sidebands. Finally, Section 4 gives conclusions and an outlook. In the Appendix A, we present a derivation of the approximate expressions for the related matrix elements, which is used for obtaining a simple approximate expression for the spin polarization of photoelectrons in two-color experiments.

2. Theory

2.1. Theoretical Description of Spin Polarization in Two-Color Multiphoton Ionization

Consider the photoionization of an atom by two spatially and temporally overlapping electromagnetic pulses of XUV and IR radiations. Both pulses are circularly polarized and collinear, propagating along the z-axis. To describe the interaction of these pulses with the atom, we use the strong field approximation (SFA) [24,25]. We suppose that the XUV pulse is comparatively weak so that its interaction with electrons can be considered in the first-order perturbation theory, and we use the rotating wave approximation. The

IR pulse is rather strong 10^{11}–10^{13} W/cm^2, but not strong enough to distort the bound atomic states. For a description of the atomic wave function, we use a single-active-electron approximation. The final continuum states of the emitted electron are described by the non-relativistic Volkov wave functions [26]. Note that we ignore the influence of the magnetic field of the IR pulse on the spin orientation of the emitted electron. As it was shown in paper [27], this influence is generally rather small. In this case, the amplitude of the photoionization can be written using the time-dependent distorted wave approximation as follows (we use atomic units throughout the paper unless otherwise indicated) [28]

$$\mathcal{A}_{\vec{k},m_s,m_j} = -i\int_{-\infty}^{\infty} dt\, \tilde{\mathcal{E}}_X(t)\langle\psi_{\vec{k}}(\vec{r},t)\chi_{m_s}\,|\,\hat{d}\,|\,\phi_{j,m_j}(\vec{r})\rangle e^{i(E_b-\omega_X)t}\,. \tag{1}$$

Here, $\tilde{\mathcal{E}}_X(t)$ is the envelope of the XUV pulse electric field, ω_X is its mean frequency, and E_b is the ionization potential. The matrix element $\langle\psi_{\vec{k}}(\vec{r},t)\chi_{m_s}\,|\,\hat{d}\,|\,\phi_{j,m_j}(\vec{r})\rangle$ describes a transition from the initial state of the atomic electron $\phi_{j,m_j}(\vec{r})$ with the total angular momentum and its projection j, m_j, to the final continuum state in the IR field $\psi_{\vec{k}}(\vec{r},t)\chi_{m_s}$ with the momentum $\vec{k}$ and spin state χ_{m_s}, with m_s being the projection of the spin on the z axis, and $\hat{d}$ being the dipole operator. Note that we ignore the spin-orbit interaction in the continuum.

For a circularly polarized XUV beam with the polarization vector $\vec{\epsilon}_X^{\,\pm}$, the dipole operator is given by

$$\hat{d}^{\pm} = (\vec{\epsilon}_X^{\,\pm}\,\vec{r}) = -\sqrt{4\pi/3}\,rY_{1,\pm1}(\hat{r})\,, \tag{2}$$

where plus and minus signs correspond to right- and left-circularly polarized XUV photons, respectively, and $Y_{l,m}$ is a spherical harmonic. The wave function $\psi_{\vec{k}}(\vec{r},t)$ in Equation (1) describes the "dressed" photoelectron in the laser field, which is characterized by the final (asymptotic) momentum $\vec{k}$. Within the SFA, the wave function of the photoelectron is represented by the non-relativistic Volkov wave function [26]:

$$\psi_{\vec{k}}(\vec{r},t) = \exp\Big\{i[\vec{k}-\vec{A}_L(t)]\vec{r} - i\Phi(\vec{k},t)\Big\}\,. \tag{3}$$

Here,

$$\Phi(\vec{k},t) = -\frac{1}{2}\int_t^{\infty} dt'\Big[\vec{k}-\vec{A}_L(t')\Big]^2 \tag{4}$$

with $\vec{A}_L(t)$ being the vector potential of the laser field, which we define as $\vec{A}_L(t) = \int_t^{\infty} dt'\vec{\mathcal{E}}_L(t')$ ($\vec{\mathcal{E}}_L(t)$ is the IR laser electric field vector). For circularly polarized IR laser light, $\vec{\mathcal{E}}_L(t)$ is

$$\vec{\mathcal{E}}_L(t) = \frac{1}{\sqrt{2}}\tilde{\mathcal{E}}_L(t)[\hat{x}\cos\omega_L t \pm \hat{y}\sin\omega_L t]\,, \tag{5}$$

where $\tilde{\mathcal{E}}_L(t)$ is the envelope of the laser pulse, ω_L is its mean frequency, $\hat{x}(\hat{y})$ is a unit vector along the x (y) axis, and the plus (minus) sign corresponds to the right- (left-) circularly polarized IR light.

Consider the matrix element in Equation (1) and uncouple the spin angular momentum in the initial state wave function using Clebsch–Gordan coefficients:

$$\langle\psi_{\vec{k}}(\vec{r},t)\chi_{m_s}\,|\,\hat{d}\,|\,\phi_{j,m_j}(\vec{r})\rangle = \langle\psi_{\vec{k}}(\vec{r},t)\chi_{m_s}\,|\,\hat{d}\,|\,\textstyle\sum_{m_0,m_s'}\Big(l_0m_0,\tfrac{1}{2}m_s'\,|\,jm_j\Big)\phi_{l_0,m_0}(\vec{r})\chi_{m_s'}\rangle\,. \tag{6}$$

Here $\phi_{\ell_0,m_0}(\vec{r})$ describes the initial state of the electron with the orbital angular momentum ℓ_0 and its projection m_0. Taking into account that the dipole operator does not act on the spin variables, the matrix element is reduced to

$$\langle\psi_{\vec{k}}(\vec{r},t)\chi_{m_s}\,|\,\hat{d}\,|\,\phi_{j,m_j}(\vec{r})\rangle = \langle\psi_{\vec{k}}(\vec{r},t)\,|\,\hat{d}\,|\,\phi_{l_0,m_0}(\vec{r})\rangle\Big(l_0m_0,\tfrac{1}{2}m_s\,|\,jm_j\Big)\,. \tag{7}$$

Substituting this expression into the amplitude (1), one obtains

$$\mathcal{A}_{\vec{k},m_s,m_j} = \left(l_0 m_0, \tfrac{1}{2} m_s \,|\, j m_j\right) \delta_{m_0, m_j - m_s} \mathcal{M}_{\vec{k},m_0}, \tag{8}$$

where $\delta_{m,m'}$ is the Kronecker symbol and

$$\mathcal{M}_{\vec{k},m_0} = -i \int_{-\infty}^{\infty} dt\, \tilde{\mathcal{E}}_X(t) \langle \psi_{\vec{k}}(\vec{r},t) \,|\, \hat{d} \,|\, \phi_{l_0 m_0}(\vec{r}) \rangle e^{i(E_b - \omega_X)t}. \tag{9}$$

The probability for the photoelectron to have a certain projection m_s is proportional to the square of the amplitude (8) averaged over projections m_j:

$$W^j_{m_s}(\vec{k}) = \frac{1}{2j+1} \sum_{m_j} |\left(l_0 m_0, \tfrac{1}{2} m_s \,|\, j m_j\right) \delta_{m_0, m_j - m_s} \mathcal{M}_{\vec{k},m_0}|^2. \tag{10}$$

In the following, we consider the z-component of the polarization vector, P^j_z which is parallel to the photon beam direction. This is the largest component in the considered energy range [3]. Additionally, this is the only non-zero component which characterizes the angle-integrated spin polarization of the total photoelectron flux. The degree of spin polarization (z-component of the polarization vector) is usually defined as the ratio

$$P^j_z(\vec{k}) = \frac{W^j_{1/2}(\vec{k}) - W^j_{-1/2}(\vec{k})}{W^j_{1/2}(\vec{k}) + W^j_{-1/2}(\vec{k})}. \tag{11}$$

In particular, for the photoionization of p atomic shell ($l_0 = 1$), $j = 1/2, m_j = \pm 1/2$ and $j = 3/2, m_j = \pm 1/2, \pm 3/2$. Then for $j = 1/2$ one has

$$W^{1/2}_{1/2}(\vec{k}) = \frac{1}{2}\left\{ |\frac{1}{\sqrt{3}} \mathcal{M}_{\vec{k},0}|^2 + |\frac{\sqrt{2}}{\sqrt{3}} \mathcal{M}_{\vec{k},-1}|^2 \right\}, \tag{12}$$

$$W^{1/2}_{-1/2}(\vec{k}) = \frac{1}{2}\left\{ |\frac{\sqrt{2}}{\sqrt{3}} \mathcal{M}_{\vec{k},1}|^2 + |\frac{1}{\sqrt{3}} \mathcal{M}_{\vec{k},0}|^2 \right\}. \tag{13}$$

For $j = 3/2$, one has

$$W^{3/2}_{1/2}(\vec{k}) = \frac{1}{4}\left\{ |\mathcal{M}_{\vec{k},1}|^2 + |\frac{\sqrt{2}}{\sqrt{3}} \mathcal{M}_{\vec{k},0}|^2 + |\frac{1}{\sqrt{3}} \mathcal{M}_{\vec{k},-1}|^2 \right\}, \tag{14}$$

$$W^{3/2}_{-1/2}(\vec{k}) = \frac{1}{4}\left\{ |\frac{1}{\sqrt{3}} \mathcal{M}_{\vec{k},1}|^2 + |\frac{\sqrt{2}}{\sqrt{3}} \mathcal{M}_{\vec{k},0}|^2 + |\mathcal{M}_{\vec{k},-1}|^2 \right\}. \tag{15}$$

Thus the spin polarization P_z for the case of $j = 1/2$ is

$$P^{1/2}_z(\vec{k}) = \frac{|\mathcal{M}_{\vec{k},-1}|^2 - |\mathcal{M}_{\vec{k},1}|^2}{|\mathcal{M}_{\vec{k},0}|^2 + |\mathcal{M}_{\vec{k},1}|^2 + |\mathcal{M}_{\vec{k},-1}|^2}, \tag{16}$$

while for $j = 3/2$ the polarization is

$$P^{3/2}_z(\vec{k}) = \frac{1}{2} \frac{|\mathcal{M}_{\vec{k},1}|^2 - |\mathcal{M}_{\vec{k},-1}|^2}{|\mathcal{M}_{\vec{k},0}|^2 + |\mathcal{M}_{\vec{k},1}|^2 + |\mathcal{M}_{\vec{k},-1}|^2}. \tag{17}$$

Note that in the considered model, $P^{3/2}_z(\vec{k}) = -\frac{1}{2} P^{1/2}_z(\vec{k})$, as it is in the single photon ionization [4].

For calculating the amplitudes $\mathcal{M}_{\vec{k},m_0}$ we expand the continuum wave function $\psi_{\vec{k}}$ in partial waves and apply the dipole selection rules. Then the matrix element of the dipole operator for circularly polarized light, Equation (2), and for a particular projection m_0 can be written as

$$d^{\pm}_{\vec{k}_0,m_0} = d_{\ell_0+1,m_0\pm1} Y_{\ell_0+1,m_0\pm1}(\theta_0,\phi_0) e^{i\delta_{\ell_0+1}} \\ + d_{\ell_0-1,m_0\pm1} Y_{\ell_0-1,m_0\pm1}(\theta_0,\phi_0) e^{i\delta_{\ell_0-1}} . \quad (18)$$

Here $d_{\ell_0\pm1,m_0\pm1}$ are the partial dipole amplitudes for the transitions from the initial state (ℓ_0, m_0), and $\delta_{\ell_0\pm1}$ are the photoionization phases. The angles (θ_0, ϕ_0) give the direction of electron emission from the atom before propagation in the optical laser field. These angles are connected with the detection angles (θ, ϕ) after propagation in the IR field by the relations:

$$\theta_0(t) = \arccos(k_z/k_0(t)) , \quad (19)$$

$$\exp(i\phi_0(t)) = \frac{(k_x - A_{Lx}(t)) + i(k_y - A_{Ly}(t))}{(k_0^2(t) - k_z^2)^{1/2}} ,$$

where $k_0^2(t) = (\vec{k} - \vec{A}_L(t))^2$.

In the particular case of p-subshell ionization ($\ell_0 = 1, m_0 = 0, \pm1$) in the absence of the IR field, s and d partial waves contribute. Then, collecting Equations (9), (18) and (19), one can obtain for a circularly polarized XUV pulse and a circularly polarized IR pulse the following expression

$$\mathcal{M}_{\vec{k},m_0} = -i \int_{-\infty}^{\infty} dt \tilde{\mathcal{E}}_X(t) \Big[d_{2,m_0\pm1} Y_{2,m_0\pm1}(\theta_0(t), \phi_0(t)) e^{i\delta_d} \\ + d_{0,0}\delta_{m_0\pm1,0} Y_{0,0} e^{i\delta_s} \Big] e^{i\Phi(\vec{k},t)} e^{i(E_b-\omega_X)t} , \quad (20)$$

where $\delta_{m,0}$ is a Kronecker symbol.

Suppose that the XUV pulse is right-circularly polarized (upper sign in Equation (18)), then

$$\mathcal{M}_{\vec{k},0} = -i \int_{-\infty}^{\infty} dt \tilde{\mathcal{E}}_X(t) d_{2,1} Y_{2,1}(\theta_0(t), \phi_0(t)) e^{i\delta_d} e^{i\Phi(\vec{k},t)} e^{i(E_b-\omega_X)t} , \quad (21)$$

$$\mathcal{M}_{\vec{k},1} = -i \int_{-\infty}^{\infty} dt \tilde{\mathcal{E}}_X(t) d_{2,2} Y_{2,2}(\theta_0(t), \phi_0(t)) e^{i\delta_d} e^{i\Phi(\vec{k},t)} e^{i(E_b-\omega_X)t} , \quad (22)$$

$$\mathcal{M}_{\vec{k},-1} = -i \int_{-\infty}^{\infty} dt \tilde{\mathcal{E}}_X(t) \Big[d_{2,0} Y_{2,0}(\theta_0(t), \phi_0(t)) e^{i\delta_d} + d_{0,0} Y_{0,0} e^{i\delta_s} \Big] e^{i\Phi(\vec{k},t)} e^{i(E_b-\omega_X)t} . \quad (23)$$

2.2. Choice of Parameters and Details of Calculations

In particular calculations for Xe atoms, the matrix elements $d_{2,2}$, $d_{2,1}$ and $d_{2,0}$ and phases δ_d and δ_s were calculated using the Herman–Skillman potential [29] in the non-relativistic single electron model. To test the results, we calculated the spin polarization of photoelectrons by expression (16) but for a negligibly small IR field. The results agree well with more advanced calculations by Cherepkov [3] within the time-independent RPAE.

The XUV pulse was assumed to be of a Gaussian shape:

$$\tilde{\mathcal{E}}_X(t) = \exp[-(t - t_0)^2/(2\tau_X^2) , \quad (24)$$

where t_0 determines the delay of the XUV pulse relative to the optical pulse (OP), and τ_X determines its duration. The duration FWHM was taken to be 23 fs. The duration of the OP is 60 fs, and we suppose that the XUV pulse is at the middle of the optical pulse.

The fine structure spitting of the 5p state in Xe is 1.3 eV. Therefore, in order to avoid overlapping of the two series of sidebands from $5p_{3/2}$ and $5p_{1/2}$ ionization, the optical photon energy is set equal to 3.1 eV as it was chosen in a recent experiment [15]. At this photon energy, the two combs of sidebands are clearly separated, which makes it possible to determine the spin polarization for each peak [15]. For the illustrations below, we chose the electron emission angle of 90°, where the number of sidebends is maximal. We also present the angle-integrated polarization.

The calculated cross section and spin polarization were convoluted with a Gaussian function which imitates the energy resolution of about 0.7 eV in a possible experiment.

In the following, we assume that the XUV pulse is right circularly polarized, while the helicity of the optical pulse is changing. Below, we present the calculated spin polarization component $P_z^{1/2}$ for the photoionization of the Xe($5p_{1/2}$) state. We remind that the spin polarization for the second component of the spin-orbit doublet Xe($5p_{3/2}$) can be easily obtained multiplying the calculated results by $-1/2$. If in an experiment the spin-orbit doublet is not resolved, the spin polarization of the photoelectrons is negligibly small [4].

3. Results for Ionization of Xe and Discussion

Figure 1a shows the results of the calculations for the Xe($5p_{1/2}$) ionization at the XUV photon energy of 38.4 eV (photoelectron energy without optical field is 25 eV) and emission angle of 90°. At this energy, the calculated photoelectron spin polarization in the absence of the optical field is $P_z^{1/2} = 0.35$. The dressing laser intensity is 5×10^{11} W/cm^2 and photon energy is 3.1 eV. At this intensity, there are two noticeable sidebands. We remind that the XUV pulse is right-circularly polarized. When the OP is also right-circularly polarized, the electron polarization at the sidebands becomes smaller (green solid line) than at the central line. On the contrary, for the left-circular polarization of optical beam, the spin polarization at the sideband is larger than at the central line. The polarization of the central line is increasing (decreasing) when the OP is right- (left-) circularly polarized in comparison with the case of the negligible optical field. We note that the calculated polarization is practically constant along the spectral line and changes abruptly to another value at the other line.

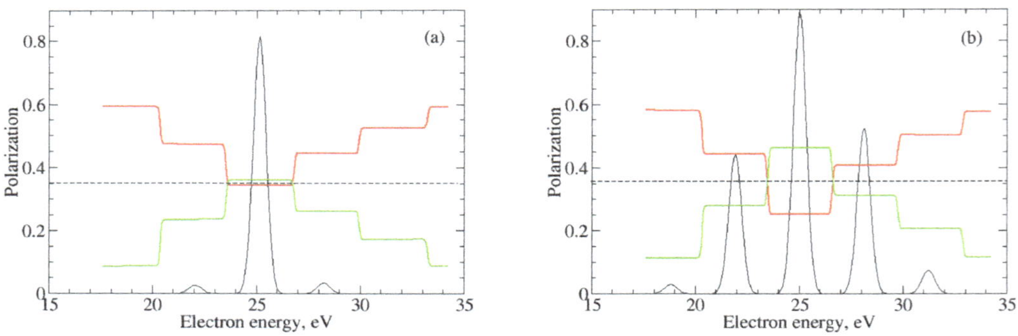

Figure 1. The photoelectron spin polarization component $P_z^{1/2}$ for the XUV photon energy $\hbar\omega_X = 38.4$ eV and the emission angle of 90°. Red line: left-circularly polarized dressing optical pulse (OP), green line: right-circularly polarized OP. The thin solid line shows electron spectrum in arbitrary units. The OP intensity is panel (**a**) 0.5×10^{12} W/cm^2, panel (**b**) 5×10^{12} W/cm^2. The dashed straight line shows the spin polarization when the OP is absent.

Figure 1b shows similar values but for larger intensity 5×10^{12} W/cm^2. With the increase in intensity, the number of sidebands is increasing. Qualitatively, the polarization results are similar to the previous case. Note that the shift of polarization for the central line increases. From the other side, the polarization of the sidebands exhibits a weak dependence on the laser field intensity.

The angular distributions of the spin polarization for the same parameters as in Figure 1b are shown in Figure 2 (upper panel) for the central line (CL), first high-energy sideband (SB1) and first low-energy sideband (SB-1). One can see that in all cases, the spin polarization component $P_z^{1/2}$ has a minimum at the emission angle of 90°. The minimal value of polarization is different for left- and right-circularly polarized optical pulses, as it is seen also in Figure 1b. In Figure 2 (lower panel), the angular distribution of the electron yield is shown for the same parameters as in the upper panel. Interestingly, the behavior of the polarization near angles 0° and 180° is drastically different for the central line, where $P_z^{1/2} \to 1$, and for the first high-energy sideband at the left-polarized IR pulse and first low-energy sideband at the right-polarized IR pulse, where $P_z^{1/2} \to 0$.

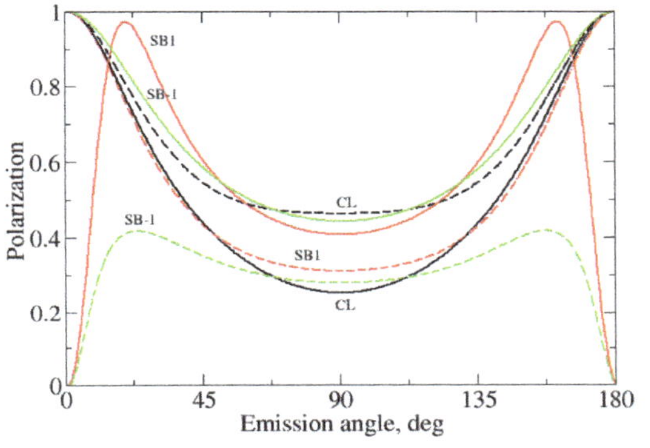

(a)

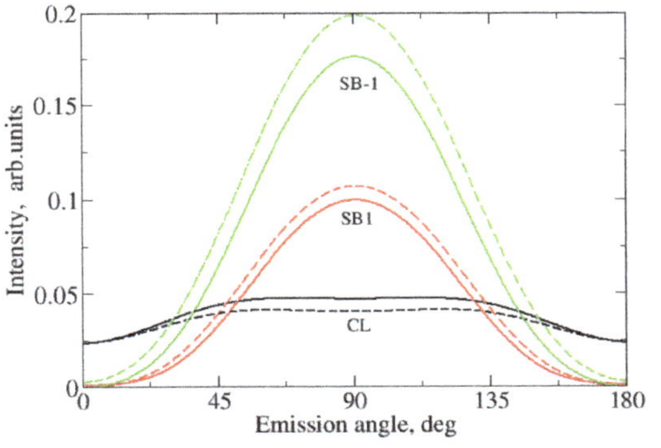

(b)

Figure 2. Angular distribution of the spin-polarization component $P_z^{1/2}$ (**a**) and of the electron yield (**b**) for the XUV photon energy of 38.4 eV (the photoelectron energy without optical field is 25 eV) and the OP intensity 5.0×10^{12} W/cm^2. Solid lines correspond to the left-circularly polarized OP, dashed lines, to the right-circularly polarized. CL denotes central line. SB1—high-energy sideband. SB-1—low-energy sideband.

This behavior may be explained using conservation of the angular momentum projection. Indeed, in our non-relativistic model, the projection of the orbital angular momentum is conserved. For the central line and right-circularly polarized XUV pulse, this gives $M_i + m_e = +1$, where M_i and m_e are projections of the ion and the electron orbital angular momenta. If the electron is emitted at zero angle, $m_e = 0$ and therefore, $M_i = 1$. Then in the final state $^2P_{1/2}$, the projection of spin S_i must be $M_s = -1/2$. Since in photoionization the spin projection is also conserved, $M_s + s_e = 0$ and therefore $s_e = 1/2$, i.e., $P_z^{1/2} = +1$. For the first high-energy sideband and left-circularly polarized optical photon, the sum of the projections of XUV and laser photons is zero, and therefore $M_i + m_e = 0$. In the forward emission, $m_e = 0$, thus $M_i = 0$ and the spin projection may be $M_s = \pm 1/2$ with equal probability, which means that $P_z^{1/2} = 0$. Similarly one can show that for the first low-energy sideband and the right- circularly polarized OP photon, the polarization is also zero in the forward direction. We note that the behavior of the spin polarization in forward and backward directions, although interesting, is practically not very important since the cross section of the sideband formation at emission angles 0° and 180° is negligibly small for circularly polarized light (see lower panel of Figure 2).

Figure 3 shows the polarization for the electron energy of 35 eV and OP intensity 1×10^{12} W/cm^2. All other parameters are the same as in Figure 1. For this energy, the electron polarization without laser field is larger [3], $P_z^{1/2} \approx 0.7$. When the optical radiation is switched on, the polarization for sidebands is changing similarly to the previous case.

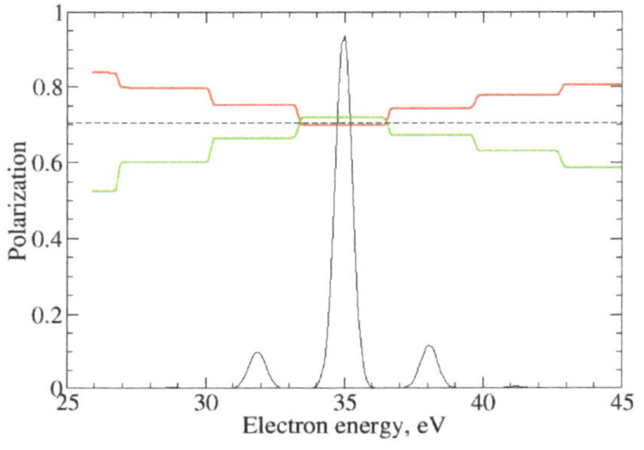

Figure 3. The same as in Figure 1 but for the XUV photon energy 48.4 eV (the photoelectron energy without optical field is 35 eV) and the OP intensity is 1.0×10^{12} W/cm^2. Red line—left-circularly polarized OP, green line—right circularly polarized OP.

The total angle-integrated polarization for sideband electrons behaves qualitatively similarly to that for the emission angle of 90°. As an example, the angle integrated polarization is shown in Figure 4 for the electron energy of 25 eV and laser intensity of 5.0×10^{12} W/cm^2.

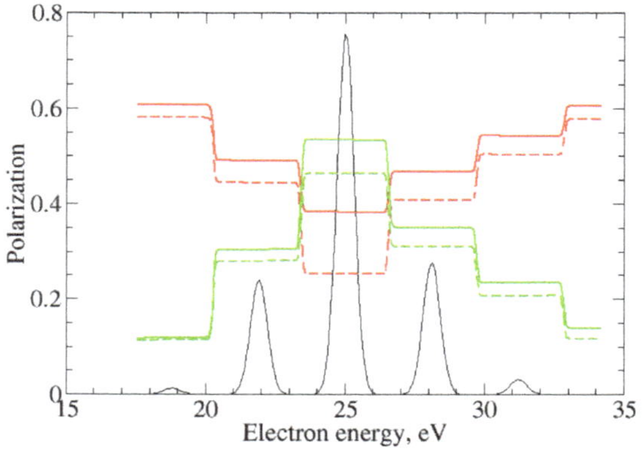

Figure 4. Total spin polarization $P_z^{1/2}$ (solid lines) and spin polarization at 90° (dashed lines) at the XUV photon energy of 38.4 eV. Red lines: left-circularly polarized OP, green lines: right-circularly polarized OP. The IR intensity is 5.0×10^{12} W/cm^2. The thin solid line shows the electron spectrum in arbitrary units.

The general behavior of the total (angle integrated) spin polarization is similar to that of the polarization at 90° emission angle. However, the total polarization is larger than at 90°, since the angular distribution of the polarization $P_z^{1/2}$ has a minimum in the direction perpendicular to the beam.

Qualitatively, the above discussed behavior of the spin polarization of the photoelectrons in the dressing laser field can be explained as follows. If one assumes that both the XUV pulse and the laser pulse are sufficiently long (i.e., contain many oscillations of the electric field), then following the procedure described in paper [30], one can obtain analytical expressions for the matrix elements $\mathcal{M}_{\vec{k},m}$. We also make additional approximations. First, we take into account that we consider electron emission at 90°. Thus we can neglect the contribution of the matrix element $\mathcal{M}_{\vec{k},0}$ (Equation (21)), containing the spherical function $Y_{2,1}$ which is small around 90°. In the matrix element $\mathcal{M}_{\vec{k},-1}$ (Equation (23)), we neglect the term containing $d_{2,0}Y_{2,0}$, since in the considered energy range $d_{2,0} << d_{0,0}$ (i.e., the contribution of the d partial wave is much smaller than the contribution of the s partial wave). Then, using the Jacobi–Anger expansion in terms of Bessel functions, the matrix elements $\mathcal{M}_{\vec{k},-1}$ and $\mathcal{M}_{\vec{k},+1}$ can be presented as a sum of contributions corresponding to different sidebands (for the derivation of these expressions, see the Appendix A)

$$\mathcal{M}_{\vec{k},-1} = -i\frac{1}{\sqrt{4\pi}}\, d_{00} \sum_{n=-\infty}^{+\infty} \tilde{\mathcal{E}}_X^{(n)} i^n J_n(q)\,, \tag{25}$$

$$\mathcal{M}_{\vec{k},1} = -i\frac{\sqrt{15}}{\sqrt{32\pi}}\, d_{22} \sum_{n=-\infty}^{+\infty} \tilde{\mathcal{E}}_X^{(n)}\, i^n \left[J_n(q) \pm \frac{2A_L}{k} J_n'(q)\right]. \tag{26}$$

Here $J_n(q)$ and $J_n'(q)$ are Bessel functions and their derivatives, $q = \frac{kA_L}{\omega_L}$, where A_L is the amplitude of the laser vector potential, and

$$\tilde{\mathcal{E}}_X^{(n)} = \int dt \tilde{\mathcal{E}}_X(t) \exp[i(E_b - \omega_X + k^2/2 + n\omega_L)t]\,. \tag{27}$$

For a long pulse, the right-hand side of the last equation is close to a delta function, which expresses the energy conservation. Each term in the sums (25) and (26) presents the contribution of the n-th sideband. If one neglects the interference between different terms and the contribution from the neighboring terms, then the squares of the matrix elements (25) and (26) for the sideband number n are

$$|\mathcal{M}_{\vec{k},-1}(n)|^2 = \frac{|d_{0,0}|^2}{4\pi}\, [J_n(q)]^2\,, \tag{28}$$

$$|\mathcal{M}_{\vec{k},1}(n)|^2 = \frac{15|d_{2,2}|^2}{32\pi}\, [J_n(q) \pm \frac{2A_L}{k} J_n'(q)]^2\,. \tag{29}$$

Ignoring small terms of the order $(\frac{A_L}{k})^2$, one can write the spin polarization $P_z^j(n)$ of the n-th sideband:

$$P_z^{1/2}(n) \approx \frac{1 - \alpha(1 \pm 4\frac{A_L}{k}\frac{J_n'(q)}{J_n(q)})}{1 + \alpha(1 \pm 4\frac{A_L}{k}\frac{J_n'(q)}{J_n q})}\,, \tag{30}$$

where $\alpha = \frac{15}{8}|d_{2,2}|^2/|d_{0,0}|^2$. Taking into account that $\alpha << 1$, one can finally obtain the following simple approximate equation for the electron spin polarization at the n-th sideband:

$$P_z^{1/2}(n) \approx 1 - 2\alpha\left[1 \pm 4\frac{A_L}{k}\frac{J_n'(q)}{J_n(q)}\right]. \tag{31}$$

Upper (lower) sign in Equation (31) corresponds to the right- (left-) circularly polarized laser pulse. (We remind that the XUV pulse is assumed to be right-circularly polarized.)

For sidebands with $n \neq 0$, and for small q, Equation (31) can be rewritten as

$$P_z^{1/2}(n) \approx 1 - 2\alpha\left[1 \pm 4\frac{|n|\omega_L}{k^2}\right]. \quad (32)$$

It follows that for the right-circularly polarized IR pulse, the electron polarization diminishes with the increase in $|n|$, while for the left polarized pulse, it increases, as we see in Figures 1 and 3. Additionally, the polarization of the sidebands does not depend on the IR laser intensity.

For the central line, the situation is different. Here, $J_0'(q)/J_0(q) = -J_1(q)/J_0(q)$ and at small q

$$P_z^{1/2}(n=0) \approx 1 - 2\alpha\left[1 \mp 2\frac{A_L^2}{\omega_L}\right]. \quad (33)$$

Thus, the electron polarization in the central line is larger for pulses with the same helicity and smaller when the pulses have different helicity, and the difference of polarizations increases with the increase in laser intensity. This is clearly seen in Figures 1 and 3.

4. Conclusions

We presented a theoretical analysis of the spin polarization of photoelectrons in the two-color ionization of atoms by short circularly polarized XUV + optical pulses. Based on the non-relativistic SFA model, we showed that spin polarization of the main line and the sidebands can be large. Basically, it is determined by the spin polarization induced by the XUV ionization. Spin polarization in the sidebands depends strongly on the helicity of the pulses and varies with the order of the sideband. Simple approximate expressions are suggested for the spin polarization of the main line and the sidebands. As an example, the spin polarization of the 5p photoelectrons from Xe atoms is calculated. Measurements of the spin polarization in two-color experiments could give important information about the photo processes in the continuum.

Author Contributions: All authors have contributed to the manuscript equally. All authors have read and agreed to the published version of the manuscript.

Funding: This research received no external funding

Acknowledgments: The authors are grateful to M. Meyer for numerous useful discussions and for critical reading the manuscript. NMK is grateful to DESY (Hamburg) for hospitality and for the financial support and to Donostia International Physics Center (DIPC) for the financial support .

Conflicts of Interest: The authors declare no conflict of interests.

Appendix A

Let derive expressions (25) and (26) for the case of a long IR pulse. Suppose that the IR pulse is circularly polarized, then the pulse electric field may be presented as

$$\mathcal{E}_L(t) = g(\alpha t)\frac{\bar{\mathcal{E}}_L}{\sqrt{2}}[\hat{x}\cos(\omega_L t) \pm \hat{y}\sin(\omega_L t)], \quad (A1)$$

where $\bar{\mathcal{E}}_L$ is the field amplitude and upper (lower) sign corresponds to right- (left-) circularly polarized IR field. The corresponding vector potential is

$$A_L(t) = g(\alpha t)A_L[\hat{x}\sin(\omega_L t) \mp \hat{y}\cos(\omega_L t)] \quad (A2)$$

with $A_L = -\bar{\mathcal{E}}_L/\sqrt{2}\omega_L$. Here, and in Equation (A1), we introduced an auxiliary function $g(x)$ which is smooth, equal to unity at small x and tends to zero limit at large $|x|$. It allows us to calculate the integral (4) when $t \to \pm\infty$, and $\alpha \to 0$. In the following, we assume that $k \gg A_L$ and ignore the quadratic term A_L^2 in Equation (4). In this approximation, taking into account that $k_x = k \sin\theta \cos\phi$ and $k_y = k \sin\theta \sin\phi$, the Volkov phase can be presented as

$$\Phi(\vec{k}, t) = \frac{k^2}{2} t + \frac{k A_L}{\omega_L} \sin\theta \cos(\phi \mp \omega_L t). \tag{A3}$$

Substituting this expression into Equations (22) and (23) and using the Jacobi–Anger expansion,

$$\exp(i\kappa \cos\alpha) = \sum_{n=-\infty}^{+\infty} i^n \exp(in\alpha) J_n(\kappa), \tag{A4}$$

where $J_n(\kappa)$ is the Bessel function, one obtains for $\mathcal{M}(\vec{k}, -1)$

$$\begin{aligned} \mathcal{M}_{\vec{k},-1} &= -i \sum_{n=-\infty}^{+\infty} \int_{-\infty}^{\infty} dt\, \tilde{\mathcal{E}}_X(t) d_{00} Y_{0,0} i^n \exp(in\phi) \\ &\times \exp(\mp in\omega_L t) J_n(q) \exp\left[i(E_b + \frac{k^2}{2} - \omega_X)t\right] \\ &= -id_{00} \frac{1}{\sqrt{4\pi}} \sum_{n=-\infty}^{+\infty} i^n \exp(\mp in\phi)\, J_n(q)\, \tilde{\mathcal{E}}_X^{(n)}, \end{aligned} \tag{A5}$$

where $q = k A_L \sin\theta / \omega_L$ and the following notation is introduced:

$$\tilde{\mathcal{E}}_X^{(n)} = \int_{-\infty}^{\infty} dt \tilde{\mathcal{E}}_X(t) \exp\left[i\left(E_b - \omega_X + \frac{k^2}{2} + n\omega_L\right)t\right]. \tag{A6}$$

Similarly for $\mathcal{M}(\vec{k}, 1)$ one obtains

$$\begin{aligned} \mathcal{M}_{\vec{k},1} &= -i \sum_{n=-\infty}^{+\infty} \int_{-\infty}^{\infty} dt\, \tilde{\mathcal{E}}_X(t) d_{2,2} Y_{2,2}(\theta_0, \phi_0) i^n \exp(in\phi) \\ &\times \exp(\mp in\omega_L t) J_n(q) \exp\left[i(E_b + \frac{k^2}{2} - \omega_X)t\right]. \end{aligned} \tag{A7}$$

The spherical harmonic $Y_{2,2}(\theta_0, \phi_0)$ can be expressed in terms of angles θ, ϕ using Equation (19) as follows:

$$\begin{aligned} Y_{2,2}(\theta_0, \phi_0) &\equiv \sqrt{\frac{15}{32\pi}} \sin^2\theta_0 \exp(i2\phi_0) \\ \approx \sqrt{\frac{15}{32\pi}} &\left[\sin^2\theta \exp(2i\phi) \pm 2i \frac{A_L}{k} \sin\theta \exp(i\phi) \exp(\pm i\omega_L t) \right. \\ &\left. + 2\frac{A_L}{k} \sin^3\theta \exp(2i\phi) \sin(\omega_L t \mp \varphi)\right]. \end{aligned} \tag{A8}$$

Here, the upper (lower) sign corresponds to the right (left) circular polarization of the IR field, and we kept only linear terms in A_L/k, which is considered to be small $A_L/k \ll 1$.

Substituting this expression into Equation (22), one obtains the matrix element $\mathcal{M}_{\vec{k},1}$

$$\begin{aligned} \mathcal{M}_{\vec{k},1} &= -id_{2,2} \sqrt{\frac{15}{32\pi}} \sum_{n=-\infty}^{+\infty} i^n \exp(\mp in\phi) J_n(q) \\ \times &\left[\sin^2\theta \exp(2i\phi) \tilde{\mathcal{E}}_X^{(n)} \pm i \frac{A_L}{k} \sin\theta (2 - \sin^2\theta) \exp(i\phi)\, \tilde{\mathcal{E}}_X^{(n\pm1)} \right. \\ &\left. \pm i \frac{A_L}{k} \sin^3\theta\, \exp(3i\phi) \tilde{\mathcal{E}}_X^{(n\mp1)}\right]. \end{aligned} \tag{A9}$$

This expression can be rewritten by rearranging the terms in the sums as

$$\mathcal{M}_{\vec{k},1} = -id_{2,2}\sqrt{\frac{15}{32\pi}} \sum_{n=-\infty}^{+\infty} \tilde{\mathcal{E}}_X^{(n)} i^n \exp[i(2\mp n)\phi] \times \left[\sin^2\theta J_n(q) + \frac{A_L}{k}(2-\sin^2\theta)\, J_{n\mp1}(q) - \frac{A_L}{k}\sin^3\theta\, J_{n\pm1}(q) \right]. \tag{A10}$$

From Equations (A5) and (A10), by setting $\theta = 90^\circ$ and $\phi = 0$ and using identity $J_{n-1}(q) - J_{n+1}(q) = 2J_n'(q)$, one obtains equations

$$\mathcal{M}_{\vec{k},-1} = -i\frac{1}{\sqrt{4\pi}} d_{00} \sum_{n=-\infty}^{+\infty} \tilde{\mathcal{E}}_X^{(n)} i^n\, J_n(q)\,, \tag{A11}$$

$$\mathcal{M}_{\vec{k},1} = -i\frac{\sqrt{15}}{\sqrt{32\pi}} d_{22} \sum_{n=-\infty}^{+\infty} \tilde{\mathcal{E}}_X^{(n)}\, i^n \left[J_n(q) \pm \frac{2A_L}{k} J_n'(q) \right], \tag{A12}$$

which coincide with Equations (25) and (26).

References

1. Fano, U. Spin orientation of photoelectrons ejected by circularly polarized light. *Phys. Rev.* **1969**, *79*, 131. [CrossRef]
2. Cherepkov, N.A. Spin polarisation of photoelectrons ejected from unpolarised atoms. *J. Phys. B At. Mol. Phys.* **1979**, *12*, 1279. [CrossRef]
3. Cherepkov, N.A. Spin polarization of atomic and molecular photoelectrons.*Adv. At. Mol. Phys.* **1983**, *19*, 395–447.
4. Heinzmann, U.; Cherepkov, N.A. Spin polarization in photoionization. In *VUV and Soft X-ray Photoionization*; Becker U., Shirley D.A., Eds.; Plenum Press: New York, NY, USA, 1996; pp. 521–559.
5. Heinzmann, U.; Dill, J.H. Spin-orbit-induced photoelectron spin polarization in angle-resolved photoemission from both atomic and condensed matter targets. *J. Phys. Condens. Matter* **2012**, *24*, 173001. [CrossRef]
6. Feder, R. *Polarized Electrons in Surface Physics*; World Scientific: Singapore, 1985.
7. Kirschner, J. *Polarized Electrons at Surfaces*; Springer: Berlin, Germany, 1985.
8. Kleinpoppen, H.; Lohmann, B.; Grum-Grzhimailo, A.N. *Perfect/Complete Scattering Experiments*, Springer Series on Atomic, Optical and Plasma Physics; Springer: Berlin/Heidelberg, Germany, 2013; Volume 75.
9. Lambropoulos, P. Spin-orbit coupling and photoelectron polarization in multiphoton ionization of atoms. *Phys. Rev. Lett.* **1973**, *30*, 413. [CrossRef]
10. Lambropoulos, P. On producing totally polarized electrons through multiphoton ionization. *J. Phys. B At. Mol. Phys.* **1974**, *7*, L33. [CrossRef]
11. Nakajima, T.; Lambropoulos, P. Electron spin-polarization in single-, two- and three-photon ionization of xenon. *Europhys. Lett.* **2002**, *57*, 25. [CrossRef]
12. Barth, I.; Smirnova, O. Spin-polarized electrons produced by strong-field ionization. *Phys. Rev. A* **2013**, *88*, 013401. [CrossRef]
13. Hartung, A.; Morales, F.; Kunitski, M.; Henrichs, K.; Laucke, A.; Richter, M.; Jahnke, T.; Kalinin, A.; Schöffler, M.; Schmidt, L.H.; et al. Electron spin polarization in strong-field ionization of xenon atoms. *Nat. Photon.* **2016**, *10*, 526. [CrossRef]
14. Liu, M.-M.; Shao, Y.; Han, M.; Ge, P.; Deng, Y.; Wu, C.H.; Gong, Q.; Liu, Y. Energy- and momentum-resolved photoelectron spin polarization in multiphoton ionization of Xe by circularly polarized fields. *Phys. Rev. Lett.* **2018**, *120*, 043201. [CrossRef]
15. Trabert, D.; Hartung, A.; Eckart, S.; Trinter, F.; Kalinin, A.; Schöffler, M.; Schmidt, L.P.H.; Jahnke, T.; Kunitski, M.; Dörner, R. Spin and Angular Momentum in Strong-Field Ionization. *Phys. Rev. Lett.* **2018**, *120*, 043202. [CrossRef]
16. Gryzlova, E.V.; Popova, M.M.; Grum-Grzhimailo, A.N. Spin polarization of photoelectrons in bichromatic extreme-ultraviolet atomic ionization. *Phys. Rev. A* **2020**, *102*, 053116. [CrossRef]
17. Popova, M.M.; Gryzlova, E.V.; Kiselev, M.D.; Grum-Grzhimailo, A.N. Symmetry violation in bichromatic ionization by a free-electron laser: Photoelectron angular distribution and spin polarization. *Symmetry* **2021**, *13*, 1015. [CrossRef]
18. Meyer, M.; Costello, J.T.; Düsterer, S.; Li, W.B.; Radcliffe, P. Two-colour experiments in the gas phase. *J. Phys. B At. Mol. Opt. Phys.* **2010**, *42*, 194006. [CrossRef]
19. Glover, T.E.; Schoenlein, R.W.; Chin, A.H.; Shank, C.V. Observation of laser assisted photoelectric effect and femtosecond high order harmonic radiation. *Phys. Rev. Lett.* **1996**, *76*, 2468. [CrossRef]
20. Meyer, M.; Cubaynes, D.; O'Keeffe, P.; Luna, H.; Yeates, P.; Kennedy, E.T.; Costello, J.T.; Orr, P.; Taïeb, R.; Maquet, A.; et al. Two-color photoionization in xuv free-electron and visible laser fields. *Phys. Rev. A* **2006**, *74*, 011401(R). [CrossRef]

21. Düsterer, S.; Radcliffe, P.; Bostedt, C.; Bozek, J.; Cavalieri, A.L.; Coffee, R.; Costello, J.T.; Cubaynes, D.; DiMauro, L.F.; Ding, Y.; et al. Femtosecond x-ray pulse length characterization at the Linac Coherent Light Source free-electron laser. *New J. Phys.* **2011**, *13*, 093024. [CrossRef]
22. Mazza, T.; Ilchen, M.; Rafipoor, A.J.; Callegari, C.; Finetti, P.; Plekan, O.; Prince, K.C.; Richter, R.; Danailov, M.B.; Demidovich, A.; et al. Determining the polarization state of an extreme ultraviolet free-electron laser beam using atomic circular dichroism. *Nat. Commun.* **2014**, *5*, 4648. [CrossRef]
23. Hartmann, G.; Lindahl, A.O.; Knie, A.; Hartmann, N.; Lutman, A.A.; MacArthur, J.P.; Shevchuk, I.; Buck, J.; Galler, A.; Glownia, J.M.; et al. Circular dichroism measurements at an x-ray free-electron laser with polarization control. *Rev. Sci. Instrum.* **2016**, *87*, 083113. [CrossRef]
24. Keldysh, L.V. Ionization in the field of a strong electromagnetic wave. *Sov. Phys. JETP* **1965**, *20*, 1307.
25. Amini, K.; Biegert, J.; Calegari, F.; Chacón, A.; Ciappina, M.F.; Dauphin, A.; Efimov, D.K.; de Morisson Faria, C.F.; Giergiel, K.; Gniewek, P. Symphony on strong field approximation. *Rep. Prog. Phys.* **2019**, *82*, 116001. [CrossRef]
26. Wolkow, D.M. Über eine Klasse von Lösungen der Diracschen Gleichung. *Z. Phys.* **1935**, *94*, 250. [CrossRef]
27. Walser, M.W.; Urbach, D.J.; Hatsagortsyan, K.Z.; Hu, S.X.; Keitel, C.H. Spin and radiation in intense laser fields. *Phys. Rev. A* **2002**, *65*, 043410. [CrossRef]
28. Kazansky, A.K.; Sazhina, I.P.; Kabachnik, N.M. Angle-resolved electron spectra in short-pulse two-color XUV+IR photoionization of atoms. *Phys. Rev. A* **2010**, *82*, 033420. [CrossRef]
29. Herman, F.; Skillman, S. *Atomic Structure Calculations*; Prentice-Hall: Englewood Cliffs, NJ, USA, 1963.
30. Kazansky, A.K.; Grigorieva, A.V.; Kabachnik, N.M. Dichroism in short-pulse two-color XUV plus IR multiphoton ionization of atoms. *Phys. Rev. A* **2012**, *85*, 053409. [CrossRef]

atoms

MDPI

Article

Spectroscopic Peculiarities at Ionization of Excited $2p^5(^2P_{J_f})3s[K]_{0,1,2}$ States of Ne: Cooper Minima and Autoionizing Resonances

Maria M. Popova [1,2], Maksim D. Kiselev [1,2,3], Sergei M. Burkov [4], Elena V. Gryzlova [2,*] and Alexei N. Grum-Grzhimailo [2,3]

1 Faculty of Physics, Lomonosov Moscow State University, 119991 Moscow, Russia
2 Skobeltsyn Institute of Nuclear Physics, Lomonosov Moscow State University, 119991 Moscow, Russia
3 School of Physics and Engineering, ITMO University, 197101 Saint Petersburg, Russia
4 Laboratory for Modeling of Quantum Processes, Pacific National University, 680035 Khabarovsk, Russia
* Correspondence: gryzlovaev@depni.sinp.msu.ru; Tel.: +7-905-763-2237

Abstract: An extensive study of photoionization from neon excited states was performed. The R-matrix approach was applied to calculate a photoionization cross-section from the metastable $2p^5(^2P_{J_f})3s[K]_{0,2}$ and dipole-allowed $2p^5(^2P_{J_f})3s[K]_1$ states. The resonance structures and Cooper minimum accessible in photoionization from the excited states by the photons with energy below 30 eV were analyzed. The parameters of the lowest autoionizing states (AISs) of even parity were extracted by fitting of the photoionization cross-section. For the dipole-allowed states, calculations are presented for unpolarized, linearly and circularly polarized radiation.

Keywords: photoionization; neon; excited states; spectroscopy; Cooper minimum; electrons correlations; R-matrix; autoionizing state; metastable state

Citation: Popova, M.M.; Kiselev, M.D.; Burkov, S.M.; Gryzlova, E.V.; Grum-Grzhimailo, A.N. Spectroscopic Peculiarities at Ionization of Excited $2p^5(^2P_{J_f})3s[K]_{0,1,2}$ States of Ne: Cooper Minima and Autoionizing Resonances. *Atoms* **2022**, *10*, 102. https://doi.org/10.3390/atoms10040102

Academic Editors: Anatoli Kheifets, Gleb Gribakin and Vadim Ivanov

Received: 30 August 2022
Accepted: 20 September 2022
Published: 26 September 2022

1. Introduction

The investigations of photoprocesses from excited states being particularly interesting for studies in non-linear optics, plasma physics and the interpretation of astrophysical data from planet and stellar atmospheres [1] are suppressed by the fact that their relaxation may occur faster than photoionization. Up to recently, there were two frameworks overcoming this obstacle: measuring photoionization of metastable states and pump-probe experiments. Both schemes are restricted in terms of states they may be applied to. The first one because only the lowest atomic states are metastable, and the second one because at least one of the pump-probe fields should be intense, and therefore is supposed to be in the optic region. The extensive review of both the experimental and theoretical research devoted to ionization from excited states is presented in [2]. With the advent of Free Electron Laser facilities (FEL), the available set of states has extended enormously. Due to such sources of high-intensity radiation, the photoionization of excited states can be studied in two-photon resonance ionization.

It was M.Ya. Amusia [3] who pointed out that, opposite to common expectations that with the increase of a shell quantum number photoionization cross-sections would tend to become hydrogen-like, ionization from an excited state may manifest even more characteristic features: "The deviations of many-electron atoms from hydrogen-like ones are still essential even for comparatively high values of the ionised level's quantum number n for any frequencies in the vicinity of the threshold and far away from it".

In particular, Cooper minimum [4,5] may appear even in the $l \to l-1$ channel [6] in photoionization cross-sections from excited states, in contrast to the photoionization from ground states [7,8]. The Cooper minimum is one of the most famous among characteristic features, which can be explained in terms of single-electron amplitudes, but its position, form and depth are very sensitive to electron–electron correlations.

Electron–electron correlations are known to be of a great importance for the explanation of characteristic features in photoionization [9–14]. They provide a very severe and challenging test for theoretical models. Nowadays, the characteristic features of noble gases have attracted a lot of attention in connection with time-delay studies [15–18].

Another important characteristic feature of a continuum is autoionizing states [19,20]. In the noble gases, odd-parity AISs with $J = 1$ can be photoexcited from the ground state and have been investigated in great detail [21–24]. AISs with $J \neq 1$, including the AISs of even parity, are populated in multiphoton processes, probably involving discrete excited states [25–27]. A multistep or multiphoton excitation allows reaching a state of any parity and angular momentum, but Cooper minima in this process has another nature [28,29]. From this point of view, the photoionization of prepared or pumped states provides different pieces of information about atomic continuum.

In noble gases, even the lowest discrete states ($\sim$10 eV) may be excited only by VUV radiation. The setup based on the joined action of synchrotron radiation and optic (IR) laser (pump-probe scheme) has been widely used for investigations of the photoionization of excited states to the near-threshold region, including the Rydberg AIS [30–33]. The ionization of the metastable states of noble gases populated by glow discharge method (optogalvanic spectroscopy) or by electron or ion impact [34] also has been a subject of numerous investigations [35–42]. To describe characteristic features in the photoionization of noble gases from excited states (Cooper minima and Rydberg AIS), different methods were applied: quantum defect theory [43], R-matrix [44], configuration interaction Pauli–Fock with a core polarization method [31,41,45] and MCHF [46].

While the Rydberg AIS of the $np^5(^2P_{1/2})ml$ configuration can be reached via VUV + IR two-photon transition, even-parity AIS with the hole-particle or doubly excited configurations can be reached only via two-photon (resonantly enhanced) ionization by VUV radiation [47–50]. Before FEL development, these AISs were investigated basically by electron- or ion-impact method [51–53].

Since intense photon pulses were obtained with the advent of FELs, and resonance few-photon ionization of atoms and ions of noble gases was observed [54–57], the problem of obtaining accurate two-photon cross-sections and cross-sections from the excited states in the XUV region has arisen.

Here, we present an extensive study of neon photoionization from excited states. The lowest excited states are metastable $2s^22p^5(^2P_{J_f})3s[K]_{0,2}$ (here and after core $1s^2$ is omitted for brevity) and the states which can be reached from the ground by dipole excitation (dipole-allowed) $2s^22p^5(^2P_{J_f})3s[K]_1$. We use the intermediate jK-coupling scheme, where a core angular momentum J_f is coupled with an electron orbital momentum l to quantum number K, and then intermediate momentum K and electron spin $1/2$ are coupled to total angular momentum J of a system. The cross-sections are obtained by the B-spline R-matrix approach in realization by O.I. Zatsarinny [58,59]. This approach provided a high quality of oscillator strength for neon ground state excitations [60] and excellent agreement for the ionization of potassium excited states [61].

The paper is organized as follows: in Section 2, we present the results obtained within different models for metastable $2s^22p^5(^2P_{J_f})3s[K]_{0,2}$ states and compare the models with each other and with available experimental data; in Section 3, we present the results obtained in a few models chosen based on Section 2 and calculated for different field polarizations for dipole-allowed $2s^22p^5(^2P_{J_f})3s[K]_1$ states; and in Section 4, we discuss the parameters of autoionizing structures.

2. PhotoIonization of Metastable $2s^22p^5(^2P_{1/2})3s[1/2]_0$ and $2s^22p^5(^2P_{3/2})3s[3/2]_2$ States

The B-spline R-matrix (BSR) approach [58] allows the orbitals of initial and target states, as well as different target states, to be non-orthogonal. With this advantage, wave functions of the initial (excited atomic) and target (ionic) states can be obtained in series of independent MCHF calculations [62]. The approach allows to account for the dependence of the individual one-electron orbitals on states of interest and accurately describe them, keeping configuration expansions compact. The R-matrix approach automatically accounts for electron correlations in the continuum, leading to essential channel interactions.

In a pure jK-coupling scheme, the allowed ionization channels are:

$$2s^22p^5(^2P_{1/2})3s[1/2]_0 + \gamma \rightarrow 2s^22p^5(^2P_{1/2}) + \varepsilon p \quad J = 1; \tag{1}$$
$$2s^22p^5(^2P_{3/2})3s[3/2]_2 + \gamma \rightarrow 2s^22p^5(^2P_{3/2}) + \varepsilon p \quad J = 1,2,3; \tag{2}$$

but due to channel interactions, the weaker channels, including εf-wave, emerge:

$$2s^22p^5(^2P_{1/2})3s[1/2]_0 + \gamma \rightarrow 2s^22p^5(^2P_{3/2}) + \varepsilon p/\varepsilon f \quad J = 1; \tag{3}$$
$$2s^22p^5(^2P_{3/2})3s[3/2]_2 + \gamma \rightarrow 2s^22p^5(^2P_{1/2}) + \varepsilon p/\varepsilon f \quad J = 1,2,3. \tag{4}$$

We explore three models with the subsequent Breit–Pauli diagonalization procedure [63] to take into account spin–orbit interaction: the model based on using real orthogonal orbitals (addressed as $\mathcal{R}$-model), the model with the use of non-orthogonal virtual orbitals ($\mathcal{V}$-model) and the approach combining both real and virtual orbitals ($\mathcal{C}$-model). For all of them, different numbers of target ionic states (see Table 1) are considered in the B-spline, bound-state close-coupling calculations. The number of accounted target states was varied: two ($2s^22p^5$ ($^2P_{3/2,1/2}$)) — to include direct photoionization only; six (+ $2s2p^6$ ($^2S_{1/2}$) and $2s^22p^43s$ ($^4P_{5/2,3/2,1/2}$)) — to allow generation of the lowest AISs, either hole-particle or doubly excited; ten (+ $2s^22p^43s$ ($^2P_{3/2,1/2}$) and $2s^22p^43s$ ($^2D_{5/2,3/2}$)); and thirteen (+ $2s^22p^43p$ ($^4P_{5/2,3/2,1/2}$)) — to shift the AIS energies closer to their experimental positions [53,64].

Table 1. Target states (named by the leading term in LSJ approximation), the leading terms in configuration mixing (in percent) and energies according to NIST database [65]. Core $1s^2$ is omitted for brevity.

Target	Energy	$\mathcal{R}$-Model	$\mathcal{V}$-Model	$\mathcal{C}$-Model
$2s^22p^5$ ($^2P_{3/2}$) $2s^22p^5$ ($^2P_{1/2}$)	21.5645 21.6613	99.92 $2s^22p^5$+0.03 $2s^22p^43p$+ 0.02 $2s^22p^33p^2$+0.01 $2s^22p^44p$	92.80 $2s^22p^5$+3.16 $2s^22p^43\bar{p}$+ 2.70 $2s2p^53\bar{s}$+0.36 0.37 $2s^22p^33\bar{p}^2$	98.24 $2s^22p^5$+0.56 $2s^22p^34\bar{p}^2$+ 0.54 $2s^22p^33\bar{d}^2$+0.35 $2s2p^53\bar{d}$
$2s2p^6$ ($^2S_{1/2}$)	48.4750	95.39 $2s2p^6$+4.03 $2s^22p^43s$+ 0.23 $2sp^53p$+0.21 $2s^22p^44s$	93.58 $2s2p^6$+2.26 $2s^22p^43\bar{d}$+ 0.77 $2s2p^43\bar{d}^2$+0.67 $2s^22p^43s$	94.70 $2s2p^6$+2.33 $2s^22p^43\bar{d}$+ 0.79 $2s2p^43\bar{d}^2$+0.75 $2s2p^44\bar{p}^2$
$2s^22p^43s$ ($^4P_{5/2}$) $2s^22p^43s$ ($^4P_{3/2}$) $2s^22p^43s$ ($^4P_{1/2}$)	48.7333 48.7975 48.8345	95.13 $2s^22p^43s$+3.76 $2s^22p^44s$ 0.79 $2s^22p^33s3p$+0.10 $2s2p^53p$ 91.80 $2s^22p^43s$+3.63 $2s^22p^44s$ 3.48 $2s2p^6$+0.76 $2s^22p^33s3p$	97.28 $2s^22p^43s$+1.24 $2s2p^43s^2$+ 0.58 $2s2p^43s3\bar{d}$+0.35 $2s^22p^23s3\bar{p}^2$	91.68 $2s^22p^43s$+4.98 $2s^22p^44\bar{s}$ 0.66 $2s^22p^33s4\bar{p}$+0.62 $2s2p^43s3\bar{d}$
$2s^22p^43s$ ($^2P_{3/2}$) $2s^22p^43s$ ($^2P_{1/2}$)	49.3478 49.4237	95.40 $2s^22p^43s$+3.41 $2s^22p^44s$+ 1.38 $2s^22p^33s4p$+0.84 $2s^22p^33s3p$	97.70 $2s^22p^43s$+0.72 $2s2p^43s^2$+ 0.59 $2s2p^43s3\bar{d}$+0.35 $2s^22p^23s3\bar{p}^2$	94.02 $2s^22p^43s$+2.50 $2s^22p^44\bar{s}$+ 1.38 $2s^22p^33s4\bar{p}$+0.74 $2s^22p^33s3p$
$2s^22p^43p$ ($^4P_{5/2}$) $2s^22p^43p$ ($^4P_{3/2}$) $2s^22p^43p$ ($^4P_{1/2}$)	52.0885 52.1161 52.1388	94.37 $2s^22p^43p$+4.70 $2s^22p^44p$+ 0.55 $2s^22p^33p^2$+0.14 $2s^22p^33p4p$	90.78 $2s^22p^43p$+6.67 $2s^22p^33p4\bar{p}$+ 1.77 $2s2p^43p3\bar{s}$+0.48 $2s2p^43p3\bar{d}$	96.73 $2s^22p^43p$+1.67 $2s^22p^33p4\bar{p}$+ 0.58 $2s^22p^43p3\bar{d}$+0.49 $2s^22p^33p^2$
$2s^22p^43s$ ($^2D_{5/2}$) $2s^22p^43s$ ($^2D_{3/2}$)	52.1135 52.1139	95.30 $2s^22p^43s$+3.71 $2s^22p^44s$+ 0.73 $2s^22p^33s3p$+0.16 $2s^22p^33s4p$	97.38 $2s^22p^43s$+1.01 $2s2p^43s^2$+ 0.62 $2s2p^43s3\bar{d}$+0.38 $2s^22p^23s3\bar{d}^2$	92.77 $2s^22p^43s$+4.03 $2s^22p^44\bar{s}$+ 1.16 $2s^22p^33s4\bar{p}$+0.66 $2s2p^43s3\bar{d}$

For the interpolation procedure, we use B-splines of the 8th order. The R-matrix radius is chosen to be 80 a.u. Moreover, we fix grid parameters: $h = 0.0125$ a.u. (step size at the origin, from 0 to 1) and $h_{max} = 0.5$ a.u. (maximum step size of the grid). These parameters were chosen in such a way that for all of the models the resulting accuracy is enough to correctly reproduce the radial part of the wave functions.

The models have the same set of configuration state functions (CSFs), including the ones with excited core to account the core–valence correlations. The initial $2s^22p^5(^2P_{J_f})3s\,[K]_{0,1,2}$ states' CSFs include all possible single and double replacements of $2s$, $2p$ and $3s$ orbitals with $n = 3,4; l = s,p,d,f$-orbitals; and target states' CSFs include single and double replacements of $2s$, $2p$ and $3s$ orbitals to each of the orbitals $n = 3,4; l = s,p,d,f$ solely. For example, in the $2s2p^6$ ($^2S_{1/2}$) decomposition, there are $2s2p^43p^2$ and $2s2p^44p^2$ configurations, but there is no $2s2p^43p4p$. In the calculation, we use the energies of target states adjusted to the experimental values [65].

In the $\mathcal{R}$-model, the initial and target states are obtained using the same real orbitals: the first set of 1*s*, 2*s* and 2*p* orbitals is optimized on the $2s^22p^5$ configuration and frozen for forthcoming calculations, then each of the real 3*s*–4*f* orbitals is optimized on the corresponding configuration with the highest weights in the CSFs set of $2s^22p^5(^2P_{3/2})3s[3/2]_1$ state, i.e., $3s$, $4s$, $3d$ and $4d$ orbitals are optimized on $2s^22p^5nl$ configuration and $3p$, $4p$ and $4f$ orbitals on $2s^22p^43snl$. The main contributions to both metastable $2s^22p^5(^2P_{J_f})3s[K]_{0,2}$ states came from

$$2s^22p^53s(99.81\%)\,,2s^22p^54s(0.07\%)\,,2s^22p^43s3p(0.05\%)\,,2s^22p^33s3p^2(0.03\%)\,,2s^22p^53d(0.02\%)$$

configurations. The main contributions to the target states are presented in Table 1.

In the $\mathcal{V}$-model, the CSFs of the initial and target states are optimized independently. Thereof, the orbitals of the leading configuration ($1s$, $2s$ and $2p$) are real; orbitals $3d-4f$ are correlation pseudo-orbitals, and 3*s* and 3*p* orbitals may be real or virtual depending on the state of interest. All orbitals were allowed to vary together. The main contributions to both $2s^22p^5(^2P_{J_f})3s[K]_{0,2}$ states came from

$$2s^22p^53s(93.09\%)\,,2s^22p^43s3p(2.86\%)\,,2s2p^53s4s(2.45\%)\,,2s^22p^33s3p^2(0.42\%)\,,2s2p^33s3d(0.37\%)$$

configurations. The main contributions to the target states are presented in Table 1. The difference between theoretical and experimental target energies in this model is less than 0.1 eV.

In the $\mathcal{C}$-model, we keep all of the orbitals presented in target states leading configurations ($1s-3p$) as they are obtained in the $\mathcal{R}$-model; therefore, they are real and orthogonal. The other ($3d-4f$) pseudo-orbitals are optimized separately for different targets and non-orthogonal. The initial states are the same as in the $\mathcal{R}$-model. The main contributions to the target states are presented in Table 1.

In Figures 1 and 2, cross-sections of photoionization from $2s^22p^5(P_{1/2})3s[1/2]_0$ and $2s^22p^5(P_{3/2})3s[3/2]_2$ metastable states are presented for these three models, with different numbers of target states in length and velocity gauges. The upper row (a,b,c) in Figure 1 and bottom row (d,e,f) in Figure 2 present results for photoionization which is forbidden within the pure *jK*-coupling scheme, the bottom row in Figure 1 and upper row in Figure 2 for allowed. In the featureless region (from 3 to 20 eV), cross-sections for allowed ionization are higher than for forbidden one.

The cross-section of allowed photoionization manifests the Cooper minimum near the threshold; the one of the forbidden photoionization does not. The energy of Cooper minimum is basically determined by the nodes of 3*s*-orbital. That is the reason why cross-sections in $\mathcal{R}$- and $\mathcal{C}$-models based on the same 3*s*-orbitals are alike at low energies. The Cooper minimum is more pronounced in the $\mathcal{V}$-model because the wave function of the initial state is more compact when virtual orbitals are included. The more target states are included, the farther Cooper minimum is pushed below the ionization threshold. A further analysis has shown that for larger CSF sets of target states, the Cooper minimum is pushed less. With the exception of the model with the fewest (two) number of target states, Cooper minima are deeper and situated at a little higher energy for calculations in the length gauge.

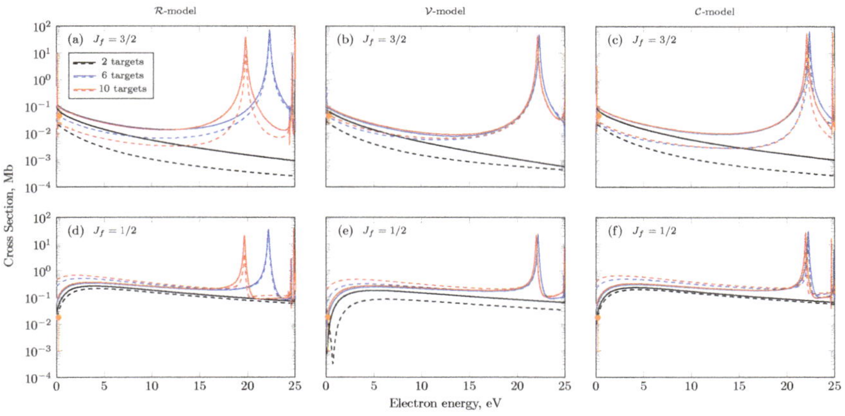

Figure 1. Photoionization cross-section of $2s^22p^5(P_{1/2})3s[1/2]_0$ state of neon to $2s^22p^5\ (^2P_{3/2})$ (upper row) and $2s^22p^5\ (^2P_{1/2})$ (bottom row). Calculations are performed within $\mathcal{R}$- (**a**,**d**), $\mathcal{V}$- (**b**,**e**) and $\mathcal{C}$-model (**c**,**f**). Solid lines are for calculations in the length gauge and dashed lines in the velocity gauge; different colors mark the number of accounted target states: two (black), six (blue) and ten (red). Experimental data points are taken from [38] and labeled by orange.

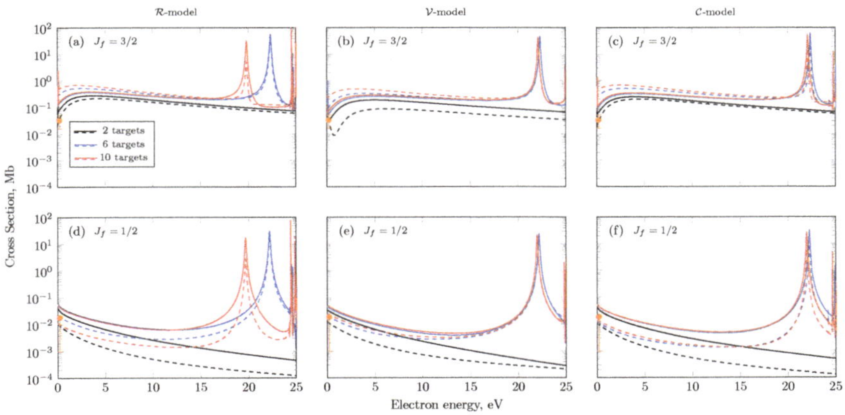

Figure 2. Photoionization cross-section of $2s^22p^5(^2P_{3/2})3s[3/2]_2$ state of neon to $2s^22p^5\ (^2P_{3/2})$ (upper row) and $2s^22p^5\ (^2P_{1/2})$ (bottom row). Calculations are performed within $\mathcal{R}$- (**a**,**d**), $\mathcal{V}$- (**b**,**e**) and $\mathcal{C}$-model (**c**,**f**). Solid lines are for calculations in the length gauge and dashed for the velocity gauge; different colors mark the number of accounted targets: two (black), six (blue) and ten (red). Experimental data points are taken from [38] and labeled by orange.

The resonance structure at $\sim$22 eV is hole-particle $2s2p^63s\ (^3S_1)$ AIS generated on the $2s2p^6\ (^2S)$ target; inclusion of the other targets may shift the position of the structure. Their influence depends on how close a target energy is to the ^2S-target. Note that only states with total momentum $J = 1$ can be excited from the $2s^22p^5(^2P_{1/2})3s[1/2]_0$ metastable state. The position of the AIS weakly depends on the model and is in accordance with that reported in [53], except for the $\mathcal{R}$-model, where the enabling of $2s^22p^43s\ (^2D_{5/2,3/2})$ target states leads to a dramatic jump in the resonance energy position (Figures 1a and 2a, red curves). In the $\mathcal{V}$- and $\mathcal{C}$-models, these ^2D targets slightly (by 0.3 eV) move the resonance position down towards the experimental value. In the $\mathcal{R}$-model, it is impossible to match the energy position of $2s2p^6\ (^2S_{1/2})$ and $2s^22p^43s\ (^2D_{5/2,3/2})$ target states, and interference between them is more essential than it is supposed to be in reality. The parameters of the resonance are discussed in Section 4.

As was pointed out in [60], introducing virtual orbitals is necessary to obtain good agreement between length and velocity gauges, and in the considered case, overall agreement is better for the $\mathcal{V}$-model, especially for the forbidden photoionization. We checked that the addition of $5\bar{l}(\bar{l} = \bar{s}, \bar{p}, \bar{d}, \bar{f})$ orbitals to the expansion of $2s^22p^5(^2P_{J_f})3s[K]_{0,2}$ states in $\mathcal{R}$- and $\mathcal{C}$-models makes the cross-sections look similar to the ones obtained within the $\mathcal{V}$-model (not shown) and pulls the Cooper minimum above the threshold. Nevertheless, it is important to notice that while the $\mathcal{V}$-model provides much faster convergence than the $\mathcal{R}$- and $\mathcal{C}$-models, the last two reproduce near-threshold behavior of angular anisotropy parameter β for the forbidden photoionization much better. Among the models with virtual orbitals, the ones including only $2s2p^5n\bar{l}^2$-type (with $n = 3,4$; $l = s,p,d,f$) pair-excitations additionally to $2s^22p^5n\bar{s} + 2s^22p^5n\bar{d}$ $(n = 3,4)$ single-excitations provide an angular anisotropy parameter similar to the one in [38] and hit the experimental values. The models allowing contributions of single excitations to $3p, 4p, 4f$ fail because configuration $2s^22p^43s3\bar{p}$ obtains the highest weight after $2s^22p^53s$. However, in the cross-sections, the difference between the use of the two sets of orbitals described above is hardly seen.

The issue is very similar to that recorded for electron scattering involving $2p^53s$ states: while there are models reproducing angle-integrated data [66], additional actions are needed to reproduce differential parameters [67].

3. Photoionization of Dipole-Allowed $2s^22p^5(^2P_{J_f})3s[K]_1$ States

In this section, we present calculations of photoionization cross-sections from dipole-allowed $2s^22p^5(^2P_{J_f})3s[K]_1$ states, which can be effectively excited by an electromagnetic field. To the best of our knowledge, there are very few updated data for them in the literature [47,50,68]. This data may be useful for the interpretation of experiments involving two-photon (probably two-color) processes at FELs [54–56]. One of the advantages of modern FELs and synchrotron facilities is that the generated radiation is polarized. Bearing in mind this possibility, we present here calculations for three cases: both exciting and ionizing fields are unpolarized, linearly polarized in the same direction and circularly polarized with equal helicities. Accounting for the dipole selection rules, the cross-section for unpolarized radiation is $\sigma^{(u)} = s_0(|D_{J=0}|^2 + |D_{J=1}|^2 + |D_{J=2}|^2)/3$, for linearly polarized $\sigma^{(l)} = s_0(|D_{J=0}|^2/3 + 2|D_{J=2}|^2/15)$ and for circularly polarized $\sigma^{(c)} = s_0|D_{J=2}|^2/5$ [46,69], where $s_0 = 4\pi\omega/3c$ and D_J is the reduced dipole matrix transition amplitude. The cross-sections by unpolarized and polarized radiation do not share spectroscopic features because different channels are involved.

In Figures 3 and 4, we present results for photoionization of $2s^22p^5(^2P_{3/2})3s[3/2]_1$ and $2s^22p^5(^2P_{1/2})3s[1/2]_1$ states, correspondingly calculated within the $\mathcal{V}$- and $\mathcal{C}$-models accounting for six and ten target states. The upper row (a,b,c) in Figure 3 and bottom row (d,e,f) in Figure 4 present allowed photoionization.

$$2s^22p^5(^2P_{1/2})3s[1/2]_1 + \gamma \rightarrow 2s^22p^5(^2P_{1/2}) + \varepsilon l \quad J = 0,1,2; \tag{5}$$

$$2s^22p^5(^2P_{3/2})3s[3/2]_1 + \gamma \rightarrow 2s^22p^5(^2P_{3/2}) + \varepsilon l \quad J = 0,1,2; \tag{6}$$

The bottom row (d,e,f) in Figure 3 and upper row (a,b,c) in Figure 4 present forbidden photoionization:

$$2s^22p^5(^2P_{1/2})3s[1/2]_1 + \gamma \rightarrow 2s^22p^5(^2P_{3/2}) + \varepsilon l/\varepsilon f \quad J = 0,1,2; \tag{7}$$

$$2s^22p^5(^2P_{3/2})3s[3/2]_1 + \gamma \rightarrow 2s^22p^5(^2P_{1/2}) + \varepsilon l/\varepsilon f \quad J = 0,1,2. \tag{8}$$

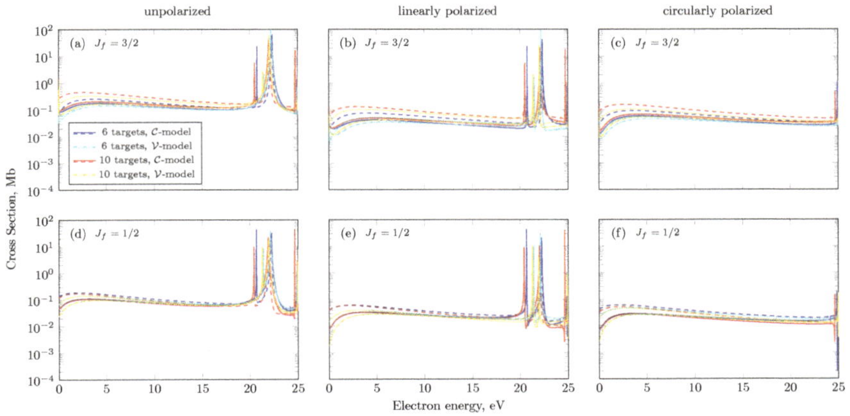

Figure 3. Cross-section calculated within the $\mathcal{V}$- (dash-dotted lines correspond to length gauge, dotted lines to velocity gauge) and $\mathcal{C}$- (solid lines correspond to length gauge, dashed lines to velocity gauge) models with six (cyan and blue) and ten (orange and red) target states for ionization of $2s^22p^5(^2P_{3/2})3s[3/2]_1$ state to $2s^22p^5\,(^2P_{3/2})$ (upper row) and $2s^22p^5\,(^2P_{1/2})$ (bottom row) ionic states for unpolarized (**a**,**d**), linearly polarized (**b**,**e**) and circularly polarized (**c**,**f**) fields.

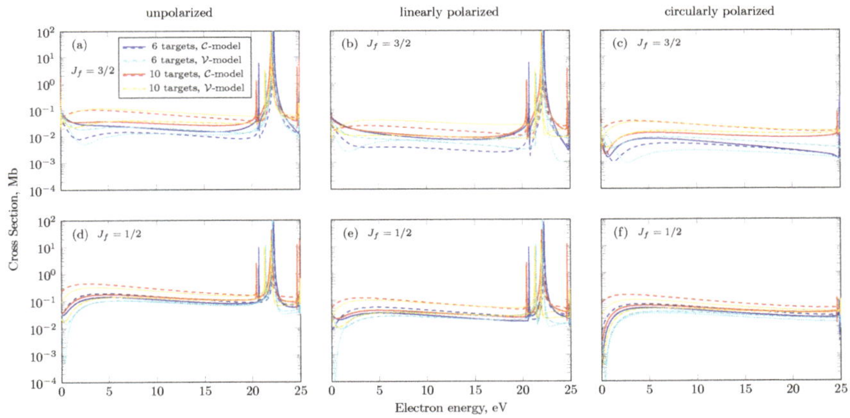

Figure 4. Cross-section calculated within the $\mathcal{V}$- (dash-dotted lines correspond to length gauge, dotted lines to velocity gauge) and $\mathcal{C}$- (solid lines correspond to length gauge, dashed lines to velocity gauge) models with six (cyan and blue) and ten (orange and red) target states for ionization of $2s^22p^5(^2P_{1/2})3s[1/2]_1$ state to $2s^22p^5\,(^2P_{3/2})$ (upper row) and $2s^22p^5\,(^2P_{1/2})$ (bottom row) ionic states for unpolarized (**a**,**d**), linearly polarized (**b**,**e**) and circularly polarized (**c**,**f**) fields. Within the $\mathcal{V}$-model, orange and cyan lines in panels (**d**,**e**,**f**) practically coincide.

Notice that in the smooth region, "allowed" and "forbidden" photoionization does not differ that much as for the metastable states, because these states are not pure, neither in the jK-scheme, nor in the LS one: in the $\mathcal{C}$-model $2s^22p^5(^2P_{1/2})3s[1/2]_1 = 0.89\,2s^22p^53s\,^1P - 0.45\,2s^22p^53s\,^3P + \ldots$ and in the $\mathcal{V}$-model $2s^22p^5(^2P_{1/2})3s[1/2]_1 = 0.89\,2s^22p^53s\,^1P - 0.38\,2s^22p^53s\,^3P + \ldots$, which differs from the pure jK-scheme where the coefficients are $2s^22p^5(^2P_{1/2})3s[1/2]_1 = 0.82\,2s^22p^53s\,^1P - 0.58\,2s^22p^53s\,^3P$.

For the allowed photoionization, the manifestation of the Cooper minimum strongly depends on polarization; in the case of circular polarization where only channels with $J = 2$ can contribute, it is placed very close to the threshold (Figures 3c and 4f). In the case of linear polarization, additional allowed channels with $J = 0$ shift the Cooper minimum to higher energy (Figures 3b and 4e). Finally, the channels with $J = 1$ allowed for unpolarized radiation smear it out (Figures 3a and 4d).

For the forbidden photoionization (Figures 3d–f and 4a–c), general tendencies are quite different from the ones observed for the metastable states (Figures 1a–c and 2d–f). While there is no Cooper minimum in the photoionization of the metastable states and the cross-section drops down from the threshold, in the photoionization of the dipole-allowed states, the cross-section may increase from threshold (Figure 3d–f), indicating that the Cooper minimum has fallen under the threshold or appears above the threshold (Figure 4a–c). In the last case, the Cooper minimum is more pronounced for circularly polarized fields.

Calculations in the velocity gauge are again more sensitive to a model than in the length gauge and more sensitive when it comes to photoionization from $2s^22p^5(^2P_{1/2})3s[1/2]_1$ than from $2s^22p^5(^2P_{3/2})3s[3/2]_1$.

In comparison with the photoionization from the metastable states, the channels with $J = 0$ bring up two more autoionizing states: hole-particle $2s2p^63s\ [^1S_0]$ AIS and doubly excited $2s^22p^43s^2\ [^3P_0]$ AIS visible in case of linearly polarized and unpolarized light (Figures 3 and 4a,b,d,e). For unpolarized fields (Figures 3 and 4a,d), channels with $J = 1$ manifest resonance $2s2p^63s\ [^3S_1]$, which dominates over the energy region. The parameters of the AISs are discussed in the next section.

4. The Resonance Structures

There are three lowest resonance structures of even parity: hole-particle resonances $2s2p^63s\ (^{1,3}S_{0,1})$ generated on $2s2p^6\ (^2S_{1/2})$ target and doubly excited resonance $2s^22p^43s^2\ (^3P_0)$ generated on $2s^22p^43s\ (^4P_{1/2})$ target. Within the jK-coupling scheme, the R-matrix [58] does not generate the AIS of configuration $2p^43s^2$ with $J = 1, 2$ on any of the target states $2p^43s(^4P_{3/2,5/2},{}^2D_{3/2,5/2})$.

In Figure 5, there is the region with AISs excited from the metastable and dipole-allowed states by unpolarized fields plotted in a more detailed way for the models with six and ten targets. The AISs' positions are different due to interchannel interaction. The inclusion of $2s^22p^4(^3P)3s\ [^2D_J]$ lowers the position of the hole-particle resonances $2s2p^63s\ [^{1,3}S_{0,1}]$ and, following the addition of higher targets, does not change them significantly. The inclusion of $2s^22p^4(^3P)3s\ [^2P_J]$ lowers the position of the doubly excited $2s^22p^4(^3P)3s^2\ [^3P_0]$ resonance and places it in experimental position [64] in $\mathcal{C}$-model. The further inclusion of $2s^22p^4(^3P)3p\ [^4P_J]$ targets (not shown) does not improve the position of the doubly excited state: in the $\mathcal{C}$-model, it is pulled 0.3 eV below the experiment, while in the $\mathcal{V}$-model, it is by 0.6 eV above. It is found that the lowest AISs of even parity in neon are excited very effectively from the excited states and dominate over direct photoionization in the corresponding energy region.

The R-matrix approach does not produce values of energy E, width Γ and Fano profile index q [19], but one can fit a resonance structure and extract these values (see Table 2). All of the resonances are excited very efficiently, whereby transitions $2s \to 2p$ and $2p \to 3s$, as a result, the AISs are practically symmetric with huge q-indexes. The large value of q makes the extraction procedure problematic, and in Table 2, we replace $|q| > 100$ by ∞ with the corresponding sign. The particle-hole AISs' width ratio is close to the ratio of their statistical weights. The Fano index may take different values and signs depending on the state it is excited from, the target state and the polarization of radiation. The latter occurs because different contributions of the channels with different angular momentum J to the cross-sections result in different backgrounds and heights of resonance.

As for the doubly excited AIS $2p^43s^2[^3P_J]$ within the jK-coupling scheme, there is only $2p^43s^2[^3P_0]$ with $\Gamma \approx 0.003$ eV. Within the LSJ-coupling scheme, all of the triplet states $2p^43s^2[^3P_J]$ are generated with $\Gamma \approx 0.001$ eV separated approximately by 0.15 eV. We checked that the hole-particle resonances are weakly affected by a coupling scheme and that they are just a little broadening ($\Gamma = 0.122$ and 0.079 eV). To the best of our knowledge, there is no experiment able to resolve the fine structure of this AIS and the choice between the jK- and LSJ-coupling schemes for them is insurmountable so far.

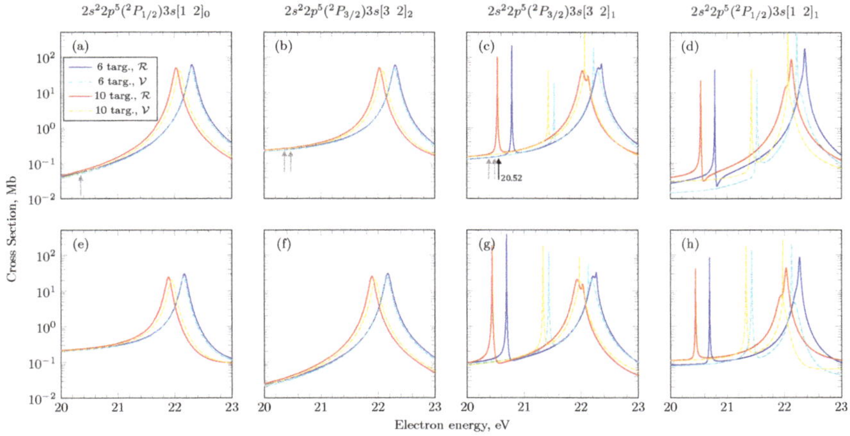

Figure 5. Cross-section calculated in the length gauge within the $\mathcal{V}$- (dash-dotted lines) and $\mathcal{C}$- (solid lines) models with six (cyan and blue) and ten (orange and red) thresholds for ionization of $2s^22p^5(^2P_{1/2})3s[1/2]_0$ **(a,e)**, $2s^22p^5(^2P_{3/2})3s[3/2]_2$ **(b,f)** and $2s^22p^5(^2P_{3/2})3s[3/2]_1$ **(c,g)**, $2s^22p^5(^2P_{1/2})3s[1/2]_1$ **(d,h)** states to $2s^22p^5\,(^2P_{3/2})$ (upper row) and to $2s^22p^5\,(^2P_{1/2})$ (bottom row) ionic states for unpolarized fields. The arrows mark positions of very narrow resonances $2p^43s^2[^2P_{J=1,2}]$ where they would appear within the LSJ-coupling scheme.

Table 2. AISs parameters in a model within ten target states: energy position E (eV), width Γ (eV) and Fano profile index q at excitation of different initial states for different polarization.

Parameter \ AIS		**$2s2p^63s\ [^3S_1]$**	**$2s2p^63s\ [^1S_0]$**
E	$\mathcal{C}$-model	22.03	22.13
Γ		0.119	0.028
q^{un}	$(^2P_{3/2})3s[3/2]_1 \to 2p^5\,(^2P_{3/2})$	−70	∞
q^{un}	$(^2P_{3/2})3s[3/2]_1 \to 2p^5\,(^2P_{1/2})$	−50	−75
q^{un}	$(^2P_{1/2})3s[1/2]_1 \to 2p^5\,(^2P_{3/2})$	∞	∞
q^{un}	$(^2P_{1/2})3s[1/2]_1 \to 2p^5\,(^2P_{1/2})$	−15	∞
q^{lin}	$(^2P_{3/2})3s[3/2]_1 \to 2p^5\,(^2P_{3/2})$	-	∞
q^{lin}	$(^2P_{3/2})3s[3/2]_1 \to 2p^5\,(^2P_{1/2})$	-	−80
q^{lin}	$(^2P_{1/2})3s[1/2]_1 \to 2p^5\,(^2P_{3/2})$	-	−∞
q^{lin}	$(^2P_{1/2})3s[1/2]_1 \to 2p^5\,(^2P_{1/2})$	-	∞
E	$\mathcal{V}$-model	22.07	22.05
Γ		0.116	0.002
q^{un}	$(^2P_{3/2})3s[3/2]_1 \to 2p^5\,(^2P_{3/2})$	−75	−∞
q^{un}	$(^2P_{3/2})3s[3/2]_1 \to 2p^5\,(^2P_{1/2})$	−50	∞
q^{un}	$(^2P_{1/2})3s[1/2]_1 \to 2p^5\,(^2P_{3/2})$	65	∞
q^{un}	$(^2P_{1/2})3s[1/2]_1 \to 2p^5\,(^2P_{1/2})$	−15	∞
q^{lin}	$(^2P_{3/2})3s[3/2]_1 \to 2p^5\,(^2P_{3/2})$	-	−∞
q^{lin}	$(^2P_{3/2})3s[3/2]_1 \to 2p^5\,(^2P_{1/2})$	-	∞
q^{lin}	$(^2P_{1/2})3s[1/2]_1 \to 2p^5\,(^2P_{3/2})$	-	−∞
q^{lin}	$(^2P_{1/2})3s[1/2]_1 \to 2p^5\,(^2P_{1/2})$	-	−∞

5. Conclusions

The photoionization cross-section of the lowest excited states of neon are presented. The R-matrix approach is applied with a variety of different parameters, and the comparison of the usability of real and virtual orbitals is carried out. Results are presented for the metastable states (total angular momentum $J = 0, 2$) ionized by unpolarized light and for the dipole-allowed states ($J = 1$) ionized by the light with the same polarization as those causing their excitation: unpolarized, linearly and circularly polarized. The spectroscopic features such as Cooper minimum, particle-hole and doubly excited autoionizing states are found to appear differently depending on polarization.

It was shown that the model based on virtual orbitals provides much better convergence in terms of length and velocity gauge, as well as in terms of target states involved. It predicts a deeper Cooper minimum placed at higher energy. On the other hand, $\mathcal{V}$-model has a tendency to become overcorrelated (unrealistically compact wave functions), which imposes the problem of a correct positioning of the narrow AISs. Based on real orbitals, the $\mathcal{C}$-model turns out to be more stable and better reproduces the AIS structures. The hole-particle AISs are found to be much broader than the doubly excited ones.

This work may serve as building blocks for the investigation of multi-photon, probably two-color ionization by VUV radiation.

Author Contributions: M.D.K. is responsible for the $\mathcal{V}$-model and M.M.P. for the $\mathcal{R}$-, $\mathcal{C}$-models. All of the authors took part in discussions of the results and preparing the manuscript. Conceptualization E.V.G. and A.N.G.-G.; investigation, formal analysis M.M.P. and M.D.K.; visualization M.M.P.; resources S.M.B.; writing—original draft preparation E.V.G., M.M.P. and M.D.K.; writing—review and editing M.M.P., M.D.K., S.M.B., E.V.G. and A.N.G.-G.; funding acquisition A.N.G.-G. and S.M.B. All authors have read and agreed to the published version of the manuscript.

Funding: The development of the $\mathcal{R}$- and $\mathcal{C}$-models and the analysis of the corresponding cross-sections' energy dependencies are supported by the Russian Foundation for Basic Research (RFBR) under project No. 20-52-12023 and by the Ministry of Science and Higher Education of the Russian Federation (grant No. 075-15-2021-1353). The preparation of the $\mathcal{V}$-model and analysis of AIS are supported by the Russian Science Foundation (project No. 21-42-04412). The calculations of the photoionization processes are performed using resources of the Shared Services "Data Center of the Far-Eastern Branch of the Russian Academy of Sciences" and supported by the Ministry of Science and Higher Education of the Russian Federation (project No. 0818-2020-0005).

Institutional Review Board Statement: Not applicable.

Informed Consent Statement: Not applicable.

Data Availability Statement: Not applicable.

Acknowledgments: The authors benefited greatly from discussions with Oleg Zatsarinny. The authors are grateful to Jon Grumer for many useful pieces of advice.

Conflicts of Interest: The authors declare no conflict of interests.

References

1. Connerade, J. *Highly Exited Atoms*; Cambridge University Press: Cambridge, UK, 1998.
2. Baig, M.A. Measurement of Photoionization Cross-Section for the Excited States of Atoms: A Review. *Atoms* **2022**, *10*, 39. [CrossRef]
3. Avdonina, N.B.; Amusia, M.Y. Characteristic features in photoionisation of excited atomic states. *J. Phys. B At. Mol. Phys.* **1983**, *16*, L543–L545. [CrossRef]
4. Cooper, J.W. Interaction of Maxima in the Absorption of Soft X Rays. *Phys. Rev. Lett.* **1964**, *13*, 762–764. [CrossRef]
5. Fano, U.; Cooper, J.W. Spectral Distribution of Atomic Oscillator Strengths. *Rev. Mod. Phys.* **1968**, *40*, 441–507. [CrossRef]
6. Msezane, A.; Manson, S.T. New Minima in Photoionization Cross Section. *Phys. Rev. Lett.* **1975**, *35*, 364–366. [CrossRef]
7. Mamsom, S.T.; Cooper, J.W. Photo-Ionization in the Soft x-Ray Range: 1Z Dependence in a Central-Potential Model. *Phys. Rev.* **1968**, *165*, 126–138. [CrossRef]
8. McGuire, E.J. Photo-Ionization Cross Sections of the Elements Helium to Xenon. *Phys. Rev.* **1968**, *175*, 20–30. [CrossRef]
9. Amusia, M.; Chernysheva, L.; Yarzhemsky, V. *Handbook of Theoretical Atomic Physics: Data for Photon Absorption, Electron Scattering, and Vacancies Decay*; Springer Science & Business Media: Berlin/Heidelberg, Germany, 2012.

10. Amusia, M.; Kheifets, A. The influence of "two-electron-two-hole excitation"s on the $3s^{-1}4p$ autoionization profile in Ar atoms. *Phys. Lett. A* **1981**, *82*, 407–411. [CrossRef]
11. Amusia, M.Y.; Ivanov, V.K.; Cherepkov, N.A.; Chernysheva, L.V. Interference effects in photoionization of noble gas atoms outer s-shells. *Phys. Lett.* **1972**, *40A*, 361–362. [CrossRef]
12. Toffoli, D.; Stener, M.; Fronzoni, G.; Decleva, P. Convergence of the multicenter B-spline DFT approach for the continuum. *Chem. Phys.* **2002**, *276*, 25–43. [CrossRef]
13. Marante, C.; Klinker, M.; Corral, I.; González-Vázquez, J.; Argenti, L.; Martín, F. Hybrid-Basis Close-Coupling Interface to Quantum Chemistry Packages for the Treatment of Ionization Problems. *J. Chem. Theory Comput.* **2017**, *13*, 499–514. [CrossRef]
14. Moitra, T.; Ponzi, A.; Koch, H.; Coriani, S.; Decleva, P. Accurate Description of Photoionization Dynamical Parameters. *J. Phys. Chem. Lett.* **2020**, *11*, 5330–5337. [CrossRef]
15. Kheifets, A.S. Time delay in valence-shell photoionization of noble-gas atoms. *Phys. Rev. A* **2013**, *87*, 063404. [CrossRef]
16. Palatchi, C.; Dahlstrom, J.M.; Kheifets, A.S.; Ivanov, I.A.; Canaday, D.M.; Agostini, P.; DiMauro, L.F. Atomic delay in helium, neon, argon and krypton. *J. Phys. B At. Mol. Opt. Phys.* **2014**, *47*, 245003. [CrossRef]
17. Alexandridi, C.; Platzer, D.; Barreau, L.; Busto, D.; Zhong, S.; Turconi, M.; Neoričić, L.; Laurell, H.; Arnold, C.L.; Borot, A.; et al. Attosecond photoionization dynamics in the vicinity of the Cooper minima in argon. *Phys. Rev. Res.* **2021**, *3*, L012012. [CrossRef]
18. Saha, S.; Mandal, A.; Jose, J.; Varma, H.R.; Deshmukh, P.C.; Kheifets, A.S.; Dolmatov, V.K.; Manson, S.T. Relativistic effects in photoionization time delay near the Cooper minimum of noble-gas atoms. *Phys. Rev. A* **2014**, *90*, 053406. [CrossRef]
19. Fano, U. Effects of Configuration Interaction on Intensities and Phase Shifts. *Phys. Rev.* **1961**, *124*, 1866–1878. [CrossRef]
20. Arimondo, E.; Clark, C.W.; Martin, W.C. Colloquium: Ettore Majorana and the birth of autoionization. *Rev. Mod. Phys.* **2010**, *82*, 1947–1958. [CrossRef]
21. Madden, R.P.; Codling, K. New Autoionizing Atomic Energy Levels in He, Ne, and Ar. *Phys. Rev. Lett.* **1963**, *10*, 516–518. [CrossRef]
22. Baig, M.A.; Connerade, J.P. Centrifugal barrier effects in the high Rydberg states and autoionising resonances of neon. *J. Phys. B At. Mol. Phys.* **1984**, *17*, 1785–1796. [CrossRef]
23. Maeda, K.; Ueda, K.; Ito, K. High-resolution measurement for photoabsorption cross sections in the autoionization regions of Ar, Kr and Xe. *J. Phys. B At. Mol. Opt. Phys.* **1993**, *26*, 1541–1555. [CrossRef]
24. Kabachnik, N.M.; Sazhina, I.P. Angular distribution and polarization of photoelectrons in the region of resonances. *J. Phys. B At. Mol. Opt. Phys.* **1976**, *9*, 1681–1697. [CrossRef]
25. Pratt, S.T.; Dehmer, P.M.; Dehmer, J.L. Three-photon excitation of autoionizing states of atomic xenon between the ${}^{2\circ}{}_{3/2}$ and ${}^{2\circ}{}_{1/2}$ fine-structure thresholds. *Phys. Rev. A* **1987**, *35*, 3793–3798. [CrossRef]
26. Blazewicz, P.R.; Stockdale, J.A.D.; Miller, J.C.; Efthimiopoulos, T.; Fotakis, C. Four-photon excitation of even-parity Rydberg states in krypton and xenon. *Phys. Rev. A* **1987**, *35*, 1092–1098. [CrossRef]
27. Koeckhoven, S.M.; Buma, W.J.; de Lange, C.A. Three-photon excitation of autoionizing states of Ar, Kr, and Xe between the ${}^2P_{3/2}$ and ${}^2P_{1/2}$ ionic limits. *Phys. Rev. A* **1994**, *49*, 3322–3332. [CrossRef]
28. Moccia, R.; Rahman, N.K.; Rizzo, A. Two-photon ionisation cross section calculations of noble gases: Results for Ne and Ar. *J. Phys. B At. Mol. Opt. Phys.* **1983**, *16*, 2737–2751. [CrossRef]
29. Saenz, A.; Lambropoulos, P. Theoretical two-, three- and four-photon ionization cross sections of helium in the XUV range. *J. Phys. B At. Mol. Opt. Phys.* **1999**, *32*, 5629–5637. [CrossRef]
30. Aloïse, S.; O'Keeffe, P.; Cubaynes, D.; Meyer, M.; Grum-Grzhimailo, A.N. Photoionization of Synchrotron-Radiation-Excited Atoms: Separating Partial Cross Sections by Full Polarization Control. *Phys. Rev. Lett.* **2005**, *94*, 223002. [CrossRef]
31. Petrov, I.; Peters, T.; Halfmann, T.; Aloıse, S.; O'Keeffe, P.; Meyer, M.; Sukhorukov, V.; Hotop, H. Lineshapes of the even $mp^5_{1/2}5n(p'/f')$ autoionizing resonances of Ar, Kr and Xe. *Eur. Phys. J. D* **2006**, *40*, 181–193. [CrossRef]
32. O'Keeffe, P.; Bolognesi, P.; Mihelic, A.; Moise, A.; Richter, R.; Cautero, G.; Stebel, L.; Sergo, R.; Pravica, L.; Ovcharenko, E.; et al. Photoelectron angular distributions from polarized Ne* atoms near threshold. *Phys. Rev. A* **2010**, *82*, 052522. [CrossRef]
33. O'Keeffe, P.; Gryzlova, E.V.; Cubaynes, D.; Garcia, G.A.; Nahon, L.; Grum-Grzhimailo, A.N.; Meyer, M. Isotopically Resolved Photoelectron Imaging Unravels Complex Atomic Autoionization Dynamics by Two-Color Resonant Ionization. *Phys. Rev. Lett.* **2013**, *111*, 243002. [CrossRef] [PubMed]
34. Bartschat, K.; Madison, D.H. Electron impact excitation of rare gases: Differential cross sections and angular correlation parameters for neon, argon, krypton and xenon. *J. Phys. B At. Mol. Opt. Phys.* **1987**, *20*, 5839–5863. [CrossRef]
35. Knight, R.D.; guo Wang, L. One-photon laser spectroscopy of the np and nf Rydberg series in xenon. *J. Opt. Soc. Am. B* **1985**, *2*, 1084–1087. [CrossRef]
36. L'Huillier, A.; Lompré, L.A.; Normand, D.; Morellec, J.; Ferray, M.; Lavancier, J.; Mainfray, G.; Manus, C. Spectroscopy of the np and nf even-parity Rydberg series in xenon by two-photon excitation. *J. Opt. Soc. Am. B* **1989**, *6*, 1644–1647. [CrossRef]
37. McCann, K.J.; Flannery, M.R. Photoionization of metastable rare-gas atoms (He*,Ne*,Ar*,Kr*,Xe*). *Appl. Phys. Lett.* **1977**, *31*, 599–601. [CrossRef]
38. Kau, R.; Petrov, I.D.; Sukhorukov, V.L.; Hotop, H. Experimental and theoretical cross sections for photoionization of metastable atoms near threshold. *J. Phys. B At. Mol. Opt. Phys.* **1996**, *29*, 5673–5698. [CrossRef]
39. Kopeika, N.S.; Shuker, R.; Yerachmiel, Y.; Gabai, Y.; Ih, C.S. Observation of Cooper minima in excited-*s*-state photoionization cross sections in neon and argon. *Phys. Rev. A* **1983**, *28*, 1517–1526. [CrossRef]

40. Petrov, I.D.; Lagutin, B.M.; Sukhorukov, V.L.; Knie, A.; Ehresmann, A. Correlation and polarization effects in two-photon photoionization of Ar. *Phys. Rev. A* **2016**, *93*, 033408. [CrossRef]
41. Petrov, I.D.; Sukhorukov, V.L.; Hotop, H. Photoionization of excited Ne* $(2p^5 3p, J = 3)$ atoms near threshold. *J. Phys. B At. Mol. Opt. Phys.* **2008**, *41*, 065205. [CrossRef]
42. Sukhorukov, V.L.; Petrov, I.D.; Schafer, M.; Merkt, F.; Ruf, M.W.; Hotop, H. Photoionization dynamics of excited Ne, Ar, Kr and Xe atoms near threshold. *J. Phys. B At. Mol. Opt. Phys.* **2012**, *45*, 092001. [CrossRef]
43. Gallagher, T.F. Quantum defect theory. In *Rydberg Atoms;* Cambridge Monographs on Atomic, Molecular and Chemical Physics; Cambridge University Press: Cambridge, UK, 1994; pp. 415–428. [CrossRef]
44. Ojha, P.C.; Burke, P.G. Photoionisation of the $3p^5 4s$ excited states of argon. *J. Phys. B At. Mol. Opt. Phys.* **1983**, *16*, 3513–3529. [CrossRef]
45. Petrov, I.D.; Sukhorukov, V.L.; Hollenstein, U.; Kaufmann, L.J.; Merkt, F.; Hotop, H. Autoionization dynamics of even Ar $(3p^5_{1/2}np', nf')$ resonances: Comparison of experiment and theory. *J. Phys. B At. Mol. Opt. Phys.* **2011**, *44*, 025004. [CrossRef]
46. Gryzlova, E.V.; O'Keeffe, P.; Cubaynes, D.; Garcia, G.A.; Nahon, L.; Grum-Grzhimailo, A.N.; Meyer, M. Isotope effects in resonant two-color photoionization of Xe in the region of the $5p^5(^2P_{1/2})4f[5/2]_2$ autoionizing state. *N. J. Phys.* **2015**, *17*, 043054. [CrossRef]
47. McKenna, C.; van der Hart, H.W. Multiphoton ionization cross sections of neon and argon. *J. Phys. B At. Mol. Opt. Phys.* **2003**, *37*, 457–470. [CrossRef]
48. van der Hart, H.W.; Greene, C.H. Multichannel photoionization spectroscopy of Ar: Total cross section and threshold photoelectrons. *Phys. Rev. A* **1998**, *58*, 2097–2105. [CrossRef]
49. Hamonou, L.; Lysaght, M.A.; van der Hart, H.W. Influence of autoionizing states on the pulse-length dependence of strong-field Ne^+ photoionization at 38.4 eV. *J. Phys. B At. Mol. Opt. Phys.* **2010**, *43*, 045601. [CrossRef]
50. Zhou, Z.; Chu, S.I. Time-dependent localized Hartree-Fock density-functional linear response approach for photoionization of atomic excited states. *Phys. Rev. A* **2009**, *79*, 053412. [CrossRef]
51. Edwards, A.K.; Rudd, M.E. Excitation of Auto-Ionizing Levels in Neon by Ion Impact. *Phys. Rev.* **1968**, *170*, 140–144. [CrossRef]
52. Olsen, J.O.; Anderson, N. Autoionizing levels in neon excited by low-energy heavy-ion impact. *J. Phys. B At. Mol. Opt. Phys.* **1977**, *10*, 101–110. [CrossRef]
53. Jureta, J.J.; Marinković, B.P.; Milosavljević, A.R.; Avaldi, L. Singly and doubly excited states in ejected electron spectra of neon at high incident electron energies. *Eur. Phys. J. D* **2015**, *69*, 74. [CrossRef]
54. Prince, K.C.; Allaria, E.; Callegari, C.; Cucini, R.; Ninno, G.D.; Mitri, S.D.; Diviacco, B.; Ferrari, E.; Finetti, P.; Gauthier, D.; et al. Coherent control with a short-wavelength free-electron laser. *Nat. Photonics* **2016**, *10*, 176. [CrossRef]
55. You, D.; Ueda, K.; Gryzlova, E.V.; Grum-Grzhimailo, A.N.; Popova, M.M.; Staroselskaya, E.I.; Tugs, O.; Orimo, Y.; Sato, T.; Ishikawa, K.L.; et al. New Method for Measuring Angle-Resolved Phases in Photoemission. *Phys. Rev. X* **2020**, *10*, 031070. [CrossRef]
56. Gryzlova, E.V.; Carpeggiani, P.; Popova, M.M.; Kiselev, M.D.; Douguet, N.; Reduzzi, M.; Negro, M.; Comby, A.; Ahmadi, H.; Waniel, V.; et al. Influence of an atomic resonance on the coherent control of the photoionization process. *Phys. Rev. Res.* **2022**, *4*, 033231. [CrossRef]
57. Gryzlova, E.V.; Ma, R.; Fukuzawa, H.; Motomura, K.; Yamada, A.; Ueda, K.; Grum-Grzhimailo, A.N.; Kabachnik, N.M.; Strakhova, S.I.; Rouzée, A.; et al. Doubly resonant three-photon double ionization of Ar atoms induced by an EUV free-electron laser. *Phys. Rev. A* **2011**, *84*, 063405. [CrossRef]
58. Zatsarinny, O. BSR: B-spline atomic R-matrix codes. *Comput. Phys. Commun.* **2006**, *174*, 273–356. [CrossRef]
59. Fischer, C.F. Towards B-Spline Atomic Structure Calculations. *Atoms* **2021**, *9*, 50. [CrossRef]
60. Zatsarinny, O.; Bartschat, K. B-spline calculations of oscillator strengths in noble gases. *Phys. Scr.* **2009**, *T134*, 014020. [CrossRef]
61. Zatsarinny, O.; Tayal, S.S. Photoionization of potassium atoms from the ground and excited states. *Phys. Rev. A* **2010**, *81*, 043423. [CrossRef]
62. Fischer, C.F.; Brage, T.; Johnsson, P. *Computational Atomic Structure. An MCHF Approach;* IOP Publishing: Bristol, UK, 1997.
63. Zatsarinny, O.; Fischer, C.F. A general program for computing angular integrals of the Breit–Pauli Hamiltonian with non-orthogonal orbitals. *Comput. Phys. Commun.* **2000**, *124*, 247–289. [CrossRef]
64. Wilden, D.G.; Hicks, P.J.; Comer, J. Electron impact studies of resonances and autoionizing states of neon. *J. Phys. B At. Mol. Phys.* **1977**, *10*, 1477–1486. [CrossRef]
65. *NIST Atomic Spectra Database (Version 5.8)*; National Institute of Standards and Technology: Gaithersburg, MD, USA, 2020. Available online: https://physics.nist.gov/asd (accessed on 18 May 2020).
66. Zeman, V.; Bartschat, K. Electron-impact excitation of the and states of neon. *J. Phys. B At. Mol. Opt. Phys.* **1997**, *30*, 4609–4622. [CrossRef]
67. Khakoo, M.A.; Wrkich, J.; Larsen, M.; Kleiban, G.; Kanik, I.; Trajmar, S.; Brunger, M.J.; Teubner, P.J.O.; Crowe, A.; Fontes, C.J.; et al. Differential cross sections and cross-section ratios for the electron-impact excitation of the neon $2p^5 3s$ configuration. *Phys. Rev. A* **2002**, *65*, 062711. [CrossRef]
68. Kheifets, A. Revealing the Target Electronic Structure with Under-Threshold RABBITT. *Atoms* **2021**, *9*, 66. [CrossRef]
69. Baier, S.; Grum-Grzhimailo, A.N.; Kabachnik, N.M. Angular distribution of photoelectrons in resonant photoionization of polarized atoms. *J. Phys. B At. Mol. Opt. Phys.* **1994**, *27*, 3363–3388. [CrossRef]

atoms

MDPI

Article

Absolute Double-Differential Cross Sections of Ultrasoft Isochromatic X-ray Radiation in Electron Scattering on Atoms

Aleksei S. Kornev [1], Boris A. Zon [1], Vladislav E. Chernov [1,*], Miron Ya. Amusia [2,†], Petr Kubelík [3] and Martin Ferus [3]

1 Physics Faculty, Voronezh State University, 394018 Voronezh, Russia
2 Racah Institute of Physics, The Hebrew University of Jerusalem, 91904 Jerusalem, Israel
3 J. Heyrovský Institute of Physical Chemistry, Academy of Sciences of the Czech Republic, Dolejškova 3, 18223 Prague, Czech Republic
* Correspondence: wladislaw.chernov@gmail.com
† Also at: A. F. Ioffe Physical-Technical Institute, 194021 St. Petersburg, Russia

Abstract: We calculate double-differential cross sections of ultrasoft X-ray bremsstrahlung in electron scattering by Ar, Kr, and Xe atoms in the soft-photon approximation. The calculations are done for the isochromatic spectra (i.e., dependence on the electron energy at a fixed photon energy of 165 and 177 eV). The results are consistent with the absolute values of the differential cross sections measured by Gnatchenko et al. (Phys. Rev. A **80**, 022707 (2009)) for the above-mentioned photon energies. For low electron energies, our theoretical isochromatic spectra are in quantitative agreement with the experimental data for Ar. For Kr, the agreement is qualitative while agreement with the Xe data is poor.

Keywords: ultrasoft X-ray bremsstrahlung; isochromatic spectra

Citation: Kornev, A.S.; Zon, B.A.; Chernov, V.E.; Amusia, M.Y.; Ferus, M.; Kubelík, P. Absolute Double-Differential Cross Sections of Ultrasoft Isochromatic X-ray Radiation in Electron Scattering on Atoms. *Atoms* **2022**, *10*, 86. https://doi.org/10.3390/atoms10030086

Academic Editor: Anatoli Kheifets, Gleb Gribakin and Vadim Ivanov

Received: 11 August 2022
Accepted: 21 August 2022
Published: 24 August 2022

1. Introduction

The long-wavelength range of the bremsstrahlung (BS) spectra has been studied for a long time because it is the region where the famous "infrared catastrophe" arises [1]. The relative intensity of BS in the long-wavelength range was first measured in [2] on thin metal targets bombarded by low-energy electrons. The experimental data were described well by the nonrelativistic Sommerfeld theory [3] which, in fact, is the theory of BS on a Coulomb center, accounting for the exponential screening of nuclei by the atomic electrons [4]. BS spectra play an important role in engineering and technical applications such as ion beam monitoring [5], plasma diagnostics [6], medical tomography imaging [7], etc.

In addition to the BS photon spectra, the dependence of BS cross sections on the scattering electron energy at a fixed photon energy (isochromatic spectra) is also of interest. These spectra were first measured for solid targets in [8], and the results are cited in a well-known monograph [9]. Korsunsky et al. [8] observed an increase of the BS probability from the zero value (at the minimal possible electron energy equal to the fixed photon energy) and a decrease of the BS probability at higher electron energy. A similar shape of the isochromatic spectra was registered for atomic targets in [10,11] in which the quasi-resonance character of the isochromatic spectra was observed. It should be noted that the detection technique for the isochromatic BS soft X-photons was proposed in [12,13]. A technique for detecting isochromatic BS photons in the UV range was proposed earlier in [14].

An analysis of the Sommerfeld [3] formula performed in [10] showed that an isochromatic spectrum maximum is achieved at the initial electron energy, E_i, equal to the following:

$$E_i^{\max} = 1.53\,\hbar\omega, \tag{1}$$

where ω is the photon frequency.

It should be noted that the knowledge of BS cross sections on atomic targets is important for studying BS generated in electron scattering on molecules [15] and clusters [16].

The "atomic" BS data are also used in Monte Carlo simulation of BS on solid targets (see, e.g., [17,18]).

All the above-cited works studied relative BS intensities only. At the same time, there exist some absolute BS cross-section value measurements. For example, high-precision ($\simeq$5.5%) absolute cross sections of BS on thin C, Al, Te, Ta, and Au targets have been recently measured in [19], which also contains references to earlier absolute BS cross-section measurements on solid targets.

To the authors' best knowledge, the first absolute BS intensity measurements on atoms (Ne, Ar, Kr, and Xe) were performed in [20] for incident electron energies ranging from 28 to 50 keV. The experimental results were shown to be in qualitative agreement with the theory that takes into account not only the traditional BS mechanism (developed by Sommerfeld [3] for the nonrelativistic case and by Bethe and Heitler [21] and Sauter [22] for the relativistic case) but also the polarization bremsstrahlung (PBS) theory [23–25], which takes into account the photons emitted by the electrons of the atomic target. However, the quantitative difference between the experimental and theoretical results was quite significant, and this discrepancy increased with the decrease of BS photon energy. Similarly, discrepancy exists between the experimental cross sections measured in [19] and the data tabulated in [26,27], which also increases with the decrease of the BS photon energy. García-Alvarez et al. [19] noted this fact as one important result of their study. Similar discrepancy between theory and experiment was also observed in recent calculations [28].

However, the authors in [19,20] did not consider BS for photon energies in the ultrasoft X-ray region. Therefore, the absolute BS cross sections measured in [12,13] in the low-energy photon energy range are of great interest. These authors recorded the BS photon spectra for 600 eV electrons scattered on Ar, Kr, and Xe atoms, as well as the isochromatic spectra for the electron energies from 0.4 to 2 keV. Their results differ by 3–4 times from the calculations by Pratt et al. [29,30], who used the radial electron wavefunctions obtained in partial-wave series by numerically integrating the radial Dirac equation with a relativistic self-consistent screened potential. In recent calculations [28], the Dirac equation for the continuum wave function was solved by the power-series method with the interaction potential obtained from the Kohn–Sham density functional theory.

In the present work, we develop an interpretation of the experimental results of [12,13] using the soft-photon approximation (SPA); the validity conditions are discussed below. The general SPA formulas are presented in Section 2. SPA was used for interpretation of the ultrasoft X-ray spectra recorded in [31,32]. The BS photon spectra were measured in [12,13] for Ar, Kr, and Xe gaseous targets. Two types of BS spectra were recorded:

(i) "Convenient" BS (or photon BS spectrum) that is the spectral and angular distribution of BS photons emitted in collisions with electrons whose energy, E_i, is fixed. Such spectra were analyzed in our recent study [33].
(ii) Isochromatic BS spectrum that is the angular distribution of BS photons as function of the colliding electron energy, E_i (which varies in some interval), at a fixed photon energy, $\hbar\omega$. The isochromatic spectra are analyzed in Section 3 of the present work.

The results are discussed in Section 4. The main conclusions are given in Section 5.

2. Methods

2.1. General Formalism of Soft-Photon Approximation (SPA)

In the nonrelativistic SPA framework, the double-differential cross section, $d\sigma_{BS}$, of BS in the electron scattering on a target can be factorized. In particular, $d\sigma_{BS}$ is the product of the differential cross section, $d\sigma_s$, of the electron scattering on the same target without γ-radiation, by the probability, dw_γ, of the photon radiation [1]:

$$\begin{aligned} d\sigma_{BS} &= d\sigma_s\, dw_\gamma, \\ dw_\gamma &= \frac{\alpha}{4\pi^2}\left[\frac{\mathbf{k}}{\omega}(\mathbf{v}_i - \mathbf{v}_f)\right]^2 \frac{d\omega}{\omega}\, d\Omega_\gamma. \end{aligned} \tag{2}$$

where $\alpha \simeq 1/137$ is the fine structure constant, $\mathbf{v}_i$, $\mathbf{v}_f$ are the electron velocity in the initial and final state, respectively; $\mathbf{k}$ is the photon wave vector, and $\mathrm{d}\Omega_\gamma$ is the element of solid angle to which the photon is radiated. The cross section in (2) is averaged over the electron spin in the initial state and summed over the photon polarizations and the electron spin in the final state.

If the ejection direction of the electron is not fixed, we should integrate Equation (2) over this direction. To do this, we choose the z axis directed along the $\mathbf{v}_i$ vector and write $\mathrm{d}\sigma_s$ in the following form:

$$\mathrm{d}\sigma_s(E_i,\theta) = \sigma_s(E_i,\theta)\,\mathrm{d}\Omega_{\mathbf{v}_f}, \tag{3}$$

where θ is the polar angle of the $\mathbf{v}_f$ vector, E_i is the energy of the impact electron. Then, Equation (2) can be written in the following form:

$$\begin{aligned}
\mathrm{d}\sigma_{\mathrm{BS}} &= \frac{\alpha}{4\pi^2\omega}\sigma_s(E_i,\theta)\left[\frac{v_i^2}{c^2}\cos^2\Theta - \frac{2v_i}{c\omega}(\mathbf{k}\mathbf{v}_f)\cos\Theta\right.\\
&\quad\left.+\frac{(\mathbf{k}\mathbf{v}_f)^2}{\omega^2}\right]\mathrm{d}\Omega_{\mathbf{v}_f}\,\mathrm{d}\omega\,\mathrm{d}\Omega_\gamma\\
&= \frac{\alpha}{4\pi^2c^2\omega}\sigma_s(E_i,\theta)\left[v_i^2\cos^2\Theta\right.\\
&\quad - 2v_iv_f\cos\Theta(\sin\Theta\sin\theta\cos\varphi+\cos\Theta\cos\theta)+\\
&\quad\left.+v_f^2(\sin\Theta\sin\theta\cos\varphi+\cos\Theta\cos\theta)^2\right]\mathrm{d}\Omega_{\mathbf{v}_f}\,\mathrm{d}\omega\,\mathrm{d}\Omega_\gamma.
\end{aligned}$$

where φ is the azimuthal angle of the $\mathbf{v}_f$ vector, Θ is the polar angle of the $\mathbf{k}$ vector (i.e., the angle between $\mathbf{v}_i$ and $\mathbf{k}$ vectors), and the azimuthal angle of the $\mathbf{k}$ vector is assumed to be zero. After integration over φ, we obtain the following:

$$\mathrm{d}\sigma_{\mathrm{BS}} = \frac{\alpha}{2\pi c^2\omega}\sigma_s(E_i,\theta)Q(\Theta,\theta)\sin\theta\,\mathrm{d}\theta\,\mathrm{d}\omega\,\mathrm{d}\Omega_\gamma, \tag{4}$$

where

$$Q(\Theta,\theta) = v_i^2\cos^2\Theta - 2v_iv_f\cos^2\Theta\cos\theta + v_f^2\left(\tfrac{1}{2}\sin^2\Theta\sin^2\theta+\cos^2\Theta\cos^2\theta\right). \tag{5}$$

2.2. SPA: Validity Conditions

SPA is valid when the irradiated photon energy is low as compared to the scattering electron energy, E_i:

$$\omega \ll E_i/\hbar. \tag{6}$$

Apart the condition (6), it is also required for SPA validity that the momentum, $\mathbf{q} = \mathbf{p}_i - \mathbf{p}_f$, transferred to the atom from the scattering electron, is much higher than the electron momentum change due to irradiation of the photon [1]. In the nonrelativistic approximation, this condition takes the following form [1]:

$$\hbar\omega/(v_iq) \ll 1. \tag{7}$$

It is seen that the condition (7) is violated at low transferred momentum values, i.e., at small scattering angles, θ. However, the contribution of the small angles into BS cross section is small due to the angular factor, $Q(\Theta,\theta)$ in Equation (4). This factor is plotted in Figure 1 as function of the scattering angle, θ, for the value of $\Theta = 83^\circ$ used experimentally in [12,13]. It is clearly seen that $Q(\Theta,\theta)$ is small for $\theta \to 0$.

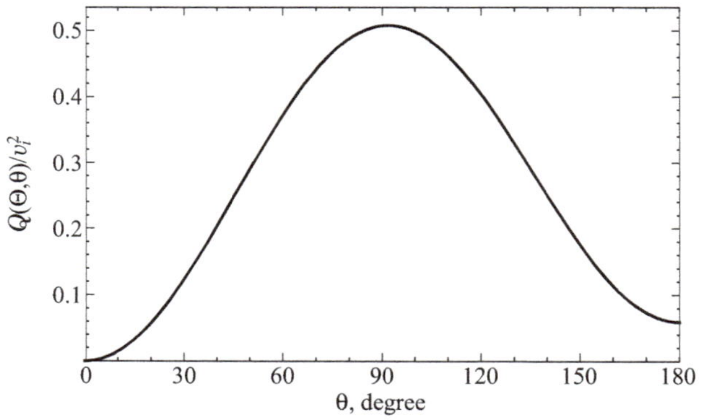

Figure 1. The angular factor, $Q(\Theta,\theta)/v_i^2$, from Equation (5) at $\Theta = 83°$ and $v_f \simeq v_i$.

3. Results on Isochromatic Spectra

Wavelength dependence of the BS intensity was obtained in experiment [13] and analyzed theoretically in [33].

Now, we proceed to the analysis of the isochromatic spectra. These spectra show the dependence of BS intensity on the electron energy, E_i, at a fixed photon energy, $\hbar\omega$. The authors in [13] list the experimental values for a modified (as compared to that in Equation (4)) differential BS cross section. This modification, as compared to Equation (4), implies calculating the angle-differential cross section of BS photon radiation into a finite small wavelength interval, $\Delta\lambda$, integrated over the scattering angle, θ,

$$\frac{d\sigma_{BS}}{d\Omega_\gamma} = \frac{\alpha}{2\pi c^2}\frac{\Delta\lambda}{\lambda}\int_0^{\pi} \sigma_s(E_i,\theta) Q(\Theta,\theta)\sin\theta\, d\theta, \tag{8}$$

where $\lambda = 2\pi\hbar/k$ is the photon wavelength, the factor $Q(\Theta,\theta)$ is defined in Equation (5). In other words, a change $d\omega/\omega \to \Delta\lambda/\lambda$ should be done in Equation (4). In the experiment presented in [13], photons were registered in the wavelength interval of $\Delta\lambda = 0.1$ nm. Equation (8) requires knowledge of the following integrals:

$$\mathcal{J}_{ab}(E_i) = \int_0^{\pi} \sigma_s(E_i,\theta)\cos^a\theta \sin^{1+b}\theta\, d\theta, \quad (ab) = (00), (10), (20), (02). \tag{9}$$

In particular, the cross section of elastic electron–atom scattering is $\sigma_{el}(E_i) = 2\pi\mathcal{J}_{00}(E_i)$.

Equation (8) implies using differential cross sections of elastic electron–atom scattering, $\sigma_s(E_i,\theta)$, tabulated as functions of the scattering angle, θ. Such data can be obtained by numerical simulation using various models for interaction of the incident electrons with atoms. For instance, the cross-section database [34] contains data for the majority of atoms of the Periodic system. These data were used in our earlier work [33] for analysis of experimental BS photon spectra at fixed energy of an electron scattering on rare gas atoms [13].

In the case of electron scattering on argon atom, the data calculated in [35] can be used as an alternative to the NIST database [34]. The results of the cross-section calculation at particular electron energies, E_i, using both datasets are presented in Table 1.

Table 1. Simulated cross section in elastic electron–atom scattering (Å^2).

E_i, keV	Ar			Kr		Xe	
	[34]	[35]	[36]	[34]	[36]	[34]	[36]
0.4	2.30	2.21	1.99	2.98	2.92	4.20	3.82
0.5	2.05	1.97	1.76	2.69	2.51	3.86	3.42
0.6	1.85	1.78	1.59	2.46	2.27	3.58	3.11
0.7	1.70	1.63	1.45	2.28	2.03	3.34	2.76
0.8	1.57	1.51	1.34	2.14	1.94	3.14	2.71
0.9	1.47	1.41	1.25	2.01	1.81	2.98	2.64
1.0	1.38	1.33	1.17	1.90	1.73	2.83	2.29
1.5	1.07	1.03	0.91	1.53	1.19	2.33	1.94
2.0	0.88	0.86	0.76	1.30	1.01	2.02	1.65

It can be seen that the cross sections of elastic electron scattering on Ar calculated in [34,35] coincide within 4%. However, the angle-differential cross sections differ significantly. As it is seen from Figure 2, the differential cross section has a sharp maximum for the forward scattering. This maximum calculated in [35] is two times higher as compared to the results of [34], and the differential cross section drops down for the scattering angles $\theta > 15^\circ$. Cross sections of electron scattering on Ar, Kr, and Xe atoms were also calculated in [36]. They are presented in Table 1 as well and their values are less as compared to those of [34]. This difference increases with the electron energy, E_i, up to 40% at $E_i = 2$ keV. Unfortunately, the angle-differential cross sections are not listed in [36].

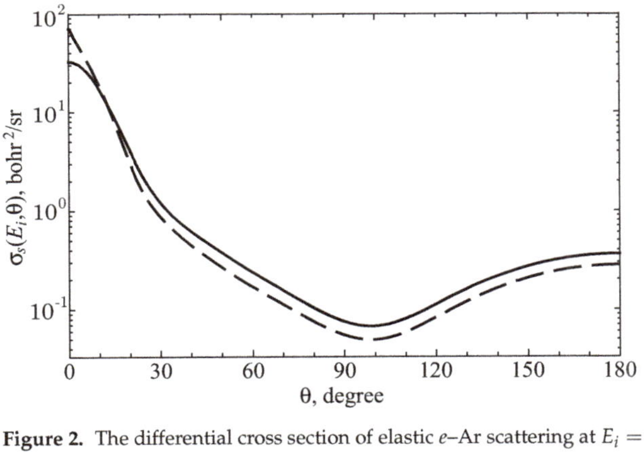

Figure 2. The differential cross section of elastic e–Ar scattering at $E_i = 0.4$ keV from [34] (solid line) and from Figure 7a of [35] (dashed line).

The results of our calculations of isochromatic BS spectra by Equation (8) are given in Figures 3–5 in comparison with the experimental data shown in Figures 5–7 of [13], respectively. We note a misprint in [13]: the cross section shown in these figures should be measured in mbarn/sr, not barn. As seen from these figures, the calculated BS cross sections increase with the initial electron energy, E_i, for Kr and Xe, and slightly decrease for Ar. We recall that calculations of the isochromatic spectra according to Pratt et al. [29,30] (as given in [13]) predict a monotonic decrease of the BS intensity as a function of E_i.

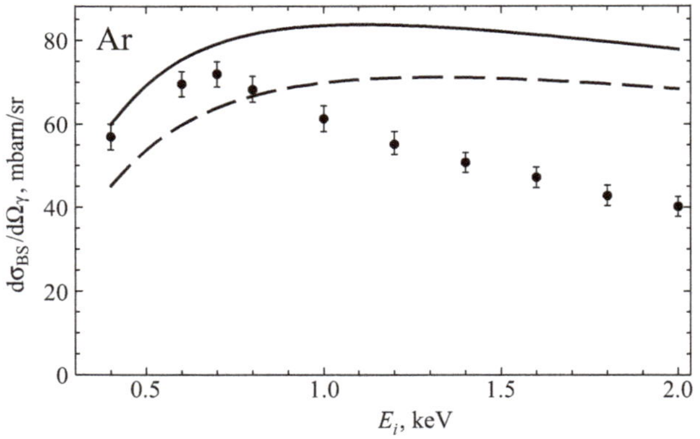

Figure 3. Isochromatic spectrum for Ar: double-differential BS cross section (8) as a function of the initial electron energy at the fixed photon energy of 177 eV. Solid line: elastic-scattering differential cross sections are taken from [34]; dashed line: from [35]. Black points with error bars: experimental values adopted from Figure 5 in [13] by manual digitization.

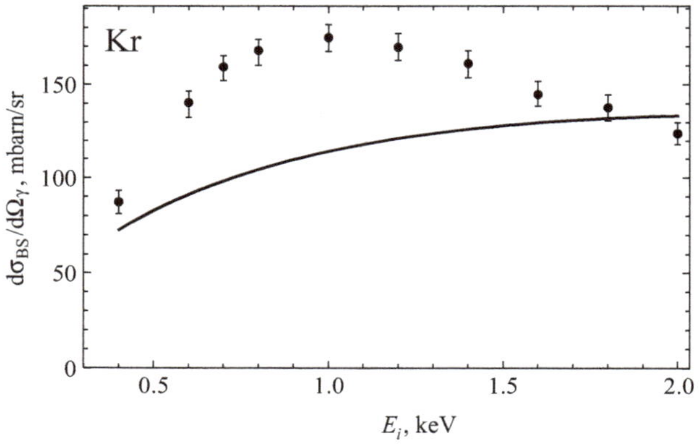

Figure 4. Same as in Figure 3 for Kr, $\hbar\omega$ = 165 eV. Experimental values adopted from Figure 6 in [13].

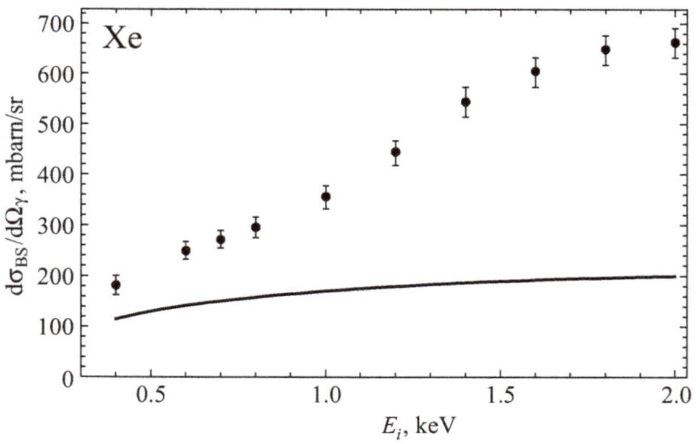

Figure 5. Same as in Figure 4 for Xe, $\hbar\omega$ = 177 eV. Experimental values adopted from Figure 7 in [13].

Note the maximum value of BS cross sections is reached at $E_i = E_i^{\max}$ (see Equation (1)) and predicted by Sommerfeld [3] in the pure Coulomb potential. Under the conditions of the experiment of [13], this maximum should be observed at $E_i \approx 0.26$ keV.

4. Discussion

In the geometry of the experiment [13] ($\Theta = 83^\circ$), according to Equation (5), the main contribution into the BS cross section (8) is made by the following term:

$$\left(\frac{d\sigma_{\mathrm{BS}}}{d\Omega_\gamma}\right)_0 = \frac{\alpha}{4\pi c^2}\frac{\Delta\lambda}{\lambda}\, v_f^2 \sin^2\Theta\, \mathcal{J}_{02}(E_i), \tag{10}$$

where the factor $\mathcal{J}_{02}(E_i)$ is defined by Equation (9). The E_i-dependence in Equation (10) is due not only to the v_f^2 factor but also to the $\mathcal{J}_{02}(E_i)$ dependence.

For the total (scattering + impact ionization) cross section $\sigma_{\mathrm{tot}}(E_i)$, this dependence on E_i in the range 0.5–2.0 keV, according to [36], can be approximated by the following equation:

$$\sigma_{\mathrm{tot}}(E_i) \sim E_i^{-B}, \tag{11}$$

where parameter B varies from 0.90 for He to 0.59 for Xe. These values make the slow dependence (11) unable to compensate the linear increase of v_f^2. However, the dependence of $\mathcal{J}_{02}$ on E_i can differ from (11). Of course, the discrepancy between theory and experiment can be also due to inaccuracy of SPA in this case.

In the case of argon (see Figure 3), the maximum in the isochromatic spectrum measured in [13] appears at the electron energy, $E_i^{\mathrm{expt}} = 0.7$ keV, which is greater compared to that given by Equation (1). For the energies less than 0.7 keV, our theoretical results are in satisfactory accord with the experiment. However, theoretical dependence of BS cross section on E_i, obtained using the differential cross section of elastic scattering from [34] has a less sharp maximum at $E_i^{\mathrm{theor}} = 1.11$ keV. When the differential cross-section data from [35] are used, the maximum appears at $E_i^{\mathrm{theor}} = 1.31$ keV. In both cases, the calculated cross section decreases with E_i slower compared to the experiment. Unfortunately, we have no explanation of such a behavior of the cross section. We note that using the elastic cross sections from [35] results in lesser BS cross section compared to using the data from [34] that make better agreement with the experiment. This can be caused by the fact that the results in [35] give (i) a more sharp maximum at the forward differential cross section and (ii) less magnitude of the cross section in the $\theta > 15^\circ$ domain, as compared to [34] (see also Figure 2). It is this angular domain that makes the main contribution into the value of $\mathcal{J}_{02}(E_i)$ in (9) due to $\sin^3\theta$ in the integral. Thus, in the case of Ar, we have quantitative agreement with the experiment at the electron energy $E_i < E_i^{\mathrm{expt}} = 0.7$ keV, and a qualitative agreement for higher energies.

In the case of krypton (see Figure 4), the experimental isochromatic spectrum [13] has the maximum at $E_i^{\mathrm{expt}} = 1.0$ keV which is higher than the value given by (1) as well. Our BS cross section calculated using the differential cross section of elastic scattering from [34], slowly increases in the studied energy range, $E_i = 0.4$–2.0 keV. Unfortunately, the authors in [35] do not provide differential cross sections of elastic *e*–Kr scattering, so our calculations for Kr can achieve only qualitative agreement with the experiment yet.

An even worse situation takes place for xenon (see Figure 5). Both experimental and theoretical isochromatic spectra increase with the electron energy, E_i, the former [13] increasing much faster than the latter. Extrapolating the Ar and Kr data, one can suppose that E_i^{expt} increases with the nuclear charge of the atomic target. Probably, one has $E_i^{\mathrm{expt}} > 2$ keV for Xe. A similar discrepancy between theoretical and experimental BS spectra in *e*–Xe scattering with fixed electron energy, $E_i = 0.6$ keV, was noted in our earlier work [33]. A possible explanation of this discrepancy could be due to greater role of so-called polarization BS [23,31,37] for Xe, as compared to the lighter rare gases. Indeed, the main atomic feature responsible for the polarization bremsstrahlung is the dipole polarizability [23]. The contribution of polarization BS into the experimental BS spectra of rare gases should increase with the nuclear charge since the polarizabilities of these gases are 11, 17 and 27 bohr^3 for Ar, Kr and Xe, respectively [38]. The discrepancy between theory and experiment for Kr isochromatic spectra can be also due to polarization BS, as well as to some inelastic processes, such as impact ionization (in Equations (4) and (8), we take into account conventional BS and elastic cross section only).

The main disadvantage of the presented theory is that it does not reproduce the BS intensity decrease for sufficient high electron energy, $E_i > E_i^{\rm expt}$, as seen from Figure 4.

5. Conclusions

The isochromatic spectra recorded in [10] and studied in detail in [13] are quasi-resonant, i.e., demonstrate a prominent maxima at some electron energy $E_i = E_i^{\rm expt}$. It should be noted that the experimental values of $E_i^{\rm expt}$ are much higher than those given by the BS theory, $E_i^{\rm max}$, in pure Coulomb potential (1). This excess of $E_i^{\rm expt}$ value increases with the target nucleus charge. Therefore, the isochromatic spectra reported in [13] for Xe did not display any maxima, since the electron energies studied in [13] did not exceed 2 keV, which is probably lower than the $E_i^{\rm max}$ value for Xe. Our model gives the first quantitative description of the increase of BS intensity at low $E_i < E_i^{\rm max}$ in the isochromatic spectra for Ar and the first qualitative description for Kr and Xe. The other theoretical models [29,30] predict a monotonic (without maxima) decrease of the BS intensity as a function of E_i.

It should be emphasized that the agreement with experiment demonstrated in this work is achieved with the help of very simple analytical expressions.

Unlike the BS spectra with the fixed electron energy [13,33], the isochromatic spectra can possibly be more sensitive indicators of the polarization component of BS since this component appears not only in xenon, but probably also in krypton.

Author Contributions: Conceptualization, B.A.Z. and M.Y.A.; methodology, B.A.Z.; visualization, A.S.K. and V.E.C.; writing—original draft preparation, B.A.Z. and A.S.K.; writing—review and editing, M.Y.A. and V.E.C.; data curation, M.F.; software, P.K.; validation, A.S.K. and V.E.C.; formal analysis, A.S.K. and V.E.C.; funding acquisition, P.K. and V.E.C. All authors have read and agreed to the published version of the manuscript.

Funding: This research was funded by joint project of Czech Science Foundation (grant number 20-10591J) and Russian Foundation of Basic Research (grant number 19-52-26006) and Ministry of Science and Higher Education of the Russian Federation (grant number FZGU-2020-0035) and ERDF/ESF "Centre of Advanced Applied Sciences" (grant number CZ.02.1.01/0.0/0.0/16_019/0000778).

Institutional Review Board Statement: Not applicable.

Informed Consent Statement: Not applicable.

Data Availability Statement: Not applicable.

Acknowledgments: The authors express their deep gratitude to A. A. Tkachenko for the detailed discussion of the experimental results presented in [12,13]. We are grateful to Prof. Fazlul Haque for kindly provided numerical data on cross section published in [35].

Conflicts of Interest: The authors declare no conflicts of interest.

Abbreviations

The following abbreviations are used in this manuscript:

BS	Bremsstrahlung
SPA	soft-photon approximation

References

1. Akhiezer, A.I.; Berestetskiĭ, V.B. *Quantum Electrodynamics; Interscience Monographs and Texts in Physics and Astronomy*; Interscience Publishers: New York, NY, USA, 1965; Volume 11.
2. Peterson, T.J.; Tomboulian, D.H. Low-Energy Continuous X-Ray Spectrum in the 80 Å to 180 Å Region. *Phys. Rev.* **1962**, *125*, 235–241. [CrossRef]
3. Sommerfeld, A. *Atombau und Spektrallinien*; Friedr. Vieweg und Sohn: Braunschweig, Germany, 1939; Volume 2.
4. Kirkpatrick, P.; Wiedmann, L. Theoretical Continuous X-Ray Energy and Polarization. *Phys. Rev.* **1945**, *67*, 321–339. [CrossRef]
5. Ralite, F.; Koumeir, C.; Guertin, A.; Haddad, F.; Mouchard, Q.; Servagent, N.; Métivier, V. Bremsstrahlung X-rays as a non-invasive tool for ion beam monitoring. *Nucl. Instrum. Methods Phys. Res. Sect. B* **2021**, *500–501*, 76–82. [CrossRef]
6. Kwak, S.; Hergenhahn, U.; Höfel, U.; Krychowiak, M.; Pavone, A.; Svensson, J.; Ford, O.; König, R.; Bozhenkov, S.; Fuchert, G.; et al. Bayesian inference of spatially resolved Z_{eff} profiles from line integrated bremsstrahlung spectra. *Rev. Sci. Instrum.* **2021**, *92*, 043505. [CrossRef]
7. Talebi, A.S.; Rajabi, H. Outcomes following transarterial radioembolization with ^{90}Y and nanoparticles loaded resin microspheres. *Appl. Radiat. Isot.* **2022**, *188*, 110405. [CrossRef]
8. Korsunsky, M.I.; Walther, A.K.; Ivanov, A.V.; Zypkin, S.I.; Ganenko, V.E. An investigation of "Bremsstrahlung" by means of excited In^{115} nuclei. *J. Phys.* **1943**, *7*, 129–137.
9. Massey, H.S.W.; Burhop, E.H.S.; Gilbody, H.B. *Electronic and Ionic Impact Phenomena. Vol. II. Electron Collisions with Molecules and Photo-Ionization*; Clarendon Press: Oxford, UK, 1969.
10. Gnatchenko, E.V.; Tkachenko, A.A.; Verkhovtseva, E.T.; Zon, B.A. Ultrasoft X-Ray Bremsstrahlung Isochromatic Spectra from 300–2000 eV Electrons on Ar and Kr. *Phys. Rev. Lett.* **2005**, *95*, 023002. [CrossRef]
11. Verkhovtseva, E.T.; Gnatchenko, E.V.; Tkachenko, A.A.; Zon, B.A. Resonance structures in polarization bremsstrahlung from electron–atom collisions. *Rad. Phys. Chem.* **2006**, *75*, 1115 – 1134. [CrossRef]
12. Gnatchenko, E.V.; Tkachenko, A.A.; Nechay, A.N. Absolute differential bremsstrahlung cross section for the scattering of 0.6-keV electrons by xenon atoms. *JETP Lett.* **2007**, *86*, 292–296. [CrossRef]
13. Gnatchenko, E.V.; Nechay, A.N.; Tkachenko, A.A. Absolute differential bremsstrahlung cross sections for 0.4–2-keV electrons scattered by Ar, Kr, and Xe atoms. *Phys. Rev. A* **2009**, *80*, 022707. [CrossRef]
14. Prince, K.C. Detector for Bremsstrahlung-Isochromatic-Spectroscopy (BIS). U.S. Patent No. 4,871,915, 3 October 1989.
15. Prajapati, S.; Singh, B.; Singh, B.; Shanker, R. Study of anisotropy of bremsstrahlung radiation emitted from 4.0 keV electrons in scattering by CH_4 molecule. *Rad. Phys. Chem.* **2018**, *153*, 92–97. [CrossRef]
16. Gnatchenko, E.V.; Nechay, A.N.; Samovarov, V.N.; Tkachenko, A.A. Polarization bremsstrahlung from xenon atoms and clusters: A cooperative effect contribution. *Phys. Rev. A* **2010**, *82*, 012702. [CrossRef]
17. Li, L.; An, Z.; Zhu, J.; Tan, W.; Sun, Q.; Liu, M. Bremsstrahlung of 5–25-keV electron impact with Al, Cu, Ag, Te, and Au thick solid targets with no polarization contribution. *Phys. Rev. A* **2019**, *99*, 052701. [CrossRef]
18. Li, L.; An, Z.; Zhu, J.; Liu, M. Experimental thick target bremsstrahlung spectra produced by 5–25 keV electrons for $6 \leq Z \leq 82$ and comparison with Monte Carlo PENELOPE simulations. *Nucl. Instrum. Methods Phys. Res. Sect. B* **2019**, *445*, 13–17. [CrossRef]
19. García-Alvarez, J.A.; Fernández-Varea, J.M.; Vanin, V.R.; Maidana, N.L. Electron-atom bremsstrahlung cross sections in the 20–100 keV energy region: Absolute measurements for $6 \leqslant Z \leqslant 79$ and comparison with theoretical databases. *J. Phys. B At. Mol. Opt. Phys.* **2018**, *51*, 225003. [CrossRef]
20. Portillo, S.; Quarles, C.A. Absolute Doubly Differential Cross Sections for Electron Bremsstrahlung from Rare Gas Atoms at 28 and 50 keV. *Phys. Rev. Lett.* **2003**, *91*, 173201. [CrossRef]
21. Bethe, H.; Heitler, W. On the stopping of fast particles and on the creation of positive electrons. *Proc. R. Soc. London Ser. A* **1934**, *146*, 83–112. [CrossRef]
22. Sauter, F. Über den atomaren Photoeffekt bei großer Härte der anregenden Strahlung. *Ann. Phys.* **1931**, *401*, 217–248. [CrossRef]
23. Amus'ya, M.Y.; Biumistrov, V.M.; Zon, B.A.; Tsytovich, V.N.; Astapenko, V.A.; Kleiman, E.B.; Korol', A.V.; Krotov, Y.A.; Kukushkin, A.B.; Lisitsa, V.S.; et al. *Polarization Bremsstrahlung*; Physics of Atoms and Molecules; Springer US: New York, NY, USA, 1992. [CrossRef]
24. Astapenko, V. *Polarization Bremsstrahlung on Atoms, Plasmas, Nanostructures and Solids*; Springer Series on Atomic, Optical, and Plasma Physics; Springer: Berlin/Heidelberg, Germany, 2013; Volume 23. [CrossRef]
25. Korol, A.V.; Solovyov, A.V. *Polarization Bremsstrahlung*; Springer Series on Atomic Optical and Plasma Physics (SSAOPP); Springer: New York, NY, USA, 2014; Volume 23. [CrossRef]
26. Kissel, L.; Quarles, C.A.; Pratt, R.H. Shape functions for atomic-field bremsstrahlung from electrons of kinetic energy 1–500 keV on selected neutral atoms $1 \leq Z \leq 92$. *At. Data Nucl. Data Tables* **1983**, *28*, 381–460. [CrossRef]
27. Seltzer, S.M.; Berger, M.J. Bremsstrahlung energy spectra from electrons with kinetic energy 1 keV–10 GeV incident on screened nuclei and orbital electrons of neutral atoms with $Z = 1$–100. *At. Data Nucl. Data Tables* **1986**, *35*, 345–418. [CrossRef]
28. Poškus, A. Shape functions and singly differential cross sections of bremsstrahlung at electron energies from 10 eV to 3 MeV for $Z = 1$–100. *At. Data Nucl. Data Tables* **2019**, *129–130*, 101277. [CrossRef]
29. Pratt, R.H.; Tseng, H.K.; Lee, C.M.; Kissel, L.; MacCallum, C.; Riley, M. Bremsstrahlung energy spectra from electrons of kinetic energy 1 keV$\leq T_1 \leq$2000 keV incident on neutral atoms 2$\leq Z \leq$92. *At. Data Nucl. Data Tables* **1977**, *20*, 175–209. [CrossRef]

30. Pratt, R.H.; Tseng, H.K.; Lee, C.M.; Kissel, L.; MacCallum, C.; Riley, M. Bremsstrahlung energy spectra from electrons of kinetic energy 1 keV$\leq T_1 \leq$2000 keV incident on neutral atoms 2$\leq Z \leq$92. *At. Data Nucl. Data Tables* **1981**, *26*, 477–481. [CrossRef]
31. Verkhovtseva, É.T.; Gnatchenko, E.V.; Zon, B.A.; Nekipelov, A.A.; Tkachenko, A.A. Bremsstrahlung in electron scattering by xenon. *Sov. Phys. JETP* **1990**, *71*, 443–448.
32. Zon, B.A. Bremsstrahlung in intermediate-energy electron scattering by noble gas atoms. *Sov. Phys. JETP* **1995**, *80*, 655–656.
33. Kornev, A.S.; Zon, B.A.; Amusia, M.Y. Absolute cross sections of ultra-soft x-ray radiation in electron scattering upon atoms and the soft-photons approximation. *Phys. Lett. A* **2019**, *383*, 2488–2491. [CrossRef]
34. Jablonski, A.; Salvat, F.; Powell, C.J.; Lee, A.Y. *NIST Electron Elastic-Scattering Cross-Section Database Version 4.0*; NIST Standard Reference Database Number 64; National Institute of Standards and Technology: Gaithersburg, MD, USA, 2016; p. 20899.
35. Haque, M.M.; Haque, A.K.F.; Jakubassa-Amundsen, D.H.; Patoary, M.A.R.; Basak, A.K.; Maaza, M.; Saha, B.C.; Uddin, M.A. $e^{\pm}$ Ar scattering in the energy range 1 eV $\leq E_i \leq$ 0.5 GeV. *J. Phys. Commun.* **2019**, *3*, 045011. [CrossRef]
36. Vinodkumar, M.; Limbachiya, C.; Antony, B.; Joshipura, K.N. Calculations of elastic, ionization and total cross sections for inert gases upon electron impact: Threshold to 2 keV. *J. Phys. B At. Mol. Opt. Phys.* **2007**, *40*, 3259–3271. [CrossRef]
37. Amusia, M.Y.; Chernysheva, L.V.; Lee, T.J.; Almlöf, J. Bremsstrahlung spectrum on Xe for 600 eV electrons. *J. Phys. B: At. Mol. Opt. Phys.* **1990**, *23*, 2899. [CrossRef]
38. Rice, J.E.; Taylor, P.R.; Korol, A.V. The Detrmination of Accurate Dipole Polarizabilities α and γ for the Noble Gases. *J. Chem. Phys.* **1991**, *94*, 4972. [CrossRef]

atoms

MDPI

Article

Satellite Excitations and Final State Interactions in Atomic Photoionization

Victor G. Yarzhemsky [1,*] and Yury A. Teterin [2,3]

1 Kurnakov Institute of General and Inorganic Chemistry of the Russian Academy of Sciences, 31 Leninsky, 119991 Moscow, Russia
2 Russian Research Center "Kurchatov Institute", 123182 Moscow, Russia; teterin_ya@nrcki.ru
3 Chemistry Department, Moscow State University, 119991 Moscow, Russia
* Correspondence: vgyar@igic.ras.ru

Abstract: Satellite excitations and final state configuration interactions appear due to the many-electron correlations and result in a photoelectron spectrum complex final state structure instead of single lines corresponding to one-hole states. In the present work, both processes are considered in a framework of the many-body perturbation theory, and two techniques, namely the spectral function and CI (configuration interaction) methods are considered. It is shown that for the calculation of satellite lineshapes and low-energy Auger decay, the spectral function method is more appropriate, but in the case of strong final state interactions, the methods of solution of Dyson equation or secular matrix are superior. The results obtained for satellites and low energy Auger decay in the Ne 1s, Ne 2p photoelectron spectra, the Co 3s, and the Th 5p photoelectron spectra are in agreement with the experimental data.

Keywords: atomic photoionization; photoelectron spectroscopy; many-electron effects; satellites; Ne 1s; Ne 2p; Co 3s; Th 5p photoelectron spectra

Citation: Yarzhemsky, V.G.; Teterin, Y.A. Satellite Excitations and Final State Interactions in Atomic Photoionization. *Atoms* **2022**, *10*, 73. https://doi.org/10.3390/atoms10030073

Academic Editors: Anatoli Kheifets, Gleb Gribakin and Vadim Ivanov

Received: 5 May 2022
Accepted: 5 July 2022
Published: 14 July 2022

1. Introduction

The interaction of a single vacancy with more complex states with two vacancies and one excited electron results in a breakdown of the one-electron picture in the 4p photoelectron spectra of lanthanides and the 5p spectra of actinides [1–5], and strong deviation of multiplet structure of the 3s spectra of 3d metals from that obtained in Hartree–Fock approximation [6–11]. The theory of these effects is essential in the investigations of materials by photoelectron spectroscopy. Satellite lines in the photoelectron spectra of noble gases have been the subject of several experimental works and their study provides detailed information on the dynamics of many electron correlations [12–17]. It was obtained theoretically [18,19] and confirmed experimentally [20] that monopole shake-up and shake-off satellites take about 20% of the intensity of the main line. Thus, the account for satellite intensities is required in using theoretical photoionization cross-sections in the elemental analysis by photoelectron spectroscopy. The first satellite calculations were made in the "overlapping" approximation [21–24], in which the satellite intensity was proportional to the square of overlap integral between ground state Hartree–Fock wave functions and relaxed final state wave functions. In addition, a combination of configuration interaction and "overlapping" methods was used for satellite calculation [25,26].

It was an idea of Miron Ya. Amusia to consider the hole potential as a perturbation potential for the ground state wave functions and to use the spectral function of the initial hole (see, e.g., [1]) to calculate the whole spectrum, i.e., main line, shake-up satellites, and shake off continuum [27]. This technique was applied for the shake-up satellites in photoelectron spectra of noble gases [28–30] and extended to valence and core Auger transitions [31–35]. Some predictions of these theoretical results for the valence Auger transitions were confirmed experimentally [36,37]. The creation of a new of HAXPES (hard x-ray photoelectron spectroscopy) experimental techniques [38,39] caused further development of the theory of many-electron effects in photoionization [40,41]. Similar theoretical approaches were used for the understanding of atomic many-electron effects in photoelectron spectra of atoms in chemical compounds, namely the 3s-spectra of Co [42], the 4p-spectra of Ba [43], the 5p-spectra of Th [44], and U [45].

Thus many-electron effects change the one-electron picture of photoionization [1] and knowledge of the nature of many-electron effects is required for the correct interpretation of XPS data on the compounds under investigation. Furthermore, in the case of noble gases, a comparison of theoretical and experimental results is required to understand photoionization and related phenomena [35]. In the present paper, the many-electron approaches to core relaxation and multiplet splitting are developed and examples of their applications are considered.

2. Theoretical Approach

The relaxation of atomic orbitals caused by ionization of the k-shell is described by the excitations of electron-hole pairs under the action of suddenly switched potential of the k-hole [27]. The corresponding Feynman diagram is shown in Figure 1a, where a wavy line stands for the interaction, determined by the Coulomb integral, right (left) directed arrows denote particles (holes), which are added to the initial atomic configuration.

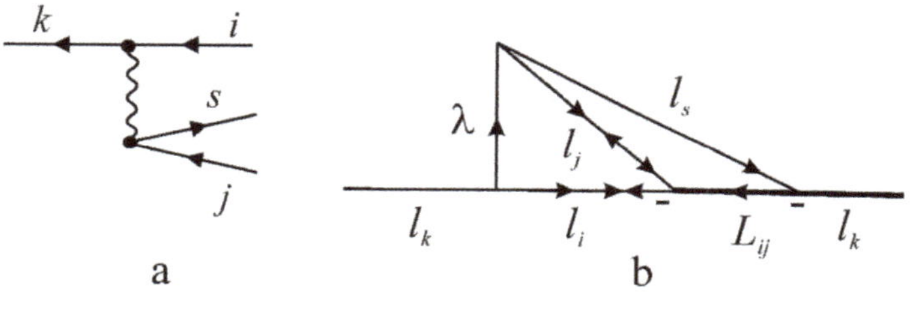

Figure 1. Graphs for the satellite's excitation. (**a**) Feynman diagram for the satellite excitation. Arrows directed from left to right (right to left) denote particles (holes), wavy line denotes Coulomb interaction. (**b**) angular momentum graph, corresponding to the Feynman diagram (**a**). Angular momentum graph for the spin part in which interaction line is removed should be included.

Coulomb integral, corresponding to a wavy line may be written as:

$$U_\gamma(kjis) = \sum_{\lambda,\mu} \alpha_\gamma^\lambda \left(kj\middle|R^\lambda\middle|is\right) + \beta_\gamma^\mu (ki|R^\mu|js), \tag{1}$$

where λ and μ are multipole values of direct and exchange parts, and angular weight factors α_γ^λ and β_γ^μ depend on the coupling scheme γ. The case $k = i$ corresponds to monopole shake-up satellites, and the case $k \neq i$ corresponds to Auger decay or to final state configuration interaction. The Coulomb integrals were calculated using the atomic wave functions $P_v(r)$, obtained by the Hartree-Fock method [46,47] using the standard formula:

$$\left(\nu_1\nu_3\middle|R^\lambda\middle|\nu_2\nu_4\right) = \int_0^\infty P_{\nu_1}(r)P_{\nu_2}(r)dr \int_0^\infty \frac{r_<^\lambda}{r_>^{\lambda+1}} P_{\nu_3}(r')P_{\nu_4}(r')dr' \tag{2}$$

To calculate the angular part of the Feynman diagram, one must plot an angular momentum graph [48], in which the interaction vertex is topologically equivalent to the Feynman diagram and the free particle and hole lines are connected in correspondence to the coupling scheme [35]. Figure 1b shows the angular momentum graph for the satellite excitation in the case of the LS-coupling scheme. The spin momentum graph is obtained by removing an interaction line [48]. Additional arrows are also added in the case of coupling of holes [49]. When reducing the angular momentum graph of Figure 1b together with the spin momentum graph and adding factors for Coulomb interaction vertex [48] one obtains:

$$\alpha^{\lambda} = f(-1)^{S_{ij}+1+\lambda} \frac{[L_{ij}S_{ij}]^{1/2}}{(2)^{1/2}[l_k]^{1/2}} \begin{Bmatrix} \lambda & l_k & l_i \\ L_{ij} & l_j & l_s \end{Bmatrix} \left\langle l_i \middle\| C^{\lambda} \middle\| l_k \right\rangle \left\langle l_j \middle\| C^{\lambda} \middle\| l_s \right\rangle, \quad (3)$$

where $f = 1$ if electrons l_i and l_j are non-equivalent, $f = \sqrt{2}$ if electrons l_i and l_j are equivalent, and $[a]$ denotes $2a + 1$.

For the exchange graph, one must interchange lines of i- and j-holes at the Coulomb interaction line, what results in additional phase factors, and one obtains the weight factor before exchange integral:

$$\beta^{\mu} = -f(-1)^{L_{ij}+l_i+l_j+\mu} \frac{[L_{ij}S_{ij}]^{1/2}}{(2)^{1/2}[l_k]^{1/2}} \begin{Bmatrix} \mu & l_k & l_j \\ L_{ij} & l_i & l_s \end{Bmatrix} \langle l_j \| C^{\mu} \| l_k \rangle \langle l_i \| C^{\mu} \| l_s \rangle \quad (4)$$

where $f = 1$, if electrons l_i and l_j are non-equivalent, and $f = 0$, if electrons l_i and l_j are equivalent. Formulas (3) and (4) differ from the formulas for Auger decay [50], obtained by coupling of initial hole's and final electron's angular momentums and spins into L_{ij} and S_{ij}, by inessential common phase factor only.

The k-vacancy spectrum, which includes the main line, discrete shake-up, continuum shake-off excitations, and Auger decay is represented by a spectral function:

$$A_k(E) = \frac{1}{\pi} \frac{\mathrm{Im}\Sigma_k(E)}{(E - \varepsilon_k - \mathrm{Re}\Sigma_k(E))^2 + \mathrm{Im}\Sigma_k(E)^2}, \quad (5)$$

where E is the energy parameter of k-hole, which runs over all relevant values and $\Sigma_k(E)-$ is the self-energy of k-vacancy.

In the second order of perturbation theory, the self-energy is represented by the Feynman diagrams of Figure 2a–c, and is defined by its real and imaginary parts as follows:

$$\mathrm{Re}\Sigma_k^{(2)}(E) = \sum_{i,j,s} \frac{\langle kj|U|is\rangle^2}{E - \varepsilon_i - \varepsilon_j + \varepsilon_s} \quad (6)$$

$$\mathrm{Im}\Sigma_k^{(2)}(E) = \pi \sum_{i,j,s} \langle kj|U|is\rangle^2 \delta(E - \varepsilon_i - \varepsilon_j + \varepsilon_s) \quad (7)$$

It is assumed, that the sums in Equations (6) and (7) include integration over continuum energies ε_s also.

The integral of the spectral function equals to the unity and the intensity of all spectral distribution is proportional the photoionization cross-section $\sigma_k(\omega)$ calculated in Hartree–Fock approximation [27].

The positions E_ν of discrete satellites are the solutions of the Dyson equation:

$$E_\nu = \varepsilon_k + \mathrm{Re}\Sigma_k(E_\nu) \quad (8)$$

The intensities of the main line and shake-up satellite relative to all spectral distribution are proportional to the spectroscopic factors:

$$f_\nu = \frac{1}{1 - \left.\frac{\partial \mathrm{Re}\Sigma_k(E)}{\partial E}\right|_{E=E_\nu}} \quad (9)$$

If the discrete line is in the continuum of another transition, its spectroscopic factor may be calculated as an integral of the spectral function. Spectroscopic factors of all main lines in photoelectron spectra calculated by spectral function [18] and "overlapping" [19] methods are approximately the same and all are close to the value 0.8, and in general agree with the experiment [20].

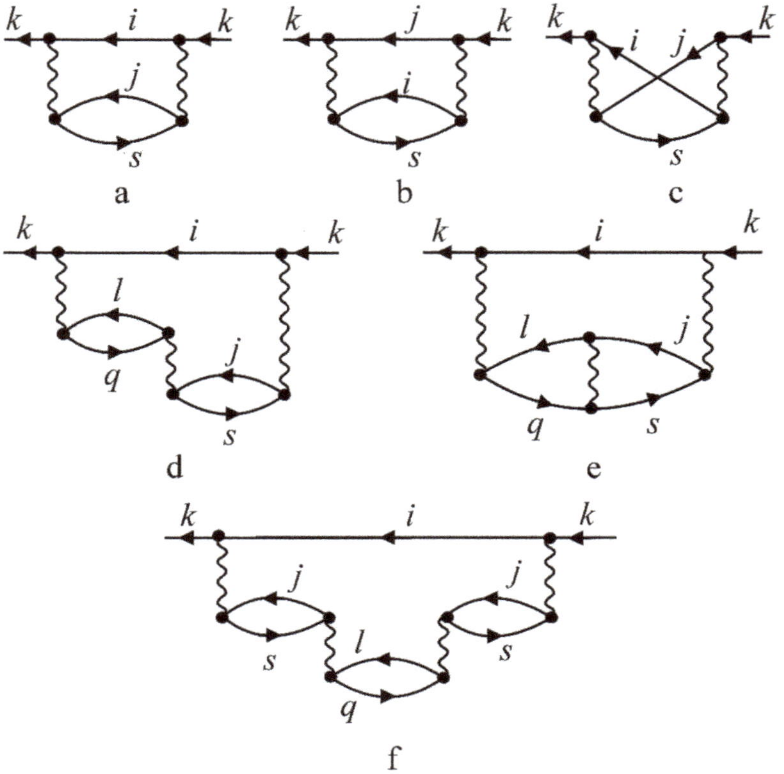

Figure 2. Feynman diagrams for the spectral function in a simplified forth-order approach. (**a–c**) Direct and exchange second-order diagrams, (**d**,**e**) direct and exchange third-order diagrams for the decay of satellite into continuum, (**f**) fourth-order diagram (exchange parts are not shown).

The second-order diagrams and Formula (7) represent the broadening of the satellite due to the direct transitions from one-hole states to the shake-off continuum $k^{-1} \to k^{-1}l^{-1}q$. If there exist at the same energy a discrete shake-up satellite $k^{-1}j^{-1}s$ it is broadened by the decay into underlying continuum $k^{-1}j^{-1}s \to k^{-1}l^{-1}q$. This process is represented by the Feynman diagram in Figure 2f and the contribution to the imaginary part of the self-energy is written as:

$$\mathrm{Im}\Sigma_k^{(4)}(E) = \frac{\pi\langle kj|U|ks\rangle^2}{(E-\varepsilon_k-\varepsilon_j+\varepsilon_s)^2}\sum_{l,q}\langle js|U|lq\rangle^2\delta(E-\varepsilon_k-\varepsilon_l+\varepsilon_q) \tag{10}$$

Figure 2d,e represent the interference between two ways of excitation of shake-off continuum, namely direct $k^{-1} \to k^{-1}l^{-1}q$, and via shake-up resonance $k^{-1} \to k^{-1}j^{-1}s \to k^{-1}l^{-1}q$. The contribution of these diagrams to the spectral function is written as:

$$\mathrm{Im}\Sigma_k^{(3)}(E) = \frac{2\pi\langle kl|U|ks\rangle}{E-\varepsilon_k-\varepsilon_j+\varepsilon_s}\sum_{l,q}\langle kl|U|kq\rangle\langle js|U|lq\rangle\delta(E-\varepsilon_k-\varepsilon_l+\varepsilon_q) \tag{11}$$

It is seen from Formulas (5), (7), (10) and (11), that contributions of decay channels to the spectral function are additive (in the numerator of Equation (5)) and the spectral functions for decay channels can be separated. Thus one can obtain the spectral function for low-energy Auger decay [31].

3. Noble Gases

Calculation of spectral functions in the simplified forth order approach (see Formulas (5)–(7), (10) and (11), and Figure 2) made it possible to reproduce strongly asymmetrical satellite lineshapes and to estimate intensities and lineshapes of low energy Auger decay. Figure 3 shows theoretical lineshape of the satellite $2s^{-1}2p^{-1}(^1P)3s(^2P)$, which is broadened by the decay into three continua $2p^{-2}(^3P)\varepsilon p(^2P)$, $2p^{-2}(^1D)\varepsilon p(^2P)$, and $2p^{-2}(^1S)\varepsilon p(^2P)$. The spectral densities of the satellite transferred to these three low energy Auger transitions are also shown. According to our calculations, the most intense is the transition to $2p^{-2}(^1D)\varepsilon p$ continuum, the intensity of transition to $2p^{-2}(^1S)\varepsilon p$ is significantly smaller, and a transitions to $2p^{-2}(^3P)\varepsilon p$ continuum is almost completely depressed. This prediction was confirmed by Kaneyashi et al. [36], who obtained the relative intensities ^1D:^1S:^3P = 1500:400:300.

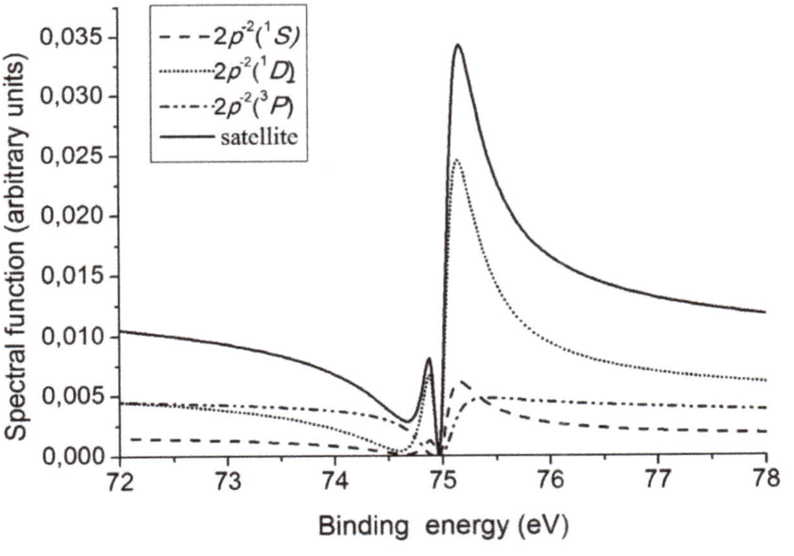

Figure 3. Theoretical lineshape of the satellite $2s^{-1}2p^{-1}(^1P)3s(^2P)$ which is broadened by the decay into three continua $2p^{-2}(^3P)\varepsilon p(^2P)$, $2p^{-2}(^1D)\varepsilon p(^2P)$ and $2p^{-2}(^1S)\varepsilon p(^2P)$ (at the satellite energy scale).

Figure 4 shows the satellite state $1s^{-1}2s^{-1}(^3S)3s(^2S)$, which is asymmetrically broadened due to decay into $1s^{-1}2p^{-1}(^3P)\varepsilon p(^2S)$ and $1s^{-1}2p^{-1}(^1P)\varepsilon p(^2S)$ continua. Theoretical curve reproduces experimental strongly asymmetrical lineshape having a Fano profile [12]. Theoretical lineshapes of low-energy Auger decay of this state are shown in Figure 5. It is seen in Figure 5, that the intensity of decay into the singlet channel $1s^{-1}2p^{-1}(^1P)\varepsilon p(^2S)$ is larger than the intensity of decay into the triplet channel $1s^{-1}2p^{-1}(^3P)\varepsilon p(^2S)$. This result is in qualitative agreement with the experimental data of Hikosaka et al. [37].

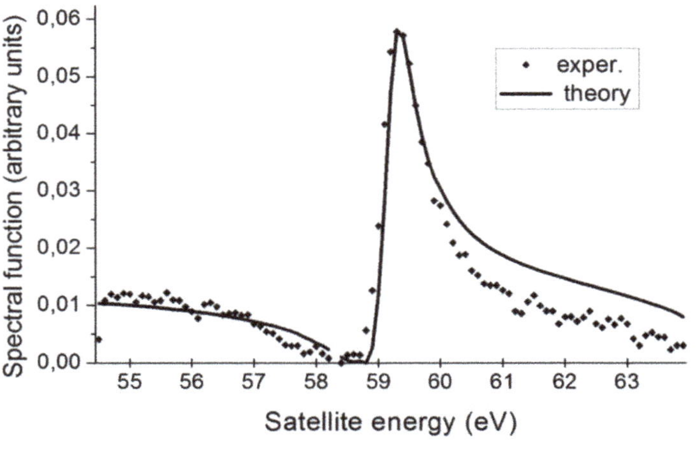

Figure 4. Theoretical lineshape of satellite $1s^{-1}2s^{-1}(^3S)3s(^2S)$ of Ne photoionization (solid line), diamonds experiment [12].

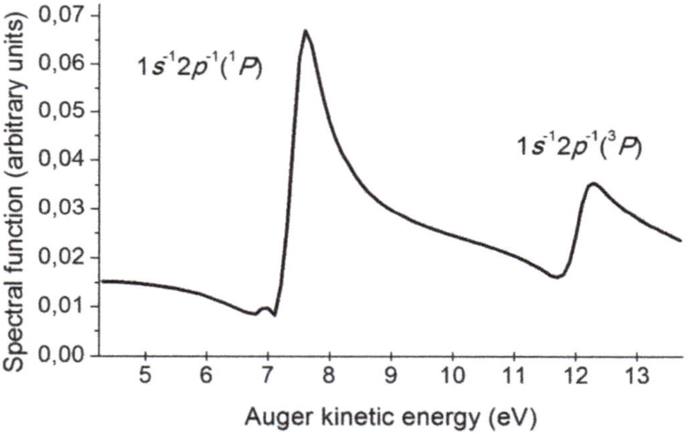

Figure 5. Theoretical lineshape of valence Auger decay of satellite state $1s^{-1}2s^{-1}(^3S)3s(^2S)$ into two continua $1s^{-1}2p^{-1}(^3P)\varepsilon p(^2S)$ and $1s^{-1}2p^{-1}(^1P)\varepsilon p(^2S)$.

4. Atoms in Chemical Compounds

The photoionization of 3*s*-levels of 3*d* elements with the configuration $3d^N(^{2S+1}L)$ due to the interaction between 3*s*-hole and unfilled 3*d*-shell results in a line split in two components. These components correspond to two states: a low spin state with total spin $S_t = S - 1/2$ and a high spin state with total spin $S_t = S + 1/2$. It can be shown that this splitting is proportional to the spin *S* of the initial $3d^N$ state. However, for $Co^{3+}(3d^6(^5D))$ using atomic Hartree–Fock wavefunctions [46,47], we obtained exchange integral $\langle 3s3s|R^2|3d3d\rangle = 0.492$ a.u., resulting in a splitting 13.4 eV between the states ^4D and ^6D, whereas experimental splitting is just 4.7 eV. Furthermore, the ratio of the spectral line intensities is not equal to a statistical ratio (2*S* + 2):2*S* = 6:4 [42]. In some works, this problem was solved by multiplication of the exchange integral by some scaling factor [10,11]. However, this formal decrease of the exchange integral is due to the interaction of the initial one-hole state $3s^{-1}3d^N$ with more complex two-hole-one-particle states $3p^{-2}3d^{N+1}$ [7–9]. This interaction is represented by the Feynman diagram shown in Figure 2a. Since holes

i and j in the final state are equivalent, the exchange part is absent. The corresponding angular momentum graphs are shown in Figure 6. In this graph, all lines correspond to holes and a black square denotes fractional parentage coefficients for the $3d^{4l+2-N}$ hole configurations. Using graphical methods for calculating angular momentum graphs [48], we obtain the following formula for the weight factor before the Coulomb integral:

$$\alpha^{\lambda} = (-1)^{L_t+L_1+L_2+l_1+l_3}\overline{G}^{LS}_{L_1S_1}\left\{\begin{array}{ccc} L_2 & l_1 & l_3 \\ \lambda & l_2 & l_2 \end{array}\right\}\left\{\begin{array}{ccc} L_t & L & l_1 \\ l_3 & L_2 & L_1 \end{array}\right\}\left\{\begin{array}{ccc} S_t & S & 1/2 \\ 1/2 & S_2 & S_1 \end{array}\right\}\times \\ \left(\begin{array}{ccc} l_1 & \lambda & l_2 \\ 0 & 0 & 0 \end{array}\right)\left(\begin{array}{ccc} l_3 & \lambda & l_2 \\ 0 & 0 & 0 \end{array}\right)(2\overline{N})^{1/2}[l_1l_3]^{1/2}[l_2][S_2SL_2L]^{1/2} \tag{12}$$

where l_1, l_2, and l_3 correspond to s-, p-, and d-electrons, respectively, $\overline{G}^{LS}_{L_1S_1}$ is the fractional parentage coefficient for the hole configurations of the $3d$ shell; $\overline{N} = 4l + 2 - N$ is the number of holes in the $3d$ shell in the initial state. In our case, we obtained coefficients α^1 for the interaction of the term $3s^{-1}3d^6(^5D)(^4D)$ of initial state with excited states $3p^{-2}(^1D)3d^7(^4F)(^4D)$ and $3p^{-2}(^1D)3d^7(^4P)(^4D)$ 0.683 and 0.446, respectively.

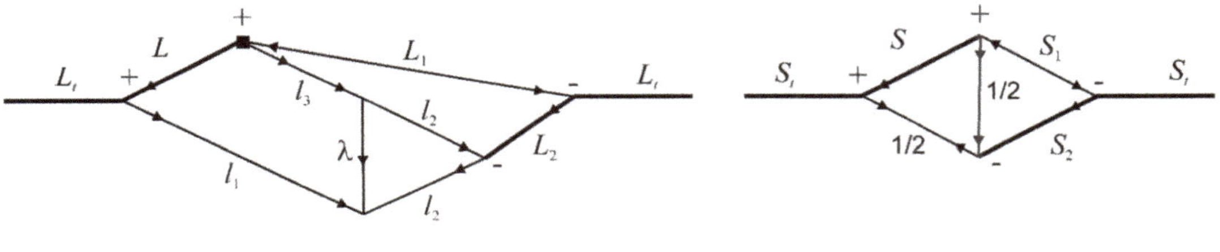

Figure 6. Angular momentum graphs for the calculation of angular weight factors for the interaction between initial state in photoionization $3s^{-1}3d^N$ and a the state $3p^{-2}3d^{N+1}$. (**a**) Orbital part, (**b**) spin part.

Formula (12) makes it possible to draw some qualitative conclusions about the configuration interaction under consideration. In the first line of the first $6j$-symbol $l_1 = 0$ and $l_3 = 2$. Thus, according to the triangle rule for the first line of this $6j$-symbol there is only one possible value for L_2, i.e., $L_2 = 2$. Therefore, only the term ^1D among three possible terms ^3P, ^1D, and ^1S, of the $3p^{-2}$ shell is involved in the configuration interaction. This circumstance makes it possible not only to simplify the calculations but also to qualitatively estimate the effect of the configuration interaction on the splitting of the $3s$-line in the photoelectron spectrum. Consider atoms with the shell more than half-filled $N \geq 2l+1$ in the state of the ground term $3d^N(^{2S+1}L)$. According to the table of fractional parentage coefficients [51], the addition of one electron to the ground term results in terms of the configuration $3d^{N+1}$ with the spin $S_1 = S - 1/2$. As shown above, the interaction of the state $3s^{-1}3d^N$ with the states $3p^{-2}3d^{N+1}$ is possible only if the $3p^{-2}$ shell has the term ^1D. For this reason and because the relation $S_1 = S - 1/2$ is fulfilled for ground terms when $N \geq 2l+1$ the high-spin state $S_t = S + 1/2$ of $3s^{-1}3d^N$ configuration does not interact with the configuration $3p^{-2}3d^{N+1}$. Therefore, the configuration interaction at $N \geq 2l+1$ affects only the position of the low-spin state, resulting in the reduction of the splitting between low-spin and high-spin states.

In the case under consideration (the ground term ^5D of the $3d^6$ shell), the interaction between the excited states involving two terms ^4P and ^4F of the configuration $3d^7$ is also possible. The Feynman diagrams of such an interaction are shown in Figure 2d,e. The corresponding formula for the self-energy part is written as:

$$\mathrm{Re}\Sigma^{(3)}(E) = \sum_{\gamma} \frac{\langle kj|U|js\rangle_{\gamma}^{2}}{E+(-\varepsilon_j-\varepsilon_j+\varepsilon_s)_{\gamma}} + \sum_{\gamma\neq\delta} \frac{2\langle kj|U|js\rangle_{\gamma}\langle kj|U|js\rangle_{\delta}\langle js|U|js\rangle_{\gamma\delta}}{\left[E+(-\varepsilon_j-\varepsilon_j+\varepsilon_s)_{\gamma}\right]\left[E+(-\varepsilon_j-\varepsilon_j+\varepsilon_s)_{\delta}\right]} \quad (13)$$

where sums run over all distinct terms γ and δ of the configuration under consideration. Since only term ^1D of $3p^{-2}$ configuration is possible, the sum runs over terms of $3d^{N+1}$ configuration with $S_1 = S_t$.

The experimental 3s-photoelectron spectrum of Co^{3+} ion in paramagnetic $BiCoO_3$ [42] is shown in Figure 7. Theoretical splitting of the 3s-line with account for configuration interaction (CI) 5.1 eV is in good agreement with experimental value 4.7 eV and with theoretical result 5.4 eV [7], obtained with account for a larger number of configurations. In the Hartree–Fock approximation (HF), the value of splitting equals 13.4 eV. The spectroscopic factor of the low-spin state calculated by Formula (9) equals 0.75. Taking into account the statistical ratio of the high-spin to low-spin component 1.5, we obtain theoretical ratio 2, which is equal to the experimental value. It was also pointed out that interaction with exited configuration 3p-4f is important [9]. However, the value of leading Coulomb integral in this case $\langle 3s3p|R^1|3p4f\rangle = -0.0088$ a.u. is significantly less then $\langle 3s3p|R^1|3p3d\rangle = 0.650$ a.u., used in the present work for the main channel of CI. That is why the main features of the spectrum can be reproduced by our method, which is equivalent to solving the secular matrix of dimension 3.

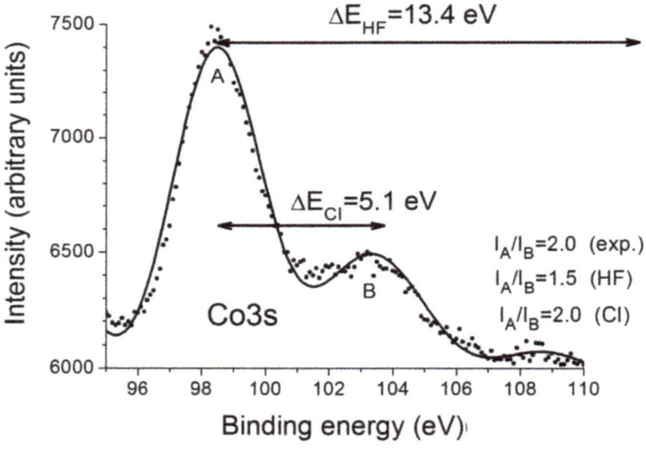

Figure 7. Experimental 3s-photoelectron spectrum of Co^{3+} ion in paramagnetic $BiCoO_3$ [42]. Theoretical line splitting in HF approximation (HF) and with account for configuration interaction (CI) are also shown.

In photoelectron spectra of the Th $5p$ [44] and U $5p$ [45] of ThO_2 and UO_2 a complex structure is observed instead of two components of the spin doublet $5p_{1/2}$ and $5p_{3/2}$. In XPS spectrum of ThO_2, shown in Figure 8 a strong satellite peak with a binding energy of about 20 eV larger than the energy of the $5p_{3/2}$ peak appears. The experimental spin-orbit splitting 55.5 eV is significantly larger than the result of our Dirac–Fock calculation of 47.9 eV and the line $5p_{1/2}$ is asymmetrically broadened. These many-electron effects appear due to the interaction of the initial hole states $5p_{3/2(1/2)}$ with the more complex two-hole-one-particle states $5d^{-2}(^{2S+1}L_J)5f(\varepsilon f)$. In this case, we used an intermediate coupling scheme and the corresponding momentum graph is shown in Figure 9. Reduction of this momentum graph results in the following formula for the weight factor:

$$\alpha^{\lambda}(L,S,J) = \sqrt{2}\begin{pmatrix} l_1 & \lambda & l_2 \\ 0 & 0 & 0 \end{pmatrix}\begin{pmatrix} l_2 & \lambda & l_3 \\ 0 & 0 & 0 \end{pmatrix}[l_1 l_2 l_2 l_3]^{1/2}[LSJj_3]^{1/2}[j_1]^{-1/2}\begin{Bmatrix} l_1 & L & l_3 \\ l_2 & \lambda & l_2 \end{Bmatrix}$$
$$\times\begin{Bmatrix} j_1 & \kappa & l_3 \\ 1/2 & j_3 & J \end{Bmatrix}\sum_{\kappa=L-1/2}^{L+1/2}(-1)^{2\kappa+L+3J+l_1+l_3+j_1+j_3+1}\begin{Bmatrix} l_3 & j_1 & \kappa \\ 1/2 & L & l_1 \end{Bmatrix}\begin{Bmatrix} J & 1/2 & \kappa \\ 1/2 & L & S \end{Bmatrix} \quad (14)$$

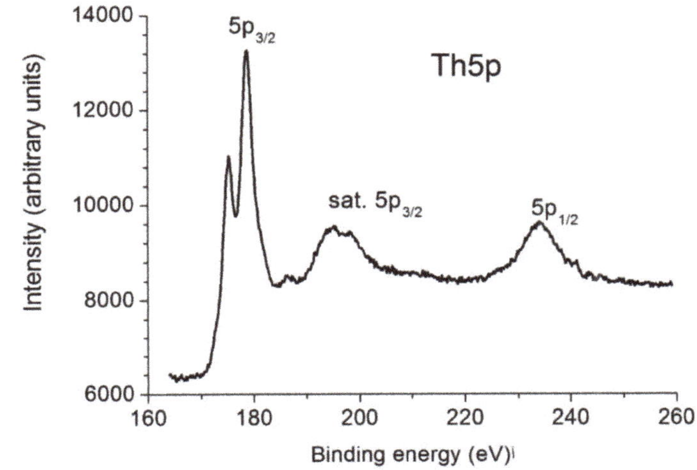

Figure 8. Experimental Th5p photoelectron spectrum of ThO_2 [44].

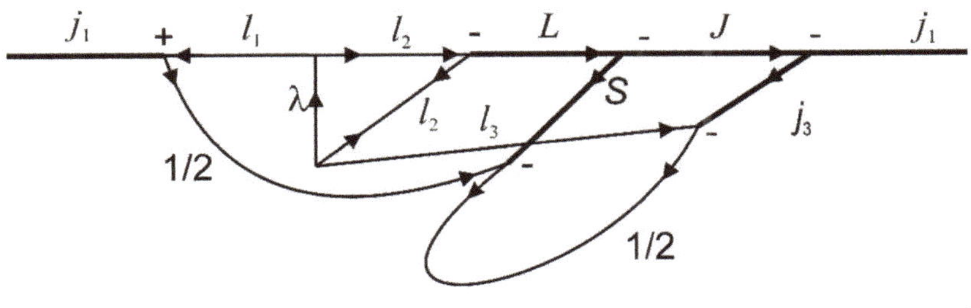

Figure 9. Momentum graph for the calculation of final state configuration interaction in intermediate coupling scheme.

Where l_1, l_2, and l_3 correspond to 5p-, 5d-holes, and 5f-electron, respectively, j_1 is the total momentum of the $5p^{-1}$ hole, L, S, and J correspond to the terms of configuration $5d^{-2}$, the summation is over all half-integer κ in the specified interval, and λ takes values 1 and 3. Formula (14) is a generalization of the non-relativistic formula for Auger decay [50] in the case of intermediate coupling. Note that the first of the 6j-symbols in Formula (14) is also present in the non-relativistic formula [43]. It follows from the triangular condition for elements of the first row of this 6j-symbol, that interaction of the initial vacancy is possible only with three terms ^1D_J, ^3F_J, and ^1G_J of the shell $5d^{-2}$.

Formula (14) was used to calculate the interaction of the initial hole $5p^{-1}$ with the excited states $5d^{-2}5f$ in the secular matrix. In addition, the interaction between two-hole-one-particle states was taken into account. The spectrum of the Th $5p_{3/2}$ electrons was calculated by solving the secular matrix. The eigenvalues of the secular matrix correspond to the positions of the spectral lines, and the squares of the elements of the eigenvector, corresponding to the energy of the main line give the intensities of all lines. In the case of ionization of the $5p_{1/2}$ subshell the levels of the excited $5d^{-2}(^{2S+1}L)5f$ configuration are far from the single-hole state, whereas Auger decay into the $5d^{-2}(^{2S+1}L)\varepsilon p,\ \varepsilon f$ continuous states is possible, which leads to the asymmetrical broadening of the Th $5p_{1/2}$ line. To calculate the spectrum of the $5p_{1/2}$ electrons, we used the spectral function method (see Formula (5)). The theoretical spectrum with overall energy shift −9.6 eV representing solid-state effects [44] is shown in Figure 10. The calculated Th $5p$ spectrum consists of three groups of lines which can be attributed to the $5p_{1/2}$ and $5p_{3/2}$ lines and satellites, i.e., the $5d^{-2}5f$ states. Many-electron effects reduce the binding energies of the $5p_{3/2}$ electrons, and as a result, spin–orbit splitting reaches the value 55 eV, which corresponds to the experimental value 56 eV. It is seen in Figure 10 that the $5p_{1/2}$ line is asymmetrically broadened with raised low binding energy side. The spectroscopic factor of the $5p_{1/2}$ line, estimated as the integral of the spectral function Equation (5) in an interval of 10 eV (which corresponds to the interval where the intensity of the line was experimentally determined) with background subtraction equals 0.69. Thus the theoretical ratio of the intensities $I(5p_{3/2})$:$I(5p_{1/2})$, when the satellites are included in $5p_{3/2}$ line, equals 2.9:1. Experiential value of this ratio depends on the background subtraction. When Shirley's method was used, this ratio equals 5:1, but in linear background subtraction, we obtained the ratio 3.6:1.

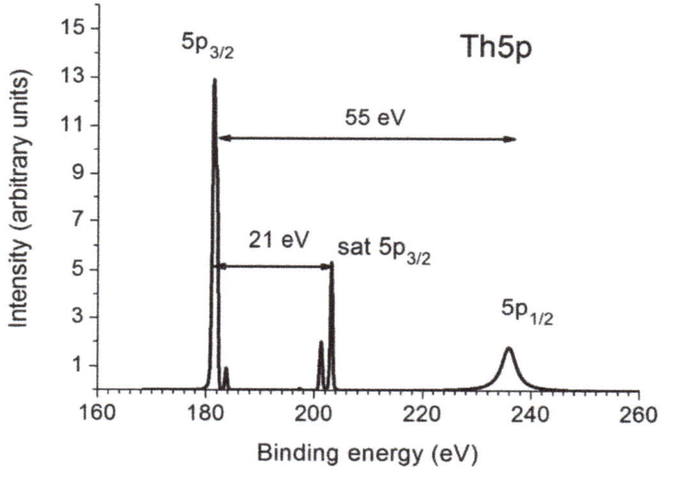

Figure 10. Theoretical Th 5p photoelectron spectrum. The satellite structures of the $5p_{3/2}$ line were calculated by the CI method. The spectral function of $5p_{1/2}$ line was calculated in the second order of perturbation theory. The lineshape $5p_{1/2}$ is slightly asymmetrical, with a raised low-binding energy side.

5. Conclusions

In the present work, it was shown that for different relations between interaction parameters in the final state, appearing after photoionization, different approaches of MBPT can be used. A spectral function approach with account for the imaginary part of the self-energy in simplified fourth-order make it possible to calculate asymmetrical lineshapes of satellites in photoelectron spectra and of low-energy Auger spectra. Dyson's equation in the third-order makes it possible to calculate the line positions in the case of weak interaction between excited states. In particular, in the case of 3s-spectra of 3d-elements accounting for a very limited number of Feynman diagrams made it possible to reach agreement with the experiment and to draw a qualitative conclusion on many-electron effects for 3d-shell more than half-filled. However, in the case of strong CI in 5p-spectra, the method of secular matrix solution is more appropriate. This series of research was inspired by Miron Ya. Amusia, who was the first person to proposed to use unified MBPT technique for CI and satellite excitations.

Author Contributions: V.G.Y., software and theoretical calculations; Y.A.T., methodology and experimental investigations. All authors have read and agreed to the published version of the manuscript.

Funding: The work was supported by IGIC RAS state assignment and RRC "KI" state assignment.

Institutional Review Board Statement: The study was conducted in accordance with the Declaration of Helsinki, and approved by the Institutional Review Board of IGIC RAS in agreement with RRC "KI" 04.05.2022.

Informed Consent Statement: Not applicable.

Data Availability Statement: Not applicable.

Conflicts of Interest: The authors declare no conflict of interest.

References

1. Wendin, G. Breakdown of one-electron pictures in photoelectron spectra. In *Structure and Bonding*; Springer: Berlin/Heidelberg, Germany, 1981; Volume 45, pp. 1–125, ISBN 978-3662157800.
2. Kowalczyk, S.P.; Ley, L.; Martin, R.L.; McFeely, F.R.; Shirley, D.A. Relaxation and final-state structure in XPS of atoms, molecules, and metals. *Faraday Discuss. Chem. Soc.* **1975**, *60*, 7–17. [CrossRef]
3. Sham, T.K.; Wendin, G. Screening and configuration-interaction effects in the 5p X-ray-photoelectron spectrum of Th metal. *Phys. Rev. Lett.* **1980**, *44*, 817–820. [CrossRef]
4. Boring, M.; Cowan, R.D. Satellite structure in the 5p and 5s X-ray-photoelectron spectra of the actinides. *Phys. Rev. B* **1981**, *23*, 445–448. [CrossRef]
5. Fadley, C.S.; Shirley, D.A. Multiplet splitting of metal-atom electron binding energies. *Phys. Rev. A* **1970**, *2*, 1109–1113. [CrossRef]
6. Kowalczyk, S.P.; Ley, L.; Pollak, R.A.; McFeely, F.R.; Shirley, D.A. New multiplet structure in photemission from MnF_2. *Phys. Rev. B* **1973**, *7*, 4009–4011. [CrossRef]
7. Viinikka, E.-K.; Ohrn, Y. Configuration mixing in the 3s-hole state of transition-metal ion. *Phys. Rev. B* **1975**, *11*, 4168–4175. [CrossRef]
8. Bagus, P.S.; Freeman, A.J.F.; Sasaki, F. Prediction of new multiplet structure in photoemission experiments. *Phys. Rev.* **1975**, *30*, 850–853. [CrossRef]
9. Bagus, P.S.; Broer, R.; Ilton, E.S. A new near degeneracy effect for photoemission in transition metals. *Chem. Phys. Lett.* **2004**, *394*, 150–154. [CrossRef]
10. Kochur, A.G.; Ivanova, T.M.; Shchukarev, A.V.; Sidorov, A.A.; Kiskin, M.A.; Novotortsev, V.M.; Eremenko, I.L. Mn3s X-ray photoelectron spectra of polynuclear trimethylacetate complexes of manganese. *Bull. Russ. Acad. Sci. Phys.* **2010**, *74*, 625–630. [CrossRef]
11. Okada, K.; Kotani, A. Interatomic and intra-atomic configuration interactions in core-level X-ray photoemission spectra of late transition metal compounds'. *J. Phys. Soc. Jpn.* **1992**, *61*, 4619–4637. [CrossRef]
12. Svensson, S.; Eriksson, B.; Martensson, N.; Wendin, G.; Gelius, U. Electron shake-up and correlation satellites and continuum shake-off distributions in X-ray photoelectron spectra of the rare gas atoms. *J. Electron Spectrosc. Relat. Phenom.* **1988**, *47*, 327–384. [CrossRef]
13. Pahler, M.; Caldwell, C.D.; Schaphorst, S.J.; Krause, M.O. Intrinsic linewidths of neon $2s2p^5$ $(^{1,3}P)nl$ ^2L correlation satellites. *J. Phys. B At. Mol. Opt. Phys.* **1993**, *26*, 1617–1625. [CrossRef]
14. Hall, R.I.; Dawber, G.K.; Ellis, K.; Zubek, M.; Avaldi, L.; King, G.C. Near-threshold study of the neon photoelectron satellites. *J. Phys. B At. Mol. Opt. Phys.* **1991**, *24*, 4133–4146. [CrossRef]
15. Yarzhemsky, V.G.; Amusia, M.Y.; Bolognesi, P.; Avaldi, L. A study of the Ne $2s2p^5(^3P)3s$ and 3p correlation satellites up to 75 eV above threshold. *J. Phys. B At. Mol. Opt. Phys.* **2010**, *43*, 185204. [CrossRef]
16. Krause, M.O.; Whitfield, S.B.; Caldwell, C.D.; Wu, J.-Z.; Schaphorst, S.J.; van der Meulen, P.; de Lange, C.A.; Hansen, R.W.C. Outer *ns*, *np* correlation satellites in the rare gases—A photoelectron spectrometry study with an undulator photon source. *J. Electron Spectrosc. Relat. Phenom.* **1992**, *58*, 79–104. [CrossRef]
17. Kikas, A.; Osborne, S.J.; Ausmees, A.; Svensson, S.; Sairanen, O.-P.; Aksela, S. High-resolution study of the correlation satellites in photoelectron spectra of the rare gases. *J. Electron Spectrosc. Relat. Phenom.* **1996**, *77*, 241–266. [CrossRef]
18. Yarzhemsky, V.G.; Nefedov, V.I.; Trzhaskovskaya, M.B.; Band, I.M.; Szargan, R. The influence of core hole relaxation on the main-line intensities in X-ray photoelectron spectra. *J. Electron Spectrosc. Relat. Phenom.* **2002**, *123*, 1–10. [CrossRef]
19. Yarzhemsky, V.G.; Trzhaskovskaya, M.B. Spectroscopic factors of atomic subshells for HAXPES applications. *At. Data Nucl. Data Tables* **2021**, *139*, 101387. [CrossRef]
20. Seah, M.P.; Gilmore, I.S. Quantitative X-ray photoelectron spectroscopy: Quadrupole effects, shake-up, Shirley background, and relative sensitivity factors from a database of true X-ray photoelectron spectra. *Phys. Rev. B* **2006**, *73*, 174113, Erratum in *Phys. Rev. B* **2007**, *75*, 149901. [CrossRef]
21. Åberg, T. Theory of X-ray satellites. *Phys. Rev.* **1967**, *156*, 35–41. [CrossRef]
22. Manne, R.; Åberg, T. Koopmans' theorem for inner-shell ionization. *Chem. Phys. Lett.* **1970**, *2*, 282–284. [CrossRef]

23. Martin, R.L.; Shirley, D.A. Theory of core-level photoemission correlation state spectra. *J. Chem. Phys.* **1976**, *64*, 3685–3689. [CrossRef]
24. Armen, G.B.; Åberg, T.; Karim, K.R.; Levin, J.C.; Crasemann, B.; Brown, G.S.; Chen, M.H.; Ice, G.E. Threshold double photoexcitation of argon with synchrotron radiation. *Phys. Rev. Lett.* **1985**, *54*, 182–185. [CrossRef] [PubMed]
25. Sukhorukov, V.L.; Lagutin, B.M.; Petrov, I.D.; Schmoranzer, H.; Ehresmann, A.; Vollweiler, F.; Schartner, K.-H. Resonance structure of the photoionization cross sections of the Kr atom in the energy region between single and double ionization thresholds. *J. Electron Spectrosc. Relat. Phenom.* **1995**, *76*, 421–426. [CrossRef]
26. Lagutin, B.M.; Petrov, I.D.; Sukhorukov, V.L.; Whitfield, S.B.; Langer, B.; Viefhaus, J.; Wehlitz, R.; Berrah, N.; Mahler, W.; Becker, U. Cross sections and angular distributions of the photoelectron correlation satellites of the Xe atom. *J. Phys. B At. Mol. Opt. Phys.* **1996**, *29*, 937–976. [CrossRef]
27. Kheifets, A.S.; Amusia MYa Yarzhemsky, V.G. On the validity of quasi-particle approximation in photoelectron spectroscopy. *J. Phys. B. At. Mol. Opt. Phys.* **1985**, *18*, L343–L350. [CrossRef]
28. Yarzhemsky, V.G.; Armen, G.B.; Larkins, F.P. Calculation of the shake-up satellites in the 1s and 2s X-ray photoelectron spectra of neon. *J. Phys. B At. Mol. Opt. Phys.* **1993**, *26*, 2785–2794. [CrossRef]
29. Kheifets, A.S. Green's function calculation of the satellite spectrum of neon. *J. Phys. B At. Mol. Opt. Phys.* **1995**, *28*, 3791–3803. [CrossRef]
30. Yarzhemsky, V.G.; Kheifets, A.S.; Armen, G.B.; Larkins, F.P. Linewidths and intensities of satellites in photoelectron spectra in the presence of an underlying continuum. *J. Phys. B At. Mol. Opt. Phys.* **1995**, *28*, 2105–2112. [CrossRef]
31. Yarzhemsky, V.G.; Larkins, F.P. The shapes of Auger decay lines in photoelectron satellite spectra. *Eur. Phys. J. D At. Mol. Opt. Plasma Phys.* **1999**, *5*, 179–184. [CrossRef]
32. Yarzhemsky, V.G.; Amusia, M.Y.; Chernysheva, L.V. Lineshape of Ne 1s photoionization satellite [1s2s](^{3}S) 3s and its valence Auger decay spectrum. *J. Electron Spectrosc. Relat. Phenom.* **2002**, *127*, 153–159. [CrossRef]
33. Yarzhemsky, V.G.; Nefedov, V.I.; Amusia MYa Chernysheva, L.V. The shapes of photoelectron satellite spectra. *Surf. Rev. Lett.* **2002**, *9*, 1209–1212. [CrossRef]
34. Yarzhemsky, V.G.; Sgamellotti, A. Auger rates of second-row atoms calculated by many-body perturbation theory. *J. Electron Spectrosc. Relat. Phenom.* **2002**, *125*, 13–24. [CrossRef]
35. Amusia, M.Y.; Chernysheva, L.V.; Yarzhemsky, V.G. *Handbook of Theoretical Atomic Physics*; Springer: Berlin/Heidelberg, Germany, 2012; p. 799. [CrossRef]
36. Kaneyasu, T.; Hikosaka, Y.; Shigemasa, E.; Penent, F.; Lablanquie, P.; Aoto, T.; Ito, K. Autoionization of the Ne^+ Rydberg states formed via valence photoemission. *J. Phys. B At. Mol. Opt. Phys.* **2007**, *40*, 4047–4060. [CrossRef]
37. Hikosaka, Y.; Aoto, T.; Lablanquie, P.; Penent, F.; Shigemasa, E.; Ito, K. Experimental investigation of core-valence double photoionization. *Phys. Rev. Lett.* **2006**, *97*, 053003. [CrossRef]
38. Püttner, R.; Goldsztejn, G.; Céolin, D.; Rueff, J.-P.; Moreno, T.; Kushawaha, R.K.; Marchenko, T.; Journel, L.; Lindle, D.W.; Piancastelli, M.N.; et al. Direct Observation of Double-Core-Hole Shake-Up States in Photoemission. *Phys. Rev. Lett.* **2015**, *114*, 093001. [CrossRef]
39. Goldsztejn, G.; Marchenko, T.; Püttner, R.; Journel, L.; Guillemin, R.; Carniato, S.; Selles, P.; Travnikova, T.; Céolin, D.; Lago, A.F.; et al. Double-Core-Hole States in Neon: Lifetime, Post-Collision Interaction, and Spectral Assignment. *Phys. Rev. Lett.* **2016**, *117*, 133001. [CrossRef]
40. Amusia, M.Y.; Chernysheva, L.V.; Yarzhemsky, V.G. On photoionization in the hard X-ray region. *JETP Lett.* **2013**, *97*, 704–707. [CrossRef]
41. Yarzhemsky, V.G.; Amusia, M.Y. Calculation of Ar photoelectron satellites in the hard-X-ray region. *Phys. Rev. A* **2016**, *93*, 063406. [CrossRef]
42. Yarzhemsky, V.G.; Teterin, Y.A.; Presnyakov, I.A.; Maslakov, K.I.; Teterin, A.Y.; Ivanov, K.E. Many-electron effects in Co 3*s* X-ray photoelectron spectra of diamagnetic $ScCoO_3$ and paramagnetic $BiCoO_3$ Cobaltites. *JETP Lett.* **2020**, *111*, 422–427. [CrossRef]
43. Yarzhemsky, V.G.; Teterin, Y.A.; Sosulnikov, M.I. Dynamic dipolar relaxation in X-ray photoelectron spectra of Ba4*p* subshell in barium compounds. *J. Electron Spectrosc. Relat. Phenomom.* **1992**, *59*, 211–222. [CrossRef]
44. Yarzhemsky, V.G.; Teterin Yu, A.; Maslakov, K.I.; Teterin, A.Y.; Ivanov, K.E. Many-electron effects in Th 5*p* and 5*s* X-ray photoelectron spectra of ThO_2. *JETP Lett.* **2021**, *114*, 609–615. [CrossRef]
45. Maslakov, K.I.; Yarzhemsky, V.G.; Teterin, Y.A.; Teterin, A.Y.; Ivanov, K.E. Complex XPS spectra structure of U5p electrons and the uranium oxidation state. *Radiochemistry* **2020**, *62*, 608–616. [CrossRef]
46. Amusia, M.Y.; Chernysheva, L.V. *Computation of Atomic Processes*; IOP Publishing: Bristol/Philadelphia, PA, USA, 1997; ISBN 9780750302296.
47. Amusia, M.Y.; Chernysheva, L.V. Computation of atomic and molecular processes. In *Introducing the ATOM-M Software Suite*; Springer: Berlin/Heidelberg, Germany, 2021; p. 456, ISBN 978-3-030-85142-2.
48. Lindgren, I.; Morrison, J. *Atomic Many-Body Theory*; Springer: Berlin/Heidelberg, Germany, 1982; p. 472. ISBN 978-3-642-96614-9.
49. Judd, B.R. *Second Quantization and Atomic Spectroscopy*; Baltimore MD Johns Hopkins press: Baltimore, MD, USA, 1967; ISBN 0801803225/9780801803222.

50. Walters, D.L.; Bhalla, C.P. Nonrelativistic auger Rates, X-ray rates, and fluorescence yields for the K shell. *Phys. Rev. A* **1971**, *3*, 1919–1926. [CrossRef]
51. Sobelman, I.I. Atomic spectra and radiative transitions. In *Chemical Physics Book Series (CHEMICAL, V. 1)*; Springer: Berlin/Heidelberg, Germany, 1979; p. 308, ISBN 978-3-662-05905-0.

atoms

MDPI

Article

Strong-Field Ionization Amplitudes for Atomic Many-Electron Targets

Stephan Fritzsche [1,2,3,*] and Birger Böning [1,2]

1 Helmholtz-Institut Jena, Fröbelstieg 3, D-07743 Jena, Germany; b.n.boening@hi-jena.gsi.de
2 GSI Helmholtzzentrum für Schwerionenforschung, D-64291 Darmstadt, Germany
3 Theoretisch-Physikalisches Institut, Friedrich-Schiller-Universität Jena, D-07743 Jena, Germany
* Correspondence: s.fritzsche@gsi.de

Abstract: The strong-field approximation (SFA) has been widely applied in the literature to model the ionization of atoms and molecules by intense laser pulses. A recent re-formulation of the SFA in terms of partial waves and spherical tensor operators helped adopt this approach to account for realistic atomic potentials and pulses of different shape and time structure. This re-formulation also enables one to overcome certain limitations of the original SFA formulation with regard to the representation of the initial-bound and final-continuum wave functions of the emitted electrons. We here show within the framework of JAC, the Jena Atomic Calculator, how the direct SFA ionization amplitude can be readily generated and utilized in order to compute above-threshold ionization (ATI) distributions for many-electron targets and laser pulses of given frequency, intensity, polarization, pulse duration and carrier–envelope phase. Examples are shown for selected ATI energy, angular as well as momentum distributions in the strong-field ionization of atomic krypton. We also briefly discuss how this approach can be extended to incorporate rescattering and high-harmonic processes into the SFA amplitudes.

Keywords: atomic; Coulomb–Volkov; direct amplitude; distorted-Volkov; electron emission; Jena Atomic Calculator; partial-wave representation; relativistic; strong-field approximation; strong-field ionization

Citation: Fritzsche, S.; Böning, B. Strong-Field Ionization Amplitudes for Atomic Many-Electron Targets. *Atoms* **2022**, *10*, 70. https://doi.org/10.3390/atoms10030070

Academic Editors: Anatoli Kheifets, Gleb Gribakin and Vadim Ivanov

Received: 3 June 2022
Accepted: 27 June 2022
Published: 30 June 2022

1. Introduction

During the past decades, strong-field ionization measurements in atoms and molecules have led to numerous insights into the electron dynamics on short time scales. In particular, several nonlinear optical processes, such as the above-threshold ionization (ATI, [1,2]), tunneling ionization, high-order harmonic generation (HHG, [3,4]), or the nonsequential double ionization (NSDI, [5]) have attracted much interest and can be readily controlled by tailoring the temporal shape and duration of ultrashort laser pulses. In ATI, for example, the energy and momentum distributions of photoelectrons are often recorded for different targets and (short) laser pulses of different frequency ω, intensity I, polarization ϵ, pulse duration (i.e., number of laser cycles, n_p), or by even steering the carrier–envelope phase $\phi^{(\mathrm{CEP})}$. In contrast to the detailed modeling of the driving laser pulse, however, the target atoms are typically described in rather a simplified manner, and especially the initial state of the *active* electron is often just taken as a hydrogenic 1s state [6,7]. Because of this and further simplifications in modeling the target atoms, many observations are still understood only *qualitatively* so far.

Figure 1a displays the prototypical geometry and observables of an ATI experiment. Here, atoms are exposed to an intense driving laser pulse with given intensity I, wavelength λ, ellipticity ϵ or, perhaps, even a superposition of such light fields. A detector D records the photoelectrons that are emitted due to the interaction of laser pulse with the target atoms. Routinely, the photoelectron energy distributions are recorded at a fixed detector position within the $x - y$ polarization plane (Figure 1b). The observed photoelectron energy spectra then exhibit ATI peaks that are spaced by the photon energy. If the detector position is varied within the polarization plane, azimuthal angular distributions can be recorded for photoelectrons of selected energy (Figure 1c). These angular distributions strongly depend not only on the shape of the driving laser pulse but also on the outgoing electron wave in the potential of the photoion. In addition, the full photoelectron momentum distributions are often measured within the polarization plane as shown in Figure 1d.

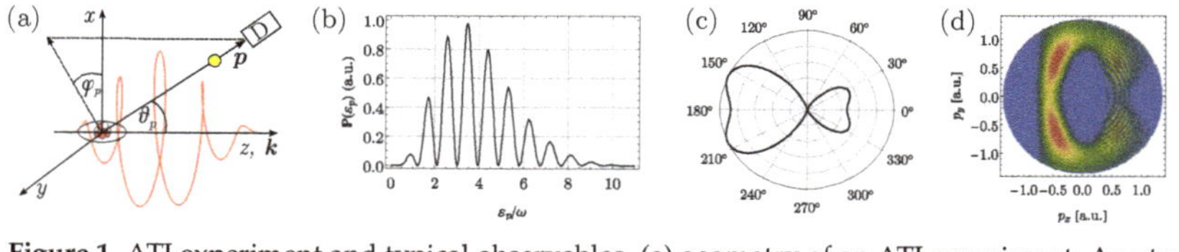

Figure 1. ATI experiment and typical observables. (**a**) geometry of an ATI experiment: An atom is irradiated by a strong laser pulse (red) of intensity I and wavelength λ that propagates along the z-axis and is polarized with ellipticity ϵ within the $x - y$ plane . Due to the interaction with the laser field, a photoelectron is emitted with momentum $\mathbf{p} = (p, \vartheta_p, \varphi_p)$ in spherical coordinates and measured at the detector D. (**b**) For a given position of the detector within the polarization plane, the measured photoelectron energy distributions exhibit several ATI peaks that are just spaced by the photon energy. (**c**) For a few-cycle driving laser pulse, the angular distribution of photoelectrons with fixed energy reveals an asymmetry which encodes details of the pulse structure and the electron continuum. (**d**) In addition, full photoelectron momentum distributions are often recorded within the polarization plane.

While any reliable theory of the strong-field (ionization) processes from above must have its roots in the time-dependent Schrödinger equation, a direct (numerical) integration of this equation becomes unfeasible already for the three-dimensional motion of a single (active) electron in a static soft-core potential, not to speak about the many-electron nature of most targets [8]. Therefore, a number of analytical methods have been developed as well and nowadays provide good insights into the underlying electron dynamics. In particular, the strong-field approximation (SFA) [9–11] provides an efficient single-electron treatment and has become, despite several limitations in its original form, a very valuable tool for computing the ATI and HHG spectra for a wide range of laser parameters and targets [12–14]. Here, however, a re-formulation of the SFA in terms of partial waves and spherical tensors [15] is applied and help adopt this method towards modern strong-field measurements. This re-formulation enables one to incorporate all central features of the incident laser pulse as well as the electronic structure of the target atoms.

To support the analysis of different strong-field measurements, this work reports an implementation of the (direct) SFA amplitude in its partial-wave representation within the framework of JAC, the Jena Atomic Calculator [16]. This toolbox, which facilitates the (relativistic) computation of atomic structures and processes [17,18], has been expanded here in order to model the initial-bound and final-Volkov states in the computation of strong-field amplitudes. Apart from the active-electron waves, however, our implementation below is flexible also in choosing the polarization, shape (envelope) and even the CEP phase of the driving laser pulse. Indeed, all these features have been found to be (very) crucial to further adopt the theoretical modeling of strong-field ionization processes to ongoing experiments.

The paper is structured as follows: After a brief discussion of the SFA and the direct amplitude in terms of partial waves in Section 2, emphasis is placed on the implementation within the framework of JAC as well as the role of appropriate data structures for simplifying the communication *with* and *within* the program. Section 3 then explains and discusses how the different energy, angular and momentum distributions can be obtained quite readily by just specifying the initial and final levels of the target atom as well as the parameters of the laser pulse. This includes the choice of atomic potential and the Volkov states in the evaluation of the amplitudes. Finally, a short summary and conclusions are given in Section 4, with emphasis on possible extensions of the code towards rescattering phenomena, or the computation of harmonic spectra.

2. Strong-Field Amplitudes and Probabilities

2.1. Brief Account of the Strong-Field Approximation

The SFA has been known as perhaps the most straight avenue for modeling strong-field ionization processes and for analyzing most of the associated ATI spectra and momentum distributions. In this approximation, a (so-called) *active* electron is assumed to undergo a transition from its initial bound state $|\psi_i(t)\rangle$ into the laser-dressed continuum $|\chi_{\mathbf{p}}(t)\rangle$ owing to its interaction with the laser pulse (cf. Figure 1a), while the motion of all other electrons of the target is typically assumed to be unaffected. Not much need to be said here about the basic SFA theory, which can be found in various texts [12,13]. In this approximation, the probability for the strong-field ionization of atoms and for finding a photoelectron with asymptotic momentum $\mathbf{p}$ at the detector,

$$P(\mathbf{p}) = p\,|T(\mathbf{p})|^2, \tag{1}$$

can then be expressed in terms of transition amplitudes as [6]

$$\begin{aligned} T(\mathbf{p}) &= T_0(\mathbf{p}) + T_1(\mathbf{p}), \\ T_0(\mathbf{p}) &= -i \int_{-\infty}^{\infty} d\tau\, \langle \chi_{\mathbf{p}}(\tau)|\, V_{\mathrm{le}}(\mathbf{r},t)\, |\psi_i(\tau)\rangle, \qquad (2) \\ T_1(\mathbf{p}) &= (-i)^2 \int_{-\infty}^{\infty} d\tau \int_{\tau}^{\infty} d\tau'\, \langle \chi_{\mathbf{p}}(\tau')|\, V(\mathbf{r})\, U_{\mathrm{le}}(\tau',\tau)\, V_{\mathrm{le}}(\mathbf{r},\tau)\, |\psi_i(\tau)\rangle, \qquad (3) \end{aligned}$$

and where $V_{\mathrm{le}}(\mathbf{r},t)$ refers to the laser–electron interaction, $U_{\mathrm{le}}(t',t)$ the time evolution and $V(\mathbf{r})$ to the *atomic* potential as seen by the outgoing electron.

Indeed, the two (strong-field) amplitudes $T_0(\mathbf{p})$ and $T_1(\mathbf{p})$ can be readily interpreted in terms of a (re-) scattered photoelectron and are often referred to as the direct and rescattering amplitudes, respectively. In this work, we shall focus especially on the direct amplitude $T_0(\mathbf{p})$ that describes those photoelectrons which are directly released from the target atom by the laser potential, $V_{\mathrm{le}}(\mathbf{r},t)\,|\psi_i(t)\rangle$, and then *freely* propagate within the laser field as Volkov solution $|\chi_{\mathbf{p}}\rangle$, until they reach the detector. Indeed, this amplitude often provides a good approximation for most strong-field ionization processes and, in particular, if the laser field is not linearly polarized. Typically, the following assumptions are made to further simplify the amplitude (2):

1. The initial-bound state $|\psi_i(t)\rangle$ is entirely determined by the atomic potential $V(\mathbf{r})$ and is not affected by the laser field.
2. The photoelectron with asymptotic momentum $\mathbf{p}$ arrives as plane-wave at the detector, i.e., $|\chi_{\mathbf{p}}(t \to \infty)\rangle = |\mathbf{p}\rangle$.
3. Once the electron is released from the atom, the atomic potential does not affect its (electronic) motion within the continuum.

Often, moreover, a Coulomb potential $V(\mathbf{r}) = -Z^{(\mathrm{eff})}/r = -\sqrt{2I_p}/r$ has been applied, in line with the ionization potential of the target atoms, and the initial state $|\psi_i(t)\rangle$ has been taken just as $1s$ ground state orbital in this (Coulomb) potential. With these assumptions in mind, the momentum distribution of the photoelectrons can then be expressed by a closed (analytical) formula. Obviously, however, these assumptions neglect both a proper representation of the initial state of the atoms as well as the (static) potential of the photoion upon the outgoing electron wave (continuum) and, hence, quite major parts of the electronic structure of the target atoms.

Several, if not most, of these assumptions can be easily released, if the initial and final states are consequently expressed in terms of partial waves as typical for atomic structure theory [15]. In such a partial wave expansion, the representation of the initial bound and (Volkov) continuum states can be incorporated along with the parameterization of the short an intense laser pulses. It is *this* representation of the (direct) transition amplitude $T_0(\mathbf{p})$ on which we shall focus in the implementation below and which paves the way for extending the strong-field theory towards the study of non-dipole contributions in light-atom interactions as well as towards many-particle correlations in strong-field ionization processes.

2.2. Partial-Wave Representation of Strong-Field Amplitudes

In the derivation of the SFA amplitude ((2) and (3)), indeed, no additional assumptions have to be made about the form of the atomic potential $V(\mathbf{r})$ or about the potentials $\Phi(\mathbf{r},t)$ and $\mathbf{A}(\mathbf{r},t)$ of the driving laser field, but which—of course—affect the Volkov states $|\chi_{\mathbf{p}}\rangle$. If, for the coupling of the radiation field, we restrict ourselves to the dipole approximation $[\,\mathbf{A}(\mathbf{r},t) \approx \mathbf{A}(t)\,]$ and the velocity gauge $[\,\Phi(\mathbf{r},t) = 0\,]$, the vector potential of an elliptically-polarized laser pulse can be written in terms of its spherical tensor components as

$$\mathbf{A}(t) \;=\; \mathrm{Re}\,\{\mathbf{A}_c(t)\} \;=\; \frac{1}{2}\,(\mathbf{A}_c(t) \,+\, \mathbf{A}_c^*(t)), \qquad \mathbf{A}_c(t) \;=\; A_o\,\mathbf{u}\,f(t)\,e^{-i\,(\omega t + \phi^{(\mathrm{CEP})})} \tag{4}$$

and where A_o denotes the (real-valued) amplitude, $f(t)$ the pulse envelope, $\omega = 2\pi\,c/\lambda$ the fundamental frequency, and where $\phi^{(\mathrm{CEP})}$ refers to the carrier–envelope phase of the laser pulse. In this notation, moreover, the (complex) polarization unit vector

$$\mathbf{u} \;=\; \frac{1}{\sqrt{1+\epsilon^2}}\,(\mathbf{e}_x \,+\, i\,\epsilon\,\mathbf{e}_y) \tag{5}$$

defines the orientation of the polarization ellipse in terms of the ellipticity $-1 \le \epsilon \le 1$. The vector potential (4) therefore implies already all the properties $\omega,\ I,\ f(t),\ \epsilon$ and $\phi^{(\mathrm{CEP})}$ of the laser field which can be controlled experimentally.

A partial-wave representation of the amplitudes ((2) and (3)) also enables one to adapt both the initial bound state $|\psi_i\rangle$ and the final continuum state to the target potential of interest [15,19]. This is readily achieved, for instance, by using (self-consistent) solutions from atomic structure theory. For the outgoing photoelectron

$$|\chi_{\mathbf{p}}(t)\rangle \;=\; \frac{1}{(2\pi)^{3/2}}\,e^{-iS_V(t)}\,\left|\phi_{\mathbf{p}}^{(A)}\right\rangle, \tag{6}$$

moreover, one only needs to replace the partial wave $\left|\phi_{\mathbf{p}}^{(A)}\right\rangle$ in the expansion of a plane-wave Volkov state by the corresponding solutions of either a Coulomb–Volkov or distorted-Volkov state in order to account for a realistic potential of the target, including the associated Coulombic and non-Coulombic phase shifts [20,21]. Here, we shall provide only a brief discussion of the theory, just enough to follow our implementation below, while all further details are given in Refs. [15,19].

For the direct SFA transition amplitude (2), Equations (23) and (24) of reference [15] display a rather lengthy formula that is written in a basis of well-defined total angular momenta for the initially bound and the final (photo-) electron. This expression depends explicitly on the Volkov phase and all the parameters of the driving laser pulse, and it also accounts for the spatial dependence of the active electron in terms of the reduced one-particle matrix elements $\langle \varepsilon_p \ell_p j_p \| \,\mathbf{p}\, \| n\ell j\rangle$ of the momentum operator, as typical for atomic structure theory. The advantage of such an expression in a spherical basis arises from the—prior and separate—integration over all radial and spherical coordinates. This expression therefore also enables one to readily incorporate and discuss different contributions from the electron–photon interaction and the representation of the active electron without any need to re-derive the transition amplitude(s) for every target potential and/or laser (pulse) configuration separately. Below, we shall focus especially upon realistic (single-electron) initial states and the improved representation of the continuum for the outgoing photoelectron. It is *this* partial-wave representation of the direct SFA amplitude which makes the present extension an integral part of the JAC toolbox and which goes well beyond of what has been used originally in the SFA. However, since the expression in Ref. [15] is still restricted to the electric-dipole approximation, it neither accounts for the magnetic field nor any spatial dependence of the electric field, though this can likely be done as well (cf. Section 4).

2.3. Implementation of the (Direct) Strong-Field Amplitude

Like in atomic structure theory, a partial-wave representation of all (strong-field) amplitudes enables one to deal quite independently with atomic potentials, the Volkov state for the outgoing electron or the laser–electron interaction in terms of the given laser parameters. Such a representation also helps identify the building blocks for computing the photoelectron spectra and/or momentum distributions, as they are observed experimentally. In the JAC toolbox, we therefore aim to distinguish between the target and laser parameters as the input of a computations, and the generated observables (spectra) as output. A simple access to all individual input parameters will enable us to compute the strong-field ionization amplitudes in quite different approximations. Although the current implementation is still restricted to the direct SFA amplitude in velocity gauge, the same or very similar building blocks will occur if other amplitudes or gauges are to be considered in the future. The use of partial waves and spherical tensors even ensures that these amplitudes can be readily combined with atomic structure codes to further include electronic correlations and relativistic contributions to the strong-field ionization studies.

Not much need to be said about JAC itself, the Jena Atomic Calculator [16] that supports atomic (structure) calculations of different kind and complexity and that has been summarized at various places [22–24]. Apart from energies and wave functions for open-shell atoms and ions, this toolbox also helps compute a good number of their excitation and decay processes. With the design and implementation of JAC [25], we moreover aim for establishing a *descriptive* language that (i) is simple enough for both, a seldom or more frequent use of this toolbox, (ii) emphasizes the underlying atomic physics and, furthermore, (iii) avoids most technical slang as common to many other electronic structure codes. The implementation of JAC is based on Julia [26,27], a recently developed programming language for scientific computing, and supports its use without much prior knowledge about neither the language nor the code itself.

Figure 2 displays a few selected atomic processes that are presently supported by JAC, and which are shown together with useful features and control parameters for calculating strong-field amplitudes. The set of parameters in the right panel of Figure 2 hereby indicates how between different pulses we shall distinguish in these computations geometries and/or gauges for the coupling of the radiation field, and as far they have been worked out until the present. In particular, the initial bound and final Volkov states of the (photo-) electron just appear in the reduced matrix elements $\langle \epsilon_p \ell_p j_p \| \, \mathbf{p} \, \| n\ell j \rangle$, and can be taken either as a hydrogenic orbital, scaled upon the ionization potential, or as realistic one-electron wave function. Here, the continuum orbitals are generated in the static potential of the photoion and with energies as measured at the detector [28].

Special care has to be taken about the envelope of the laser pulses. In a spherical-wave expansion, this envelope enters the direct amplitude in terms of (so-called) pulse-shape integrals $\mathcal{F}_1[\pm\omega; f; \mathbf{p}]$ and $\mathcal{F}_2[f; \mathbf{p}]$, cf. Ref. [15]. These one-dimensional, (time) integrals are often obtained numerically but can be evaluated also analytically for continuous beams and a few other forms of the envelope. In our implementation below, the envelope (shape) of the laser pulse is accessed by a proper (abstract) data type, quite similar to the frequency, intensity, number of cycles and the polarization of the incident pulses. In typical applications of JAC, one needs to select these parameters based on the given setup of the experiment and different practical considerations in order to keep the computations feasible.

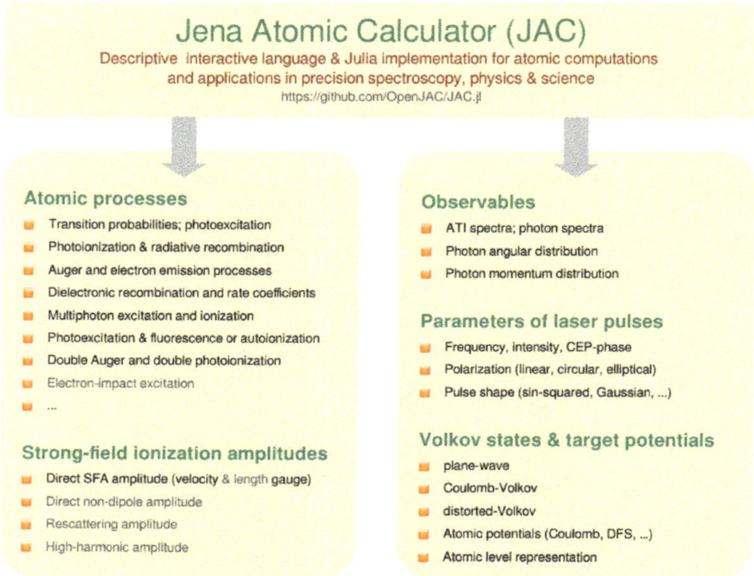

Figure 2. Selected applications of the JAC toolbox that help calculate atomic structures and processes as well as strong-field ionization amplitudes in various approximations. Apart from choosing between typical strong-field observables, the Volkov states and the parameterization of the laser pulses can be controlled rather flexibly. See Refs. [16,25] for a detailed account of all other features of this toolbox.

2.4. Data Types for Modeling Photoelectron Distributions and Above-Threshold Experiments

From a physics viewpoint, we normally wish to trace back the simulation of different spectra and photoelectron distributions to just computing the (direct) SFA amplitude from above, though for specifically selected target atoms, approximation of wave functions and parameters of the laser pulse. Obviously, this requires simple access to all these data as well as special care to bring them together with the internal calls of the program. To facilitate the communication *with* and the data transfer *within* the program, the JAC toolbox is built upon a large number of data structures in order to specify useful and frequently recurring objects in such computations, and which also establish their language elements. Two prominent examples for such data structures, that frequently appear in atomic structure theory, are an `Orbital` for specifying the quantum numbers and radial components of single-electron orbital functions, or a `Level` for the full representation of an approximate initial or final bound state of the target atoms, and which itself comprises all information about the orbitals, the coupling of the angular momenta and the mixing of the many-electron target states. These target states are typically obtained self-consistently in a Dirac–Fock–Slater potential and hence are based on orbitals in line with the given target. JAC's explicit set of data structures has been enlarged for the present update of the code by several types and now helps compute, analyze and explore the desired photoelectron spectra for different laser pulses and targets. In total, there are about 250 of these data structures in JAC, though most of them remain hidden to the user.

To model different strong-field ionization measurements, we wish (and need) to characterize especially the incident laser pulse in terms of its frequency, intensity, shape and polarization as well as the observables (spectra) to be simulated. In addition, we wish to control the target potentials and representation of the Volkov states in the strong-field amplitude. All this input is very central to the implementation and must be readily accessible by the given hierarchy of data structures. While we shall not explain these structures in all detail here, Figure 3 displays a few of them from JAC's `Pulse` and `StrongField` modules. The abstract type `Pulse.AbstractEnvelope` (middle panel), for example, just deals with the envelope of the laser pulse and comprises various concrete types for specifying a particular shape, pulse duration or number of cycles. Similarly, the data type `StrongField.AbstractSFAObservable` enables one to specify the observable of choice and its resolution. All this information about the observable, target and pulse parameters finally define (an instance of) a `StrongField.Computation` (lower panel), and which can be utilized in JAC analogue to the previously implemented `Atomic.Computation` [16,29] or `Cascade.Computation` [30,31].

```
abstract type StrongField.AbstractSFAObservable ... defines an abstract as well as a number of
  concrete data types for observables that can be computed by means of the implemented SFA
  amplitude.

+ struct SfaEnergyDistribution               ... to compute an energy spectrum of the photoelectrons.
+ struct SfaMomentumDistribution             ... to compute a momentum distribution.
+ struct SfaAzimuthalAngularDistribution     ... to compute an angular distribution of photoelectrons
                                                 as function of the azimuthal angle phi but for a
                                                 fixed energy and polar angle.
+ struct SfaPolarAngularDistribution         ... to compute the angular distribution as function of
                                                 theta but for fixed energy and phi.

StrongField.SfaEnergyDistribution <: StrongField.AbstractSFAObservable, etc.
```

```
abstract type Pulse.AbstractEnvelope ... defines an abstract type to deal with the envelope
  of a laser pulse; it comprises the following (concrete) subtypes:

+ InfiniteEnvelope          ... to represent an infinte (plane-wave) pulse.
+ RectangularEnvelope       ... to represent a finite rectangular pulse.
+ SinSquaredEnvelope        ... to represent a finite sin^2 pulse.
+ GaussianEnvelope          ... to represent a Gaussian light pulse.

Pulse.RectangularEnvelope <: Pulse.AbstractEnvelope, etc.
```

```
struct StrongField.Computation ... defines a type for the computation of strong-field amplitudes
  and observables; it enables the user to specify all the parameters as they are typically known
  from strong-field ionization experiments.

+ observable          ::AbstractSFAObservable          ... SFA obserable to be calculated in this run.
+ nuclearModel        ::Nuclear.Model                  ... Model, charge and parameters of the nucleus.
+ grid                ::Radial.Grid                    ... Radial grid to be used for the computation.
+ initialLevel        ::Level                          ... Initial level of the atom.
+ finalLevel          ::Level                          ... Final level of the atom.
+ beam                ::Pulse.AbstractBeam             ... Type and properties of the incident pulse.
+ envelope            ::Pulse.AbstractEnvelope         ... Envelope of the incident light pulse.
+ polarization        ::Basics.AbstractPolarization    ... Polarization of the pulse.
+ volkov              ::AbstractVolkovState            ... Approach for the Volkov states.
+ settings            ::StrongField.Settings           ... Settings to control the SFA computation.
```

Figure 3. Selected data structures of the JAC toolbox that help specify and perform a `StrongField.Computation` (lower panel). Apart from the observable of interest (upper panel), the nuclear model, radial grid as well as the initial and final level of the target atom, one needs to specify the properties of the laser pulse in terms of its beam type, envelope (middle panel) and the polarization of the incident light. Moreover, the user can select the Volkov state approach and a number of further settings. See Table 1 for other data types that are closely related to `StrongField.Computations`.

Finally, Table 1 displays several other data structures that are relevant as well for the computation and analysis of strong-field photoelectron distributions. They are explained only in brief, while further details can be obtained from JAC's User Guide [25] or by just using Julia's help facilities [32]. The definition and hierarchy of these data structures however nicely illustrate how they help implement different strong-field ionization scenarios and, hence, a wide range of potential applications in atomic and atto-second physics. In the next section, we make use of these data types to simulate various energy, angular and momentum distributions for a krypton target.

Table 1. Selected data structures of the JAC toolbox that are relevant for `StrongField.Computations`. Here, only a brief explanation is given, while further details can be found by using Julia's help facilities.

Struct & Brief Explanation.
`Basics.AbstractPolarization`: defines an abstract type to deal with different polarizations of light and electron beams; it presently comprises the concrete subtypes: `LinearPolarization`, `LeftCircular`, `RightCircular`, `LeftElliptical`, `RightElliptical`.
`ManyElectron.Level`: defines a type for an atomic level in terms of its quantum number, energy and with regard to an explicitly specified relativistic basis.
`Pulse.AbstractBeam`: defines an abstract type to deal with various basic laser pulses as they often characterized in terms of their amplitude, frequency, carrier–envelope phase, etc. In general, the basic beam properties are independent of the (pulse) envelope and the polarization properties which are handled and communicated separately (to and within the program).
`Pulse.AbstractEnvelope`: defines an abstract type to deal with different envelopes of the laser pulses with regard to their shape, pulse duration or number of cycles; it comprises the concrete subtypes: `InfiniteEnvelope`, `RectangularEnvelope`, `SinSquaredEnvelope` and `GaussianEnvelope`.
`StrongField.AbstractVolkovState`: defines an abstract type to specify the Volkov states in the computation of SFA amplitudes; it comprises the concrete types: `FreeVolkov`, `CoulombVolkov`, `DistortedVolkov`.
`StrongField.SphericalAmplitude`: to compute and store the amplitude at a given energy-angular point (energy, theta, phi) in the momentum space of the outgoing electron.
`StrongField.Settings`: specifies further settings for the computation of SFA amplitudes and observables, including the choice of the multipole field, gauge and several others.

3. Energy and Momentum Distributions for Atomic Krypton

In the literature, the SFA has been frequently applied for comparing the energy and momentum distributions with experiments and for studying pulses and targets of quite a different sort. In these computations, more often than not, the active electron has initially been assumed to be in a hydrogenic 1s state, and by just matching the ionization potential to the target of interest. However, such a simple approach provides only little insight into the role that the target atoms play in strong-field ionization. Here, we wish to demonstrate that our partial-wave representation of the SFA amplitude enables us to adopt the initial-bound and final-Volkov states to realistic target potentials. We also show how the ATI spectra and momentum distributions can be obtained for pulses of different intensity, polarization and pulse duration. All these computations are performed by applying the JAC toolbox [16], which integrates the electronic structure and a good deal of atomic processes within a single computational framework, and which has now been expanded to facilitate the simulation of strong-field ionization distributions. For the sake of convenience, all simulations below are performed for krypton (I_p = 14 eV) and a right-circularly polarized, $n_p = 8$ cycle driving laser pulse with wavelength $\lambda = 800$ nm, intensity $I = 10^{14}$ W/cm^2 and carrier–envelope phase $\phi^{(\mathrm{CEP})} = 0$. Here, we shall not compare our implementation with experiment or previous computations, which have been done recently for a number of other targets [19].

3.1. *Above-Threshold Energy Spectra*

Often, the observed ATI spectra can be *qualitatively* reproduced by simply using the SFA and plane-wave Volkov continuum states, since the peak structure of these photoelectron spectra itself arises from the interaction of the (quasi-) free electron with the electric field of the ionizing laser pulse. For these reasons, most energy spectra also exhibit distinct peaks, which are just spaced by the photon energy of the incident laser beam. These peaks become easily visible by measuring the photoelectron energy for a fixed azimuthal angle φ_o along some line in the $p_x - p_y$ polarization plane. Besides the selected laser parameters, these energy spectra depend of course also on the target atoms as well as on how the photoelectrons are described on their way to the detector, including the Volkov continuum and, possibly, even a re-scattering of the photoelectrons.

The ATI energy spectra of the strong-field photoelectrons can be modeled also by the present extension to the JAC toolbox. Figure 4 displays the (Julia) input which needs to be prepared by the user and which enables one to calculate such spectra for different targets and pulses. In this input, we have assumed that the $4p^6\ ^1S_0$ ground level of atomic krypton and the $4p^5\ ^2P_{1/2,3/2}$ final levels of the photoion have been computed before by the JAC toolbox and are just provided by the variables `initialLevel` and `finalLevel`. Here, we make use of a slightly larger charge of the nucleus in order to adopt the $4p$ ionization potential to experiment. To characterize the laser pulse, moreover, we provide the wavelength, intensity and carrier–envelope phase and assume a $\sin^2$ envelope as well as a right-circularly polarized plane-wave beam. Some of these given parameters first need to be converted to atomic units in order to be applicable in the computation of the field amplitude. We also specify here the velocity gauge and the electric-dipole (E1) interaction, even if these parameters must not be changed in the present implementation. The choice of a hydrogenic orbital with scaled nuclear charge can be made by a boolean in the `StrongField.Settings()`.

```
nuclearModel     = Nuclear.Model(36.53)
grid             = Radial.Grid(Radial.Grid(false), rnt = 4.0e-6, h = 5.0e-2, hp = 2.0e-3, rbox = 20.0)

# Choose electron continuum in the laser field
volkov           = StrongField.FreeVolkov()

# Define laser beam parameter
wavelength       = 800.;       intensity = 1e14;      CEP = 0.;
envelope         = Pulse.SinSquaredEnvelope(8)
polarization     = Basics.RightElliptical(1)

omega            = convertUnits("energy: from wavelength [nm] to atomic", wavelength)
intensity        = convertUnits("intensity: from W/cm^2 to atomic", intensity)
A0               = Pulse.computeFieldAmplitude(intensity, omega)
beam             = Pulse.PlaneWaveBeam(A0, omega, CEP)

# Spectral observations and settings
observable       = StrongField.SfaEnergyDistribution(pi/2, 0.0, 200, 10*omega)
sfaSettings      = StrongField.Settings([E1], "VelocityGauge", true, true, false, false, true)

comp             = StrongField.Computation(observable, nuclearModel, grid, initialLevel, finalLevel,
                                           beam, envelope, polarization, volkov, sfaSettings)
StrongField.perform(comp, output=true)
```

Figure 4. Julia input for generating the black-solid ATI spectrum in the left panel of Figure 5 for a krypton target, if irradiated by an $n_p = 8$ cycle $\sin^2$ laser pulse with a central wavelength of $\lambda = 800$ nm and intensity $I = 10^{14}$ W/cm^2. The laser pulse is right-circularly polarized and has a carrier–envelope phase $\phi^{(\mathrm{CEP})} = 0$. This input describes the complete strong-field computation, but where the $4p^6\ ^1S_0$ ground and the final $4p^5\ ^2P_{1/2,3/2}$ levels of krypton are assumed to be generated before by the JAC toolbox. Although no attempt is made to explain this input in all detail, this figure nicely demonstrates how readily JAC can be utilized to generate rather different spectra and distributions. See text for further explanations.

In the input above, we finally also specify as observable an `SfaEnergyDistribution()`, and which is to be calculated for $\vartheta = \pi/2$ and $\varphi = 0$ (i.e., along the x-axis), and for 200-electron energies in the interval $0 \leq \varepsilon_{\mathrm{kin}} \leq 10\,\omega \approx 15$ eV. All this input together determines the (strong-field) computation `comp::StrongField.Computation` and can be readily adopted to many other experimental scenarios. All that is needed in JAC is to `perform(comp, output=true)` this computation, and where the optional parameter `output=true` just tells JAC to return the calculated data (amplitudes) to the user for printing and post-processing.

Figure 5 displays the photoelectron energy spectra, emitted along the x-axis, for a krypton target and a right-circularly polarized laser pulse. The left panel shows the spectra as obtained for a computed with a hydrogenic $1s$ initial wave function with adopted ionization potential and for a plane-wave Volkov continuum (black-solid curve) as well as a Coulomb–Volkov continuum (red dashed curve). On the right panel, in contrast, the spectra are computed for an initial $4p$ orbital of neutral krypton and a plane-wave Volkov continuum (black-solid curve), a Coulomb–Volkov continuum (red-dashed curve) as well as a distorted-Volkov continuum (blue-dotted curve). In all these computations, a right-circularly polarized $\sin^2$ pulse of wavelength $\lambda = 800$ nm, intensity $I = 10^{14}$ W/cm^2, carrier–envelope phase $\phi^{(\text{CEP})} = 0$ and with just $n_p = 8$ cycles has been utilized.

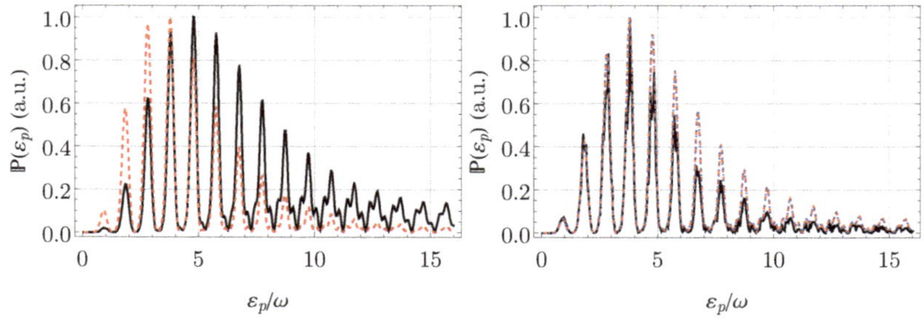

Figure 5. Photoelectron energy spectra, emitted along the x-axis within the polarization plane, for a neutral krypton target and a right-circularly polarized laser pulse. The left panel shows the spectra as computed with a hydrogenic 1s initial wave function with adopted ionization potential and for a plane-wave Volkov continuum (black-solid curve) as well as a Coulomb–Volkov continuum (red-dashed curve). On the right panel, in contrast, the spectra are computed for an initial $4p$ orbital of neutral krypton and a plane-wave Volkov continuum (black-solid curve), a Coulomb–Volkov continuum (red-dashed curve) as well as a distorted-Volkov continuum (blue-dotted curve). In all these computations, a right-circularly polarized $\sin^2$ pulse of wavelength $\lambda = 800$ nm, intensity $I = 10^{14}$ W/cm^2, carrier–envelope phase $\phi^{(\text{CEP})} = 0$ and with $n_p = 8$ cycles has been utilized.

Input quite similar to Figure 4 can be employed also for studying the angle and momentum distributions of photoelectrons for different laser pulses and targets. While no further input data will be shown below, we refer the reader for details to the User Guide and the online documentation of the JAC program. Moreover, rather moderate changes to this input will be sufficient in the future to expand the `StrongField` module to other gauges, amplitudes or many-electron features. While such an expansion of the code indeed appears straightforward, major effort will still be needed for its implementation and testing.

3.2. Photoelectron Angular Distribution for Elliptically-Polarized Laser Pulses

In the electric-dipole (E1) approximation, the angular distribution of the photoelectrons is restricted to the $x - y$ polarization plane and just reflects at fixed photoelectron energy the ionization probability in Equation (1) for different azimuthal angles $0 \leq \varphi \leq 360°$. If, moreover, the laser field dominates the electron dynamics in the continuum, the observed photoelectron angular distribution (PAD) should also reflect the symmetry of the vector potential of the laser beam. In practice, however, a *Coulomb asymmetry* in the PAD was (first) observed by Goreslavski et al. [33] in the ATI of xenon gas targets and, since then, has been found to be a valuable testbed for improving theory. For lithium, argon and xenon, for example, the SFA theory was shown to reproduce this asymmetry, if a target-specific initial orbital function is chosen along with a distorted-Volkov continuum for the active electron [34].

Figure 6 displays different photoelectron angular distributions in the polarization plane ($\vartheta_p = \pi/2$) for a krypton target. Angular distributions are shown for elliptically-polarized laser pulses with $\epsilon = 0.36$ (left panel) and $\varepsilon = 0.56$ (right panel) at fixed photoelectron energy $\varepsilon_p \approx 2.9\,\omega$ according to the third ATI peak in Figure 5. Different approximations are compared for these angular distributions: a hydrogenic 1s initial orbital together with a plane-wave Volkov continuum (black-solid curves); a self-consistent initial $4p$ orbital of neutral krypton together with a Coulomb–Volkov continuum (red long-dashed curves); the same initial $4p$ orbital but together with a distorted-Volkov continuum (blue-dashed curves). All these distributions are normalized on their maximum, while all other laser parameters are the same as in Figure 5. Indeed, a self-consistent $4p$ orbital of neutral krypton together with a Coulomb–Volkov continuum (red long-dashed curves) leads to a clear rotation of the PAD as mentioned above. Moreover, the PAD no longer exhibits an inversion symmetry with regard to the origin because of the short duration of the laser pulse. If, in addition, the Coulomb–Volkov continuum is replaced by an distorted-Volkov continuum (blue-dashed curves), and which accounts for an outgoing electron in the potential of the Kr^{+} photoion, the rotation angle still changes rather remarkably. In Ref. [19], it was shown that such a distorted-Volkov continuum (often) leads for different targets to better agreement with experiment.

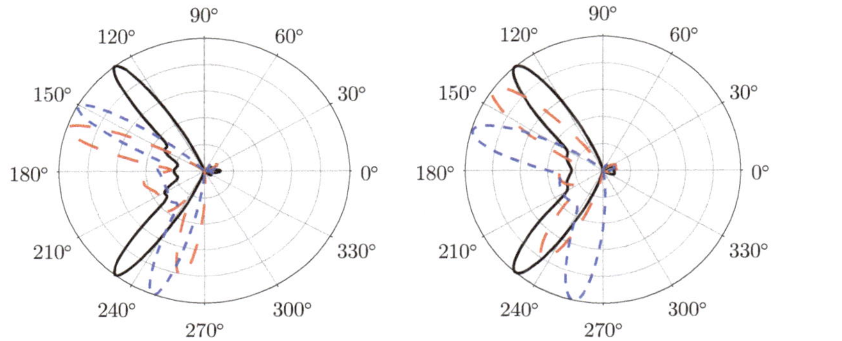

Figure 6. Photoelectron angular distributions in the polarization plane ($\vartheta_p = \pi/2$) for a krypton target. Angular distributions are shown for elliptically-polarized laser beams with $\epsilon = 0.36$ (**left panel**) and $\epsilon = 0.56$ (**right panel**) at fixed photoelectron energy $\varepsilon_p \approx 2.9\,\omega$ according to the third ATI peak in Figure 5. Different approximations are compared for these angular distributions: a hydrogenic 1s initial orbital together with a plane-wave Volkov continuum (black-solid curves); a self-consistent initial $4p$ orbital of neutral krypton together with a Coulomb–Volkov continuum (red long-dashed curves); the same initial $4p$ orbital but together with a distorted-Volkov continuum (blue-dashed curves). All distributions are normalized on their maximum, while all other laser parameters are the same as in Figure 5.

3.3. *Photoelectron Momentum Distribution for Few-Cycle Laser Pulses*

Theoretical photoelectron momentum distributions (PMD) have been calculated in the literature by means of quite different methods, and by making use of even a larger number of case-specific modifications to these methods. Generally, the experimentally observed symmetries of the PMD cannot be explained so readily by just applying a plane-wave Volkov continuum [33], but can be improved further if the Coulomb potential of the residual ion is taken into account. In our implementation of the SFA direct amplitude, this is achieved by replacing the plane-wave Volkov continuum by either Coulomb–Volkov or distorted-Volkov states. For the low-energy photoelectrons with (say) $\varepsilon_p \lesssim 2\omega$, the ionization probability is then often enhanced by up to an order of magnitude, if the ionic charge just increases from $Z = 0$ to 1. This has been explained by the attractive Coulomb potential of the residual ion that pulls the electron back to the ion and hence reduces its kinetic energy. The low-energy part of the ATI spectra can be further improved by adding a short-range potential to the (long-range) Coulomb potential and by making use of distorted-Volkov states.

Figure 7 shows the photoelectron momentum distributions in the polarization plane ($\vartheta_p = \pi/2$) for the strong-field ionization of a krypton target. Momentum distributions are shown for circularly-polarized laser beams with three different CEP phases: $\phi^{(\mathrm{CEP})} = 0^\circ$ (left panel), $\phi^{(\mathrm{CEP})} = 45^\circ$ (middle panel) and $\phi^{(\mathrm{CEP})} = 90^\circ$ (right panel) and by applying a self-consistent initial $4p$ orbital of neutral krypton together with a plane-wave Volkov continuum. All further laser parameters are the same as in Figure 5. Obviously, the PMD in this figure exhibits a (very) clear rotation since the photoelectrons are preferably emitted in the polarization plane along the maximum of the vector potential [6], and which changes with the carrier envelope phase $\phi^{(\mathrm{CEP})}$. It will be interesting to explain with JAC in future work how the Coulomb asymmetry and the CEP dependence act together upon the angular or momentum distributions, and, especially of the initial-bound and distorted-Volkov continuum states of different atomic targets, are taken properly into account. In these studies, both the Coulomb and short-range interactions can be easily incorporated into the continuum by just replacing the radial wave functions of the active electron.

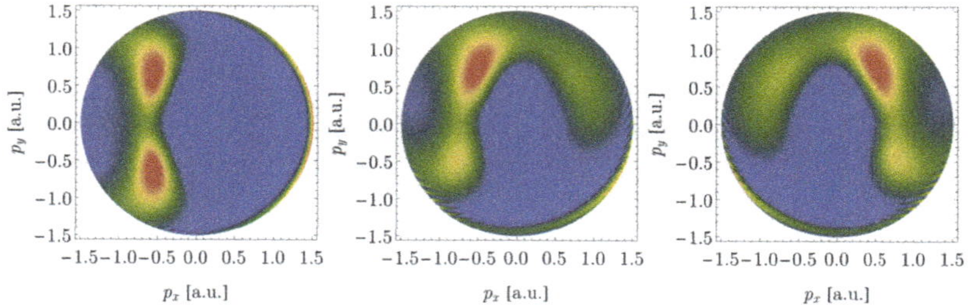

Figure 7. Photoelectron momentum distributions in the polarization plane ($\vartheta_p = \pi/2$) for a krypton target. Momentum distributions are shown for circularly-polarized laser beams with three different CEP phases: $\phi^{(\mathrm{CEP})} = 0^\circ$ (**left panel**), $\phi^{(\mathrm{CEP})} = 45^\circ$ (**middle panel**) and $\phi^{(\mathrm{CEP})} = 90^\circ$ (**right panel**) and by applying a self-consistent initial $4p$ orbital of neutral krypton together with a plane-wave Volkov continuum. All further laser parameters are the same as in Figure 5.

4. Conclusions and Outlook

Up to the present, the SFA has been found as perhaps the most powerful method for predicting or analyzing the electron dynamics in strong-field ionization. Often, this approximation helps describe features in the observed electron distributions even *quantitatively*, if the initial-bound and final-continuum states of the photoelectron are well adopted to the target atoms, and if combined with a proper parameterization of the laser field. With the present implementation of the direct SFA amplitude into the JAC toolbox, this method can now be applied to different targets and strong-field scenarios. In particular, the implementation of the SFA in the partial-wave representation enables us to readily control (and replace) the wave functions and various details about the laser–electron interaction. It also enables us to extend this implementation for incorporating further interactions and mechanisms into the modeling.

Detailed calculations are performed for a krypton target as well as for different ATI spectra and PMD. These examples clearly show how the target potential affects the photoelectrons on their way to the detector and, hence, all the observed spectra. In particular, we have demonstrated how the electronic structure of the atomic targets can be taken into account in the representation of the active electron and how the dynamics of the outgoing electron can be readily controlled by applying different approximations for the Volkov continuum. Moreover, the use of partial waves and spherical tensor operators facilitates a simpler comparison of different pulse shapes and how they influence the observed ATI spectra and PMD.

Several extensions to the SFA are still desirable and appear feasible within a framework, which is based on a partial-wave representation of the associated strong-field amplitudes. While further effort will be needed to decompose these amplitudes into a form, suitable for computations, a few useful extensions concern:

- **Non-dipole interactions:** For spatially-structured light fields, non-dipole contributions to the Volkov continuum usually arise from the spatially dependent Volkov phase [35–37], and which need first to be expressed into a partial-wave representation in order to become applicable within JAC. These *non-dipole* terms beyond the widely used E1 approximation capture the combined-electric *and* magnetic-fields upon the electron dynamics [38,39]. Their implementation into the JAC toolbox will help predict the energy and momentum shifts at long wavelengths of the driving fields.

- **Coupling of the radiation field:** Apart from the (commonly applied) velocity gauge, the direct amplitude can be also implemented in length gauge. This leads to more complicated pulse-shape integrals that also comprise the reduced matrix elements of the momentum operator, since the kinetic momentum then needs to be replaced by the (time-dependent) canonical momentum. While such an implementation requires further work, the direct SFA amplitude in length gauge was shown to provide more accurate results for the ionization of non-spherical np electrons [40].
- **Rescattering amplitude:** For laser pulses with proper polarization, the electrons are known to be partly *rescattered* by the photoion, which then leads to processes, such as high-order ATI, the non-sequential double ionization, or even to high-order harmonics beyond the well-known *cut-off* law [41]. A partial-wave representation of the rescattering amplitude (2) is currently worked out and can be applied to account for realistic rescattering potentials.
- **High-harmonic generation:** Similar to the rescattering ATI amplitude above, a recombination amplitude needs to be computed in order to obtain the dipole moment of emitted high-harmonic radiation. For modeling HHG, again, we expect to benefit from a re-formulation of the dipole amplitude in terms of partial waves and from including realistic initial and continuum orbitals [42,43].
- **Role of bound states:** The coupling of the ground and continuum states to other excited (bound) states has been analyzed in the literature for just a (very) few selected HHG spectra [44]. A partial-wave representation of the SFA amplitudes facilitates the coupling to excited states of the target and may help explain the formation and influence of (autoionizing) resonances in the HHG plateau.
- **Many-electron effects:** A consequent partial-wave decomposition of all strong-field amplitudes enables one to incorporate many-electron contributions beyond a (spherical) short-range potential into the formalism. Apart from the self-consistent field and the mixing of important configurations, this also refers to the treatment of the multipole contributions (higher than E1), if the corresponding many-electron matrix elements are utilized [45,46].
- **Nonsequential double ionization (NSDI):** When the photoelectron returns to the photoion, the electron can scatter inelastically under the ionization of a second electron. Theoretically, the NSDI is typically described semi-classically by using excitation and/or ionization cross sections for the second (ionizing) step of the process [47,48]. A partial-wave representation of all associated quantum SFA amplitude facilitates a coherent treatment of this nonlinear ionization process.

For all these desirable extensions, the partial-wave representation of the SFA [15], and its implementation in JAC provides a straight and conceivably the best way to advance theory and the light–atom interaction in strong fields.

Author Contributions: Methodology, S.F. and B.B.; software, S.F. and B.B.; writing—review and editing, S.F. and B.B. All authors have read and agreed to the published version of the manuscript.

Funding: This work has been funded by the Deutsche Forschungsgemeinschaft (DFG, German Research Foundation)—440556973.

Institutional Review Board Statement: Not applicable.

Informed Consent Statement: Not applicable.

Data Availability Statement: Not applicable.

Conflicts of Interest: The authors declare no conflict of interest.

References

1. Agostini, P.; Fabre, F.; Mainfray, G.; Petite, G.; Rahman, N.K. Free-free transitions following six-photon ionization of xenon atoms. *Phys. Rev. Lett.* **1979**, *42*, 1127. [CrossRef]
2. Paulus, G.G.; Nicklich, W.; Huale, X.; Lambropoulos, P.; Walther, H. Plateau in above threshold ionization spectra. *Phys. Rev. Lett.* **1994**, *72*, 2851. [CrossRef] [PubMed]
3. McPherson, A.; Gibson, G.; Jara, H.; Johann, U.; Luk, T.S.; McIntyre, I.A.; Boyer, K.; Rhodes, C.K. Studies of multiphoton production of vacuum-ultraviolet radiation in the rare gas. *J. Opt. Soc. Am. B* **1987**, *4*, 595–601. [CrossRef]
4. Ferray, M.; L'Huillier, A.; Li, X.F.; Lompre, L.A.; Mainfray, G.; Manus, C. Multiple-harmonic conversion of 1064 nm radiation in rare gases. *J. Phys. B* **1988**, *21*, L31. [CrossRef]
5. L'Huillier, A.; Lompre, L.A.; Mainfray, G.; Manus, C. Multiply charged ions formed by multiphoton absorption processes in the continuum. *Phys. Rev. Lett.* **1982**, *48*, 1814. [CrossRef]
6. Milosevic, D.B.; Paulus, G.G.; Bauer, D.; Becker, W. Above-threshold ionization by few-cycle pulses. *J. Phys. B* **2006**, *39*, R203. [CrossRef]
7. Böning, B.; Abele, P.; Paufler, W.; Fritzsche, S. Above-threshold ionization of Ba^+ with realistic initial states in the strong-field approximation. *J. Phys. B* **2021**, *54*, 025602. [CrossRef]
8. Ivanov, I.A.; Kheifets, A.S. Angle-dependent time delay in two-color XUV+IR photoemission of He and Ne. *Phys. Rev. A* **2017**, *96*, 013408. [CrossRef]
9. Keldysh, L.V. Ionization in the field of a strong electromagnetic wave. *Sov. Phys. JETP* **1964**, *20*, 1307.
10. Faisal, F.H.M. Multiple absorption of laser photons by atoms. *J. Phys. B At. Mol. Phys.* **1973**, *6*, L89. [CrossRef]
11. Reiss, H.R. Effect of an intense electromagnetic field on a weakly bound system. *Phys. Rev. A* **1980**, *22*, 1786. [CrossRef]
12. Ivanov, M.Y.; Spanner, M.; Smirnova, O. Anatomy of strong field ionization. *J. Mod. Opt.* **2005**, *52*, 165–184. [CrossRef]
13. Amini, K.; Biegert, J.; Calegari, F.; Chacón, A.; Ciappina, M.F.; Dauphin, A.; Efimov, D.K.; de Morisson Faria, C.F.; Giergiel, K.; Gniewek, P.L.; et al. Symphony on strong field approximation. *Rep. Progr. Phys.* **2019**, *82*, 116001. [CrossRef]
14. Kheifets, A. Revealing the target electronic structure with under-threshold RABBITT. *Atoms* **2021**, *9*, 66. [CrossRef]
15. Böning, B.; Fritzsche, S. Partial-wave representation of the strong-field approximation. *Phys. Rev. A* **2020**, *102*, 053108. [CrossRef]
16. Fritzsche, S. A fresh computational approach to atomic structures, processes and cascades. *Comp. Phys. Commun.* **2019**, *240*, 1–14. [CrossRef]
17. Schippers, S.; Martins, M.; Beerwerth, R.; Bari, S.; Holste, K.; Schubert, K.; Viefhaus, J.; Savin, D.W.; Fritzsche, S.; Müller, A. Near L-edge single and multiple photoionization of singly charged iron ions. *Astrophys. J.* **2017**, *849*, 5. [CrossRef]
18. Fritzsche, S. Level structure and properties of open f-shell elements. *Atoms* **2022**, *10*, 7. [CrossRef]
19. Böning, B.; Fritzsche, S. Partial-wave representation of the strong-field approximation. Atomic states and Coulomb asymmetry. *Phys. Rev. A* **2022**, *submitted*.
20. Faisal, F.H.M. Strong-field S-matrix theory with final-state Coulomb interaction in all orders. *Phys. Rev. A* **2016**, *94*, 031401. [CrossRef]
21. Milosevic, D.B.; Becker, W. Atom-Volkov strong-field approximation for above-threshold ionization. *Phys. Rev. A* **2019**, *99*, 043411. [CrossRef]
22. Schippers, S.; Beerwerth, R.; Ábrók, L.; Bari, S.; Buhr, T.; Martins, M.; Ricz, S.; Viefhaus, J.; Fritzsche, S.; Müller, A. Prominent role of multielectron processes in K-shell double and triple photodetachment of oxygen anions. *Phys. Rev. A* **2016**, *94*, 041401(R). [CrossRef]
23. Beerwerth, R.; Buhr, T.; Perry-Sassmannshausen, A.; Stock, S.O.; Bari, S.; Holste, K.; Kilcoyne, A.D.; Reinwardt, S.; Ricz, S.; Savin, D.W.; et al. Near L-edge single and multiple photoionization of triply charged iron ions. *Astrophys. J.* **2019**, *887*, 189. [CrossRef]
24. Perry-Sassmannshausen, A.; Buhr, T.; Borovik, A., Jr.; Martins, M.; Reinwardt, S.; Ricz, S.; Stock, S.O.; Trinter, F.; Müller, A.; Fritzsche, S.; et al. Multiple photodetachment of carbon anions via single and double core-hole creation. *Phys. Rev. Lett.* **2020**, *124*, 083203. [CrossRef]
25. Fritzsche, S. JAC: User Guide, Compendium & Theoretical Background. *Unpublished*. Available online: https://github.com/OpenJAC/JAC.jl (accessed on 10 February 2022).
26. Julia 1.7 Documentation. Available online: https://docs.julialang.org/ (accessed on 10 May 2022).
27. Bezanson, J.; Chen, J.; Chung, B.; Karpinski, S.; Shah, V.B.; Vitek, J.; Zoubritzky, L. Julia: Dynamism and performance reconciled by design. *Proc. ACM Program. Lang.* **2018**, *2*, 120. [CrossRef]
28. Fritzsche, S.; Fricke, B.; Sepp W.D. Reduced L_1 level-width and Coster-Kronig yields by relaxation and continuum interactions in atomic zinc. *Phys. Rev. A* **1992**, *45*, 1465. [CrossRef]
29. Gaigalas, G.; Fritzsche, S. Angular coefficients for symmetry-adapted configuration states in jj-coupling. *Comput. Phys. Commun.* **2021**, *267*, 108086. [CrossRef]
30. Fritzsche, S.; Palmeri, P.; Schippers, S. Atomic cascade computations. *Symmetry* **2021**, *13*, 520. [CrossRef]
31. Fritzsche, S. Dielectronic recombination strengths and plasma rate coefficients of multiply-charged ions. *Astron. Astrophys.* **2021**, *656*, 163. [CrossRef]
32. Julia Comes with a Full-Featured Interactive and Command-Line REPL (Read-Eval-print Loop) that Is Built into the Executable of the Language. Available online: https://docs.julialang.org/en/v1/stdlib/REPL/ (accessed on 10 May 2022).

33. Goreslavski, S.P.; Paulus, G.G.; Popruzhenko, S.V.; Shvetsov-Shilovski, N.I. Coulomb asymmetry in above-threshold ionization. *Phys. Rev. Lett.* **2004**, *93*, 233002. [CrossRef]
34. Böning, B.; Fritzsche, S. Steering the longitudinal photoelectron momentum in above-threshold ionization with not quite collinear laser beams. *Phys. Rev. A* **2022**, *accepted.*
35. Böning, B.; Paufler, W.; Fritzsche, S. Nondipole strong-field approximation for spatially structured laser fields. *Phys. Rev. A* **2019**, *99*, 053404. [CrossRef]
36. Wolkow, D.M. Über eine Klasse von Lösungen der Diracschen Gleichung. *Z. Phys.* **1935**, *94*, 250–260. [CrossRef]
37. Rosenberg, L.; Zhou, F. Generalized Volkov wave functions: Application to laser-assisted scattering. *Phys. Rev. A* **1993**, *47*, 2146. [CrossRef]
38. Böning, B.; Fritzsche, S. Above-threshold ionization driven by Gaussian laser beams: Beyond the electric dipole approximation. *J. Phys. B At. Mol. Phys.* **2021**, *54*, 144002. [CrossRef]
39. Böning, B.; Paufler, W.; Fritzsche, S. Polarization-dependent high-intensity Kapitza-Dirac effect in strong laser fields. *Phys. Rev. A* **2020**, *101*, 031401(R).
40. Bauer, D.; Milosevic, D.B.; Becker, W. Strong-field approximation for intense-laser–atom processes: The choice of gauge. *Phys. Rev. A* **2005**, *72*, 023415. [CrossRef]
41. Dionissopoulou, S.; Lyras, A.; Mercouris, T.; Nicolaides, C.A. High-order above threshold ionization spectrum of hydrogen and photoelectron angular distributions. *J. Phys. B At. Mol. Phys.* **1995**, *28*, L109. [CrossRef]
42. Paufler, W.; Böning, B.; Fritzsche, S. High harmonic generation with Laguerre-Gaussian beams. *J. Opt.* **2019**, *21*, 094001. [CrossRef]
43. Paufler, W.; Böning, B.; Fritzsche, S. Tailored orbital angular momentum by high-harmonic generation from counterrotating bi-circular Laguerre-Gaussian beams. *Phys. Rev. A* **2018**, *98*, 011401(R).
44. Klaiber, M.; Hatsagortsyan, K.Z.; Keitel, C.H. Sub-barrier pathways to Freeman resonances. *Phys. Rev. A* **2020**, *102*, 053105. [CrossRef]
45. Johnson, W.R. *Atomic Structure Theory: Lectures on Atomic Physics*; Springer: Berlin/Heidelberg, Germany, 2007.
46. Fritzsche, S. The RATIP program for relativistic calculations of atomic transition, ionization and recombination properties. *Comp. Phys. Commun.* **2012**, *183*, 1525. [CrossRef]
47. Chen, Z.; Wang, Y.; Morishita, T.; Hao, X.; Chen, J.; Zatsarinny, O.; Bartschat, K. Revisiting the recollisional excitation-tunneling process in strong-field nonsequential double ionization of helium. *Phys. Rev. A* **2019**, *100*, 023405. [CrossRef]
48. Liu, F.; Chen, Z.; Morishita, T.; Bartschat, K.; Böning, B.; Fritzsche, S. Single-cycle versus multicycle nonsequential double ionization of argon. *Phys. Rev. A* **2021**, *104*, 013105. [CrossRef]

atoms

MDPI

Article

Accurate Exponential Representations for the Ground State Wave Functions of the Collinear Two-Electron Atomic Systems

Evgeny Z. Liverts * and Nir Barnea

Racah Institute of Physics, The Hebrew University, Jerusalem 91904, Israel; nir@phys.huji.ac.il
* Correspondence: Evgeny.Liverts@mail.huji.ac.il

Abstract: In the framework of the study of helium-like atomic systems possessing the collinear configuration, we propose a simple method for computing compact but very accurate wave functions describing the relevant S-state. It is worth noting that the considered states include the well-known states of the electron–nucleus and electron–electron coalescences as a particular case. The simplicity and compactness imply that the considered wave functions represent linear combinations of a few single exponentials. We have calculated such model wave functions for the ground state of helium and the two-electron ions with nucleus charge $1 \leq Z \leq 5$. The parameters and the accompanying characteristics of these functions are presented in tables for number of exponential from 3 to 6. The accuracy of the resulting wave functions are confirmed graphically. The specific properties of the relevant codes by Wolfram Mathematica are discussed. An example of application of the compact wave functions under consideration is reported.

Keywords: two-electron atoms; wave functions; collinear configuration; Fock expansion; Wolfram Mathematica

Citation: Liverts, E.Z.; Barnea, N. Accurate Exponential Representations for the Ground State Wave Functions of the Collinear Two-Electron Atomic Systems. *Atoms* **2022**, *10*, 4. https://doi.org/10.3390/atoms10010004

Academic Editors: Anatoli Kheifets, Gleb Gribakin and Vadim Ivanov

Received: 5 November 2021
Accepted: 20 December 2021
Published: 29 December 2021

1. Introduction

In this paper, we present a technique for building compact and simple wave functions of high accuracy, describing two-electron atomic systems such as H^-, He, Li^+, Be^{2+} and B^{3+} with the *collinear* arrangement of the particles [1]. The study of mechanism of double photoionization of the helium-like atomic systems by high energy photons [2,3] can serve as an example of possible application (see the details in the next Section).

Methods enabling us to calculate the relevant wave function (WF) and the corresponding non-relativistic energy differ from each other by the calculation technique, spatial variables and basis sets. It is well-known that the S-state WF, $\Psi(r_1, r_2, r_{12})$, is a function of three variables: the distances $r_1 \equiv |\mathbf{r}_1|$ and $r_2 \equiv |\mathbf{r}_2|$ between the nucleus and electrons, and the interelectron distance $r_{12} \equiv |\mathbf{r}_1 - \mathbf{r}_2|$, where $\mathbf{r}_1$ and $\mathbf{r}_2$ represent radius-vectors of the electrons. We shall pay special attention to the bases that differ from each other both in the kind of the basis functions and in its number (basis size). The Hartree atomic units are used throughout the paper.

It would be useful to give some examples of basis sets intended for describing the relevant S states. The correlation function hyperspherical harmonic method (CFHHM) [4,5] employs the basis representing the product of the hyperspherical harmonic (HH) as an angular part, and the *numerical* radial part. The corresponding basis size N equals (as a rule) 625. The Pekeris-like method (PLM) [6–8] is used intensively in the current work. The basis size of the PLM under consideration is $N = 1729$ (for the number of shells $\Omega = 25$), and the basis functions can be finally reduced to the form $\exp(\alpha r_1 + \beta r_2 + \gamma r_{12}) r_1^l r_2^m r_{12}^n$, where α, β and γ are the real constants and l, m, n are non-negative integers. Hylleraas [9] (see also [10,11]) was the first who employed the same basis but with $\gamma = 0$. The authors of Ref. [12] have performed variational calculations on the helium isoelectronic sequence using modification of the basis set that employed by Frankowski and Pekeris [13]. They managed to get very accurate results using the reduced basis of the size $N = 230$. The variational basis functions of the type $\exp(\tilde{\alpha}r_1 + \tilde{\beta}r_2 + \tilde{\gamma}r_{12})$ with complex exponents were used in the works of Korobov [14] ($N = 1400$–2200) and Frolov [15] for $N = 600$–2700 (see also references therein). Application of the Gaussian bases of the size $N > 100$ can be found in the book [16]. The reviews on the helium-like atomic system and the methods of their calculations can be found, e.g., in the handbook [17].

In this paper, we propose a simple method of calculation of the compact but very accurate WFs describing the two-electron atom/ion with collinear configuration. The results and example of application of the relevant technique are presented in the next sections.

2. Calculation Technique

The simplicity of the WFs under consideration implies that the form

$$f_N(r) = \sum_{k=1}^{N} C_k \exp(-b_k r) \tag{1}$$

represents the sum of a few single exponentials, whereas the compactness means that their number $3 \leq N \leq 6$ in Equation (1), unlike the basis sizes mentioned in the introduction. The relevant accuracy will be discussed later. It is seen that the RHS of Equation (1) includes N linear parameters C_k and N nonlinear parameters b_k with $k = 1, 2, \ldots, N$.

The *collinear* arrangement of the particles consisting of the nucleus and two electrons can be described by a single scalar parameter λ as follows [1]:

$$r_1 = r, \qquad r_2 = |\lambda| r, \qquad r_{12} = (1-\lambda) r, \tag{2}$$

where $\lambda \in [-1, 1]$, and r is the distance between the nucleus and the electron most distant from it. Clearly $\lambda = 0$ corresponds to the *electron–nucleus* coalescence, and $\lambda = 1$ to the *electron–electron* coalescence. The boundary value $\lambda = -1$ corresponds to the collinear **e-n-e** configuration with the same distances of both electrons from the nucleus. In general, $0 < \lambda \leq 1$ corresponds to the collinear arrangement of the form **n-e-e** where both electrons are on the same side of the nucleus. Accordingly, $-1 \leq \lambda < 0$ corresponds to the collinear arrangement of the form **e-n-e** where the electrons are on the opposite sides of the nucleus. The absolute value $|\lambda|$ measures the ratio of the distances of the electrons from the nucleus.

Thus, for the particles with collinear arrangement we can introduce the *collinear* WF of the form

$$\Phi(r, \lambda) \equiv \Psi(r, |\lambda| r, (1-\lambda) r)/\Psi(0,0,0). \tag{3}$$

It should be emphasized that, e.g., the PLM WF with *collinear* configuration reduces to the form

$$\Phi_{PLM}(r, \lambda) = \exp(-\delta_\lambda r) \sum_{p=0}^{\Omega} c_p(\lambda) r^p, \tag{4}$$

where $\Omega = 25$ for the current (standard) consideration, as it was mentioned earlier.

We can give an example of the physical problem where the collinear WF of the form (4) cannot be applied, but the quite accurate WF of the form (1) is required instead. In Refs. [2,3], the mechanism of photoionization in the two-electron atoms is investigated. Calculations of various differential characteristics (cross sections) of ionization are based on computation of the triple integral of the form

$$\int d^3\mathbf{r}\, e^{i\mathbf{q}\mathbf{r}}\, {}_1F_1(i\xi_1, 1, ip_1 r - i\mathbf{p}_1\mathbf{r})\, {}_1F_1(i\xi_2, 1, ip_2 r - i\mathbf{p}_2\mathbf{r}) \Phi(r, 1), \tag{5}$$

where $\mathbf{p}_j$ ($j = 1, 2$) are the momenta of photoelectrons, $\mathbf{q}$ is the recoil momentum, $\xi_j = Z/p_j$, i is the imaginary unit, and ${}_1F_1(\ldots)$ is the confluent hypergeometric function of the first kind. The most important for our consideration is the fact that integral (5) contains the *collinear* WF $\Phi(r, 1)$ describing the case of the electron–electron coalescence ($\lambda = 1$) in the helium-like atom/ion with the nucleus charge Z. It is clear that the numerical computation of the triple integral (5) is not impossible, but rather a difficult problem, especially for building the relevant graphs. Fortunately, already in 1954 [18], the explicit expression for the triple integral which is very close to integral (5) was derived. In fact, integral (5) can be calculated by simple differentiation (with respect to a parameter) of the explicit form for the integral mentioned above, but only under condition that the WF, $\Phi(r,1)$ is represented by a single exponential of the form $\exp(-br)$ (with positive parameter b, of course).

According to the Fock expansion [19,20] (see also [21,22]), we have:

$$\Psi(r_1,r_2,r_{12})/\Psi(0,0,0) \underset{R\to 0}{=} 1 - Z(r_1+r_2) + \frac{1}{2}r_{12} - Z\left(\frac{\pi-2}{3\pi}\right)\left(R^2 - r_{12}^2\right)\ln R + O(R^2), \tag{6}$$

where $R = (r_1^2 + r_2^2)^{1/2}$ is the hyperspherical radius. Using Equation (6) and the collinear conditions (2), we obtain the Fock expansion for the collinear WF in the form:

$$\Phi(r,\lambda) \underset{r\to 0}{=} 1 + \eta_\lambda r + \zeta_\lambda r^2 \ln r + \xi_\lambda r^2 + \dots \tag{7}$$

where

$$\eta_\lambda = -Z(1+|\lambda|) + \frac{1-\lambda}{2}, \tag{8}$$

$$\zeta_\lambda = -\frac{2Z\lambda(\pi-2)}{3\pi}, \tag{9}$$

and the general form of the coefficient ξ_λ being rather complicated will be discussed later. The necessity of the equivalent behavior of the model WF, (1) and the variational WF, $\Phi(r,\lambda)$ near the nucleus ($r \to 0$) results in the following two coupled equations for $2N$ parameters $\mathbf{C}_N \equiv \{C_1, C_2, \dots, C_N\}$ and $\mathbf{b}_N \equiv \{b_1, b_2, \dots, b_N\}$ of the model WF:

$$\sum_{k=1}^{N} C_k = 1, \tag{10}$$

$$\sum_{k=1}^{N} C_k b_k = Z(1+|\lambda|) + \frac{\lambda-1}{2}. \tag{11}$$

Equation (10) follows from the condition $\Phi(0,\lambda) = 1$, whereas Equation (11) is obtained by equating the linear (in r) coefficients of the power series expansion of the model WF (1) and the Fock expansion (7).

As it was mentioned above, to obtain the fully defined model WF of the form (1) one needs to determine $2N$ coefficients. To solve the problem with given Equations (10) and (11), we need to find extra $2(N-1)$ coupled equations for parameters of the exponential form (1). To this end, we propose to use the definite integral properties of the collinear WF (3).

A number of numerical results presenting expectation values of Dirac-delta functions $\langle\delta(\mathbf{r}_1)\rangle$, $\langle\delta(\mathbf{r}_{12})\rangle \equiv \langle\delta(\mathbf{r}_1-\mathbf{r}_2)\rangle$ and $\langle\delta(\mathbf{r}_1)\delta(\mathbf{r}_2)\rangle$ for the helium-like atoms can be found in the proper scientific literature (see, e.g., [15,17,23] and references therein). It was shown [1] that expectation values mentioned above represent the particular cases of the more general expectation value

$$\langle\delta(\mathbf{r}_1-\lambda\mathbf{r}_2)\rangle = 4\pi\langle\delta(\mathbf{r}_1)\delta(\mathbf{r}_2)\rangle\int_0^\infty |\Phi(r,\lambda)|^2 r^2 dr, \tag{12}$$

where

$$\langle\delta(\mathbf{r}_1)\delta(\mathbf{r}_2)\rangle = \Psi^2(0,0,0)/\int \psi^2(\mathbf{r}_1,\mathbf{r}_2)d^3\mathbf{r}_1 d^3\mathbf{r}_2 \tag{13}$$

is a square of the normalized WF taken at the nucleus. It is seen that the expectation value (12) is fully defined by the *collinear* WF, $\Phi(r,\lambda)$.

We propose to use the integrals of the form

$$S_n = \int_0^{\infty} |\Phi(r,\lambda)|^2 r^n dr, \qquad (n = 0,\ 1,\ 2,\dots) \tag{14}$$

for deriving $2(N-1)$ extra coupled equations required, in its turn, for determining $2N$ coefficients defining the model WF, (1). Replacing $\Phi(r,\lambda)$ in the RHS of Equation (14) by the model WF (1) and using the closed form of the corresponding integral, one obtains n equation of the form

$$S_n = n! \sum_{j=1}^{N} \sum_{k=1}^{N} \frac{C_j C_k}{(b_j + b_k)^n}, \tag{15}$$

where, in fact, $\mathbf{C}_N \equiv \mathbf{C}_N(Z,\lambda)$, $\mathbf{b}_N \equiv \mathbf{b}_N(Z,\lambda)$ are the coefficients we are requested, whereas the integrals $S_n \equiv S_n(Z,\lambda)$ can be computed using, for example, the PLM WFs according to definition (14). The technique proposed, in fact, represents a variant of the "Method of Moments" (see, e.g., [24]) supplemented by the boundary conditions (10) and (11).

The problem is that it is necessary to select a set (sample) of integers $\{n_1, n_2, \dots, n_{2(N-1)}\}$ describing Equations (14) and (15) for each triple of numbers (Z,λ,N). Those selected samples are presented in Tables 1–4, along with the corresponding parameters of the model WFs.

Table 1. Parameters of the model WFs $f_3(r)$.

λ	Z	b_1	b_2	b_3	C_1	C_2	C_3	n_1, n_2, n_3, n_4	$10^4 R_3$
−1	1	1.31221085126	2.30773912084	10.6895455708	1.21535118385	−0.203680351269	−0.0116708325859	1, 2, 3, 4	5.1
	2	3.30331779151	5.37300989029	23.5666840162	1.10218667377	−0.0971397085718	−0.00504696520154	1, 2, 3, 4	2.8
	3	5.29600197779	8.57050886387	37.1879160606	1.06342095849	−0.0603343891163	−0.00308656937009	1, 2, 3, 4	1.8
	4	7.29186771909	11.8104087791	51.8308243021	1.04554901504	−0.0433987968749	−0.00215021816402	1, 2, 3, 4	1.3
	5	9.28979602694	14.9934154050	64.5399905936	1.03580989481	−0.0340832362749	−0.00172665853579	1, 2, 3, 4	1.0
−0.5	1	0.34947029202	0.917929731034	3.34452782210	0.00104484900062	1.06791415059	−0.0689589995860	2, 4, 6, 8	26.1
	2	2.42281226939	4.84214960756	20.8946624090	1.05104853479	−0.0479768260983	−0.00307170869253	2, 3, 4, 5	0.82
	3	3.92179097813	7.24103626268	28.1658903414	1.03561982532	−0.0330601878282	−0.00255963748729	2, 3, 4, 5	0.38
	4	5.42080777652	9.74980079227	37.6087443483	1.02691550123	−0.0249667281386	−0.00194877309539	2, 3, 4, 5	0.31
	5	6.92015063432	12.2629734705	46.2827741201	1.02161443408	−0.0200074716239	−0.00160696245286	2, 3, 4, 5	0.25
0	1	0.298919116361	0.595029813767	7.19650438253	0.297067039246	0.704003189159	−0.00107022840522	0, 2, 3, 4	34.8
	2	1.37487894240	1.79415111548	6.51734657710	0.692940198275	0.307826353261	−0.000766551535934	2, 3, 4, 5	0.16
	3	2.38544848722	2.98500782211	11.9948030706	0.806354227022	0.193817876564	−0.000172103585808	2, 3, 4, 5	0.06
	4	3.39003538779	4.18238632004	44.1576144385	0.860366370158	0.139650495901	−0.0000168660592834	2, 3, 4, 5	0.05
	5	4.39300951665	5.38932092160	77.5849424817	0.892332808812	0.107671063544	$-3.87235633559 \times 10^{-6}$	2, 3, 4, 5	0.05
0.5	1	0.866272795833	1.35400690862	4.90509052234	0.445709671715	0.522361791814	0.0319285364710	2, 3, 4, 5	13.7
	2	2.43745271333	3.52128593757	13.1425881206	0.790450763794	0.200669891111	0.00887934509580	2, 3, 4, 5	3.06
	3	3.94735833664	5.62554439840	21.0951038899	0.867392165040	0.127429856680	0.00517797827994	2, 3, 4, 5	1.72
	4	5.45119270055	7.71384461880	28.7740927992	0.902403168528	0.0938941380922	0.00370269337946	2, 3, 4, 5	1.18
	5	6.95360810088	9.81345881494	36.9476130590	0.923088330095	0.0740947033493	0.00281696655580	2, 3, 4, 5	0.91
1	1	1.53675502654	2.33712280638	8.25913664369	0.596515660760	0.379791535994	0.0236928032457	2, 3, 4, 5	9.5
	2	3.58172751587	5.24357995810	19.2222658045	0.819528550521	0.172004590265	0.00846685921408	2, 3, 4, 5	2.85
	3	5.58931707310	8.08649238609	29.9307024541	0.881008315694	0.113793976078	0.00519770822784	2, 3, 4, 5	1.7
	4	7.59237815227	10.9111334599	40.2659747119	0.910825396831	0.0853703529884	0.00380425018111	2, 3, 4, 5	1.19
	5	9.59453605830	13.7549847587	51.2908698042	0.929005400236	0.0680615517954	0.00293304796837	2, 3, 4, 5	0.94

Table 2. Parameters of the model WFs $f_4(r)$.

λ	Z	b_1 C_1	b_2 C_2	b_3 C_3	b_4 C_4	$n_1, n_2, n_3, n_4, n_5, n_6$	$10^5 R_4$
−1	1	1.32020535772 1.26227358934	2.02749050880 −0.227201776295	4.54443362311 −0.0326343898304	23.5929861083 −0.00243742321816	1, 2, 3, 4, 5, 6	2.7
	2	3.32040412399 1.14400570872	4.62849779909 −0.128974625641	10.9532235110 −0.0142762170106	59.9204389330 −0.000754866072272	0, 1, 2, 3, 4, 5	4.24

Table 2. *Cont.*

λ	Z	b_1 C_1	b_2 C_2	b_3 C_3	b_4 C_4	$n_1, n_2, n_3, n_4, n_5, n_6$	$10^5 R_4$
	3	5.31424074104 1.09045506589	7.36821061275 −0.0810843733976	17.2777611913 −0.00889378923384	91.9038190157 −0.000476903253974	0, 1, 2, 3, 4, 5	2.64
	4	7.30994394442 1.06441904628	10.1897566947 −0.0579854208721	24.4040098169 −0.00616816635722	148.641527280 −0.000265459051415	0, 1, 2, 3, 4, 5	2.18
	5	9.30868327139 1.05138164659	12.8750916787 −0.0460610340035	29.9556776119 −0.00504641308811	155.981269889 −0.000274199500577	0, 1, 2, 3, 4, 5	1.48
−0.5	1	0.760415630622 0.170778873309	0.980090816301 0.969455001129	1.92593212146 −0.130694209709	8.20874671468 −0.00953966472852	2, 3, 4, 6, 7, 8	11.0
	2	0.454124268500 $3.82312760898 \times 10^{-6}$	2.42341297045 1.05207275919	4.76754499162 −0.0486110885984	19.5795924347 −0.00346549372265	2, 3, 4, 5, 6, 7	8.6
	3	0.436440152659 $-7.40656697931 \times 10^{-7}$	3.92140440426 1.03517571156	7.30105365639 −0.0327434647393	28.9040922033 −0.00243150616122	2, 3, 4, 5, 6, 7	8.3
	4	5.42161394245 1.02782877628	9.42798545298 −0.0249202692879	25.0471247671 −0.00283800779973	233.462987854 $-7.04991926150 \times 10^{-5}$	1, 2, 3, 4, 5, 6	0.94
	5	6.92122844867 1.02259238932	11.7547534041 −0.0200065945557	29.3462817580 −0.00247182415422	174.466927481 $-1.13970613965 \times 10^{-4}$	1, 2, 3, 4, 5, 6	0.55
0	1	0.262567575399 0.124958793217	0.389274177978 0.334742657170	0.649724323343 0.549255567787	2.23082077059 −0.00895701817486	2, 3, 4, 5, 10, 12	8.6
	2	0.863128362371 0.000151625897771	1.37696102972 0.700462341914	1.80094039452 0.300247696739	6.23029650467 −0.000861664550097	2, 3, 4, 5, 10, 12	0.85
	3	1.99107587529 0.00268122936659	2.39179071871 0.817445974254	3.00926579184 0.180106425521	10.6422108618 −0.000233629141750	2, 3, 4, 5, 10, 12	0.16
	4	2.81017418133 0.00171603685935	3.39598731473 0.868965245468	4.21438633241 0.129353924670	27.4015188295 $-3.52069972580 \times 10^{-5}$	2, 3, 4, 5, 10, 12	0.18
	5	4.38456769518 0.865405670253	5.14428012476 0.120564394928	6.14092966448 0.0141051979746	16.8182459760 $-7.52631553724 \times 10^{-5}$	2, 3, 4, 5, 10, 12	0.047
0.5	1	0.813357864690 0.270203654224	1.16216119635 0.601608930427	2.11035956249 0.118017239524	8.06302367772 0.0101701758248	2, 3, 4, 5, 6, 7	12.8
	2	2.39591108217 0.666977476198	3.04778032570 0.298527484633	5.77051889963 0.0320148612526	23.1405644612 0.00248017791627	2, 3, 4, 5, 6, 7	2.7
	3	3.90776167717 0.781821062131	4.88664625816 0.198012796435	9.27871837715 0.0187434533393	37.4614080792 0.00142268809457	2, 3, 4, 5, 6, 7	1.5
	4	5.41235086929 0.836936916554	6.71460919468 0.148699140499	12.7195616951 0.0133242070845	50.2714813824 0.00103973586269	2, 3, 4, 5, 6, 7	1.0
	5	6.91531456746 0.870584918595	8.55334168563 0.118467258140	16.2935883030 0.0101878055295	66.2413275150 0.000760017736373	2, 3, 4, 5, 6, 7	0.81
1	1	1.45895844248 0.374761964378	1.94229863142 0.526163652625	3.64144369441 0.0926648508651	14.6402400267 0.00640953213255	2, 3, 4, 5, 6, 7	11.9
	2	3.52660868338 0.702939199818	4.46313533469 0.262807685971	8.41366993352 0.0319120324665	33.9863982377 0.00234108174421	2, 3, 4, 5, 6, 7	2.9
	3	5.53691125164 0.800429839233	6.92690040645 0.178804526892	13.1173116907 0.0193437507870	53.3038611455 0.00142188308737	2, 3, 4, 5, 6, 7	1.7
	4	7.54061400125 0.848381505855	9.36808709974 0.136508216332	17.6989833742 0.0140382215614	70.3315968499 0.00107205625181	2, 3, 4, 5, 6, 7	1.1
	5	9.54371206725 0.878994353474	11.8356042739 0.109401348385	22.5015053896 0.0108088103065	91.8749584959 0.000795487835141	2, 3, 4, 5, 6, 7	0.91

Table 3. Parameters of the model WFs $f_5(r)$.

λ	Z	b_1 C_1	b_2 C_2	b_3 C_3	b_4 C_4	b_5 C_5	$n_1, n_2, n_3, n_4,$ n_5, n_6, n_7, n_8	$10^6 R_5$
−1	2	3.33403942234 1.19779177816	4.15321766999 −0.165381648981	7.22771669209 −0.0277431203764	18.0394750352 −0.00439609964953	98.9124103713 −0.000270909151645	0, 1, 2, 3, 4, 5, 6, 7	4.5
	3	5.32933696681 1.12638430126	6.59886648653 -0.105910712803	11.4228762079 −0.0175600965385	28.2096051133 −0.00273317456436	145.866483798 −0.000180317352143	0, 1, 2, 3, 4, 5, 6, 7	2.8
	4	7.32573160412 1.09052624934	9.10153474687 −0.076177131758	15.9377755826 −0.0124353779117	40.8969472027 −0.00183726599787	290.857815110 $-7.64736726593\times10^{-5}$	0, 1, 2, 3, 4, 5, 6, 7	2.4
	5	9.32452700647 1.07219627060	11.5219291480 −0.060471688997	19.8392221432 −0.0100638089132	48.4980928929 −0.00155216378212	239.735778051 -0.000108608909614	0, 1, 2, 3, 4, 5, 6, 7	1.5
−0.5	1	0.782723673384 0.247875527976	1.00579499131 0.932587577289	1.67516053163 −0.156445025546	3.84347571684 −0.0223795177635	20.7036985208 −0.00163856195617	1, 2, 3, 4, 6, 7, 8, 9	12.3
	2	2.01254813456 0.0111225319077	2.43474345293 1.05138001690	4.24085899493 −0.0543797090181	9.91825838265 −0.00759679366215	49.9226239551 −0.000526046127036	0, 2, 3, 4, 5, 6, 7, 9	6.3
	3	2.66771135582 0.000323581113314	3.92527807057 1.03908286278	6.68858179039 −0.0341431378917	15.1815880350 −0.00487131736096	69.4634730215 −0.000391988635926	1, 2, 3, 4, 5, 6, 7, 8	3.6
	4	5.02658424937 0.0176544037925	5.43330301186 1.01381753461	8.82243334116 −0.0270777601306	19.5615991527 −0.00410558781973	96.7309045026 −0.000288590450437	1, 2, 3, 4, 5, 6, 7, 8	3.6
	5	6.76854109699 0.08493484927950	6.93768261467 0.939048336705	11.4098098294 −0.0209035403321	26.4501588397 −0.00289062795813	130.894355406 −0.000189017693978	0, 1, 2, 3, 4, 5, 6, 7	3.7
0.5	1	0.790902643508 0.196607406672	1.06035478911 0.508558045232	1.50532224818 0.247655549880	3.03736727252 0.0423479201908	11.1410180677 0.00483107802464	2, 3, 4, 5, 6, 7, 8, 9	20.9
	2	2.37639295406 0.590329529453	2.86725870274 0.341324441817	4.29974756696 0.0576037324822	8.77965927735 0.00973365189967	35.0370610367 0.00100864434805	2, 3, 4, 5, 6, 7, 8, 9	3.8
	3	3.88798021172 0.721150604506	4.60274029470 0.238651951852	6.92722664338 0.0340403271804	14.1907934142 0.00558868261345	57.3767081273 0.000568433848049	2, 3, 4, 5, 6, 7, 8, 9	2.1
	4	5.39378534978 0.791446509718	6.35210931219 0.181080013887	9.65638723913 0.0233325486312	19.9037177229 0.00373563625064	76.9969095748 0.000405291512930	2, 3, 4, 5, 6, 7, 8, 9	1.2
	5	6.89631641077 0.831962265527	8.08529427507 0.146815487037	12.3139186919 0.0180923679551	25.6710836429 0.00285602492664	107.258963467 0.000273854553497	2, 3, 4, 5, 6, 7, 8, 9	1.1
1	1	1.40532919228 0.228872138883	1.76542107311 0.564008891655	2.67094532497 0.177278010589	5.73653600501 0.0281992012650	28.8596269922 0.00164175760899	2, 3, 4, 5, 6, 7, 8, 9	26.2
	2	3.49162225456 0.600886634251	4.12954660811 0.328953425818	6.24917362659 0.0598766127109	13.0406927609 0.00945702883157	55.6660835808 0.000826298388269	2, 3, 4, 5, 6, 7, 8, 9	5.2
	3	5.50808155281 0.736257943800	6.48741655765 0.224466915903	10.0193133315 0.0337303661233	21.4634677872 0.00511826651080	95.2150755643 0.000426507663627	2, 3, 4, 5, 6, 7, 8, 9	2.8
	4	7.51732319016 0.807735699959	8.88501166054 0.166935972380	14.1919981920 0.0220982375951	32.7751210282 0.00303510193517	162.408084363 0.000194988130580	1, 2, 3, 4, 5, 6, 7, 8	2.5
	5	9.52058905892 0.845567446616	11.2258539229 0.134819493378	17.9584770013 0.0171241354063	41.6932712805 0.00236103746563	236.717424813 0.000127887134412	1, 2, 3, 4, 5, 6, 7, 8	2.0

Table 4. Parameters of the model WFs $f_6(r)$ for the negative ion of hydrogen ($Z = 1$).

λ	b_1 C_1	b_2 C_2	b_3 C_3	b_4 C_4	b_5 C_5	b_6 C_6	$n_1, n_2,$ $\ldots, n_{10}$	$10^6 R_6$
−1	0.616882428065 0.0000220356978505	1.32144483672 1.27099937960	1.98116133623 −0.227846680046	3.89714002309 −0.0372255443263	10.7703678294 −0.00577032259649	117.105539565 −0.000178868331372	0, 1, 2, …, 9	7.7
−0.5	0.784443648962 0.255668095458	1.00933647226 0.933467043932	1.62815973825 −0.159437110775	3.28191702383 −0.0255522308928	9.07710968927 −0.00407564245776	175.255656318 −0.0000701552652768	0, 1, 2, …, 9	3.5
0.5	0.787145306516 0.183792703068	1.03714737333 0.466750597846	1.42318385820 0.288402624695	2.66484832743 0.0530568221124	7.69976806889 0.00798238877269	534.086213753 0.0000148635064310	0, 1, 2, …, 9	7.2
1	1.36357313996 0.134755301707	1.649819114196 0.5137129189258	2.20488118195 0.272937433228	3.61727560292 0.0657008289814	8.21954358047 0.0126953713973	125.741386557 0.000198145760468	2, 3, 4, …,11	4.8

To solve the set of Equations (10), (11) and $2(N-1)$ nonlinear equations of the form (15) we apply, as the first step, the built-in function **NSolve[. . .]** of the Wolfram *Mathematica*. The additional conditions (inequalities) $\mathbf{b}_N > 0$ are used. The program **NSolve** generates all possible solutions. However, only one of them represents the nodeless solution that corresponds to the ground-state WF. We have computed and presented the parameters of the model WFs for $3 \leq N \leq 6$. It was mentioned above that the **NSolve** is used only at the first step. The reason is that this program works normally (with no problems) only for $N \leq 3$, that is for number of equations $2N \leq 6$. Even for $2N = 6$ computer freezes for a few second capturing 100 % of CPU time, and then normal operation is restored. However, for $2N = 8$ *Mathematica* (through **NSolve**) takes all CPU time, and computer freezes for an indefinite time. This is happened for any settings of *Mathematica*, e.g., for any settings in "Parallel Kernel Configuration". We checked that this problem persists in different computers and for different version of *Mathematica* (9, 10.3, 11.0, 12.1). Therefore, to solve the relevant set of nonlinear equations for the number of exponentials $N > 3$ we employed the built-in (*Mathematica*) program **FindRoot[...]**. Unlike **NSolve** this program generates only one solution (if it exists, of course) starting its search from some initial values $\mathbf{C}_N^{(0)}, \mathbf{b}_N^{(0)}$ for which we take the values $\mathbf{C}_{N-1}, \mathbf{b}_{N-1}$ of the corresponding calculation on the $N-1$ exponentials. The conditions of the positive exponents and the WF nodeless are certainly preserved.

To estimate the accuracy of the model WF we employ the following integral representation

$$R_N = \int_0^\infty r|f_N(r) - \Phi(r,\lambda)|dr \left(\int_0^\infty r|\Phi(r,\lambda)|dr \right)^{-1}. \tag{16}$$

Note that the function $r\Phi$ is more indicative than Φ, at least, for the ground state.

3. Results

The two-exponential representations (excepting the case of $Z = 1$) for the two-particle coalescences only (corresponding to the particular cases $\lambda = 0$ and $\lambda = 1$) were reported in Ref. [25]. In the current paper, we calculate the parameters $\mathbf{C}_N$ and $\mathbf{b}_N$ of the model WFs, $f_N(r) \equiv f_N(r;\lambda)$ for the number of exponentials $3 \leq N \leq 6$. Our calculations are represented for various *collinear* configurations including in particular the two-particle coalescences and the boundary case $\lambda = -1$. The results are presented in Tables 1–4 together with the corresponding accuracy estimations R_N and the sets $\{n_1, n_2, \dots, 2(N-1)\}$ of integers included into the integrals (14). It is seen from all tables that the more exponentials generate the higher accuracy of the model WF.

One should note that for $\lambda = 0$, describing the case of the electron–nucleus coalescence, we were able to calculate the model WFs, $f_N(r)$ represented by three and four exponentials only ($N = 3$, 4). However, at least the case of $N = 4$ shows very high accuracy, which is confirmed by the following. Recall that the integral R_N characterizes the general accuracy of $f_N(r)$. In order to track changes in accuracy with distance r we used the logarithmic function of the form

$$L_N^{(\lambda)}(r) = \log_{10}|1 - f_N(r)/\Phi_{PLM}(r,\lambda)|. \tag{17}$$

It is seen from Tables 1 and 2 that at least for $\lambda = 0$ and given N the minimal accuracy (represented by maximum R_N) is demonstrated by the negative ion H^-, whereas the maximum accuracy (represented by minimum R_N) is demonstrated by the positive ion B^{3+}. The logarithmic functions $L_N^{(0}(r)$ are shown in Figures 1 and 2 for these two-electron ions with boundary (under consideration) nucleus charges $Z = 1$ and $Z = 5$. It is seen that the deviations of the model WF from the PLM WF are practically uniform along the r-axis, and that one extra exponential improves accuracy by 1-2 (decimal) orders. Regarding the accuracy of the model WF, $f_4(r)$ we would like to emphasize the following. In Ref. [1] (see Fig. 3(b) therein) it was displayed the logarithmic function $\mathcal{L}(r)$ of the form (17), which describes the difference between the PLM WF and the CFHHM WF for the $\lambda = 0$ *collinear* configuration of the H^- ion. The so called correlation function hyperspherical harmonic method (CFHHM) [4,5] with the maximum HH indices $K_m = 128$ (1089 HH basis functions) was used for calculation of the fully (3-dimensional) WF of the negative ion H^-. Comparison of the logarithmic estimations $L_4^{(0}(r)$ and the corresponding $\mathcal{L}(r)$ shows that the model WF $f_4(r)$ is even more close to the PLM WF than the CFHHM WF for all values of r, which indicates the *extremely high accuracy* of the model WF (at least for $\lambda = 0$ and $Z = 1$) represented by *four exponentials* only. It is seen (see Figure 2) that the accuracy of the model WF $f_4(r)$ for B^{3+} is higher by about 2 decimal orders than the 4-exponential WF for H^-. The logarithmic estimation $\mathcal{L}(r)$ for B^{3+} is not presented in Ref. [1]. However, the relevant calculations show that for this case ($\lambda = 0$ and $Z = 5$), the model WF $f_4(r)$ is more close to the PLM WF than the CFHHM WF, as well.

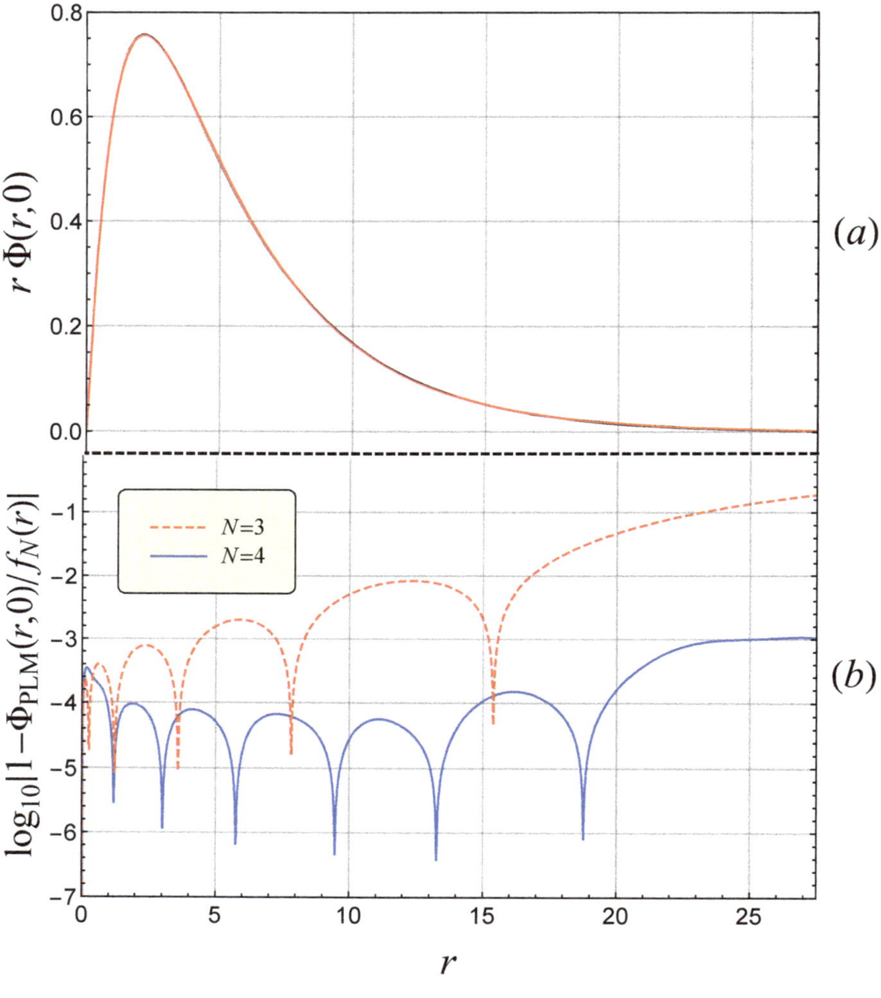

Figure 1. Negative ion of hydrogen $H^-(Z=1)$: (**a**) the WF, $\Phi(r,0)$ at the electron-nucleus coalescence (the *collinear* configuration with $\lambda = 0$) times r; (**b**) the logarithmic estimates $\mathcal{L}_4^{(0}(r)$ and $\mathcal{L}_3^{(0}(r)$ of the difference (see Equation (17)) between the model WF, $f_4(r)$ and the PLM WF (solid curve, blue online), and between the model WF, $f_3(r)$ and the PLM WF (dashed curve, red online), respectively.

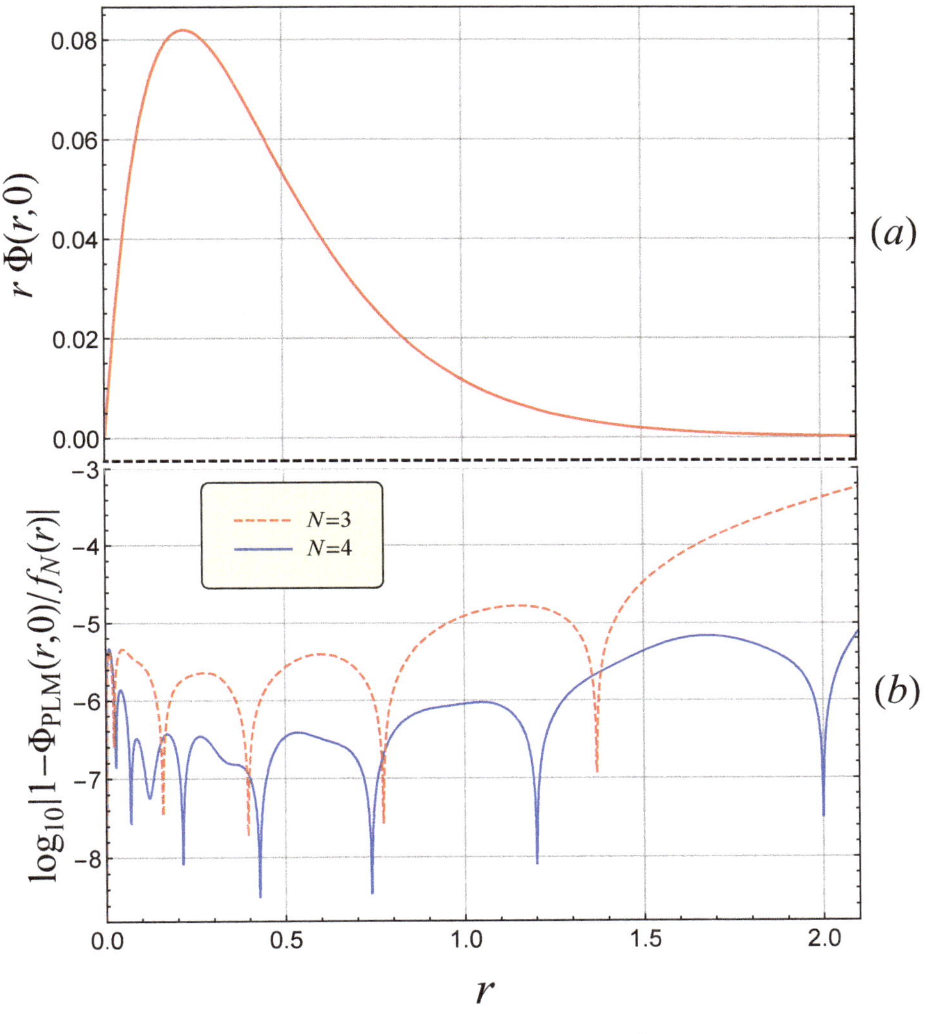

Figure 2. Ground state of the positive ion of boron $B^{3+}(Z=5)$: (**a**) the WF, $\Phi(r,0)$ at the electron–nucleus coalescence (the *collinear* configuration with $\lambda = 0$) times r; (**b**) the logarithmic estimates $\mathcal{L}_4^{(0}(r)$ and $\mathcal{L}_3^{(0}(r)$ of the difference between the model WF, $f_4(r)$ and the PLM WF (solid curve, blue online), and between the model WF, $f_3(r)$ and the PLM WF (dashed curve, red online), respectively.

It was mentioned earlier that the behavior of the two-electron atomic WF near the nucleus is described by the Fock expansion (6), which reduces to expansion (7) for the *collinear* arrangement of the particles. The most compact model WFs represented by the sum of three or four exponentials were obtained for the case of the electron-nucleus coalescence corresponding to the *collinear* parameter $\lambda = 0$. Tables 1 and 2 together with Figures 1 and 2 demonstrate the high accuracy of those model WFs. It should be emphasized that the accuracy of $f_4(r)$ for $\lambda = 0$ is close to the accuracy of the variational PLM WF, $\Phi_{PLM}(r,0)$ for all $r > 0$. Furthermore, the relevant calculations show that the model WF $f_4(r)$ mentioned above is, in fact, more accurate than $\Phi_{PLM}(r,0)$ in the vicinity of nucleus ($r \to 0$). We can argue this because the leading terms of the series expansion of $f_4(r)$ (for $\lambda = 0$) are more close to the corresponding terms of the Fock expansion than the ones for $\Phi_{PLM}(r,0)$. Actually, Equations (10) and (11) provide by definition the condition $f_4(0)=1$ and $f_4'(0)=-Z+1/2$, corresponding exactly to the Fock expansion. Moreover, it is seen from Equation (9)) that for $\lambda=0$ the logarithmic term of the Fock expansion is annihilated because $\zeta_0=0$, and hence $F''(0)/2=\xi_0$, where we denoted $F(r)\equiv\Phi(r,0)$. One should notice that $\lambda=0$ is, in fact, the single case of the *collinear* arrangement when the explicit expression for the angular Fock coefficient ξ_λ can be derived in the form [1,22]

$$\xi_0=\frac{1-2E}{12}-Z\left(\frac{3-\ln 2}{6}\right)+\frac{1}{3}Z^2, \tag{18}$$

where E is the non-relativistic energy of the two-electron atom/ion under consideration. It is seen from Table 5 that (besides $f_4'(0)$) the values of $f_4''(0)/2$ is much closer to the theoretical values (18) than $F_{PLM}''(0)/2$ for all Z. These results confirm the above conclusion about the accuracy of the model WF near the nucleus.

Table 5. The first and second derivatives of the *collinear* WF with $\lambda=0$ at the nucleus. The PLM WF, $F(r)\equiv\Phi_{PLM}(r,0)$ at the electron–nucleus coalescence is introduced.

Z	$F'(0)$	$-Z+1/2$	$F''(0)/2$	$f_4''(0)/2$	ξ_0
1	−0.506379	−0.5	0.169101	0.123314	0.12015
2	−1.50228	−1.5	1.20558	1.13429	1.13167
3	−2.50175	−2.5	3.24348	3.14574	3.14323
4	−3.50323	−3.5	6.38221	6.15306	6.15469
5	−4.50140	−4.5	10.3165	10.1691	10.1661

Author Contributions: All authors contributed equally. All authors have read and agreed to the published version of the manuscript.

Funding: This research received no external funding.

Acknowledgments: This work was supported by the PAZY Foundation, Israel.

Conflicts of Interest: The authors declare no conflict of interest.

References

1. Liverts, E.Z.; Krivec, R.; Barnea, N. Collinear configuration of the helium atom and two-electron ions. *Ann. Phys.* **2020**, *422*, 168306. [CrossRef]
2. Amusia, M.Y.; Drukarev, E.G.; Liverts, E.Z.; Mikhailov, A.I. Effects of small recoil momenta in one-photon two-electron ionization. *Phys. Rev. A* **2013**, *87*, 043423. [CrossRef]
3. Amusia, M.Y.; Drukarev, E.G.; Liverts, E.Z. Small recoil momenta double ionization of He and two-electron ions by high energy photons. *Eur. Phys. J. D* **2020**, *74*, 173. [CrossRef]
4. Haftel, M.I.; Mandelzweig, V.B. Fast Convergent Hyperspherical Harmonic Expansion for Three-Body Systems. *Ann. Phys.* **1989**, *189*, 29–52. [CrossRef]
5. Haftel, M.I.; Krivec, R.; Mandelzweig, V.B. Power Series Solution of Coupled Differential Equations in One Variable. *J. Comp. Phys.* **1996**, *123*, 149–161. [CrossRef]
6. Pekeris, C.L. Ground State of Two-Electron Atoms. *Phys. Rev.* **1958**, *112*, 1649–1658. [CrossRef]
7. Liverts, E.Z.; Barnea, N. *S*-states of helium-like ions. *Comp. Phys. Comm.* **2011**, *182*, 1790–1795. [CrossRef]
8. Liverts, E.Z.; Barnea, N. Three-body systems with Coulomb interaction. Bound and quasi-bound *S*-states. *Comp. Phys. Comm.* **2013**, *184*, 2596–2603. [CrossRef]
9. Hylleraas, E.A. Neue Berechnung der Energie des Heliums im Grundzustande, sowie des tiefsten Terms von Ortho-Helium. *Z. Phys.* **1929**, *54*, 347–366. [CrossRef]
10. Chandrasekhar, S.; Herzberg, G. Energies of the Ground States of He, Li^+, and 0^{6+}. *Phys. Rev.* **1955**, *98*, 1050–1054. [CrossRef]
11. Kinoshita, T. Ground State of the Helium Atom. *Phys. Rev.* **1957**, *105*, 1490–1502. [CrossRef]
12. Freund, D.E.; Huxtable, B.D.; Morgan, J.D., III. Variational calculations on the helium isoelectronic sequence. *Phys. Rev. A* **1984**, *29*, 980–982. [CrossRef]
13. Frankowski, K.; Pekeris, C.L. Logarithmic Terms in the Wave Functions of the Ground State of Two-Electron Atom. *Phys. Rev.* **1966**, *146*, 46–49. [CrossRef]

14. Korobov, V.I. Coulomb three-body bound-state problem: Variational calculations of nonrelativistic energies. *Phys. Rev. A* **2000**, *61*, 064503. [CrossRef]
15. Frolov, A.M. Multibox strategy for constructing highly accurate bound-state wave functions for three-body systems. *Phys. Rev. E* **2001**, *64*, 036704. [CrossRef] [PubMed]
16. Suzuki, Y.; Varga, K. *Stochastic Variational Approach to Quantum-Mechanical Few-Body Problems*; Springer: Berlin, Germany; New York, NY, USA; London, UK; Milan, Italy; Paris, France; Tokyo, Japan, 1998.
17. Drake, G.W.F. High Precision Calculations for Helium, Section 11. In *Handbook of Atomic, Molecular, and Optical Physics*; Drake, G.W.F., Ed.; AIP Press: New York, NY, USA, 1996.
18. Nordsieck, A. Reduction of an Integral in the Theory of Bremsstrahlung. *Phys. Rev.* **1954**, *93*, 785–787. [CrossRef]
19. Fock, V.A. On the Schrodinger Equation of the Helium Atom. *Izv. Akad. Nauk SSSR Ser. Fiz.* **1954**, *18*, 161–174.
20. Fadeev, L.D.; Khalfin, L.A.; Komarov, I.V. (Eds.) *VA Fock-Selected Works: Quantum Mechanics and Quantum Field Theory*; CRC Press: London, UK; Washington, DC, USA, 2004; p. 525.
21. Abbott, P.C.; Maslen, E.N. Coordinate systems and analytic expansions for three-body atomic wavefunctions: I. Partial summation for the Fock expansion in hyperspherical coordinates. *J. Phys. A Math. Gen.* **1987**, *20*, 2043–2075. [CrossRef]
22. Liverts, E.Z.; Barnea, N. Angular Fock coefficients. Refinement and further development. *Phys. Rev. A* **2015**, *92*, 042512. [CrossRef]
23. Frolov, A.M. On the Q-dependence of the lowest-order QED corrections and other properties of the ground 1^1S-states in the two-electron ions. *Phys. Rev. E* **2001**, *64*, 036704-6. [CrossRef]
24. Watkins, J.C. *An Introduction to the Science of Statistics: From Theory to Implementation*, 1st ed.; Topic 13: Method of Moments; 2016. Available online: https://www.math.arizona.edu/~jwatkins/statbook.pdf (accessed on 4 November 2021).
25. Liverts, E.Z.; Amusia, M.Y.; Krivec, R.; Mandelzweig, V.B. Boundary solutions of the two-electron Schrodinger equation at two-particle coalescences of the atomic systems. *Phys. Rev. A* **2006**, *73*, 012514-9. [CrossRef]

atoms

MDPI

Article

Time Delay in Electron Collision with a Spherical Target as a Function of the Scattering Angle

Miron Ya. Amusia [1,2,†], Arkadiy S. Baltenkov [3,†] and Igor Woiciechowski [4,*,†]

1 Racah Institute of Physics, The Hebrew University, Jerusalem 91904, Israel
2 Ioffe Physical-Technical Institute, 194021 St. Petersburg, Russia
3 Arifov Institute of Ion-Plasma and Laser Technologies, Tashkent 100125, Uzbekistan; arkbalt@mail.ru
4 Mathematics Department, Alderson Broaddus University, 101 College Hill Drive, Philippi, WV 26416, USA
* Correspondence: woiciechowskiia@ab.edu; Tel.: +1-304-457-6254
† These authors contributed equally to this work.

Abstract: We have studied the *angular time delay* in slow-electron elastic scattering by spherical targets as well as the *average time delay* of electrons in this process. It is demonstrated how the *angular time delay* is connected to the Eisenbud–Wigner–Smith (EWS) time delay. The specific features of both angular and energy dependencies of these time delays are discussed in detail. The potentialities of the derived general formulas are illustrated by the numerical calculations of the time delays of slow electrons in the potential fields of both absolutely hard-sphere and delta-shell potential well of the same radius. The conducted studies shed more light on the specific features of these time delays.

Keywords: electron scattering from atoms and molecules; Eisenbud–Wigner–Smith (EWS) time delay; angular time delay

Citation: Amusia, M.Y.; Baltenkov, A.S.; Woiciechowski, I. Time Delay in Electron Collision with a Spherical Target as a Function of the Scattering Angle. *Atoms* **2021**, *9*, 105. https://doi.org/10.3390/atoms9040105

Academic Editors: Anatoli Kheifets, Gleb Gribakin and Vadim Ivanov

Received: 9 November 2021
Accepted: 28 November 2021
Published: 1 December 2021

1. Introduction

In the first experiments, the purpose of which was to study the time delays of electrons in atomic photoeffect, electrons with the wave vector **k** emitted along the polarization vector **e** of the absorbed photon were recorded [1–3]. With this experimental technique, the delay times of the electrons escaping at an arbitrary angle to the vector **e** were unknown. Now, investigations of time delays as a function of the emission angle θ have become available [4–7], and the corresponding calculations have been able to reproduce this dependence for different atoms [8–13]. The electron delay time is a function depending on both the photoelectron emission angle θ with respect to the radiation polarization vector **e** and the photoelectron energy E. In most calculations of the time delay, its dependence on the energy E is analyzed at fixed values of the angle θ, revealing the pronounced angle dependence for large emission angles.

The angular dependence of the time delay of the wave packet scattered (or emitted) by a spherical target was obtained by Froissard, Goldberger, and Watson in [14], where the following expression for the angular time delay of the packet scattered in the direction θ was derived:

$$\Delta t(k,\theta) = \hbar \frac{\partial}{\partial E} \arg f(k,\theta), \ \theta \neq 0. \tag{1}$$

Here, $f(k,\theta)$ denotes the amplitude of electron elastic scattering by a target [15]

$$f(k,\theta) = \frac{1}{2ik} \sum_{l} (2l+1)(\exp 2i\delta_l - 1) P_l(\cos\theta), \tag{2}$$

where $\delta_l(k)$ is the partial scattering phase shifts and $P_l(\cos\theta)$ are the Legendre polynomials. According to (1), the forward scattering $\theta = 0$ must be excluded due to the interference effects between the forward scattered wave and the incident wave that give rise to the optical theorem [15].

The domain of applicability of the *angular time delay* $\Delta t(k,\theta)$ (1) is considerably broader than that of the Eisenbud–Wigner–Smith (EWS) partial-wave time delay [16–18]

$$\tau_l(k) = 2\hbar\frac{\partial\delta_l}{\partial E}. \tag{3}$$

In particular, Equation (1) serves as the basis for describing the temporal picture of atomic photoionization processes [17–25]. Equation (1) in this case needs not to be modified to exclude $\theta = 0$ as the problem of the interference with the unscattered wave does not exist in the case of photoionization. The scattering amplitude $f(k,\theta)$ for this process must be replaced in Equation (1) by the photoionization amplitude $f^{ph}(\omega,\theta)$, where ω is the photon energy

$$\tau(\omega,\theta) = \hbar\frac{\partial}{\partial\omega}\arg f^{ph}(\omega,\theta). \tag{4}$$

The dipole selection rules in photoionization of l-states of atom A lead to emission into the continuum of the pair of electronic spherical waves $Y_{l+1,m}(\mathbf{k})$ and $Y_{l-1,m}(\mathbf{k})$, propagating in the potential field of the atomic residue A^+ with the phase shifts $\delta_{l+1}(k)$ and $\delta_{l-1}(k)$, correspondingly, where k is the linear photoelectron momentum. The function $f^{ph}(\omega,\theta)$, therefore, is a linear combination of these spherical functions, the coefficients of which are determined by the corresponding dipole matrix elements $D_{l\pm1}(\omega)$. The energy derivative of the function (1) implicitly includes the derivatives of both phase shifts $\delta'_{l\pm1}(k)$ and matrix elements $D'_{l\pm1}$. The prime sign here and further denotes differentiation with respect to the electron kinetic energy E.

The time delay (4) at some electron emission angles θ was studied in the series of works on *photoionization* [19–24]. To the best of our knowledge, the angular dependence of the time delay in *elastic electron scattering* (1) has received no attention so far. Our goal in this article is to close somewhat the gap in the area of investigation of the angular time delay in electron scattering (1) by spherical targets.

We will see further that when only one scattering phase is different from zero in the scattering amplitude (2), the angular time delay (1) does not depend on the scattering angle. Here, we analyze the scattering amplitude $f(k,\theta)$ containing two Legendre polynomials only, i.e., we will consider model targets, in which, as in the case of the dipole photoelectric effect, only one pair of phase shifts is different from zero.

In Section 2, the angle dependence of the angular time delays $\Delta t(k,\theta)$ for some fixed electron momenta k is investigated. In Section 3, the time delay is studied as a function of k for some fixed polar angles θ of the scattering of an incident plane wave train. Finally, the function $\Delta t(k,\theta)$ is averaged over the distance of the order of the de Broglie wavelength, and the *average angular time delay* $\langle\Delta t(k)\rangle$ is obtained in Section 4.

2. Angular θ-Dependence of the Function $\Delta t(k,\theta)$

The argument of the amplitude $f(k,\theta)$ is determined by the ratio of the imaginary part of the function (2) $\Im f(k,\theta)$ to its real part $\Re f(k,\theta)$

$$\arg f(k,\theta) = \arctan\frac{\Im f(k,\theta)}{\Re f(k,\theta)} \tag{5}$$

whereas the angular time delay (1) is described by the general expression

$$\Delta t(k,\theta) = \frac{\mathrm{d}}{\mathrm{d}E}\arg f(k,\theta) = \frac{(\Im f)'(\Re f) - (\Re f)'(\Im f)}{|f|^2} \tag{6}$$

Here, and everywhere below, we use the atomic system of units. Let us first consider the case when all the phase shifts in (2), with the exception of $\delta_l(k)$, are equal to zero. In this case,

$$\begin{aligned} f(k,\theta) &= \frac{1}{2k}(2l+1)P_l(\cos\theta)\sin 2\delta_l + i\frac{1}{k}P_l(\cos\theta)\sin^2\delta_l, \\ \arg f(k,\theta) &= \arctan(\tan\delta_l), \\ \Delta t(k,\theta)_l &= \frac{d\delta_l}{dE} \equiv \delta_l'. \end{aligned} \tag{7}$$

It is seen that the angular time delay does not depend on the scattering angle θ and it is equal to half of the EWS-partial time delay (3).

Suppose that only two scattering phases $\delta_0(k)$ and $\delta_1(k)$ are nonzero. In this case, the scattering amplitude and its argument are represented as

$$\begin{aligned} f(k,\theta) &= \frac{1}{2k}(P_0\sin 2\delta_0 + 3P_1\sin 2\delta_1) + i\frac{1}{k}(P_0\sin^2\delta_0 + 3P_1\sin^2\delta_1) \\ \arg f(k,\theta) &= \arctan\frac{2(P_0\sin^2\delta_0 + 3P_1\sin^2\delta_1)}{(P_0\sin 2\delta_0 + 3P_1\sin 2\delta_1)},\ P_{0,1} \equiv P_{0,1}(\cos\theta). \end{aligned} \tag{8}$$

Differentiating the argument of the scattering amplitude (8), we obtain the expression for the time delay

$$\Delta t(k,\theta)_{01} = \frac{P_0[P_0\sin^2\delta_0 + 3P_1\sin(2\delta_0-\delta_1)\sin\delta_1]\delta_0' + 3P_1[3P_1\sin^2\delta_1 + P_0\sin(2\delta_1-\delta_0)\sin\delta_0]\delta_1'}{P_0^2\sin^2\delta_0 + 6P_0P_1\sin\delta_0\sin\delta_1\cos(\delta_0-\delta_1) + 9P_1^2\sin^2\delta_1} \tag{9}$$

as a function of both scattering angle θ and electron momentum $k = \sqrt{2E}$.

Repeating the calculations similar to those in formulae (8), we obtain the expression for the time delay in the case of nonzero phases $\delta_0(k)$ and $\delta_2(k)$

$$\Delta t(k,\theta)_{02} = \frac{P_0[P_0\sin^2\delta_0 + 5P_2\sin(2\delta_0-\delta_2)\sin\delta_2]\delta_0' + 5P_2[5P_2\sin^2\delta_2 + P_0\sin(2\delta_2-\delta_0)\sin\delta_0]\delta_2'}{P_0^2\sin^2\delta_0 + 10P_0P_2\sin\delta_0\sin\delta_2\cos(\delta_0-\delta_2) + 25P_2^2\sin^2\delta_2}. \tag{10}$$

It is easy to demonstrate that when only two scattering phases $\delta_l(k)$ and $\delta_{l'}(k)$ are nonzero in the electron scattering amplitude (2), the angular delay time (5) is determined by the following combination of the Legendre polynomials $P_l(\cos\theta)$ and $P_{l'}(\cos\theta)$:

$$\Delta t(k,\theta)_{ll'} = \frac{\sum_{i=l,l'}\sum_{j=l,l'}[(2i+1)(2j+1)P_iP_j\sin(2\delta_i-\delta_j)\sin\delta_j]\delta_i'}{\sum_{i=l,l'}\sum_{j=l,l'}(2i+1)(2j+1)P_iP_j\sin\delta_i\sin\delta_j\cos(\delta_i-\delta_j)}. \tag{11}$$

Explicit expressions for the time delays for selected nonzero scattering phase pairs (11) are given in [26], where the results of the calculations of the θ- and E-dependencies of the corresponding angular time delays are also given. We use both hard-sphere and delta-shell potentials as potential functions for the model targets. For these potentials, the analytical expressions for the scattering phases are known. When an electron is scattered by the model target in the form of an ideally repulsive solid sphere of radius R, the phase shifts of the electron are determined by the formula [27]

$$\tan\delta_l(k) = \frac{j_l(kR)}{n_l(kR)}, \tag{12}$$

where $j_l(kR)$ and $n_l(kR)$ are the spherical Bessel functions.

The scattering phase shifts of an electron for another model target taken in the form of an attractive delta-shell (delta-shell potential well [28]) are determined by the expression (see Equation (10) in [29])

$$\tan\delta_l = \frac{xj_l^2(x)}{xj_l(x)n_l(x) - 1/R\Delta L}, \tag{13}$$

where the variable $x = kR$. The parameter ΔL in (13) is the jump of the logarithmic derivative of the electron wave functions at the point $r = R$ where the delta-shell potential $U(r) = -U_0\delta(r - R)$ is infinitely negative. In the numerical calculations of phase shifts (12) and (13), the radii R and the parameter ΔL have the same values as those used in our article [29], where the EWS time delay of slow electrons scattered by a C_{60} cage was calculated.

Figure 1 shows the results of the calculation by formula (9) of the angular time delay $\Delta t(k,\theta)_{01}$ as a function of the scattering angle θ for some fixed electron momenta k. The left panel corresponds to the scattering on the solid sphere. The right panel corresponds to the delta-shell target. The angular time delays in these figures are given in atomic units. The atomic unit of time is equal to 24.2 attoseconds. Despite the different scales of the graphs on both panels, they show qualitatively similar behavior. The only exceptions are for the curves at $k = 0.68$. The graph of the angular dependence for the hard-sphere is almost a straight line passing from a positive to a negative half-plane at the angle of about 60°, whereas on the right panel, this curve almost coincides with the x-axis. According to both panels, at low electron energies ($k = 0.17$ and 0.34), the time delay of the scattering packet is negative at all the scattering angles. The rest of the curves (except the hard-sphere target at $k = 0.85$) are alternating for both targets. At the momenta $k = 0.51$ and $k = 1.0$, the time delays on the right panel reach its maximum ($\sim$ 298 atomic units (au) at $\theta = 95°$ in the first case and $\sim$140 au at the same angle in the second one). The appearance of these sharp peaks in the curves in Figure 1 is due to the almost vanishing of the denominator in the expression (9). The curves at $k = 0.85$ and $k = 1.0$ on the left panel cross the x-axis into the positive half-plane in the region of 90°, forming a peak with a height of ~30 atomic units.

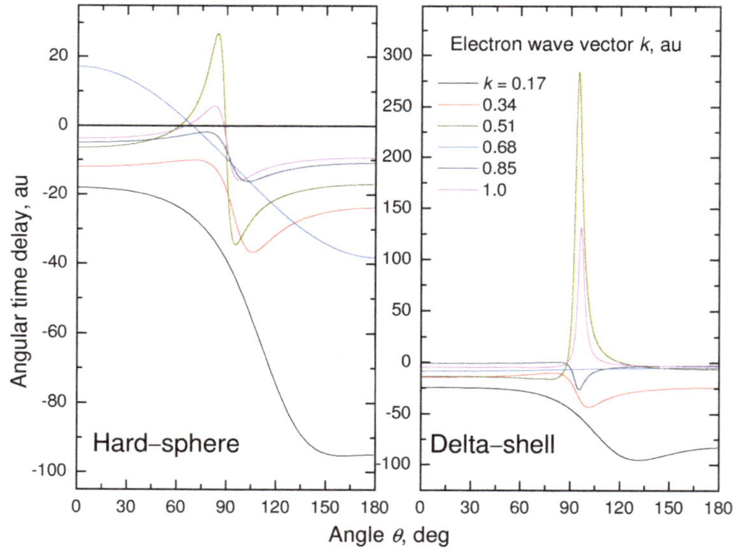

Figure 1. Angular time delay $\Delta t(k,\theta)_{01}$ (9) as a function of the polar angle θ for fixed electron wave vectors k. The functions $P_0(\cos\theta)$ and $P_1(\cos\theta)$ used in (9) is the pair of Legendre polynomials in the amplitude of electron elastic scattering $f(k,\theta)$ (2).

Figure 2 depicts the curves corresponding to the pair of polynomials $P_0(\cos\theta)$ and $P_2(\cos\theta)$. We see here the results of the calculation with formula (10) of the angular time delay $\Delta t(k,\theta)_{02}$ as a function of the scattering angle θ. As the sum of the orbital moments (indices of the Legendre polynomials) is an even number, the curves $\Delta t(k,\theta)_{02}$ in Figure 2 are symmetric relative to the angle $\theta = 90°$. The curves on the left panel, except for the curve at $k = 0.34$, lie entirely in the lower half-plane. The situation is quite different when the wave packet scatters by the delta-shell target. The behavior of the curve at $k = 1.0$ on the right panel is particularly interesting. This curve lies entirely in the positive half-plane, which allows it to be depicted in polar coordinates (see the inset in the right panel). The 3D-picture of the function $\Delta t(k,\theta)_{02}$ is a figure of rotation of this curve around the polar axis z, along which the incident plane wave train hits the target. The "wings of the star" shown there correspond to the polar scattering angles $\theta = 57°$ and $123°$. The qualitative similarity of the curves on both panels of Figure 2 is obvious.

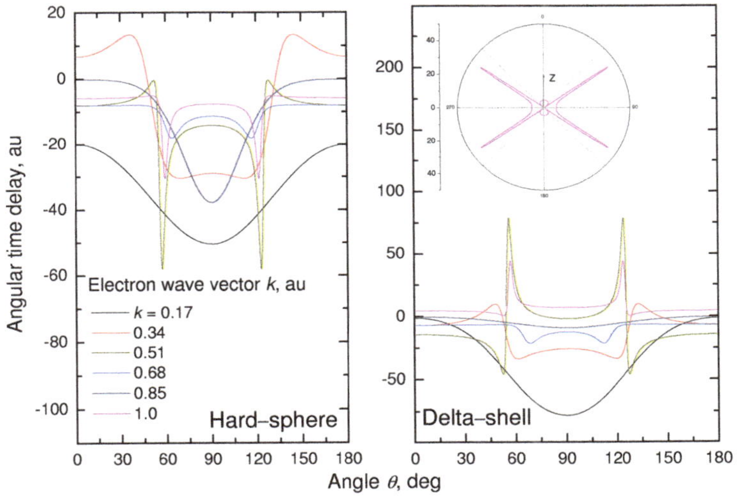

Figure 2. Angular time delay $\Delta t(k,\theta)_{02}$ (10) as a function of the polar angle θ for fixed electron wave vectors k. The functions $P_0(\cos\theta)$ and $P_2(\cos\theta)$ used in (10) is the pair of Legendre polynomials in the amplitude of electron elastic scattering $f(k,\theta)$ (2). The inset in the right panel is the plot at $k = 1.0$ in a polar coordinate system. The 3D-plot of the function $\Delta t(k,\theta)_{02}$ is a figure of rotation of this curve around the polar axis z, along which the incident plane wave train hits the target.

Note the similarity of the curves in Figure 2 and the angular spectra in Figure 1a,b of the article in [10] (devoted to the study of angular resolved time delays in photoemission from different atomic sub-shells of noble gases). A direct comparison of the function $\Delta t(k,\theta)$ for the processes of photoionization and elastic scattering cannot be conducted. An exclusion is the case when the dipole matrix element of photo-transitions varies slightly with the radiation frequency, and their derivatives with respect to the photon energy are negligible. Nevertheless, photoelectron spectra are similar to the scattering spectra in that they are symmetric relative to the angle $\theta = 90°$. Qualitative behavior of the scattering spectrum on the delta-shell at $k = 0.51$ in Figure 2 and the photoelectron spectrum in panel (a) of Figure 1 is similar. The same is to be for the curves at $k = 0.85$ in Figure 1 and those in panel (b) of Figure 1 in [10].

Summarizing, we note that according to Figures 1 and 2, the angular θ-dependencies of the function $\Delta t(k,\theta)$ are represented by nontrivial rapidly oscillating curves lying at low electron energies mainly in the negative half-plane. The situation changes with increasing the electron energy where the dependencies become smooth.

3. k-Dependence of Function $\Delta t(k,\theta)$

We now investigate the angular time delay $\Delta t(k,\theta)$ as a function of the electron energy E for some fixed values of polar angles θ. The calculation results by formulas (9) and (10) are shown in Figure 3.

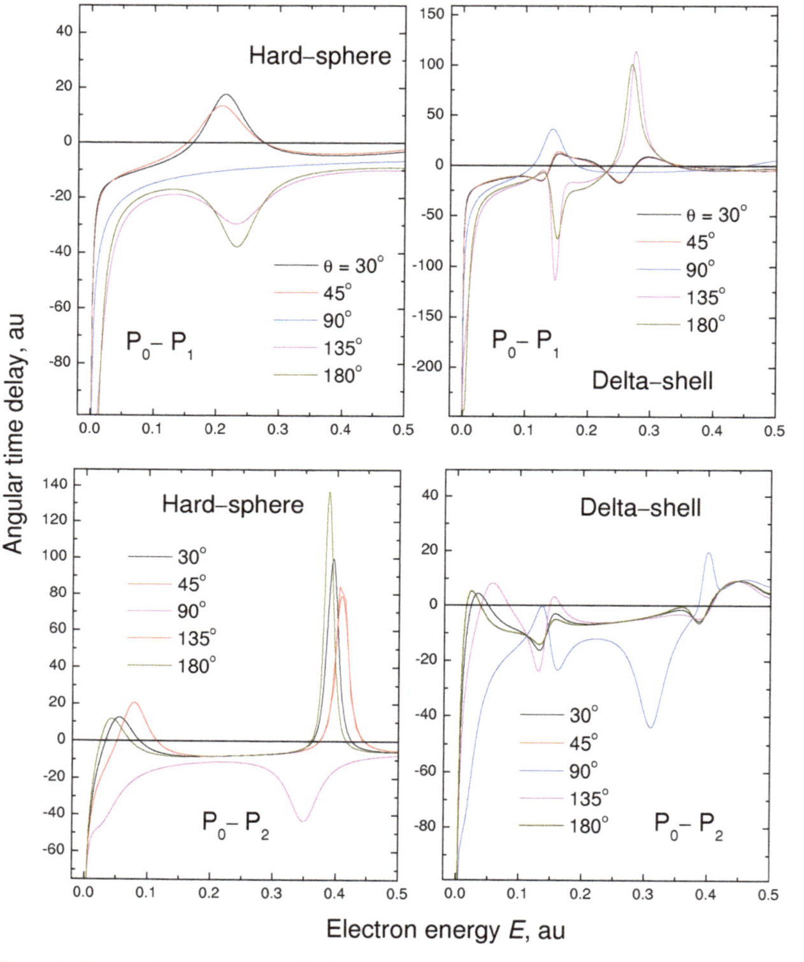

Figure 3. The angular time delay $\Delta t(k,\theta)$ as a function of the electron energy E for some fixed values of the polar angle θ. $P_0(\cos\theta)$-$P_1(\cos\theta)$ is the pair of Legendre polynomials in the upper panels. $P_0(\cos\theta)$ and $P_2(\cos\theta)$ are the polynomials used in both lower panels.

All curves in this figure tend to infinity at small electron momenta. The reason for this is that the scattering phase shift in short-range potentials must follow the Wigner threshold law $\delta_l(E) \propto E^{l+1/2}$ [30]. In the case of s-phase shift, we have $\delta_0(E \to 0) \propto \pi - E^{1/2}$. The time delay $\Delta t(k,\theta)_{01}$ and $\Delta t(k,\theta)_{02}$, that contain the derivative of the s-phase shift, for $k \to 0$ tends to infinity: $\delta_0'(E \to 0) \propto -E^{-1/2}$. For the orbital moments $l > 0$ the derivative of the phase shifts does vanish at the threshold as $\delta_0'(E \to 0) \propto -E^{l-1/2}$.

The left column of the figures corresponds to the electron scattering by the hard-sphere potential. The figures in the right column correspond to scattering by the delta-shell potential. In the upper right panel of Figure 3, the curves practically coincide with each other at small scattering angles θ, up to the angle of 45°. The graphs corresponding to the angles of 135° and 180° have alternating signs, and they are characterized by the peaks in both positive and negative half-planes of the coordinate system. We see a qualitatively similar picture in the lower panel of this column where the curves for $\Delta t(k,\theta)_{02}$ are presented. The presence of the derivative of the s-phase shift in formula (10) also leads this function to infinity at small electron energies. The curves for angles 30° and 180° almost coincide in this figure. The curve at $\theta = 90°$ is characterized by the maximum negative amplitude of oscillations. In the lower-left panel of Figure 3, we observe strong resonance behavior of all curves, except for the one at $\theta = 90°$ and energy $E \approx 0.4$ atomic units.

In the second and third sections, we limited ourselves to the specific examples of two nonzero phases in the expansion of the wave function of a scattered electron (2) into partial waves. It is very difficult to interpret rapidly oscillating dependence of the time delays upon the energy E and scattering angle θ even for this simple example. An increase in the number of included essential scattering phases significantly affects the picture of the angular time delays. The increase makes the time delays rapidly oscillating when they are averaged over the energy of incident electrons. As a consequence, the scattering angle becomes inevitable to make the angular time delay $\Delta t(k,\theta)$ observable in an experiment.

4. Average Time Delay of Scattering Process

The average angular time delay $\langle \Delta t(k) \rangle$ is obtained from (1) by averaging over the energy spectrum of the incident wave packet, as well as over the directions weighted by the differential cross section $|f(k,\theta)|^2$. This averaging is reduced to the calculation of the integral of the product $|f(k,\theta)|^2 \Delta t(k,\theta)$ over all angles of electron scattering by the target and division of the obtained result by the total cross section of elastic electron scattering $\sigma_{tot}(k)$. The calculation of the integral is complicated by the fact that, according to (1), the function $\Delta t(k,\theta)$ is not defined at $\theta = 0$. It was shown in [31] that the contribution to the integral from the forward scattering of an electron is determined by the real part of the scattering amplitude at zero angles. As a result of such averaging, Nussenzweig [31–33] obtained the expression

$$\langle \Delta t(k) \rangle = \frac{1}{\sigma_{tot}(k)} \left(\int |f(k,\theta)|^2 \Delta t(k,\theta) d\Omega + \frac{2\pi}{k^2} \frac{d}{dE} [k \Re f(k,0)] \right) = \\ = \frac{\pi}{\sigma_{tot} k^2} \sum_l (2l+1) 2\delta_l' = \frac{\pi}{\sigma_{tot} k^2} \sum_l (2l+1) \tau_l(k). \quad (14)$$

The second term on the left-hand side of Equation (14) eliminates the contribution of the forward scattering into the average angular time delay. Thus, the average time delay for the plane wave train $\langle \Delta t(k) \rangle$ is a linear combination of the EWS time delays $\tau_l(k)$ (3). The results of the calculation of the function $\langle \Delta t(k) \rangle$ (14) in the case of electrons scattered by the hard-sphere target are shown in Figure 4.

Figure 4 also shows the dependencies calculated under the assumption that the statistical weight of $\tau_l(k)$ in the sum (14) is not equal to $\pi(2l+1)/\sigma_{tot}k^2$. Instead, it is the ratio of the electron elastic scattering partial cross section $\sigma_l(k)$ to the total cross section $\sigma_l(k)/\sigma_{tot}(k)$. For more information about this assumption see, for example, Equation (10) in [10] or Equation (8) in [29]. The deep peak of the curve corresponding to the combination of the Legendre polynomials P_0 and P_2 is due to the resonant behavior of curves at $E \sim 0.4$ au in Figure 3.

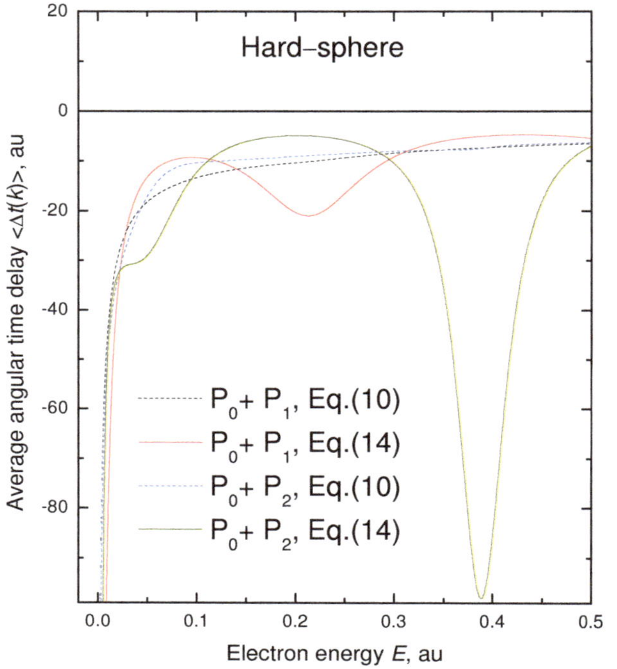

Figure 4. The function (14) in the case of electrons scattered by the hard-sphere target. $P_0 + P_1$ and $P_0 + P_2$ are the pairs of Legendre polynomials $P_l(\cos\theta)$. Note that Equation (10) corresponds to Formula (10) in [10].

5. Concluding Remarks

Using the instructive soluble example of electron scattering by the hard-sphere potential and delta-shell potential well, we for the first time explicitly obtained the angular time-delay $\Delta t(k,\theta)$ in terms of the scattering phase shifts $\delta_l(k)$ and their energy derivatives $\delta_l'(k)$. We demonstrated the complexity of $\Delta t(k,\theta)$ as a function of the incoming electron energy E and the scattering angle θ. We saw that $\Delta t(k,\theta)$ and the function $\langle\Delta t(k)\rangle$, even averaged over proper intervals of E and θ, are more sensitive to the scattering phases than the absolute cross section $\sigma_{tot}(k)$ and even the differential in angle scattering cross section that is proportional to $|f(k,\theta)|^2$. This is because the time delay functions depend not only on the cross section phases, but also upon their energy derivatives. This makes theoretical and experimental investigation of time delays a promising direction of research in the area of atomic scattering.

Author Contributions: M.Y.A., A.S.B. and I.W. contributed equally. All authors have read and agreed to the published version of the manuscript.

Funding: This research received no external funding.

Institutional Review Board Statement: Not applicable.

Informed Consent Statement: Not applicable.

Data Availability Statement: Data are available upon request.

Acknowledgments: In this section you can acknowledge any support The authors are thankful to Eric Morrow for proofreading the manuscript.

Conflicts of Interest: The authors declare no conflicts of interest.

References

1. Schultze, M.; Fieß, M.; Karpowicz, N.; Gagnon, J.; Korbman, M.; Hofstetter, M.; Neppl, S.; Cavalieri, A.L.; Komninos, Y.; Mercouris, T.; et al. Delay in photoemission. *Science* **2010**, *328*, 1658–1662. [CrossRef] [PubMed]
2. Cavalieri, A.L.; Muller, N.; Uphues, T.; Yakovlev, V.; Baltuska, A.; Horvath, B.; Schmidt, B.; Blumel, L.; Holzwarth, R.; Hendel, S.; et al. Attosecond spectroscopy in condensed matter. *Nature* **2007**, *449*, 1029–1032. [CrossRef] [PubMed]
3. Kheifets, A.S. Time delay in valence-shell photoionization of noble-gas atoms. *Phys. Rev. A* **2013**, *87*, 063404. [CrossRef]
4. Heuser, S.; Galan, A.J.; Cirelli, C.; Marante, C.; Sabbar, M.; Boge, R.; Lucchini, M.; Gallmann, L.; Ivanov, I.; Kheifets, A.; et al. Angular dependence of photoemission time delay in helium. *Phys. Rev. A* **2016**, *94*, 063409. [CrossRef]
5. Busto, D.; Vinbladh, J.; Zhong, S.; Isinger, M.; Nandi, S.; Maclot, S.; Johnsson, P.; Gisselbrecht, M.; L'Huillier, A.; Lindroth, E.; et al. Fano's propensity rule in angle-resolved attosecond pump-probe photoionization. *Phys. Rev. Lett.* **2019**, *123*, 133201. [CrossRef] [PubMed]
6. Fuchs, J.; Douguet, N.; Donsa, S.; Martin, F.; Burgdörfer, L.; Argenti, L.; Cattaneo, L.; Keller, U. Time delays from one-photon transitions in the continuum. *Optica* **2020**, *7*, 154–161. [CrossRef]
7. Cirelli, C.; Marante, C.; Heuser, S.; Petersson, L.; Galan, A.J.; Argenti, L.; Zhong, S.; Busto, D.; Isinger, M.; Nandi, S.; et al. Anisotropic photoemission time delays close to a Fano resonance. *Nat. Commun.* **2018**, *9*, 955. [CrossRef]
8. Ivanov, I.A. Time delay in strong-field photoionization of a hydrogen atom. *Phys. Rev. A* **2011**, *83*, 023421. [CrossRef]
9. Dahlström, J.M.; Lindroth, E. Study of attosecond delays using perturbation diagrams and exterior complex scaling. *J. Phys. B At. Mol. Opt. Phys.* **2014**, *47*, 124012. [CrossRef]
10. Wätzel, J.; Moskalenko, A.S.; Pavlyukh, Y.; Berakdar, J. Angular resolved time delay in photoemission. *J. Phys. B At. Mol. Opt. Phys.* **2015**, *48*, 025602. [CrossRef]
11. Ivanov, I.A.; Kheifets, A.S. Angle-dependent time delay in two-color XUV+ IR photoemission of He and Ne. *Phys. Rev. A* **2017**, *96*, 013408. [CrossRef]
12. Bray, A.W.; Naseem, F.; Kheifets, A.S. Simulation of angular-resolved RABBITT measurements in noble-gas atoms. *Phys. Rev. A* **2018**, *97*, 063404. [CrossRef]
13. Hockett, P. Angle-resolved RABBITT: Theory and numerics. *J. Phys. B At. Mol. Opt. Phys.* **2017**, *50*, 154002. [CrossRef]
14. Froissart, M.; Goldberger, M.L.; Watson, K.M. Spatial Separation of Events in S-Matrix Theory. *Phys. Rev.* **1963**, *131*, 2820–2826. [CrossRef]
15. Landau, L.D.; Lifshitz, E.M. *Quantum Mechanics, Non-Relativistic Theory*, 3rd ed.; Pergamon Press: Oxford, UK, 1977.
16. Eisenbud, L.E. The Formal Properties of Nuclear Collisions. Ph.D. Thesis, Princeton University, Princeton, NJ, USA, 1948.
17. Wigner, E.P. Lower limit for the energy derivative of the scattering phase shift. *Phys. Rev.* **1955**, *98*, 145–147. [CrossRef]
18. Smith, F.T. Lifetime matrix in collision theory. *Phys. Rev.* **1960**, *118*, 349–356. [CrossRef]
19. Pazourek, R.; Nagele, S.; Bugdörfer, J. Time-resolved photoemission on the attosecond scale: opportunities and challenges. *Faraday Discuss.* **2013**, *163*, 353–377. [CrossRef] [PubMed]
20. Pazourek, R.; Nagele, S.; Bugdörfer, J. Attosecond chronoscopy of photoemission. *Rev. Mod. Phys.* **2015**, *87*, 765–802. [CrossRef]
21. Deshmukh, P.C.; Banerjee, D. Time delay in atomic and molecular collisions and photoionisation/photodetachment. *Int. Rev. Phys. Chem.* **2020**, *40*, 127–153. [CrossRef]
22. Deshmukh, P.C.; Mandal, A.; Saha, S.; Kheifets, A.S.; Dolmatov, V.K.;Manson, S.T. Attosecond time delay in the photoionization of endohedral atoms A@C_{60}: A probe of confinement resonances. *Phys. Rev. A* **2014**, *89*, 053424. [CrossRef]
23. Hockett, P.; Frumker, E.; Villeneuve, D.M.; Corkum, P.B. Time delay in molecular photoionization. *J. Phys. B At. Mol. Opt. Phys.* **2016**, *49*, 095602. [CrossRef]
24. Baykusheva, D.; Wörner, H.J. Theory of attosecond delays in molecular photoionization. *J. Chem. Phys.* **2017**, *146*, 124306. [CrossRef] [PubMed]
25. Amusia, M.Y.; Chernysheva, L.V. Time delay of photoionization by Endohedrals. *JETP Lett.* **2020**, *112*, 219–224. [CrossRef]
26. Amusia, M.Y.; Baltenkov, A.S.; Woiciechowski, I.A. ArXive. Available online: https://arxiv.org/abs/2103.08528 (accessed on 2 March 2021).
27. Schiff, L.I. *Quantum Mechanics*, 3rd ed.; McGraw-Hill: New York, NY, USA, 1968.
28. Amusia, M.Y.; Baltenkov, A.S.; Krakov, B.G. Photodetachment of negative C_{60} ions. *Phys. Lett. A* **1998**, *243*, 99–105. [CrossRef]
29. Amusia, M.Y.; Baltenkov, A.S. Time delay in electron-C_{60} elastic scattering in a Dirac bubble potential model. *J. Phys. B At. Mol. Opt. Phys.* **2019**, *52*, 015101. [CrossRef]
30. Wigner, E.P. On the behavior of cross sections near thresholds. *Phys. Rev.* **1948**, *73*, 1002–1009. [CrossRef]
31. Nussenzveig, H.M. Causality in nonrelativistic quantum scattering. *Phys. Rev.* **1968**, *177*, 1848–1856. [CrossRef]
32. Nussenzveig, H.M. Time delay in quantum scattering. *Phys. Rev. D* **1972**, *6*, 1534–1542. [CrossRef]
33. De Carvalho, C.A.A.; Nussenzveig, H.M. Time delay. *Phys. Rep.* **2002**, *364*, 83–174. [CrossRef]

atoms

MDPI

Article

Fock Expansion for Two-Electron Atoms: High-Order Angular Coefficients

Evgeny Z. Liverts [1,*] and Rajmund Krivec [2]

[1] Racah Institute of Physics, The Hebrew University, Jerusalem 91904, Israel
[2] Department of Theoretical Physics, J. Stefan Institute, 1000 Ljubljana, Slovenia
* Correspondence: evgeny.liverts@mail.huji.ac.il

Abstract: The Fock expansion, which describes the properties of two-electron atoms near the nucleus, is studied. The angular Fock coefficients $\psi_{k,p}(\alpha,\theta)$ with the maximum possible value of subscript p are calculated on examples of the coefficients with $5 \leq k \leq 10$. The presented technique makes it possible to calculate such angular coefficients for any arbitrarily large k. The mentioned coefficients being leading in the logarithmic power series representing the Fock expansion, they may be indispensable for the development of simple methods for calculating the helium-like electronic structure. The theoretical results obtained are verified by other suitable methods. The Wolfram Mathematica is used extensively.

Keywords: two-electron atom/ion; wave function; Fock expansion; angular coefficients; Wolfram Mathematica

Citation: Liverts, E.Z.; Krivec, R. Fock Expansion for Two-Electron Atoms: High-Order Angular Coefficients. *Atoms* **2022**, *10*, 135. https://doi.org/10.3390/atoms10040135

Academic Editors: Anatoli Kheifets, Gleb Gribakin and Vadim Ivanov

Received: 23 September 2022
Accepted: 2 November 2022
Published: 7 November 2022

1. Introduction

The properties of a two-electron atomic (helium-like) system with an infinitely massive nucleus of charge Z and nonrelativistic energy E are defined by the wave function (WF) $\Psi(r_1, r_2, r_{12})$, where r_1 and r_2 are the electron–nucleus distances, and r_{12} is the distance between the electrons. The behavior of the ground state WF in the vicinity of the nucleus located at the origin is determined by the Fock expansion [1]

$$\bar{\Psi}(r_1, r_2, r_{12}) \equiv \Psi(R, \alpha, \theta) = \sum_{k=0}^{\infty} R^k \sum_{p=0}^{[k/2]} \psi_{k,p}(\alpha, \theta) \ln^p R, \tag{1}$$

where the hyperspherical coordinates R, α and θ are defined by the relations:

$$R = \sqrt{r_1^2 + r_2^2}, \quad \alpha = 2\arctan\left(\frac{r_2}{r_1}\right), \quad \theta = \arccos\left(\frac{r_1^2 + r_2^2 - r_{12}^2}{2r_1 r_2}\right). \tag{2}$$

The convergence of expansion (1) was proven in Ref. [2]. The angular Fock coefficients (AFC) $\psi_{k,p}$ satisfy the Fock recurrence relation (FRR)

$$\left[\Lambda^2 - k(k+4)\right]\psi_{k,p}(\alpha, \theta) = h_{k,p}(\alpha, \theta) \tag{3}$$

with the RHS of the form [3,4]:

$$h_{k,p} = 2(k+2)(p+1)\psi_{k,p+1} + (p+1)(p+2)\psi_{k,p+2} - 2V\psi_{k-1,p} + 2E\psi_{k-2,p}. \tag{4}$$

The dimensionless Coulomb potential representing the electron–electron and electron–nucleus interactions is

$$V \equiv \frac{R}{r_{12}} - Z\left(\frac{R}{r_1} + \frac{R}{r_2}\right) = \frac{1}{\xi} - \frac{2Z\eta}{\sin\alpha}, \tag{5}$$

where we have introduced the important (in what follows) angular quantities:

$$\xi = \sqrt{1 - \sin\alpha\cos\theta}, \quad \eta = \sqrt{1 + \sin\alpha}. \tag{6}$$

The hyperspherical angular momentum operator, projected on S states, is defined as

$$\Lambda^2 = -4\left[\frac{\partial^2}{\partial\alpha^2} + 2\cot\alpha\frac{\partial}{\partial\alpha} + \frac{1}{\sin^2\alpha}\left(\frac{\partial^2}{\partial\theta^2} + \cot\theta\frac{\partial}{\partial\theta}\right)\right]. \tag{7}$$

It is clear that all circumnuclear features of the two-electron atoms (ions) are defined by the Fock expansion (1). There are a large number of methods for calculating the electronic structure of the two-electron atomic systems. An excellent review on this topic can be found in Refs. [3,5–8]. However, we know only one technique that correctly represents the WF $\Psi(r_1, r_2, r_{12})$ near the nucleus. It is the so-called correlation function hyperspherical harmonic method (CFHHM) [9–11]. The expansion in hyperspherical harmonics (HHs) provides the correct representation of the AFCs. However, the HH expansion is known to converge very slowly. Although this method makes it possible to increase the convergence of the HH expansion, a sufficiently good accuracy requires a large HHs' basis size, which, in turn, creates great computational difficulties.

Thus, it would be extremely useful to develop a much simpler method for calculating the WF with correct behavior near the nucleus. In this regard, we would like to emphasize the following important peculiarities of the Fock expansion (FE). It follows from definition (1) that the FE can be split into individual power series (lines) associated with definite power of ln$R$. In other words, the FE can be represented in the form:

$$\begin{aligned}\Psi = (\ln R)^0&\left(\psi_{0,0} + R\psi_{1,0} + R^2\psi_{2,0} + \ldots\right)\\ &+(\ln R)^1R^2\left(\psi_{2,1} + R\psi_{3,1} + R^2\psi_{4,1} + \ldots\right)\\ &\quad+(\ln R)^2R^4\left(\psi_{4,2} + R\psi_{5,2} + R^2\psi_{6,2} + \ldots\right)\\ &\qquad+(\ln R)^3R^6\left(\psi_{6,3} + R\psi_{7,3} + R^2\psi_{8,3} + \ldots\right)\\ &\qquad\quad+(\ln R)^4R^8\left(\psi_{8,4} + R\psi_{9,4} + R^2\psi_{10,4} + \ldots\right) + \ldots\end{aligned} \tag{8}$$

It is seen that the leading term of each line represents the product $(\ln R)^{k/2}R^k\psi_{k,k/2}(\alpha,\theta)$ with even k. The first AFCs ($\psi_{0,0} = 1$) corresponding to $k = 0, 2, 4$ are well-known (see, e.g., [3,4]):

$$\psi_{1,0} = \frac{1}{2}\xi - Z\eta, \tag{9}$$

$$\psi_{2,1} = -\frac{Z(\pi-2)}{3\pi}(1-\xi^2), \tag{10}$$

$$\psi_{3,1} = \frac{Z(\pi-2)}{36\pi}\left[6Z\eta(1-\xi^2) + \xi(5\xi^2-6)\right], \tag{11}$$

$$\psi_{4,2} = \frac{Z^2(\pi-2)(5\pi-14)}{540\sqrt{\pi}}\left[Y_{40}(\alpha,\theta) + \sqrt{2}\,Y_{42}(\alpha,\theta)\right]. \tag{12}$$

The normalized HHs are

$$Y_{40}(\alpha,\theta) = \pi^{-3/2}(4\cos^2\alpha - 1), \quad Y_{42}(\alpha,\theta) = 2\sqrt{2}\pi^{-3/2}\sin^2\alpha P_2(\cos\theta), \tag{13}$$

where the $P_n(x)$ denote the Legendre polynomials.

In this paper, we present the theoretical calculations of the AFCs $\psi_{5,2}(\alpha,\theta)$, $\psi_{6,3}(\alpha,\theta)$, $\psi_{7,3}(\alpha,\theta)$, $\psi_{8,4}(\alpha,\theta)$ and $\psi_{9,4}(\alpha,\theta)$ included into the $k = 4$, $k = 6$ and $k = 8$ "lines" of the expansion (8), and also the AFC $\psi_{10,5}(\alpha,\theta)$ representing the leading term of the $k = 10$ "line". It is important to note that all mentioned angular coefficients represent the AFCs $\psi_{k,p}$ with the maximum possible p for a given k.

2. Derivation of the Angular Fock Coefficient $\psi_{5,2}(\alpha,\theta)$

The FRR (3) and (4) for $k = 5$ and $p = 2$ reduce to the form

$$\left(\Lambda^2 - 45\right)\psi_{5,2}(\alpha,\theta) = h_{5,2}(\alpha,\theta), \tag{14}$$

where

$$h_{5,2}(\alpha,\theta) = -2V\psi_{4,2}(\alpha,\theta). \tag{15}$$

Using Equations (5), (6), (12) and (13), it is convenient to represent the RHS of Equation (14) in the form

$$h_{5,2}(\alpha,\theta) = -\frac{Z^2(\pi-2)(5\pi-14)}{270\sqrt{\pi}}\left[3\pi^{-3/2}(2h_1+h_2) - 2Z(h_3+\sqrt{2}h_4)\right], \tag{16}$$

where

$$h_1 = \frac{(1-\xi^2)^2}{\xi}, \quad h_2 = \frac{\cos(2\alpha)}{\xi}, \quad h_3 = \frac{\eta Y_{40}(\alpha,\theta)}{\sin\alpha}, \quad h_4 = \frac{\eta Y_{42}(\alpha,\theta)}{\sin\alpha}. \tag{17}$$

Accordingly, we obtain the solution of Equation (14) in the identical form

$$\psi_{5,2}(\alpha,\theta) = -\frac{Z^2(\pi-2)(5\pi-14)}{270\sqrt{\pi}}\left[3\pi^{-3/2}(2f_1+f_2) - 2Z(f_3+\sqrt{2}f_4)\right], \tag{18}$$

where the AFC-components f_i satisfy the individual Fock recurrence relations (IFRRs)

$$\left(\Lambda^2 - 45\right)f_i = h_i. \qquad (i = 1,2,3,4) \tag{19}$$

We sequentially find solutions to each of the IFRRs (19) using various methods presented in Ref. [4].

2.1. Solution of the IFRR $(\Lambda^2 - 45)f_3 = \eta Y_{40}/\sin\alpha$

Moving from simpler to more complex solutions, let us start with IFRR

$$\left(\Lambda^2 - 45\right)f_3 = h_3. \tag{20}$$

The RHS $h_3 \equiv h_3(\alpha)$ represents the function of only one angle variable α. It was shown [4] that the solution of the corresponding IFRR (20) reduces to the solution $g(\rho) = f_3(\alpha)$ of the inhomogeneous differential equation

$$(\rho^2+1)^2 g''(\rho) + 2\rho^{-1}(\rho^2+1)g'(\rho) + 45g(\rho) = -\mathrm{h}(\rho), \tag{21}$$

where $\rho = \tan(\alpha/2)$, and

$$\mathrm{h}(\rho) \equiv h_3(\alpha) = \frac{(4\cos^2\alpha - 1)\sqrt{1+\sin\alpha}}{\pi^{3/2}\sin\alpha} = \frac{(\rho+1)(3\rho^4 - 10\rho^2 + 3)}{2\pi^{3/2}\rho(\rho^2+1)^{3/2}}. \tag{22}$$

For convenience, we solve the Equation (21) with the RHS $\mathrm{h}(\rho)$ not containing the multiplier $\pi^{-3/2}$. The final solution f_3 will be multiplied by this factor.

Using the method of variation of parameters, one obtains [4] the particular solution of Equation (21) in the form

$$g^{(p)}(\rho) = v_{50}(\rho)\int\frac{u_{50}(\rho)\mathrm{h}(\rho)d\rho}{(\rho^2+1)^2 W_0(\rho)} - u_{50}(\rho)\int\frac{v_{50}(\rho)\mathrm{h}(\rho)d\rho}{(\rho^2+1)^2 W_0(\rho)}, \tag{23}$$

where

$$W_0(\rho) = -(\rho^2+1)/\rho^2. \tag{24}$$

The independent solutions of the homogeneous equation associated with Equation (21) are [4]:

$$u_{50}(\rho)=\frac{(\rho^2+1)^{9/2}}{\rho}\,{}_2F_1\left(4,\frac{7}{2};\frac{1}{2};-\rho^2\right)=\frac{1-7\rho^2(3-5\rho^2+\rho^4)}{\rho(\rho^2+1)^{5/2}},\tag{25}$$

$$v_{50}(\rho)=(\rho^2+1)^{9/2}\,{}_2F_1\left(4,\frac{9}{2};\frac{3}{2};-\rho^2\right)=\frac{1-35\rho^2+21\rho^4-\rho^6}{7(\rho^2+1)^{5/2}},\tag{26}$$

where ${}_2F_1(\ldots)$ is the Gaussian hypergeometric function. The substitution of Equations (24)–(26) into the general representation (23) yields

$$g^{(p)}(\rho)=\frac{-7\rho\{\rho[5\rho(3\rho-4)(3\rho+5)-24]+23\}-23}{420\rho(\rho^2+1)^{5/2}}.\tag{27}$$

The general solution of the inhomogeneous equation can be expressed as the sum of the general solution of the associated homogeneous (complementary) equation and the particular solution of the inhomogeneous equation, whence

$$g(\rho)=g^{(p)}(\rho)+c_u u_{50}(\rho)+c_v v_{50}(\rho),\tag{28}$$

where the coefficients c_u and c_v are currently undetermined. To choose these coefficients, it is necessary to determine the behavior of all independent solutions on the boundaries of the domain $[0,\infty]$. We easily obtain:

$$g^{(p)}(\rho)\underset{\rho\to 0}{=}-\frac{23}{420\rho}-\frac{23}{60}+\frac{451\rho}{840}+O(\rho^2),\tag{29}$$

$$g^{(p)}(\rho)\underset{\rho\to\infty}{=}-\frac{3}{4\rho}-\frac{1}{4\rho^2}+\frac{85}{24\rho^3}+O(\rho^{-4}),\tag{30}$$

$$u_{50}(\rho)\underset{\rho\to 0}{=}\frac{1}{\rho}-\frac{47\rho}{2}+O(\rho^3),\tag{31}$$

$$u_{50}(\rho)\underset{\rho\to\infty}{=}-7+\frac{105}{2\rho^2}+O(\rho^{-4}),\tag{32}$$

$$v_{50}(\rho)\underset{\rho\to 0}{=}1-\frac{15\rho^2}{2}+O(\rho^4),\tag{33}$$

$$v_{50}(\rho)\underset{\rho\to\infty}{=}-\frac{\rho}{7}+\frac{47}{14\rho}+O(\rho^{-3}).\tag{34}$$

It is seen that the particular solution $g^{(p)}(\rho)$ is divergent at $\rho=0$, whereas the solutions of the homogeneous equation associated with Equation (21) are divergent, at $\rho=0$ and $\rho=\infty$ for $u_{50}(\rho)$ and $v_{50}(\rho)$, respectively. Thus, to avoid the divergence on the whole range of definition, one should set $c_u=23/420$ and $c_v=0$ in the general solution (28). Then, the final physical solution becomes

$$\begin{aligned}f_3(\alpha)&=-\frac{(\rho+1)(23\rho^4+22\rho^3-122\rho^2+22\rho+23)}{60\pi^{3/2}(\rho^2+1)^{5/2}}=\\&=-\frac{1}{60\pi^{3/2}}[11\sin\alpha+21\cos(2\alpha)+2]\sqrt{1+\sin\alpha}.\end{aligned}\tag{35}$$

2.2. Solution of the IFRR $(\Lambda^2-45)f_4=\eta Y_{42}/sin\alpha$

It was shown in Ref. [4] that the solution $f_4\equiv f_4(\alpha,\theta)$ of the IFRR

$$\left(\Lambda^2-45\right)f_4=2\pi^{-3/2}\eta\sqrt{2}\sin\alpha P_2(\cos\theta)\tag{36}$$

can be found in the form

$$f_4 = 2\sqrt{2}\,\pi^{-3/2}\sin^2\alpha P_2(\cos\theta)g_4(\rho), \tag{37}$$

where the function $g_4(\rho)$ satisfies the equation

$$(\rho^2+1)^2 g_4''(\rho) + 2\rho^{-1}[1+\rho^2+2(1-\rho^4)]g_4'(\rho) + 13g_4(\rho) = -\mathrm{h}_4(\rho), \tag{38}$$

with

$$\mathrm{h}_4(\rho) = \frac{\eta}{\sin\alpha} = \frac{(\rho+1)\sqrt{\rho^2+1}}{2\rho}. \tag{39}$$

Using the method of variation of parameters, one obtains [4] the particular solution of Equation (38) in the form

$$g_4^{(p)}(\rho) = v_{52}(\rho)\int\frac{u_{52}(\rho)\mathrm{h}_4(\rho)d\rho}{(\rho^2+1)^2W_2(\rho)} - u_{52}(\rho)\int\frac{v_{52}(\rho)\mathrm{h}_4(\rho)d\rho}{(\rho^2+1)^2W_2(\rho)}, \tag{40}$$

where

$$W_2(\rho) = -\frac{5}{\rho}\left(\frac{\rho^2+1}{\rho}\right)^5. \tag{41}$$

The independent solutions of the homogeneous equation associated with Equation (38) are:

$$u_{52}(\rho) = \frac{(\rho^2+1)^{13/2}}{\rho^5}\,{}_2F_1\left(4,\frac{3}{2};-\frac{3}{2};-\rho^2\right) = \frac{1+11\rho^2+99\rho^4-231\rho^6}{\rho^5\sqrt{\rho^2+1}}, \tag{42}$$

$$v_{52}(\rho) = (\rho^2+1)^{13/2}\,{}_2F_1\left(4,\frac{13}{2};\frac{7}{2};-\rho^2\right) = \frac{231-99\rho^2-11\rho^4-\rho^6}{231\sqrt{\rho^2+1}}. \tag{43}$$

Thus, the particular solution (40) reduces to the form:

$$g_4^{(p)}(\rho) = -\frac{11(21\rho^5+9\rho^4+\rho^2)+1}{2772\rho^5\sqrt{\rho^2+1}}. \tag{44}$$

Considering the series expansions for $g_4^{(p)}(\rho)$, $u_{52}(\rho)$ and $v_{52}(\rho)$ on the boundaries of the range of definition ($\rho\in[0,\infty]$), it can be shown that function

$$g_4^{(p)}(\rho) + \frac{1}{2772}u_{52}(\rho) = -\frac{\rho+1}{12\sqrt{\rho^2+1}} = -\frac{1}{12}\sqrt{1+\sin\alpha} \tag{45}$$

represents the physical (finite) solution of Equation (38), and hence we finally obtain:

$$f_4(\alpha,\theta) = -\frac{\sqrt{2}}{6\pi^{3/2}}(\sin\alpha)^2\sqrt{1+\sin\alpha}\;P_2(\cos\theta). \tag{46}$$

2.3. *Solution of the IFRR* $(\Lambda^2-45)f_1 = (1-\xi^2)^2/\xi$

It is important to note that the RHS h_1 (see Equation (17)) of the IFRR

$$\left(\Lambda^2-45\right)f_1 = h_1(\xi) \tag{47}$$

is a function of ξ (only) defined by Equation (6). For this case [4,12], the solution of Equation (47) coincides with the solution of the inhomogeneous differential equation

$$(\xi^2-2)f_1''(\xi) + \xi^{-1}(5\xi^2-4)f_1'(\xi) - 45f_1(\xi) = h_1(\xi). \tag{48}$$

A particular solution of Equation (48) can be found by the method of variation of parameters in the form [12]

$$f_1^{(p)}(\xi) = \frac{1}{7\sqrt{2}}\left[u_5(\xi)\int v_5(\xi)w(\xi)d\xi - v_5(\xi)\int u_5(\xi)w(\xi)d\xi\right], \tag{49}$$

where

$$w(\xi) = h_1(\xi)\xi^2\sqrt{2-\xi^2}. \tag{50}$$

The linearly independent solutions of the homogeneous equation associated with Equation (48) are defined by the relations

$$u_5(\xi) = \frac{P^{1/2}_{13/2}(\xi/\sqrt{2})}{\xi\sqrt[4]{2-\xi^2}} = \frac{2^{1/4}(8\xi^6 - 28\xi^4 + 28\xi^2 - 7)}{\sqrt{\pi(2-\xi^2)}}, \tag{51}$$

$$v_5(\xi) = \frac{Q^{1/2}_{13/2}(\xi/\sqrt{2})}{\xi\sqrt[4]{2-\xi^2}} = \frac{\sqrt{\pi}(-8\xi^6 + 20\xi^4 - 12\xi^2 + 1)}{2^{3/4}\xi}. \tag{52}$$

where $P^\mu_\nu(x)$ and $Q^\mu_\nu(x)$ are the associated Legendre functions of the first and second kind, respectively. A substitution of the representations (50)–(52) into (49) yields the particular solution:

$$f_1^{(p)}(\xi) = -\frac{1}{60}\xi(13\xi^4 - 30\xi^2 + 15). \tag{53}$$

It can be verified that the particular solution $f_1^{(p)}(\xi)$ is finite on the whole range of definition ($\xi \in [0, \sqrt{2}]$), whereas the solutions of the homogeneous equation associated with Equation (48) are divergent at $\xi = \sqrt{2}$ ($\alpha = \pi/2, \theta = \pi$) and $\xi = 0$ ($\alpha = \pi/2, \theta = 0$) for $u_5(\xi)$ and $v_5(\xi)$, respectively. Thus, we can conclude that the final physical solution of the IFFR (47) coincides with the particular solution (53), whence

$$f_1 = -\frac{1}{60}\xi(13\xi^4 - 30\xi^2 + 15) = -\frac{1}{60}\sqrt{1-\sin\alpha\cos\theta}[\sin\alpha\cos\theta(4 + 13\sin\alpha\cos\theta) - 2]. \tag{54}$$

2.4. Solution of the IFRR $(\Lambda^2 - 45)f_2 = cos(2\alpha)/\xi$

To solve the IFRR

$$\left(\Lambda^2 - 45\right)f_2 = h_2, \tag{55}$$

with the RHS h_2 defined by Equation (17), first of all, it is necessary to recall Sack's representation [13] (see also [3,4]) for ξ^ν with $\nu = -1$:

$$\xi^{-1} = \sum_{l=0}^{\infty} P_l(\cos\theta)\left(\frac{\sin\alpha}{2}\right)^l F_l(\rho), \tag{56}$$

where

$$F_l(\rho) = {}_2F_1\left(\frac{l}{2} + \frac{1}{4}, \frac{l}{2} + \frac{3}{4}; l + \frac{3}{2}; \frac{4\rho^2}{(\rho^2+1)^2}\right) = \begin{cases} \mathfrak{F}_l(\rho) & 0 \le \rho \le 1 \\ \mathfrak{F}_l(1/\rho) & \rho \ge 1 \end{cases} \tag{57}$$

with

$$\mathfrak{F}_l(\rho) = (\rho^2 + 1)^{l+\frac{1}{2}}. \tag{58}$$

This enables us to present the RHS of Equation (55) in the form

$$h_2 \equiv \frac{\cos(2\alpha)}{\xi} = \sum_{l=0}^{\infty} P_l(\cos\theta)(\sin\alpha)^l \mathrm{h}_l(\rho), \tag{59}$$

where

$$\mathrm{h}_l(\rho) = 2^{-l}F_l(\rho)\cos(2\alpha) = 2^{-l}F_l(\rho)\left[1 - \frac{8\rho^2}{(\rho^2+1)^2}\right]. \tag{60}$$

In turn, it was shown in Ref. [4] that in the case where the RHS is determined by Equation (59), the solution of the corresponding IFRR (55) can be found in the form

$$f_2(\alpha,\theta) = \sum_{l=0}^{\infty} P_l(\cos\theta)(\sin\alpha)^l \sigma_l(\rho), \tag{61}$$

where the function $\sigma_l(\rho)$ satisfies the inhomogeneous differential equation

$$(\rho^2+1)^2\sigma_l''(\rho) + 2\rho^{-1}[1+\rho^2+l(1-\rho^4)]\sigma_l'(\rho) + (5-2l)(9+2l)\sigma_l(\rho) = -\mathrm{h}_l(\rho). \tag{62}$$

The linearly independent solutions of the homogeneous equation associated with Equation (62) are:

$$u_{5l}(\rho) = \frac{(\rho^2+1)^{l+9/2}}{\rho^{2l+1}}\,{}_2F_1\left(4, \frac{7}{2}-l; \frac{1}{2}-l; -\rho^2\right) = \frac{(\rho^2+1)^{l-5/2}}{\rho^{2l+1}} \times$$
$$\left[\rho^6\left(1+\frac{120}{2l-5}-\frac{120}{2l-3}+\frac{24}{2l-1}\right) + 3\rho^4\left(1+\frac{40}{2l-3}-\frac{24}{2l-1}\right) + 3\rho^2\left(1+\frac{8}{2l-1}\right)+1\right], \tag{63}$$

$$v_{5l}(\rho) = (\rho^2+1)^{l+9/2}\,{}_2F_1\left(4, \frac{9}{2}+l; \frac{3}{2}+l; -\rho^2\right) = (\rho^2+1)^{l-5/2} \times$$
$$\left[\rho^6\left(1-\frac{24}{2l+3}+\frac{120}{2l+5}-\frac{120}{2l+7}\right) + 3\rho^4\left(1+\frac{24}{2l+3}-\frac{40}{2l+5}\right) + 3\rho^2\left(1-\frac{8}{2l+3}\right)+1\right]. \tag{64}$$

The method of variation of parameters enables us to obtain the particular solution of the inhomogeneous differential Equation (62) in the form

$$\sigma_l^{(p)}(\rho) = v_{5l}(\rho)\int \frac{u_{5l}(\rho)\mathrm{h}_l(\rho)d\rho}{(\rho^2+1)^2 W_l(\rho)} - u_{5l}(\rho)\int \frac{v_{5l}(\rho)\mathrm{h}_l(\rho)d\rho}{(\rho^2+1)^2 W_l(\rho)}, \tag{65}$$

where

$$W_l(\rho) = -\frac{2l+1}{\rho}\left(\frac{\rho^2+1}{\rho}\right)^{2l+1}. \tag{66}$$

Note that due to different representations for the function $F_l(\rho)$ (see Equation (57)) at values of ρ less and greater than 1, we obtain special representations for a particular solution in these two regions:

$$\sigma_l^{(0)}(\rho) = \frac{(\rho^2+1)^{l-5/2}}{2^{l+1}(2l-1)(2l-3)}\left[(2l-3)\rho^4 - 4(l-2)\rho^2 - \frac{4l^2+4l-27}{3(2l-5)}\right], \quad 0 \le \rho \le 1 \tag{67}$$

$$\sigma_l^{(1)}(\rho) = \frac{\rho^{-2l-1}(\rho^2+1)^{l-5/2}}{2^{l+1}(2l+3)(2l+5)}\left[\frac{4l^2+4l-27}{3(2l+7)} + 4(l+3)\rho^2 - (2l+5)\rho^4\right]. \quad \rho \ge 1 \tag{68}$$

It can be verified that both functions (67) and (68) have no singularities on their domains of definition. On the other hand, function $u_{5l}(\rho)$ is singular at $\rho = 0$, whereas $v_{5l}(\rho)$ is singular at $\rho = \infty$. This means that one should search the general solution of Equation (62) in the form:

$$\sigma_l(\rho) = \sigma_l^{(0)}(\rho) + c_{5l}^{(v)} v_{5l}(\rho), \qquad 0 \le \rho \le 1 \tag{69}$$

$$\sigma_l(\rho) = \sigma_l^{(1)}(\rho) + c_{5l}^{(u)} u_{5l}(\rho). \qquad \rho \ge 1 \tag{70}$$

Note that two coefficients $c_{5l}^{(v)}$ and $c_{5l}^{(u)}$ are presently undetermined. To calculate them, we need to find two equations relating these coefficients. The first equation is quite obvious. It follows from the condition that the representations (69) and (70) are coincident at the common point $\rho = 1$, that is

$$\sigma_l^{(0)}(1) + c_{5l}^{(v)} v_{5l}(1) = \sigma_l^{(1)}(1) + c_{5l}^{(u)} u_{5l}(1). \tag{71}$$

This relationship reduces to the first desired equation:

$$c_{5l}^{(u)}(2l+3)(2l+5)(2l+7) = c_{5l}^{(v)}(2l-5)(2l-3)(2l-1) + \frac{27-4l-4l^2}{3\times 2^{l+1}}. \tag{72}$$

It can be verified that $\mathrm{h}_l(0) = \mathrm{h}_l(\infty) = 2^{-l}$. It follows from these relations that $\sigma_l^{(0)}(0) + c_{5l}^{(v)} v_{5l}(0) = \sigma_l^{(1)}(\infty) + c_{5l}^{(u)} u_{5l}(\infty)$. It can be assumed that the last equation represents the second desired equation. However, this assumption turns out to be false, because it again leads to Equation (72).

We propose the following method to find the second desired equation. Recall that any suitable function of the angles α and θ may be expanded into HHs since they form a complete set:

$$f(\alpha,\theta) = \sum_{n=0(2)}^{\infty} \sum_{l=0}^{n/2} f_{n,l} Y_{n,l}(\alpha,\theta), \tag{73}$$

where (see, e.g., [3])

$$f_{n,l} = \int f(\alpha,\theta) Y_{n,l}(\alpha,\theta) d\Omega \tag{74}$$

with

$$d\Omega = \pi^2 \sin^2\alpha \sin\theta d\alpha d\theta. \qquad \alpha \in [0,\pi],\ \theta \in [0,\pi] \tag{75}$$

For the function $f(\alpha,\theta) = f_2(\alpha,\theta)$ represented by Equation (61), the expansion coefficient with $n = 2l$ becomes

$$f_{2l,l} = \pi^2 \int_0^{\pi}\int_0^{\pi} f_2(\alpha,\theta) Y_{2l,l}(\alpha,\theta) \sin^2\alpha \sin\theta d\alpha d\theta = \frac{2\pi^2 N_{2l,l}}{2l+1}\Big[K_0(l) + c_l^{(v)} K_v(l) + c_l^{(u)} K_u(l)\Big], \tag{76}$$

where

$$\begin{aligned} K_0(l) &= \int_0^{\pi/2} (\sin\alpha)^{2l+2} \sigma_l^{(0)}(\rho) d\alpha + \int_{\pi/2}^{\pi} (\sin\alpha)^{2l+2} \sigma_l^{(\infty)}(\rho) d\alpha = \\ &= \frac{\sqrt{2}(-24l^3 - 100l^2 + 198l + 249)}{(2l-5)(2l-3)(2l+3)(2l+5)(2l+7)(2l+9)}, \end{aligned} \tag{77}$$

$$K_v(l) = \int_0^{\pi/2} (\sin\alpha)^{2l+2} v_{5l}(\rho) d\alpha = \frac{2^{l+3/2}(2l-1)}{(2l+5)(2l+9)}, \tag{78}$$

$$K_u(l) = \int_{\pi/2}^{\pi} (\sin\alpha)^{2l+2} u_{5l}(\rho) d\alpha = \frac{2^{l+3/2}(2l+3)(2l+7)}{(2l-5)(2l-3)(2l+9)}. \tag{79}$$

To derive the results (76)–(79), we use the representation $Y_{2l,l}(\alpha,\theta) = N_{2l,l} \sin^l \alpha P_l(\cos\theta)$ for the particular case of the HHs, and the orthogonality property for the Legendre polynomials. It should be noted that the explicit form of the normalization constant $N_{2l,l}$ is not required.

On the other hand, expanding $f_2(\alpha,\theta)$ in HHs, and inserting this expansion into the LHS of the IFRR (55), we obtain

$$\left(\Lambda^2 - 45\right) f_2(\alpha,\theta) = \sum_{n=0(2)}^{\infty} \sum_{l=0}^{n/2} f_{n,l}[n(n+4) - 45] Y_{n,l}(\alpha,\theta). \tag{80}$$

To derive the last equation, we use the fact that $Y_{n,l}(\alpha,\theta)$ is an eigenfunction of the operator Λ^2 with an eigenvalue equal to $n(n+4)$, that is

$$\Lambda^2 Y_{n,l}(\alpha,\theta) = n(n+4)Y_{n,l}(\alpha,\theta). \tag{81}$$

The HH expansion of the RHS of Equation (55) is

$$h_2(\alpha,\theta) = \sum_{n=0(2)}^{\infty} \sum_{l=0}^{n/2} h_{n,l} Y_{n,l}(\alpha,\theta). \tag{82}$$

Hence,

$$f_{2l,l} = \frac{h_{2l,l}}{4l(l+2)-45}. \tag{83}$$

Using again Sack's representation (56) and (57) and Equation (74), we obtain the expansion coefficient $h_{2l,l}$ in explicit form:

$$h_{2l,l} = \pi^2 \int_0^{\pi} \int_0^{\pi} h_2(\alpha,\theta) Y_{2l,l}(\alpha,\theta) \sin^2\alpha \sin\theta d\alpha d\theta = -\frac{2^{7/2}\pi^2 N_{2l,l}(4l^2+24l+19)}{(2l+1)(2l+3)(2l+5)(2l+7)}. \tag{84}$$

Thus, inserting (84) into the RHS of Equation (83) and equating the result to the RHS of Equation (76) we obtain the desired *second equation* in the form:

$$c_{5l}^{(u)}(2l+3)(2l+5)(2l+7) = -c_{5l}^{(v)}(2l-5)(2l-3)(2l-1) - 2^{-l-1}(2l+1). \tag{85}$$

Solving the system of two linear Equations (72) and (85) gives the desired coefficients:

$$c_{5l}^{(u)} = -\frac{2^{-l-1}(l+4)(2l-3)}{3(2l+3)(2l+5)(2l+7)}, \quad c_{5l}^{(v)} = \frac{2^{-l-1}(l-3)(2l+5)}{3(2l-5)(2l-3)(2l-1)}. \tag{86}$$

It should be noted that the method described above for calculating the coefficients $c_{5l}^{(u)}$ and $c_{5l}^{(v)}$ is very reliable, but quite complex. A much simpler method is based on the statement that the point $\rho = 1$ represents the match point for the functions defined by Equations (69) and (70). This means that not only these functions, but also their first (at least) derivatives must coincide at this point. Thus, the second required equation relating the coefficients $c_{5l}^{(u)}$ and $c_{5l}^{(v)}$ is:

$$\left.\frac{d\sigma_l^{(0)}(\rho)}{d\rho}\right|_{\rho=1} + c_{5l}^{(v)} \left.\frac{dv_{5l}(\rho)}{d\rho}\right|_{\rho=1} = \left.\frac{d\sigma_l^{(1)}(\rho)}{d\rho}\right|_{\rho=1} + c_{5l}^{(u)} \left.\frac{du_{5l}(\rho)}{d\rho}\right|_{\rho=1}. \tag{87}$$

The solution of the system of two Equations (72) and (87) again gives the coefficients defined by Equation (86). Substituting these coefficients into the representations (69) and (70), we finally obtain:

$$f_2(\alpha,\theta) = \frac{1}{6}\sum_{l=0}^{\infty} \frac{\zeta_l(\rho)P_l(\cos\theta)}{(2l-1)(2l+3)}, \tag{88}$$

where

$$\zeta_l(\rho) = \begin{cases} \chi_l(\rho), & 0 \le \rho \le 1 \\ \chi_l(1/\rho), & \rho \ge 1 \end{cases} \tag{89}$$

with

$$\chi_l(\rho) = \frac{\rho^l}{(\rho^2+1)^{5/2}}\left[\frac{(l-3)(2l-1)\rho^6}{2l+7} + 9l\rho^4 - 9(l+1)\rho^2 - \frac{(l+4)(2l+3)}{2l-5}\right]. \tag{90}$$

It is clear that only the function (69) is required for calculating the function $\chi_l(\rho)$. Thus, in fact, we need to calculate only one coefficient $c_{5l}^{(v)}$ to define this function. In this regard, it is important to emphasize that the representation (89) reflects the fact that the WF of a two-electron atomic system must preserve its parity when interchanging the electrons. For the singlet S-states (which include the ground state) this means that the AFC and/or its component preserves its form (including the sign) under the transformation $\alpha \rightleftarrows \pi - \alpha$. For the AFC-component $f_2(\alpha,\theta)$, represented by the series expansion (61), this property corresponds (in terms of variable ρ) to the relationship:

$$\sigma_l^{(0)}(\rho^{-1}) + c_{5l}^{(v)} v_{5l}(\rho^{-1}) = \sigma_l^{(1)}(\rho) + c_{5l}^{(u)} u_{5l}(\rho). \tag{91}$$

The elimination of the RHSs between Equations (71) and (91) for $\rho = 1$ yields the identity, whereas the use of Equation (87) instead of Equation (71) yields the required equation:

$$\frac{d\sigma_l^{(0)}(\rho)}{d\rho}\bigg|_{\rho=1} + c_{5l}^{(v)} \frac{dv_{5l}(\rho)}{d\rho}\bigg|_{\rho=1} = \frac{d\sigma_l^{(0)}(\rho^{-1})}{d\rho}\bigg|_{\rho=1} + c_{5l}^{(v)} \frac{dv_{5l}(\rho^{-1})}{d\rho}\bigg|_{\rho=1}. \tag{92}$$

The solution of the last equation gives the coefficients $c_{5l}^{(v)}$ presented by Equation (86). Note that the coefficient $c_{5l}^{(u)}$ can then be calculated by the use of Equation (71) if needed.

In the general case, we cannot sum the infinite series (88) to obtain the function $f_2(\alpha,\theta)$ in an explicit closed form. However, this can be done for some special angles α and/or θ. For example, it is worth noting that the angles $\theta = 0, \pi$ correspond to the collinear configuration [14] of the two-electron atomic system in question. For these cases we obtain

$$f_2(\alpha, 0) = \pm \frac{(\rho-1)(12\rho^4 - 13\rho^3 - 88\rho^2 - 13\rho + 12)}{90(\rho^2+1)^{5/2}}, \tag{93}$$

$$f_2(\alpha, \pi) = -\frac{(\rho+1)(12\rho^4 + 13\rho^3 - 88\rho^2 + 13\rho + 12)}{90(\rho^2+1)^{5/2}}. \tag{94}$$

Sign "+" in Equation (93) corresponds to $0 \le \alpha \le \pi/2$ $(0 \le \rho \le 1)$, whereas "−" to $\pi/2 \le \alpha \le \pi$ $(\rho \ge 1)$. The list of special θ-angles can be supplemented with an intermediate angle $\theta = \pi/2$:

$$f_2\left(\alpha, \frac{\pi}{2}\right) = -\frac{2(\rho^4 - 3\rho^2 + 1)}{15(\rho^2+1)^2}. \tag{95}$$

It is worth noting that for the important cases of the nucleus–electron and electron–electron coalescence, Equation (88), respectively, reduces to:

$$f_2(0,\theta) = -\frac{2}{15}, \qquad f_2\left(\frac{\pi}{2}, 0\right) = 0. \tag{96}$$

To derive the results (93)–(96) we used the relationships:

$$P_n(0) = \sqrt{\pi}\,\Gamma^{-1}\left(\frac{1-n}{2}\right)\Gamma^{-1}\left(\frac{n}{2}+1\right), \quad P_n(1) = 1, \quad P_n(-1) = (-1)^n, \tag{97}$$

where $\Gamma(x)$ is the gamma function.

3. Derivation of the Angular Fock Coefficient $\psi_{6,3}(\alpha,\theta)$

We start this section by considering the FRR (3) and (4) for $k = 6$ and $p = 2$:

$$\left(\Lambda^2 - 60\right)\psi_{6,2} = 48\psi_{6,3} - 2V\psi_{5,2} + 2E\psi_{4,2}. \tag{98}$$

Next, let us expand each function in Equation (98) into HHs, using Equation (73). This gives

$$\psi_{k,p} = \sum_{n=0(2)}^{\infty} \sum_{l=0}^{n/2} c_{nl}^{(kp)} Y_{n,l}(\alpha,\theta), \tag{99}$$

with $\{k,p\} = \{6,3\}, \{6,2\}, \{4,2\}$ and

$$V\psi_{5,2} = \sum_{n=0(2)}^{\infty} \sum_{l=0}^{n/2} \mathrm{f}_{nl} Y_{n,l}(\alpha,\theta), \tag{100}$$

where the dimensionless potential V is defined by Equation (5), whereas the expansion coefficient f_{nl} can be calculated by the formula

$$\mathrm{f}_{nl} = \int V\psi_{5,2}Y_{n,l}(\alpha,\theta)d\Omega, \tag{101}$$

according to Equation (74).

It follows from Equation (4) that $h_{k,k/2} = 0$ for even k. Using additionally Equation (81), we can conclude that the AFC $\psi_{k,k/2}$ (with even k) represents the linear combination of the HHs, $Y_{k,l}(\alpha,\theta)$. Hence,

$$c_{nl}^{(63)} = 0 \quad for \quad n \neq 6, \tag{102}$$

$$c_{nl}^{(42)} = 0 \quad for \quad n \neq 4. \tag{103}$$

Equating the coefficients for the HHs, $Y_{6,l}(\alpha,\theta)$ in both sides of Equation (98), we obtain:

$$0 = 48c_{6l}^{(63)} - 2\mathrm{f}_{6l}. \tag{104}$$

Hence, (using additionally Equation (101)),

$$c_{6l}^{(63)} = \frac{1}{24}\int V\psi_{5,2}Y_{6,l}(\alpha,\theta)d\Omega. \tag{105}$$

Note that the LHS of Equation (104) equals zero, because $(\Delta^2 - 60)Y_{6,l} = 0$ as follows from Equation (81).

Thus, according to Equations (99) and (102), the AFC $\psi_{6,3}(\alpha,\theta)$ represents a linear combination of four HHs, $Y_{6,l}(\alpha,\theta)$ with $l = 0,1,2,3$. The contribution of each HH is determined by the coefficient $c_{6l}^{(63)}$ given by Equation (105). However, it is easy to prove that only the coefficients with odd values of l are nonzero for $\psi_{6,3}(\alpha,\theta)$. Indeed, this has already been mentioned in Section 2.4 that the WF of a two-electron atom/ion must preserve its parity when interchanging the electrons. For the singlet S-states this means that only the HHs, which preserve the sign under transformation $\alpha \leftrightarrows \pi - \alpha$, differ from zero in the expansion of the WF, and hence in the expansion of any AFC. In turn, it is easy to show that only $Y_{n,l}(\alpha,\theta)$ with even values of $(n/2 - l)$ satisfy the above property. Hence, the AFC in question, becomes

$$\psi_{6,3}(\alpha,\theta) = a_{61}Y_{6,1}(\alpha,\theta) + a_{63}Y_{6,3}(\alpha,\theta), \tag{106}$$

where we denoted $a_{6l} \equiv c_{6l}^{(63)}$ $(l = 1,3)$ for convenience and simplicity, and where the normalized HHs are

$$Y_{6,1}(\alpha,\theta) = \frac{2[\sin\alpha + 3\sin(3\alpha)]\cos\theta}{\pi^{3/2}\sqrt{5}}, \quad Y_{6,3}(\alpha,\theta) = \frac{8\sin^3\alpha P_3(\cos\theta)}{\pi^{3/2}\sqrt{5}}. \tag{107}$$

Using Formula (105) and taking into account the representations (5) and (18) for the dimensionless potential V and the AFC $\psi_{5,2}$, respectively, we can represent the desired coefficients in the form

$$a_{6l} = -\frac{(\pi-2)(5\pi-14)}{6480}\left(I_{l,4}Z^4 + I_{l,3}Z^3 + I_{l,2}Z^2\right), \quad (l = 1,3) \tag{108}$$

where

$$I_{l,4} = 4\pi^{3/2}\int_0^\pi\int_0^\pi\left[f_3(\alpha,\theta) + \sqrt{2}f_4(\alpha,\theta)\right]\eta Y_{6,l}(\alpha,\theta)\sin\alpha\sin\theta d\alpha d\theta, \tag{109}$$

$$I_{l,3} = -2\int_0^\pi\int_0^\pi\left\{\frac{3\eta[2f_1(\alpha,\theta) + f_2(\alpha,\theta)]}{\sin\alpha} + \frac{\pi^{3/2}\left[f_3(\alpha,\theta) + \sqrt{2}f_4(\alpha,\theta)\right]}{\xi}\right\}\times \times Y_{6,l}(\alpha,\theta)\sin^2\alpha\sin\theta d\alpha d\theta, \tag{110}$$

$$I_{l,2} = 3\int_0^\pi \int_0^\pi [2f_1(\alpha,\theta) + f_2(\alpha,\theta)]\xi^{-1} Y_{6,l}(\alpha,\theta)\sin^2\alpha \sin\theta d\alpha d\theta. \tag{111}$$

To calculate the integral (110), it is useful to separate the contributions which include the functions f_1, f_3 and f_4, represented by the explicit closed expressions, and the function f_2, represented by the infinite series (88). We obtain:

$$I_{l,3} = I_{l,3}^{(134)} - 6I_{l,3}^{(2)}, \tag{112}$$

where

$$I_{l,3}^{(134)} = -2\int_0^\pi \int_0^\pi \left\{ \frac{6\eta f_1(\alpha,\theta)}{\sin\alpha} + \frac{\pi^{3/2}\left[f_3(\alpha,\theta) + \sqrt{2}f_4(\alpha,\theta)\right]}{\xi} \right\} Y_{6,l}(\alpha,\theta)\sin^2\alpha \sin\theta d\alpha d\theta, \tag{113}$$

$$I_{l,3}^{(2)} = \int_0^\pi \int_0^\pi f_2(\alpha,\theta)\eta Y_{6,l}(\alpha,\theta)\sin\alpha \sin\theta d\alpha d\theta. \tag{114}$$

The integrals (113) can be taken in an explicit (closed) form that gives:

$$I_{1,3}^{(134)} = \frac{3(45\pi - 122)}{35\pi^{3/2}\sqrt{5}}, \qquad I_{3,3}^{(134)} = \frac{245\pi - 816}{70\pi^{3/2}\sqrt{5}}. \tag{115}$$

The problem of calculating the integrals (114) is that the corresponding integrands contain the function $f_2(\alpha,\theta)$ represented by the infinite series (88). Fortunately, using the orthogonality relationship for the Legendre polynomials, we can get these integrals also in explicit form. Changing the order of summation and integration, we easily obtain:

$$\begin{aligned} I_{1,3}^{(2)} &= \frac{\pi^{-3/2}}{3\sqrt{5}} \sum_{l=0}^\infty \int_0^\pi [\sin\alpha + 3\sin(3\alpha)]\,\eta \left[\frac{\zeta_l(\rho)}{(2l-1)(2l+3)}\right] \sin\alpha d\alpha \int_0^\pi P_l(\cos\theta)\cos\theta\sin\theta d\theta \\ &= \frac{2\pi^{-3/2}}{45\sqrt{5}} \int_0^\pi [\sin\alpha + 3\sin(3\alpha)]\,\eta\,\zeta_1(\rho)\sin\alpha d\alpha = \frac{7\pi + 22}{210\pi^{3/2}\sqrt{5}}, \end{aligned} \tag{116}$$

$$\begin{aligned} I_{3,3}^{(2)} &= \frac{4\pi^{-3/2}}{3\sqrt{5}} \sum_{l=0}^\infty \int_0^\pi \eta \left[\frac{\zeta_l(\rho)}{(2l-1)(2l+3)}\right] \sin^4\alpha\, d\alpha \int_0^\pi P_l(\cos\theta)P_3(\cos\theta)\sin\theta d\theta \\ &= \frac{8\pi^{-3/2}}{945\sqrt{5}} \int_0^\pi \eta\,\zeta_3(\rho)\sin^4\alpha\, d\alpha = \frac{3\pi - 32}{180\pi^{3/2}\sqrt{5}}. \end{aligned} \tag{117}$$

Recall that $\eta \equiv \eta(\alpha)$ is defined by Equation (6) and $\rho = \tan(\alpha/2)$.

It can be shown (using fairly long nontrivial derivations) that the integrals $I_{l,2}$ and $I_{l,4}$ vanish both for $l = 1$ and $l = 3$. This means that (according to the representations (106) and (108)) the AFC, $\psi_{6,3}(\alpha,\theta)$ is proportional to the third power of the nucleus charge Z (only), which is in full agreement with Formula (13) from Ref. [4].

Thus, combining the results of this section, we obtain the nonzero coefficients $a_{6,l}$ in the simple final form:

$$a_{61} = -\frac{(\pi - 2)(5\pi - 14)(32\pi - 97)}{56700\pi^{3/2}\sqrt{5}}Z^3, \tag{118}$$

$$a_{63} = -\frac{(\pi - 2)(5\pi - 14)(357\pi - 1112)}{680400\pi^{3/2}\sqrt{5}}Z^3. \tag{119}$$

4. Derivation of the Angular Fock Coefficients $\psi_{7,3}(\alpha,\theta)$ and $\psi_{8,4}(\alpha,\theta)$

In Sections 2 and 3 we detailed the derivation of the AFCs $\psi_{5,2}(\alpha,\theta)$ and $\psi_{6,3}(\alpha,\theta)$, respectively. Therefore, for the AFCs $\psi_{7,3}(\alpha,\theta)$ and $\psi_{8,4}(\alpha,\theta)$, we give only abbreviated derivations, and include extended explanations only in case of significant differences.

4.1. The AFC $\psi_{7,3}(\alpha,\theta)$

The FRR (3) and (4) for $k = 7$ and $p = 3$ reduces to the form

$$\left(\Lambda^2 - 77\right)\psi_{7,3}(\alpha,\theta) = h_{7,3}(\alpha,\theta), \tag{120}$$

where

$$h_{7,3}(\alpha,\theta) = -2V\psi_{6,3}(\alpha,\theta). \tag{121}$$

Using Equations (106), (107) and (5) the RHS of Equation (120) can be represented in the form:

$$h_{7,3}(\alpha,\theta) = \frac{(\pi-2)(5\pi-14)Z^3}{340200\sqrt{5}\pi^{3/2}}\left\{\frac{\bar{h}_1+\bar{h}_2}{\sqrt{5}\pi^{3/2}} - 2Z\left[12(32\pi-97)\bar{h}_3 + (357\pi-1112)\bar{h}_4\right]\right\}, \tag{122}$$

where

$$\bar{h}_1 = 20\xi^{-1}\left[12(32\pi-97)(1-\xi^2) + (357\pi-1112)(1-\xi^2)^3\right], \tag{123}$$

$$\bar{h}_2 = 60(688-255\pi)\xi^{-1}\sin^3\alpha\cos\theta, \tag{124}$$

$$\bar{h}_3 = \eta(\sin\alpha)^{-1}Y_{6,1}(\alpha,\theta), \quad \bar{h}_4 = \eta(\sin\alpha)^{-1}Y_{6,3}(\alpha,\theta). \tag{125}$$

Accordingly, the solution of the FRR (120) can be found in the form:

$$\psi_{7,3}(\alpha,\theta) = \frac{(\pi-2)(5\pi-14)Z^3}{340200\sqrt{5}\pi^{3/2}}\left\{\frac{\bar{f}_1+\bar{f}_2}{\sqrt{5}\pi^{3/2}} - 2Z\left[12(32\pi-97)\bar{f}_3 + (357\pi-1112)\bar{f}_4\right]\right\}, \tag{126}$$

where the AFC components $\bar{f}_i$ satisfy the IFRRs

$$\left(\Lambda^2 - 77\right)\bar{f}_i = \bar{h}_i. \qquad (i = 1,2,3,4) \tag{127}$$

Note that the components $\bar{h}_i$ of the RHS $h_{7,3}$ of the FRR (120) for the AFC $\psi_{7,3}$ are reasonably close to the components h_i of the RHS $h_{5,2}$ of the FRR (14). Therefore, we only briefly dwell on the conclusions of the corresponding results, as we noted earlier.

It is seen from Equation (123) that the RHS $\bar{h}_1$ is a function of a single variable ξ defined by Equation (6). The solution of the corresponding IFRR was described in Section IV of Ref. [4] (see also Section II of Ref. [12]) and illustrated (among others) in Section 2.3 of the current article. Thus, following the technique mentioned above, we obtain:

$$\bar{f}_1 = \left(\frac{41437\pi}{12} - \frac{74342}{7}\right)\xi^7 + \left(36476 - \frac{35588\pi}{3}\right)\xi^5 + \\ + \frac{5}{2}(4931\pi - 15156)\xi^3 + 5(2276 - 741\pi)\xi. \tag{128}$$

It can be verified that the RHSs $\bar{h}_3$ and $\bar{h}_4$ represent functions of the form $f(\alpha)P_l(\cos\theta)$ with l equals 1 and 3, respectively. The solution of the corresponding IFRR was described in Section V of Ref. [4] and illustrated in Sections 2.1 and 2.2 of the current article. This enables us to obtain:

$$\bar{f}_3 = -\frac{\rho(1+\rho)(29+\rho\{16+\rho[\rho(16+29\rho)-114]\})\cos\theta}{9\sqrt{5}\,\pi^{3/2}(\rho^2+1)^{7/2}}, \tag{129}$$

$$\bar{f}_4 = -\frac{\sin^3\alpha\sqrt{1+\sin\alpha}}{2\sqrt{5}\,\pi^{3/2}}P_3(\cos\theta). \tag{130}$$

Recall that the ρ variable was defined previously in Section 2.1.

The RHS $\bar{h}_2$ represented by Equation (124) is slightly more complicated than h_2 discussed in Section 2.4. In this regard, it would be useful to clarify two points.

First, using representation (56) for ξ^{-1}, we can rewrite Equation (124) in the form:

$$\bar{h}_2 = 60(688 - 225\pi)\tilde{h}_2 \tag{131}$$

where

$$\tilde{h}_2 = \sum_{l=0}^{\infty} 2^{-l}(\sin\alpha)^{l+3}F_l(\rho)\cos\theta P_l(\cos\theta), \tag{132}$$

and where $F_l(\rho)$ is defined by Equations (57) and (58). In order to apply the solution of the corresponding IFRR by the method described in Section 2.4 (see also [4]), the θ-dependent l-component in the series expansion of $\tilde{h}_2$ must be pure $P_l(\cos\theta)$. To solve the problem, one could use the general formula representing the Clebsch–Gordan series for a product of two *spherical* harmonics. However, in our simple case, it is easier to use the recurrence relation for the Legendre polynomials

$$(l+1)P_{l+1}(x) - (2l+1)xP_l(x) + lP_{l-1}(x) = 0, \tag{133}$$

which enables us to represent $\tilde{h}_2$ in the desired form:

$$\tilde{h}_2 = \sum_{l=0}^{\infty} \mathfrak{h}_l(\rho)(\sin\alpha)^l P_l(\cos\theta), \tag{134}$$

where

$$\mathfrak{h}_l(\rho) = \frac{l}{2^{l-1}(2l-1)}\sin^2\alpha F_{l-1}(\rho) + \frac{l+1}{2^{l+1}(2l+3)}\sin^4\alpha F_{l+1}(\rho). \tag{135}$$

The second point is related to the calculation of the coefficient

$$\tilde{h}_{2l,l} = \pi^2 \int_0^{\pi}\int_0^{\pi} \tilde{h}_2(\alpha,\theta)Y_{2l,l}(\alpha,\theta)\sin^2\alpha\sin\theta d\alpha d\theta \tag{136}$$

in the HH expansion of $\tilde{h}_2$ (see the corresponding Equation (84) for calculation of $\psi_{5,2}(\alpha,\theta)$). Of course, we can use representation (134) and (135) and then apply the orthogonality condition for the Legendre polynomials. However, the simpler way is to use the original representation (132) taking into account that $\cos\theta \equiv P_1(\cos\theta)$. In this case, we can apply the well-known formula for the integral of three Legendre polynomials

$$\int_{-1}^{1} P_l(x)P_L(x)P_{l'}(x)dx = 2\begin{pmatrix} l & L & l' \\ 0 & 0 & 0 \end{pmatrix}^2, \tag{137}$$

where the RHS represents twice the square of the Wigner 3-j symbol.

Thus, applying the methodologies outlined in Section 2.4, and given the above features, one obtains

$$\bar{f}_2 = \frac{1}{48}\sum_{l=0}^{\infty}\frac{\bar{\zeta}_l(\rho)P_l(\cos\theta)}{(2l-1)(2l+3)}, \tag{138}$$

where

$$\bar{\zeta}_l(\rho) = \begin{cases} \bar{\chi}_l(\rho), & 0 \le \rho \le 1 \\ \bar{\chi}_l(1/\rho), & \rho \ge 1 \end{cases} \tag{139}$$

with

$$\begin{aligned} \bar{\chi}_l(\rho) = &-\frac{\rho^l}{(\rho^2+1)^{7/2}}\left\{\frac{(32l^2+26l-25)\rho^6}{2l+5}\left[\frac{(2l-1)\rho^2}{2l+9}+4\right]+\right. \\ &\left.+\frac{1}{2l-3}\left[\frac{6(84l^2+84l-95)\rho^4}{2l+5} - (32l^2+38l-19)\left(\frac{2l+3}{2l-7}+4\rho^2\right)\right]\right\}. \end{aligned} \tag{140}$$

Recall that the component $\bar{f}_2$ in the RHS of Equation (126) is equal to $60(688-225\pi)\bar{f}_2$ according to representation (131).

As in the case of the AFC $\psi_{5,2}(\alpha,\theta)$, there are combinations of special hyperspherical angles α and θ for which the component $\bar{f}_2 \equiv \bar{f}_2(\alpha,\theta)$ of the AFC $\psi_{7,3}(\alpha,\theta)$ can be obtained in closed form. In particular, one obtains:

$$\bar{f}_2(\alpha,0) = \mp\frac{(\rho-1)(95\rho^6+1166\rho^5-1879\rho^4-8844\rho^3-1879\rho^2+1166\rho+95)}{5040(\rho^2+1)^{7/2}}, \quad (141)$$

$$\bar{f}_2(\alpha,\pi) = \frac{(\rho+1)(95\rho^6-1166\rho^5-1879\rho^4+8844\rho^3-1879\rho^2-1166\rho+95)}{5040(\rho^2+1)^{7/2}}, \quad (142)$$

$$\bar{f}_2(\alpha,\frac{\pi}{2}) = \frac{19\rho^4+10\rho^2+19}{1008(\rho^2+1)^2}. \quad (143)$$

Sign "$-$" in Equation (141) corresponds to $0 \le \alpha \le \pi/2$ $(0 \le \rho \le 1)$, whereas "+" corresponds to $\pi/2 \le \alpha \le \pi$ $(\rho \ge 1)$.

For the important cases of the nucleus–electron and electron–electron coalescence, representation (138)–(140) is simplified to:

$$\bar{f}_2(0,\theta) = \frac{19}{1008}, \qquad \bar{f}_2\left(\frac{\pi}{2},0\right) = 0. \quad (144)$$

4.2. The AFC $\psi_{8,4}(\alpha,\theta)$

Having at our disposal the AFC $\psi_{7,3} \equiv \psi_{7,3}(\alpha,\theta)$, we can calculate the AFC $\psi_{8,4} \equiv \psi_{8,4}(\alpha,\theta)$ using the FRR (3) and (4) for $k=8$ and $p=3$:

$$\left(\Lambda^2-96\right)\psi_{8,3} = 80\psi_{8,4} - 2V\psi_{7,3} + 2E\psi_{6,3}. \quad (145)$$

It follows from Equation (81) and the FRR (3) and (4) for $k=8$ and $p=4$ that the AFC $\psi_{8,4}$ is a linear combination of the HHs, $Y_{8,l} \equiv Y_{8,l}(\alpha,\theta)$. Moreover, given that only $Y_{n,l}(\alpha,\theta)$ with even values of $n/2-l$ are suitable for singlet S-states, we obtain:

$$\psi_{8,4} = a_{80}Y_{8,0} + a_{82}Y_{8,2} + a_{84}Y_{8,4}. \quad (146)$$

For further derivations, it is advisable to represent the HHs in the form

$$Y_{8,l}(\alpha,\theta) = y_{8l}(\alpha)P_l(\cos\theta), \quad (147)$$

where

$$y_{80}(\alpha) = \pi^{-3/2}[2\cos(4\alpha)+2\cos(2\alpha)+1], \quad (148)$$

$$y_{82}(\alpha) = \frac{2}{\pi^{3/2}}\sqrt{\frac{10}{7}}\sin^2\alpha[4\cos(2\alpha)+3], \quad (149)$$

$$y_{84}(\alpha) = \frac{8}{\pi^{3/2}}\sqrt{\frac{2}{7}}\sin^4\alpha. \quad (150)$$

It was found in Section 3 that $\psi_{6,3} \equiv \psi_{6,3}(\alpha,\theta)$ is the linear combination of the HHs $Y_{6,l}(\alpha,\theta)$. Thus, expanding each function of Equation (145) in HHs, and equating the coefficients for $Y_{8,l}$, we obtain (see the corresponding result (105) for a_{6l})

$$a_{8l} = \frac{1}{40}\int V\psi_{7,3}Y_{8,l}d\Omega, \quad (151)$$

where the potential V is defined by Equation (5). When deriving the last equation, it was taken into account that $(\Delta^2-96)Y_{8,l}=0$, as follows from Equation (81).

A direct substitution of the representations (5), (126) and (147) into the RHS of Equation (151) yields:

$$a_{8l} = \frac{(\pi-2)(5\pi-14)Z^3}{13608000\sqrt{5}\,\pi^{3/2}} \times$$
$$\times \int \left(\frac{1}{\xi} - \frac{2Z\eta}{\sin\alpha}\right)\left\{\frac{\check{f}_1+\check{f}_2}{\sqrt{5}\pi^{3/2}} - 2Z\Big[12(32\pi-97)\check{f}_3 + (357\pi-1112)\check{f}_4\Big]\right\}Y_{8,l}(\alpha,\theta)d\Omega. \tag{152}$$

It follows from Equation (13) of Ref. [4] that only the coefficients at Z^4 are nonzero on the RHS of the last equation. Hence, Equation (152) reduces to the form:

$$a_{8l} = -\frac{(\pi-2)(5\pi-14)Z^4}{6804000\sqrt{5}\,\pi^{3/2}}\left[\frac{S_{1l}+S_{2l}}{\sqrt{5}\pi^{3/2}} + 12(32\pi-97)S_{3l} + (357\pi-1112)S_{4l}\right], \tag{153}$$

where

$$S_{1l} = \pi^2\int_0^\pi\int_0^\pi \check{f}_1(\xi)Y_{8,l}(\alpha,\theta)\eta\sin\alpha\sin\theta d\alpha d\theta, \tag{154}$$

$$S_{2l} = \pi^2\int_0^\pi\int_0^\pi \check{f}_2(\alpha,\theta)Y_{8,l}(\alpha,\theta)\eta\sin\alpha\sin\theta d\alpha d\theta =$$
$$= \frac{5\pi^2(688-225\pi)}{(2l-1)(2l+1)(2l+3)}\int_0^{\pi/2}\bar{\chi}_l(\rho)y_{8l}(\alpha)\eta\sin\alpha d\alpha, \tag{155}$$

$$S_{nl} = \pi^2\int_0^\pi\int_0^\pi \check{f}_n(\alpha,\theta)Y_{8,l}(\alpha,\theta)\xi^{-1}\sin^2\alpha\sin\theta d\alpha d\theta. \qquad (n=3,4) \tag{156}$$

The identifiers ξ and η are defined by Equation (6), whereas functions $\bar{\chi}_l(\rho)$ can be calculated by Formula (140). When deriving Equation (155), we applied the orthogonality condition for the Legendre polynomials. Fortunately, all integrals (154)–(156) can be taken in closed form. Thus, by collecting these results and substituting them into the RHS of Equation (153), we finally obtain the desired coefficients in the form:

$$a_{8l} = \frac{Z^4(\pi-2)(5\pi-14)}{\pi^{5/2}}b_{8l}, \tag{157}$$

with

$$b_{80} = \frac{\pi(150339\pi-927292)+1430792}{19289340000}, \quad b_{82} = \frac{\pi(751965\pi-4654046)+7200976}{1928934000\sqrt{70}},$$
$$b_{84} = \frac{\pi(3190317\pi-19828996)+30802176}{25719120000\sqrt{14}}. \tag{158}$$

5. Results and Discussions

The angular Fock coefficients $\psi_{k,p} \equiv \psi_{k,p}(\alpha,\theta)$ with the *maximum possible* value of subscript p were calculated on examples of the coefficients with $5 \le k \le 10$. The results obtained in Sections 2–4 are summarized in Appendices A and B. The AFCs $\psi_{9,4}$ and $\psi_{10,5}$ are presented in Appendix C without derivations. To find the latter AFCs, the methods described in the main sections were used. The presented technique makes it possible to calculate such AFCs for any arbitrarily large k. These coefficients are leading in the logarithmic power series representing the Fock expansion (see Equation (8)). As such, they may be indispensable for the development of simple methods for calculating the helium-like electronic structure.

The proposed technique, as well as the final results, are quite complex. Therefore, both require verification. We are aware of two ways for the above-mentioned verification. The first one is to use the Green's function (GF) approach (see Ref. [1] and also Ref. [15], Section

4) which enables us to calculate (at least, numerically) the AFC (or its component) by the following integral representation:

$$\psi_{k,p}(\alpha,\theta) = \frac{1}{8\pi}\int_0^\pi d\alpha' \sin^2\alpha' \int_0^\pi d\theta' \sin\theta' \, h_{k,p}(\alpha',\theta') \int_0^\pi \frac{\cos\left[\left(\frac{k}{2}+1\right)\omega\right]}{\sin\omega}(1-\lambda)d\varphi, \tag{159}$$

where ω is an angle defined by the relation

$$\cos\omega = \cos\alpha\cos\alpha' + \sin\alpha\sin\alpha'(\cos\theta\cos\theta' + \sin\theta\sin\theta'\cos\varphi), \tag{160}$$

whereas

$$\lambda = \begin{cases} 0 & k \ \ odd \\ \omega/\pi & k \ \ even \end{cases}. \tag{161}$$

For even k and maximum value of $p = k/2$, the RHS $h_{k,k/2}$ of the FRR (3) equals zero. This implies that the GF formula (159) cannot be applied in this case. Hence, only the AFCs $\psi_{k,p}$ with odd values of k (and maximum p) can be verified with the GF method. Thus, numerically calculating (for various combinations of angles α and θ) the triple integrals (159) representing the AFCs $\psi_{5,2}(\alpha,\theta)$, $\psi_{7,3}(\alpha,\theta)$ and $\psi_{9,4}(\alpha,\theta)$, we verified that the representations obtained for them in Sections 2 and 4.1 and in Appendix C were correct.

The second verification method considered, covering all possible combinations of angles, being quite complex, is the only method known to us that correctly displays the WF near the nucleus. This is the CFHH method mentioned in the Introduction. It is based on decomposing the full WF into a form

$$\Psi^{\mathrm{CFHH}}(r_1,r_2,r_{12}) = \exp[f(r_1,r_2,r_{12})]\Phi^{\mathrm{CFHH}}(R,\alpha,\theta), \tag{162}$$

where the so-called correlation function f can be taken in a simple linear form

$$f(r_1,r_2,r_{12}) = c_1 r_1 + c_2 r_2 + c_{12} r_{12}. \tag{163}$$

The so-called "cusp parametrization"

$$c_1 = c_2 = -Z, \qquad c_{12} = 1/2 \tag{164}$$

is used as a rule. For a small enough hyperspherical radius R, the function Φ is represented as

$$\Phi^{\mathrm{CFHH}}(R,\alpha,\theta) = \frac{1}{d_{0,0}(\alpha,\theta)}\sum_{k=0}^{K}(2\kappa R)^k \sum_{p=0}^{[k/2]} d_{k,p}(\alpha,\theta)\ln^p(2\kappa R), \tag{165}$$

where $\kappa = \sqrt{-2E}$, and functions $d_{k,p}(\alpha,\theta)$ are expanded in N (basis size) HHs. It follows from representation (165) that the AFCs $\psi_{k,p}(\alpha,\theta)$ can be expressed in terms of the functions $d_{k',p'}(\alpha,\theta)$ calculated by the CFHHM. For example, for the AFCs in question, one obtains:

$$\psi^{\mathrm{CFHH}}_{k,k/2}(\alpha,\theta) = \frac{(2\kappa)^k d_{k,k/2}(\alpha,\theta)}{d_{0,0}(\alpha,\theta)}. \tag{166}$$

We calculated all AFCs discussed in this article using CFHHM with $K = 18$ and $N = 1600$. The angles $0 \le \alpha \le \pi$ and $0 \le \theta \le \pi$ with step $\pi/6$ were considered. The relative difference $|1 - \psi_{k,p}(\alpha,\theta)/\psi^{\mathrm{CFHH}}_{k,p}(\alpha,\theta)|$ was less than 10^{-7} for all examined cases, including $1 \le Z \le 5$. This indicates that all our theoretical calculations were correct.

Author Contributions: Conceptualization, E.Z.L.; Methodology, E.Z.L.; Software, E.Z.L. and R.K.; Validation, R.K.; Writing—original draft, E.Z.L. All authors have read and agreed to the published version of the manuscript.

Funding: This research received no external funding.

Conflicts of Interest: The authors declare no conflict of interest.

Appendix A

Summarizing the results of Section 2, one obtains.

$$\psi_{5,2}(\alpha,\theta) = -\frac{Z^2(\pi-2)(5\pi-14)}{270\sqrt{\pi}}\left[3\pi^{-3/2}(2f_1+f_2)-2Z(f_3+\sqrt{2}f_4)\right], \quad \text{(A1)}$$

where

$$f_1 = -\frac{1}{60}\sqrt{1-\sin\alpha\cos\theta}[\sin\alpha\cos\theta(4+13\sin\alpha\cos\theta)-2], \quad \text{(A2)}$$

$$f_3(\alpha) = -\frac{1}{60\pi^{3/2}}[11\sin\alpha+21\cos(2\alpha)+2]\sqrt{1+\sin\alpha}, \quad \text{(A3)}$$

$$f_4(\alpha,\theta) = -\frac{\sqrt{2}}{6\pi^{3/2}}(\sin\alpha)^2\sqrt{1+\sin\alpha}\,P_2(\cos\theta), \quad \text{(A4)}$$

$$f_2(\alpha,\theta) = \frac{1}{6}\sum_{l=0}^{\infty}\frac{\zeta_l(\rho)P_l(\cos\theta)}{(2l-1)(2l+3)}. \quad \text{(A5)}$$

The ζ function is defined as follows:

$$\zeta_l(\rho) = \begin{cases} \chi_l(\rho), & 0 \le \rho \le 1 \\ \chi_l(1/\rho), & \rho \ge 1 \end{cases} \quad \text{(A6)}$$

where

$$\chi_l(\rho) = \frac{\rho^l}{(\rho^2+1)^{5/2}}\left[\frac{(l-3)(2l-1)\rho^6}{2l+7}+9l\rho^4-9(l+1)\rho^2-\frac{(l+4)(2l+3)}{2l-5}\right]. \quad \text{(A7)}$$

Recall that $\rho = \tan(\alpha/2)$, and special cases of the function $f_2 \equiv f_2(\alpha,\theta)$, when they can be obtained in closed form, are represented by Equations (93)–(96).

Summarizing the results of Section 3, one obtains:

$$\psi_{6,3}(\alpha,\theta) = \frac{Z^3(\pi-2)(5\pi-14)}{56700\pi^{3/2}\sqrt{5}}\left[(97-32\pi)Y_{6,1}(\alpha,\theta)+\frac{(1112-357\pi)}{12}Y_{6,3}(\alpha,\theta)\right], \quad \text{(A8)}$$

where $Y_{n,l}(\alpha,\theta)$ are the normalized hyperspherical harmonics.

Appendix B

Summarizing the results of Section 4.1, one obtains:

$$\psi_{7,3}(\alpha,\theta) = \frac{(\pi-2)(5\pi-14)Z^3}{340200\sqrt{5}\pi^{3/2}}\left\{\frac{\bar{f}_1+\bar{f}_2}{\sqrt{5}\pi^{3/2}}-2Z\left[12(32\pi-97)\bar{f}_3+(357\pi-1112)\bar{f}_4\right]\right\}, \quad \text{(A9)}$$

where

$$\bar{f}_1 = \left(\frac{41437\pi}{12} - \frac{74342}{7}\right)\xi^7 + \left(36476 - \frac{35588\pi}{3}\right)\xi^5 + \\ + \frac{5}{2}(4931\pi - 15156)\xi^3 + 5(2276 - 741\pi)\xi. \tag{A10}$$

$$\bar{f}_3 = -\frac{\rho(1+\rho)(29+\rho\{16+\rho[\rho(16+29\rho)-114]\})\cos\theta}{9\sqrt{5}\,\pi^{3/2}(\rho^2+1)^{7/2}}, \tag{A11}$$

$$\bar{f}_4 = -\frac{\sin^3\alpha\sqrt{1+\sin\alpha}}{2\sqrt{5}\,\pi^{3/2}}P_3(\cos\theta), \tag{A12}$$

$$\bar{f}_2 = \frac{1}{48}\sum_{l=0}^{\infty}\frac{\bar{\zeta}_l(\rho)P_l(\cos\theta)}{(2l-1)(2l+3)}. \tag{A13}$$

The $\bar{\zeta}$ function is defined as follows:

$$\bar{\zeta}_l(\rho) = \begin{cases} \bar{\chi}_l(\rho), & 0 \le \rho \le 1 \\ \bar{\chi}_l(1/\rho), & \rho \ge 1 \end{cases}, \tag{A14}$$

where

$$\bar{\chi}_l(\rho) = -\frac{\rho^l}{(\rho^2+1)^{7/2}}\left\{\frac{(32l^2+26l-25)\rho^6}{2l+5}\left[\frac{(2l-1)\rho^2}{2l+9}+4\right]+ \right. \\ \left. + \frac{1}{2l-3}\left[\frac{6(84l^2+84l-95)\rho^4}{2l+5} - (32l^2+38l-19)\left(\frac{2l+3}{2l-7}+4\rho^2\right)\right]\right\}. \tag{A15}$$

Recall that variable ξ is defined by Equation (6), and special cases of the function $\bar{f}_2 \equiv \bar{f}_2(\alpha,\theta)$, when they can be obtained in closed form, are represented by Equations (141)–(144).

Summarizing the results of Section 4.2, one obtains:

$$\psi_{8,4} = \frac{Z^4(\pi-2)(5\pi-14)}{\pi^{5/2}}[b_{80}Y_{8,0}(\alpha,\theta) + b_{82}Y_{8,2}(\alpha,\theta) + b_{84}Y_{8,4}(\alpha,\theta)], \tag{A16}$$

where

$$b_{80} = \frac{\pi(150339\pi - 927292) + 1430792}{19289340000}, \quad b_{82} = \frac{\pi(751965\pi - 4654046) + 7200976}{1928934000\sqrt{70}}, \\ b_{84} = \frac{\pi(3190317\pi - 19828996) + 30802176}{25719120000\sqrt{14}}. \tag{A17}$$

Appendix C

In Sections 2 and 3, the AFCs $\psi_{5,2}(\alpha,\theta)$ and $\psi_{6,3}(\alpha,\theta)$ were calculated with detailed derivations. In Section 4, the AFCs $\psi_{7,3}(\alpha,\theta)$ and $\psi_{8,4}(\alpha,\theta)$ were presented with a very brief derivations. The corresponding results were summarized in Appendices A and B. The current Appendix presents the AFCs $\psi_{9,4} \equiv \psi_{9,4}(\alpha,\theta)$ and $\psi_{10,5} \equiv \psi_{10,5}(\alpha,\theta)$ without derivations. To find the latter AFCs, the methods described in the main sections were used.

So, the first AFC under consideration can be represented as:

$$\psi_{9,4} = 2Z^4[2ZX_1(\alpha,\theta) - X_2(\alpha,\theta)], \tag{A18}$$

where

$$X_1(\alpha,\theta) = \check{a}_{80}\check{f}_1 + \check{a}_{82}\check{f}_2 + \check{a}_{84}\check{f}_3, \tag{A19}$$

$$X_2(\alpha,\theta) = \frac{35}{\pi^{3/2}}\sqrt{\frac{2}{7}}\,\check{a}_{84}\check{f}_4 + \frac{(\pi-2)(5\pi-14)}{123451776000\pi^4}\Big[c_5\check{f}_5 + c_6\check{f}_6 + c_7\check{f}_7 + 16(c_8\check{f}_8 + c_9\check{f}_9)\Big]. \tag{A20}$$

Here, $\check{a}_{8l} = Z^{-4}a_{8l}$, where the coefficients a_{8l} are defined by Equations (157) and (158), whereas the other coefficients are:

$$\begin{gathered} c_5 = \pi(29757524 - 4780401\pi) - 46286848, \quad c_6 = \pi(9581100\pi - 59458928) + 92239360, \\ c_7 = \pi(28060 + 10149\pi) - 167168, \quad c_8 = 9\pi(134543\pi - 828732) + 11488128, \\ c_9 = \pi(4804833\pi - 29773780) + 46119680. \end{gathered} \tag{A21}$$

The functions $\check{f}_i \equiv \check{f}_i(\alpha,\theta)$ are:

$$\check{f}_1 = -\frac{(\rho+1)(563\rho^8 + 1012\rho^7 - 8932\rho^6 - 3668\rho^5 + 23954\rho^4 - 3668\rho^3 - 8932\rho^2 + 1012\rho + 563)}{1260\pi^{3/2}(\rho^2+1)^{9/2}}, \tag{A22}$$

$$\check{f}_2 = -\left(\frac{8}{\pi^{3/2}}\sqrt{\frac{10}{7}}\right)\frac{\rho^2(1+\rho)(126 + 49\rho - 424\rho^2 + 49\rho^3 + 126\rho^4)}{300(\rho^2+1)^{9/2}}P_2(\cos\theta), \tag{A23}$$

$$\check{f}_3 = -\frac{2}{5\pi^{3/2}}\sqrt{\frac{2(1+\sin\alpha)}{7}}\sin^4\alpha\, P_2(\cos\theta), \tag{A24}$$

$$\check{f}_4 = -\frac{\xi(315 - 1680\xi^2 + 2814\xi^4 - 1854\xi^6 + 419\xi^8)}{1260}, \tag{A25}$$

$$\check{f}_5 = -\frac{\xi}{60}(2\xi^2-3)(2\xi^2-1)(4\xi^4 - 10\xi^2 + 5). \tag{A26}$$

The remaining $\check{f}$ functions are represented by series:

$$\check{f}_j = \frac{(\rho^2+1)^{-9/2}}{k_j}\sum_{l=0}^{\infty}\frac{\rho^l\check{\zeta}_{jl}(\rho)}{(2l-1)(2l+3)}P_l(\cos\theta), \qquad (j = 6,7,8,9) \tag{A27}$$

where

$$k_6 = 6, \quad k_7 = 60, \quad k_8 = 24, \quad k_9 = 40, \tag{A28}$$

and the corresponding $\check{\zeta}$ functions are:

$$\begin{aligned} \check{\zeta}_{6l}(\rho) = {} & \frac{(2l-15)(2l-1)(l+1)\rho^{10}}{(2l+7)(2l+11)} + \frac{(22l^2-5l-12)\rho^8}{(2l+7)} + \frac{10(2l^2+11l+3)\rho^6}{2l+7} - \\ & -\frac{10(2l^2-7l-6)\rho^4}{2l-5} - \frac{(22l^2+49l+15)\rho^2}{2l-5} - \frac{l(2l+3)(2l+17)}{(2l-9)(2l-5)}, \end{aligned} \tag{A29}$$

$$\begin{aligned} \check{\zeta}_{7l}(\rho) = {} & \frac{(2l-1)(4l^2+160l-189)\rho^{10}}{(2l+7)(2l+11)} + \frac{35(4l^2+40l-9)\rho^8}{(2l+7)} - \frac{350(4l^2+16l+3)\rho^6}{2l+7} + \\ & +\frac{350(4l^2-8l-9)\rho^4}{2l-5} - \frac{35(4l^2-32l-45)\rho^2}{2l-5} - \frac{(2l+3)(4l^2-152l-345)}{(2l-9)(2l-5)}, \end{aligned} \tag{A30}$$

$$\check{\zeta}_{8l}(\rho) = -\frac{(2l-1)(56l^3+250l^2+338l+171)\rho^{10}}{(2l+5)(2l+7)(2l+11)} - \frac{(136l^3+314l^2-110l-153)\rho^8}{(2l+5)(2l+7)} - \frac{2(80l^4+652l^3+566l^2-1824l-873)\rho^6}{(2l-3)(2l+5)(2l+7)} + \frac{2(80l^4-332l^3-910l^2+1320l+945)\rho^4}{(2l-5)(2l-3)(2l+5)} + \frac{(136l^3+94l^2-330l-135)\rho^2}{(2l-5)(2l-3)} + \frac{(2l+3)(56l^3-82l^2+6l-27)}{(2l-9)(2l-5)(2l-3)}, \quad \text{(A31)}$$

$$\check{\zeta}_{9l}(\rho) = \frac{(2l-1)(24l^3-150l^2-670l-439)\rho^{10}}{(2l+5)(2l+7)(2l+11)} + \frac{5(72l^3+162l^2-70l-103)\rho^8}{(2l+5)(2l+7)} + \frac{10(16l^4+220l^3+222l^2-804l-423)\rho^6}{(2l-3)(2l+5)(2l+7)} - \frac{10(16l^4-156l^3-342l^2+652l+399)\rho^4}{(2l-5)(2l-3)(2l+5)} - \frac{5(72l^3+54l^2-178l-57)\rho^2}{(2l-5)(2l-3)} - \frac{(2l+3)(24l^3+222l^2-298l-57)}{(2l-9)(2l-5)(2l-3)}, \quad \text{(A32)}$$

It is important to emphasize that the representations (A27)–(A32) are valid only for $0 \leq \rho \leq 1$. For values $\rho > 1$, one should replace ρ with $1/\rho$, which is equivalent to simply redefining ρ as $\cot(\alpha/2)$.

The second AFC under consideration is of the form:

$$\psi_{10,5} = -\frac{Z^5(\pi-2)(5\pi-14)}{\pi^{7/2}}[b_{10,1}Y_{10,1}(\alpha,\theta) + b_{10,3}Y_{10,3}(\alpha,\theta) + b_{10,5}Y_{10,5}(\alpha,\theta)], \quad \text{(A33)}$$

where

$$b_{10,1} = \frac{\pi[3\pi(6840010557\pi - 63828704998) + 595609133656] - 617517605744}{401025378600000\sqrt{105}}, \quad \text{(A34)}$$

$$b_{10,3} = \frac{\pi[\pi(9194460432\pi - 85833963053) + 267084629592] - 277009842768}{100256344650000\sqrt{30}}, \quad \text{(A35)}$$

$$b_{10,5} = \frac{\pi[\pi(622341848670\pi - 5812646794643) + 18095537797140] - 18776793358080}{10025634465000000\sqrt{42}}, \quad \text{(A36)}$$

and $Y_{10,l}(\alpha,\theta)$ with $l = 1,3,5$ are the normalized HHs.

References

1. Fock, V.A. On the Schrödinger Equation of the Helium Atom. *Izv. Akad. Nauk SSSR Ser. Fiz.* **1954**, *18*, 161–174.
2. Morgan, J.D., III. Convergence properties of Fock's expansion for S-state eigenfunctions of the helium atom. *Theor. Chim. Acta* **1986**, *69*, 181–223. [CrossRef]
3. Abbott, P.C.; Maslen, E.N. Coordinate systems and analytic expansions for three-body atomic wavefunctions: I. Partial summation for the Fock expansion in hyperspherical coordinates. *J. Phys. A Math. Gen.* **1987**, *20*, 2043–2075. [CrossRef]
4. Liverts, E.Z.; Barnea, N. Angular Fock coefficients. Refinement and further development. *Phys. Rev. A* **2015**, *92*, 042512. [CrossRef]
5. Nakashima, H.; Nakatsuji, H. Solving the Schrödinger equation for helium atom and its isoelectronic ions with the free iterative complement interaction (ICI) method. *J. Chem. Phys.* **2007**, *127*, 224104. [CrossRef]
6. Rodriguez, K.V.; Gasaneo, G.; Mitnik, D.M. Accurate and simple wavefunctions for the helium isoelectronic sequence with correct cusp conditions. *J. Phys. B* **2007**, *40*, 3923–3939. [CrossRef]
7. Forrey, R.C. Compact representation of helium wave functions in perimetric and hyperspherical coordinates. *Phys. Rev. A* **2004**, *69*, 022504. [CrossRef]
8. Drake, G.W.F. (Ed.) High Precision Calculations for Helium. In *Atomic, Molecular, and Optical Physics Handbook*; AIP Press: New York, NY, USA, 1996; Section 11.

9. Haftel, M.I.; Mandelzweig, V.B. Exact Solution of Coupled Equations and the Hyperspherical Formalism: Calculation of Expectation Values and Wavefunctions of Three Coulomb-Bound Particles. *Ann. Phys.* **1983**, *150*, 48–91. [CrossRef]
10. Haftel, M.I.; Mandelzweig, V.B. Fast Convergent Hyperspherical Harmonic Expansion for Three-Body Systems. *Ann. Phys.* **1989**, *189*, 29–52. [CrossRef]
11. Haftel, M.I.; Krivec, R.; Mandelzweig, V.B. Power Series Solution of Coupled Differential Equations in One Variable. *J. Comp. Phys.* **1996**, *123*, 149–161. [CrossRef]
12. Liverts, E.Z. Analytic calculation of the edge components of the angular Fock coefficients. *Phys. Rev. A* **2016**, *94*, 022504. [CrossRef]
13. Sack, R.A. Generalization of Laplace's expansion to arbitrary powers and functions of the distance between two points. *J. Math. Phys.* **1964**, *5*, 245–251. [CrossRef]
14. Liverts, E.Z.; Krivec, R.; Barnea, N. Collinear configuration of the helium atom two-electron ions. *Ann. Phys.* **2020**, *422*, 168306. [CrossRef]
15. Liverts, E.Z. Co-spherical electronic configuration of the helium-like atomic systems. *Ann. Phys.* **2022**, *436*, 168669. [CrossRef]

MDPI
St. Alban-Anlage 66
4052 Basel
Switzerland
Tel. +41 61 683 77 34
Fax +41 61 302 89 18
www.mdpi.com

Atoms Editorial Office
E-mail: atoms@mdpi.com
www.mdpi.com/journal/atoms

www.ingramcontent.com/pod-product-compliance
Lightning Source LLC
Chambersburg PA
CBHW061617220926
51850CB00005B/530
9783036565989